COMMUNICATION THEORY AND SIGNAL PROCESSING FOR TRANSFORM CODING

Khamies El-Shennawy
Arab Academy for Science and Technology
and Maritime Transport AASTMT

Content

About the Author

Dr. Khamies El-Shennawy is a Prof. of Marine Communication in the Arab Academy for Science and Technology and Maritime Transport, Alexandria, Egypt. He is the President Assistant of the Academy for technology transfer. He was born in Alexandria, in March 1945. His B.Sc., Diploma of Higher studies, M.Sc., Ph.D. , in 1968, 1976, 1980, 1987, respectively, all from Alexandria University, Faculty of Engineering, Department of Communication and Electronics, Alexandria, Egypt, also Diploma of Higher studies from the Soviet Union in Radio Engineering in the period 1972-1973 when he was officer in the Naval Forces of Egypt. Prof. El-Shennawy is Senior member IEEE (S`83, M`87, SM`92), reviewer and member in Technical Committee TC-10 of the Instrumentation and Measurement IM society, editor/reviewer in the International Journal of Computer Science and Communication Security IJCSCS. Prof. Dr. El-Shennawy published about 90 papers in IEEE Transactions and Conferences, other international and local conferences and journals. Prof. El-Shennawy is member of the computer scientific society, Alexandria, Egypt, member of the Engineering syndicate association of Egypt, member of the Veterans and War VICTMS association of Egypt, member of the International Association of Science and Technology for Development IASTED, member of the International Frequency Sensor Association IFSA, R & D Projects, Japanese International Cooperation Agency JICA awards for his contributions, Nov.1990. Also, Honors include Two Naval Forces Outstanding achievement medals in Science and Engineering 1973 and 1998.

During the periods (1999-2014) El-Shennawy is Prof. Dr., (1996-1999) is associated Prof., (1992-1996) is assistant Prof., and (1988-1992) is a Dr. Eng., all in the Arab Academy for Science and Technology and Maritime Transport AASTMT, teaching and researches, Alexandria, Egypt, and in the period (1987-1988) is a Dr. Eng. in the electrical department, faculty of Engineering, Gharyounis university, Benghazi, Libya. Another periods of practical work, where in the period 1984-1987, he was service manager in international communication companies: Ford aerospace, Thomson CSF, and Emerson. where local projects were established in Egypt, and also in the period 1968-1984, was service manager officer in the Egyptian naval forces, Alexandria, Egypt, where international special training in German wireless communication companies: Hagenuk company in Kiel and Hamburg, and Becker company in Baden-Baden, Siemens company in Berlin, and Racal Communication Security COM-SEC, wireless communication company in England. In the field of Surface Acoustic Wave Filters, Prof. El-Shennawy has contributed a new design criterion for improving the performance of the band-pass signal processing, defined by Kham-Shen criteria for Ideal Band-pass Systems which should be taken into consideration during the design procedures, the criteria relates the transmission time delay of the filter and the periodic time of the carrier wave.

The fields of interest to Prof. El-Shennawy are: distributed networks, Charge Coupled Devices CCD and Surface Acoustic Wave SAW devices for modern communication systems, data computer communication, speech coding, speech enhancement, Voice over Internet Protocol VoIP, communication security systems (Encryption, DES, AES, public key techniques, scrambling), speech and image watermarking, audio and video compression (transform coding), acoustics, Ultra Wide Band UWB wireless communication systems, Multi Carrier Direct Sequence Code Division Multiple Access MC-DS-CDMA, Worldwide interoperability for Microwave Access WiMAX and Vehicular AdHoc Networks VANETS, Electronic Chart Display Information Systems ECDIS, Global Positioning Systems GPS, Global Maritime Distress and Safety Systems GMDSS, air-borne and space-borne remote sensing.

PREFACE

Prof. Dr. Khamies Mohammed Ali El-Shennawy, the author, believes that the book: "Communication Theory and Signal Processing for Transform Coding", is tailored for the requirements of the individual area of the signal processing in communication systems. The students of the undergraduate studies in the institutes, colleges, universities, and academies and want to specialize in the field of communication systems and signal processing, this book, is their innovation, and is more essential to them before the entrance of their specialized work in communication systems, in order to get the talent and the ability to have the faster solution for all the problems in analog and digital communication and their applications. Prof. El-Shennawy teach to the students of the undergraduate studies: circuit theory, communication theory, communication systems, data communication, electro-magnetic, antennas, and acoustics, and supervise the graduation projects as applications of Surface Acoustic Wave SAW devices in communication systems, communication security systems (encryption and decryption techniques, stream cipher, Data Encryption Standard DES, public key encryption, factorization and logarithmic and elliptic curve encryption and signature techniques, advanced encryption standard), global maritime distress and safety systems, electronic chart display and information systems, global positioning systems, air-borne and space-borne remote sensing. Also supervise and teach to the students of the graduate and post graduate studies for diploma of high studies, M.Sc. and Ph.D. courses: communication intelligence, data computer communication, surface acoustic wave devices and charge coupled devices in modern communication systems, ultra wide band technique, speech and digital coding, voice over internet protocol, audio and video compression (transform coding), source coding techniques, speech and digital watermarking, in the Arab Academy for Science and Technology and Maritime Transport AASTMT, College of Engineering and Technology, Department of Communication & Electronics and Computer Studies, Alexandria, Egypt, since 1988 to 2014 and still, three semesters every year.

This book contains a great number of numerous examples and solved problems and exercises, to explain the methodology of Fourier analysis, Fourier series, Fourier transform and properties, Discrete Fourier Transform DFT, Fast Fourier Transform FFT, Discrete Cosine Transform DCT, Discrete Wavelet Transform DWT, Contourlet Transform CT. The book proves that the students need mathematics in communication more than you may think, and make the student has the ability to deal with the DFT, FFT, DCT, DWT, and CT, utility computer programs. The advantage of this book is the simplicity, attract the student, easy to solve the problems using different ways, and with its wider contents in communication theory, applied in communication systems.

Also, this book is beneficial to the engineers of the graduate and post graduate studies, and to the researchers in the research centers because the book contains a great number of mathematical operations and is considered very important in the research results, solving their problems. The book is a big jump to the students and engineers in understanding, realization, and makes the understanding of their prediction fields wider. The book is a very good chance to the students and the engineers for verifying their predictive results in the communication problems and give them more trust. The book is considered, the first step, mathematically solving the communication problems.

Chapter I is an introduction to the model of communication system, signal contamination, why modulation and demodulation, Shannon-Hartley theorem, some basics concepts of signals and classification of signals waveforms: periodic and unperiodic, deterministic and random, Dirac delta function, unit step function, power and energy, causal and non-causal, analog and digital, and low-pass and band-pass signals, and five solved problems, as well as numerical examples.

Chapter II is review of the classical methods for the spectral analysis of the Fourier series and power spectra, Fourier series real coefficients and complex exponential coefficient methods, orthogonality, spectrum of periodic signals (discrete spectrum), sinc function, Parseval`s power theorem, power spectral density, and eleven solved problems, as well as numerical examples.

Chapter III is devoted to the spectral analyses of Fourier transform and energy spectra, spectrum of unperiodic signals (continuous spectrum), spectrum of some important integrable signals, rectangular pulse and sinc spectra, triangle pulse and sinc squared spectra, Gaussian pulse and Gaussian spectra, radio frequency pulse and two sinc spectra, decaying and rising single sided exponential pulses, Fourier transform properties, linearity, duality, scaling, shifting, differentiation, integration property of the zero boundary condition functions, convolution, area, and conjugate properties, Rayleigh`s energy theorem, energy spectral density, Fourier transform of the real and imaginary parts of a time function, and twenty eight solved problems, as well as numerical examples.

Chapter IV presents the Fourier transform of the special functions (non-integrable signals), Dirac delta function, exponential and sinusoidal functions, signum function, and unit step function. Also the chapter presents the Fourier transform integration property of the non-zero boundary condition functions, the relation between the Dirac delta function and the unit step function, Fourier transform of the error function, and twenty one solved problems, as well as numerical examples.

Chapter V evaluates the spectral analysis of the periodic signals using Fourier transformation, periodic Dirac delta functions, periodic rectangular functions, periodic triangle functions and six solved problems.

Chapter VI analysis the correlation function and spectral density, energy spectral density, power spectral density, autocorrelation function of energy signals, Fourier transform of autocorrelation function for energy signals, evaluation of energy content in terms of autocorrelation function, autocorrelation function of power signals, periodicity of autocorrelation function for power signals, Fourier transform of autocorrelation function for power signals, evaluation of average power in terms of autocorrelation function, cross-correlation function, cross-correlation function of energy signals, Fourier transform of cross-correlation function for energy signals, cross spectral density of energy signals, orthogonal energy signals in terms of cross-correlation function, cross-correlation function of power signals, Fourier transform of cross-correlation function for power signals, cross spectral density of power signals, orthogonal power signals in terms of cross-correlation function, and twenty three solved problems.

Chapter VII shows the signal transmission and systems, impulse response, transfer function, cascaded systems, causal and non-causal systems, stable and non-stable systems, bandwidth of low-pass and band-pass systems, relation between input and output energy spectral densities, distortionless system, ideal low-pass filter, ideal band-pass filter, distortion systems, amplitude distortion, phase distortion, uniformly distributed resistance capacitance interconnects systems, and sixteen solved problems.

Chapter VIII describes the Hilbert transform, Hilbert transform of sinusoidal functions, Hilbert transform and orthogonality, Hilbert transform and convolution principle, Hilbert transform of narrow band-pass signals, some important Hilbert transforms, and four solved problems, as well as numerical examples.

Chapter IX explains different analysis of the narrow band-pass signals and systems, pre-envelope, complex envelope, natural envelope, band-pass systems, equivalent low-pass technique, new design criterion for band-pass systems (Kham-Shen Criteria), input/output pre-envelope technique, dispersive systems, and envelope delay (group delay), and seven solved problems, as well as numerical examples.

Chapter X illustrates the numerical computation of the Fourier transform, sampling theorem, discrete Fourier transform and properties (linearity, shifting, and circular convolution), fast Fourier transform, sine and cosine transforms, discrete cosine transform, drawbacks of Fourier transform, short time Fourier transform, wavelet transform, discrete wavelet transform, contourlet transform, some application of compression techniques, lossless and lossy coding, Huffman encoding, run length encoding, Lempel-Ziv-Wekh encoding, predictive encoding, delta encoding, drawbacks of compression techniques, audio compression, MPEG layers I, II, III of audio compression, video compression, Joint Photographic Experts Group JPEG, JPEG initiative (JPEG 2000), Moving Picture Experts Group MPEG, MPEG-2, MPEG-4, principles behind compression, MPEG-4 International Standard (MP4), Transform-domain weighted interleave Vector Quantization TwinVQ in MPEG-4, comparison of MPEG-4 (H.264) and JPEG-2000 video compression, and fourteen solved problems, as well as numerical examples.

The ten chapters of the book are essentially suited for two semesters. The first semester on communication theory (from chapter one to chapter nine). It is expected that the reader has knowledge of mathematics, electronics, and circuit theory. The second semester on signal processing, audio and image processing, numerical computation, transform coding and compression techniques (chapter ten). The book is characterized by three directions, the mathematical point of view, the communication theory point of view, and the utility computer programs. The make up of the material for each course may be determined only by the backgrounds and interests, thereby allowing considerable flexibility in making up the course material. As an aid to the teacher of each course, a detailed solutions manual for all the unsolved problems which at the end of the chapters, is available from the publisher.

Prof. Dr. Khamies M. A. El-Shennawy
President Assistant
Arab Academy for Science and
Technology and Maritime Transport.
P.O.Box 1029, Alexandria, Egypt.
khamies@ieee.org

Acknowledgments

I would like to express my deep gratitude to Prof. Dr. Richard L. Magin, Department of Bioengineering, University of Illinois at Chicago, and to Prof. Dr. Ashfaq Khokhar, Department of Electrical and Computer Engineering, University of Illinois at Chicago, and to Prof. Dr. Ehab Sabry, Department of Electronics and Communication, Arab Academy for Science and Technology and Maritime Transport AASTMT, Alexandria, Egypt, for their helpful inputs and suggestions.

I wish to thank Mr. El-Sayed Barakat, AASTMT, for his tireless effort in drawing so many different versions of the manuscript.

Conflict of Interest

The author confirms that this eBook content have no conflict of interest.

CHAPTER 1

Communication System and Signals

Abstract: Communication is the process where the information (message signal may be speech or text or picture), is transferred from point in space and time (source) to another point, through transmission channel (communication medium), separating the transmitter from the receiver. Communication enters our daily lives in many different ways, telephones, radios, televisions, and computer terminals. Communication also provides the senses for ships on the seas, aircraft in flights, and satellite in space.
Keywords: Model of communication system; Shannon-Hartley law; Classification of signals.

1.1. Model of Communication System

The main parts of the communication system, Fig.1.1, are: the input transducer, the transmitter, the transmission medium (channel), the receiver, and the output transducer.

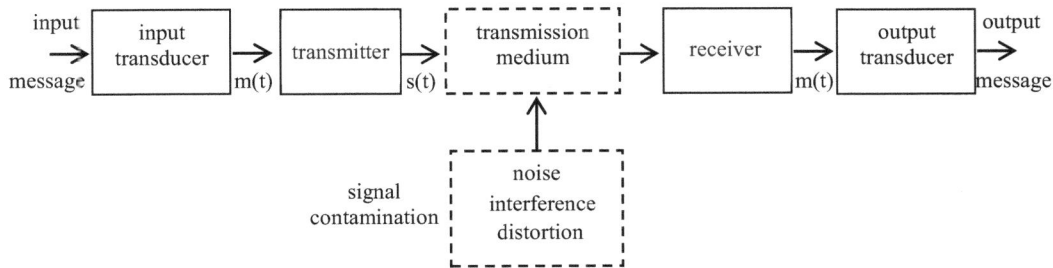

Fig.1.1. Model of communication system.

The input transducer converts the input message into a time-varying electrical baseband signal m(t) which designates the band of low frequencies delivered by the source of information, such as:

  i. The microphone in the telephony transmission.
 ii. The Morse key in the low speed telegraphy transmission (Morse code).
iii. The teleprinter in the high speed telegraphy transmission (Telex).
 iv. The facsimile in the video transmission (Fax).
  v. The Laptop or Desktop in the data transmission (E-mail, Internet, Voice over Internet Protocol VoIP, and audio video systems protocols).

The transmitter prepares the baseband signal m(t), Fig.1.1, into suitable form to be transmitted in the transmission medium (channel), by performing the modulation processing for efficient transmission. Efficient transmission requires a shift of the range of the baseband low frequencies components of m(t) into other high frequency ranges suitable for transmission in the channel, in the form of a transmitted modulated signal s(t).

The transmission medium (channel) achieves the connection between the transmitter and the receiver such as:

  i. Transmission Line TL may be pair of two wires or coaxial cable where the transmitted signal s(t) propagates in a form of electrical energy. The coaxial cable is preferred due to minimum losses and achieves higher data rates.
 ii. Free space (ether), where the transmitted signal s(t) is converted into Electro-Magnetic energy by means of radiating antenna. The vector product of the Electric field and magnetic field, gives the direction of propagation.
iii. Waveguide, where the radiating antenna converts the transmitted signal s(t) into Electro-Magnetic energy to propagate inside the waveguide which has rectangular or square or circular cross-section area and cupper made.
 iv. Optical fiber, where the transmitted signal s(t) is converted into light energy using special electro-optical converters such as Light Emitting Diodes LED and laser diodes. Fiber optics achieves very high data transfer rates and provides greater capacity.

The receiver extracts the desired high frequency range of the transmitted modulated signal s(t) from the channel and recreates the original baseband signal m(t) by performing demodulation processing.

The output transducer converts the output baseband signal m(t) into the original message (speech or text or picture) such as:

  i. The loud speaker in the telephony and Morse code (low speed telegraphy) transmissions.

  ii. The teleprinter in the high speed telegraphy transmission (Telex).

  iii. The facsimile in the video transmission (Fax).

  iv. The Laptop or Desktop in the data transmission (E-mail, Internet, Voice over Internet Protocol VoIP, and audio video systems protocols).

1.2. Signal Contamination

Signal transmission inside the communication system is exposed to contamination of distortion, interference and noise, causing fading at the input signal of the receiver, and consequently some physical changes occur in the original baseband signal m(t) at the output of the receiver. Transmitters, receivers and channels are carefully designed to minimize the signal contamination such as:

  i. The distortion is the signal shape alteration, because the response of the system to the signal, is not perfect due to the system design and the nonlinearities of the electronic devices. Distortion may be amplitude distortion or frequency distortion (delay distortion). Distortion is avoided by good communication system design.

  ii. The interference is the undesirable received external signals similar to the desired received signal such as the image frequency.

  iii. The noise is the unpredicted natural signals, mixed with the desired transmitted signal such as the external noise (atmospheric noise, galactic noise, and man made noise), and the internal noise (shot noise and thermal noise). Noise disturb the processing of the original signals in the communication system. Some noise is reduced using noise filters and others are never eliminated such as thermal noise.

1.3. Modulation and Demodulation

The baseband signal m(t) (low-pass signal) designates the band of low frequencies components which can not propagate very long distances such as audio frequencies from 16 Hz to 20 KHz, the baseband signal of the voice channel from 300 Hz to 3400 Hz (the bandwidth of the voice channel is 3.1 KHz), and the low frequency components of the video signal range.

In the transmitter, a carrier signal c(t) has high frequency f_c is generated by an oscillator inside the transmitter ($f_c \geq 30$ KHz). The carrier frequency is much higher than the maximum frequency of the baseband signal m(t). For efficient transmission, the carrier signal c(t) is modulated by the baseband signal m(t) in the modulator inside the transmitter forming the modulated signal s(t) (band-pass signal), Fig.1.1. This band-pass signal s(t) is a modulated signal contains certain band of high frequency components around or adjacent to the carrier frequency f_c and can be transmitted in the transmission medium (channel) over a much longer distance convoying the information m(t).

An example, the sound is a baseband signal m(t) propagates in the space with sound speed 343 mt/sec, while the modulated Radio Frequency RF wave s(t) propagates in the space with the speed of light 3×10^8 mt/sec. In the receiver, the received modulated band-pass signal s(t) is demodulated to recreate the original baseband low-pass signal m(t).

In the Continuous Wave CW modulation, the carrier wave and the baseband signal m(t) are sinusoidal waves, with two families of modulation systems, namely, Amplitude Modulation AM and Angle Modulation systems. The waveform of the carrier wave c(t) is given by

$$c(t) = A_c \cos(2\pi f_c t + \varphi_c) \qquad \text{volt}$$

, where A_c, f_c, φ_c are the magnitude, frequency, and phase of the carrier wave. In the amplitude modulation, the carrier amplitude varies in accordance with the baseband signal m(t) while the carrier frequency f_c and the carrier phase φ_c are kept constants. Different kinds of amplitude modulation are used: Double Side Band-Transmitted Carrier DSB-TC, Double Side Band-Suppressed Carrier DSB-SC, Quadrature Amplitude Modulation QAM, and Single Side Band SSB modulation techniques. In the angle modulation, either the carrier frequency varies in accordance with the baseband signal m(t) while the carrier amplitude A_c and the carrier phase φ_c are kept constants, this is defined by Frequency Modulation FM, or the carrier phase varies in accordance with the baseband signal m(t) while the carrier amplitude A_c and the carrier frequency f_c are kept constants, and this is defined by Phase Modulation PM. The validity of these modulation techniques are demonstrated in the continuous wave communication [1].

In the digital modulation, the carrier wave is a train of pulses, has pulse amplitude, pulse width, and pulse position, while the baseband signal m(t) is sampled with sampling frequency f_s which must be higher than or at least twice the maximum frequency of m(t), in order to, in the receiver, the recovered original signal m(t) can be reconstructed with vanishingly small distortion (sampling theorem), (chapter X). Three families of digital modulation systems, namely, pulse analog modulation, pulse code modulation, and delta modulation. In the pulse analog modulation, a Pulse Amplitude Modulation PAM where the pulse amplitude varies in accordance with the baseband signal m(t) while the pulse width and the pulse position are kept constants, a Pulse Width Modulation PWM where the pulse width varies in accordance with the baseband signal m(t) while the pulse amplitude and the pulse position are kept constants, and a Pulse Position Modulation PPM where the pulse position varies in accordance with the baseband signal m(t) while the pulse amplitude and pulse width are kept constants, the former of the pulse analog modulation is analogous to the continuous wave modulation. In the Pulse Code Modulation PCM, the samples amplitude of the baseband signal m(t) are quantized and the standard quantization causes inherent some round-off error associated with the PCM systems due to the mathematical approximation, the number of the standard quantization levels occupy the total amplitude range of the baseband signal m(t), and then the quantization levels are coded, where the Amplitude Shift Keying ASK, the Frequency Shift Keying FSK, and the Phase Shift Keying PSK modulations. The Delta Modulation DM may be described as a type of PCM where only one bit encoding is used with two quantization levels only, expressing the baseband signal as a train of delta functions $\pm \delta(t)$, the positive delta function means the sampled amplitude of m(t) is increased while the negative delta function means the sampled amplitude of m(t) is decreased. The validity of these modulation techniques are demonstrated in the digital communication systems [2].

In communication system, there are two primary resources: the transmitting Radio Frequency RF power and the channel frequency bandwidth. These two resources should be used as efficiently as possible. In most channels, one resource may be considered than the other, so the communication channels are classified into: either band-limited system such as the telephony circuit channel where the bandwidth of the voice channel is 3.1 KHz, or power-limited system such as space satellite communication link channel where the wide band and the ultra-wide band. These primary resources are related by the following Shannon-Hartley law:

$$C = B \log_2(1 + \frac{S}{N_{th}}) \qquad \text{bits/sec} \qquad (1.1)$$

, where C is the maximum channel capacity or the rate of message transmission in bits per second, B is the channel frequency bandwidth in Hz and S/N_{th} is the channel signal to noise power ratio. For a given channel capacity C, the channel bandwidth B is increased on the account of the reduction of the channel S/N_{th} ratio (the received S/N_{th} ratio). For optimum receiver, the output S/N_{th} ratio of the receiver equals the S/N_{th} ratio of the channel. The Shannon-Hartley theorem is too complex to drive here, however, its validity is demonstrated in [3], pp.421–423.

An example of the band limited system, consider a voice signal channel being used, via modulator/demodulator (modem), to transmit digital data. The bandwidth of the voice channel is 3100 Hz. A typical value of signal to noise power ratio is 30 dB (ratio of 1000:1). Then the maximum capacity of this voice channel achievable, is given by

$$C = 3100 \log_2 (1 + 1000) = 30,894 \qquad \text{bits/sec}$$

Another example of the ultra-wide band system, consider an Ultra Wide Band UWB signal channel being used to transmit digital data. The bandwidth of the UWB channel is 7 GHz. A typical value of signal to noise power ratio is 0.1 dB (ratio of 1.023:1). Then the maximum capacity of this UWB channel achievable, is given by

$$C = 7\times10^9 \log_2 (1 + 1.023) = 7.116 \qquad \text{Giga bits/sec}$$

These two examples illustrate that, for the Ultra Wide Band UWB channel, the channel capacity C and the bandwidth B are converge, where the signal level may be under the noise level, while for the narrow band voice channel, the channel capacity C and the bandwidth B are diverge where the signal level is much higher than the noise level. Shannon-Hartley law, Eq.(1.1), is considered the central theorem of the information theory. It is evident from this theorem that the bandwidth and the signal power can be exchanged for one another. To transmit the information at a given rate, the transmitted signal power is reduced provided that the bandwidth is increased correspondingly. Similarly, the bandwidth may be reduced if the signal power is willing to be increased. The process of modulation is really a means of affecting this exchange between the bandwidth and the signal to noise power ratio.

The available thermal noise power N_{th} generated by a resister is proportional to its absolute temperature, in addition to being proportional to the bandwidth B, which are related by the following formula

$$N_{th} = k_B T_{emp} B \qquad\qquad \text{watt} \qquad (1.2)$$

, where k_B is the Boltzmann's constant (1.38×10^{-23} Joule/Kelvin), T_{emp} is the absolute temperature (Kelvin degree = 273 + °C), B Hz is the bandwidth, and °C is degree Celsius.

Problem 1.1
A voice signal channel being used, via modulator/demodulator (modem), to transmit digital data. The bandwidth of the voice channel is 4000 Hz. A typical value of signal to noise power ratio is 28 dB, (ratio of 631:1). Evaluate:
  i. The capacity of the voice channel achievable.
  ii. The capacity of the voice channel achievable if the bandwidth is doubled while the transmitted signal power remains constant.
Solution
  i. The capacity of the voice channel achievable, is given by

$$C_1 = 4000 \log_2(1 + 631) = 37{,}216 \qquad\qquad \text{bits/sec}$$

  ii. When the bandwidth is doubled while the transmitted signal power remains constant, and according to Eq.(1.2), the noise power will be doubled, and the capacity of the voice channel achievable yields

$$C_2 = 8000 \log_2[1 + (631/2)] = 66{,}448 \qquad\qquad \text{bits/sec}$$

, when the bandwidth is doubled while the transmitted signal power remains constant, the capacity of the voice channel achievable is increased by a ratio C_2/C_1 of 1.785.

In order to study the communication system, you must be familiar with various ways of representing signals and classification of signals, in the time domain and in the frequency domain, to show the waveforms and the spectral contents of the different communication signals.

1.4. Signals and Classification of Signals
The signal x(t) is a single-valued waveform function dependent of the time t, and can be seen by the oscilloscopes. It may be real or complex and convoys the information, the time t is real independent variable. In the frequency domain, the signal x(t) contains a certain spectrum of frequency components. Fourier analysis achieves the conformal mapping between the time domain of the signal x(t) and its frequency domain. The signal representation in time domain and frequency domain, hinges on the particular type of the signal being considered, depending on the feature of the interest.

Signals may be classified into six different classes:
  i. Periodic and unperiodic signals.
  ii. Deterministic and random signals.
  iii. Power and energy signals.
  iv. Causal and non-causal signals
  v. Analog and digital signals.
  vi. Low-pass and band-pass signals.

1.4.1. Periodic and Unperiodic Signals
The waveform equation of the periodic signal $x_p(t)$, Fig.1.2b, is expressed by

$$x_p(t) = \sum_{m=-\infty}^{\infty} x(t - mT_o) \qquad \text{for all t,} \qquad m = 0,1,2,3,\dots \quad (1.3)$$

, where x(t) is the unperiodic signal, Fig.1.2a, and defines one cycle of the periodic signal $x_p(t)$, so x(t) is considered the generating function of the periodic signal $x_p(t)$ in the period T_o, and T_o defines the periodic time of $x_p(t)$. The future behavior of the periodic signal is certainty and known because it exists from $-\infty$ to $+\infty$. Therefore, in the limit, the period T_o of the unperiodic signal becomes infinitely large, and x(t) yields

$$x(t) = \lim_{T_o \to \infty} x_p(t) \qquad\qquad (1.4)$$

, generally the future behavior of the unperiodic signals may be uncertain and unknown.

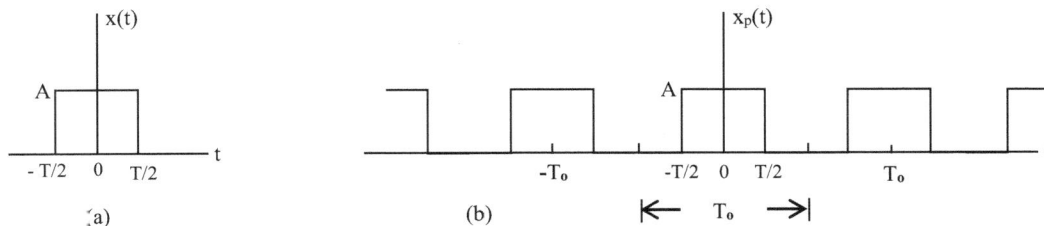

Fig.1.2. a) Unperiodic rectangular signal, b) Periodic rectangular signal.

1.4.2. Deterministic and Random Signals

The deterministic signal is a completely specified function in time and amplitude, and has not uncertainty with respect to value at any time such as the rectangular signals, Fig.1.2, where the unperiodic rectangular signal, Fig.1.2a, is denoted by

$$x(t) = A \, rect \left(\frac{t}{T}\right)$$

, and the periodic rectangular signal, Fig.1.2b, is expressed by

$$x_p(t) = \sum_{m=-\infty}^{\infty} A \, rect \left[\frac{t - mT_o}{T}\right] \qquad \text{for all t,} \qquad m = 0, 1, 2, 3, \dots$$

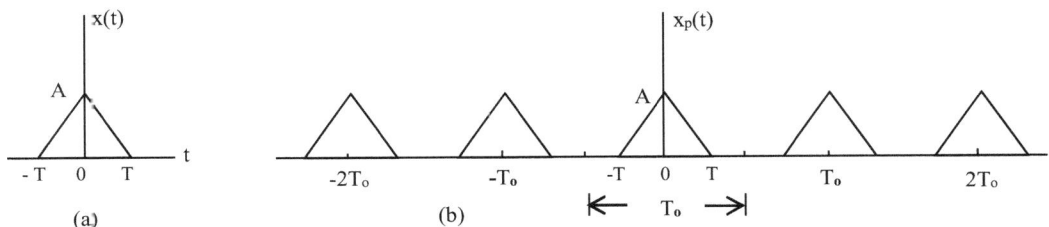

Fig.1.3. a) Unperiodic triangle signal, b) Periodic triangle signal.

Also the triangle signals, Fig.1.3, where the unperiodic triangle signal, Fig.1.3a, is denoted by

$$x(t) = A \, tri \left(\frac{t}{T}\right)$$

, and the periodic triangle signal, Fig.1.3b, is expressed by

$$x_p(t) = \sum_{m=-\infty}^{\infty} A \, tri \left[\frac{t - mT_o}{T}\right] \qquad \text{for all t,} \qquad m = 0, 1, 2, 3, \dots$$

, and also the unit step function $u(t)$, the signum function $sgn(t)$, and the Dirac delta function $\delta(t)$ (unit impulse), Fig.1.4. are deterministic signals.

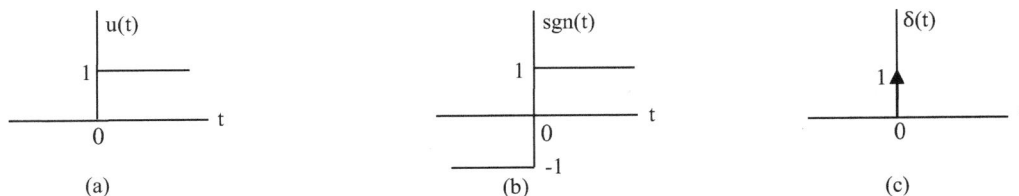

Fig.1.4. a) Unit step function, b) Signum function, c) Dirac delta function.

While the random signal (stochastic waveform) is not completely specified function in time and amplitude, and has some degree of uncertainty before the signal actually occurs, so it may be viewed as collection or ensemble of unknown signals, Fig.1.5. The future behavior of the random signal is uncertainty and unknown, so it must be modeled probabilistically. A typical example of the random signal is the voice waveform.

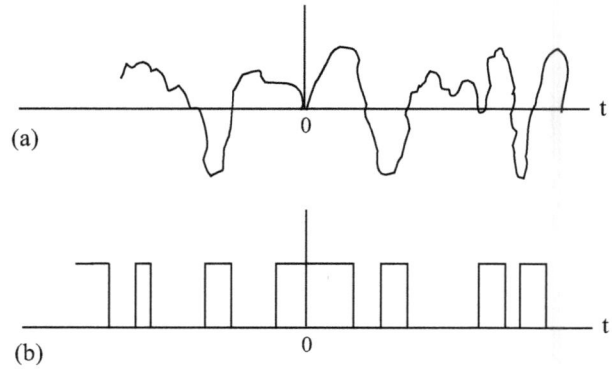

Fig.1.5. a) Continuous random signal, b) Digital random signal.

1.4.2.1. Dirac Delta Function

Dirac Delta function (unit impulse) is a pulse with duration equals zero, given at the origin, its pulse amplitude goes to infinity with unit strength (unit area), Fig.1.4c, and denoted by $\delta(t)$, where

$$\int_{-\infty}^{\infty} \delta(t)dt = 1 \qquad \text{at} \quad t = 0 \qquad (1.5)$$

$$= 0 \qquad \text{at} \quad t \neq 0$$

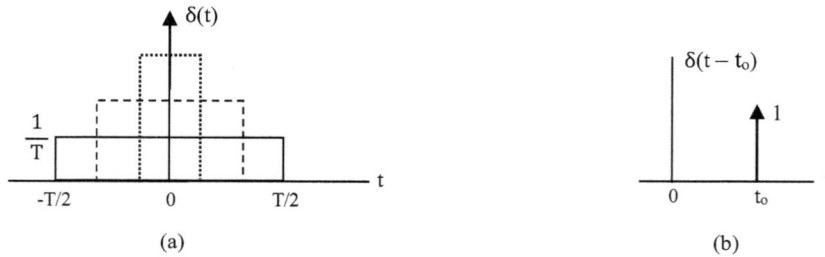

Fig.1.6. a) Unit area rectangular pulse, with duration T variable parameter, when the limit T tends to zero, the pulse tends to the Dirac delta function, b) A shifted Dirac delta function at $t = t_o$.

, consider a rectangular pulse x(t) with duration T and amplitude 1/T, Fig.1.6a, x(t) is expressed by

$$x(t) = \frac{1}{T} \ \text{rect}(\frac{t}{T})$$

, assume that the pulse duration T is a variable parameter, and in the limit T tends to zero. This unit area rectangular pulse becomes infinitely narrow in duration and infinitely large in amplitude, remaining its area finite and fixed at unity. Then the Dirac delta function is given by

$$\delta(t) = \lim_{T \to 0} \frac{1}{T} \ \text{rect}(\frac{t}{T}) \qquad (1.6)$$

, the multiplication of an arbitrary function x(t) by the delta function $\delta(t)$, is given by

$$x(t) \ \delta(t) = \ x(0) \ \delta(t) \qquad \text{at} \ \ t = 0 \qquad (1.7)$$

$$= 0 \qquad \text{otherwise}$$

, also the multiplication of x(t) by a shifted delta function $\delta(t - t_o)$, Fig.1.6b, is given by

$$x(t)\,\delta(t - t_o) = x(t_o)\,\delta(t - t_o) \qquad \text{at } t = t_o \qquad (1.8)$$
$$= 0 \qquad\qquad\qquad \text{otherwise}$$

, where t_o is the shifting time, x(0) and $x(t_o)$ are the values of the arbitrary function x(t) at the origin and at t_o respectively. Then the integrations for a function x(t) multiplied by a delta function, using Equations (1.5), (1.7) and (1.8), yield

$$\int_{-\infty}^{\infty} x(t)\,\delta(t)\,dt = x(0) \qquad (4.4a)$$

, and

$$\int_{-\infty}^{\infty} x(t)\,\delta(t - t_o)\,dt = x(t_o) \qquad (4.4b)$$

, since the definition of the Dirac delta function, Eq.(1.6), is the Limit T→0 of a rectangular pulse having duration T, and amplitude 1/T, of constant area (unity). Then the time scaling of the Dirac delta function $\delta(at)$, where the "a" is positive real constant (a ≥ 0), is given by

$$\delta(at) = \frac{1}{|a|}\,\delta(t) \qquad (4.4c)$$

Dirac Delta function $\delta(t)$ is the differentiation of the unit step function u(t), Eq.(4.21b). $\delta(t)$ is an even function and is not a function in the usual sense, but it is very important concept or tool which allows the easy solution of many problems in the communication theory that would otherwise quite difficult.

Numerical examples

i. $\int_{-\infty}^{\infty} (t^4 + t^2 + t - 1)\,\delta(t - 1)\,dt = 2$, where $\delta(t - 1)$ occurs at $t = 1$

ii. $\int_{-\infty}^{\infty} (t^3 + 3)\,\cos[2\pi(t + \tfrac{1}{2})]\,\delta(8 - 4t)\,dt = -\frac{11}{4}$, where $\delta(8 - 4t) = \tfrac{1}{4}\delta(t - 2)$

1.4.2.2. Unit step function
The unit step function u(t), Fig.1.4a, is defined by

$$u(t) = 1 \qquad t > 0$$
$$= \tfrac{1}{2} \qquad t = 0$$
$$= 0 \qquad t < 0$$

, where the origin value of the step function, point of discontinuity, is the mean of the values of the function, also u(at) = u(τ), where τ = at, and the "a" is positive real constant (a ≥ 0). The unit step function u(t) is the integration of the Dirac Delta function $\delta(t)$, Eq,(4.21a).

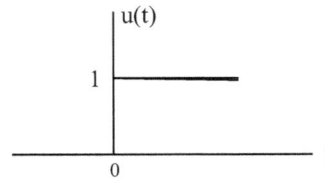

Fig.1.4a. Unit step function.

Numerical examples

     The product $\delta(t)\,u(t) = \tfrac{1}{2}\delta(t)$

,   the product $\delta(t - 2)\,u(t - 2) = \tfrac{1}{2}\delta(t - 2)$

,   the product $\delta(t - 3)\,u(t - 2) = \delta(t - 3)$

,   the product $\delta(t + 6)\,u(t + 4) = 0$

, and   the product $\delta(t + 5)\,u(3 - t) = \delta(t + 5)\,u[-(t - 3)] = \delta(t + 5)$

1.4.3. Power and Energy Signals
The instantaneous dissipated power in an electrical element of pure resistance R, is defined by

$$p_{in}(t) \;=\; \frac{|v(t)|^2}{R} \;=\; |i(t)|^2\,R \qquad\qquad watt$$

, where $v(t)$ and $i(t)$ represent the electrical voltage and current. The total energy consumed by the resistance R, is defined by the area under the instantaneous power $p_{in}(t)$, Fig.1.7, and is given by

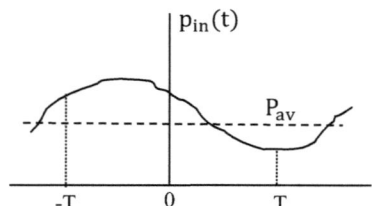

Fig.1.7. The instantaneous dissipated power versus time t.

$$E \;=\; \lim_{T\to\infty} \int_{-T}^{T} p_{in}(t)\,dt \qquad\qquad joule$$

, and the average power will be

$$P_{av} \;=\; \lim_{T\to\infty} \frac{1}{2T} \int_{-T}^{T} p_{in}(t)\,dt \qquad\qquad watt$$

In the signal analysis, it is customary to work within one ohm resistor, regardless if whether the given signal $x(t)$ represent a voltage $v(t)$ or current $i(t)$, and then the instantaneous dissipated power is expressed by

$$p_{in}(t) = |x(t)|^2 \qquad\qquad watt/ohm$$

The power signal (periodic and random) of periodic time T_o, has finite average power and has total energy equals infinity because it extends from $-\infty$ to $+\infty$. Since the periodic signal $x_p(t)$ is a special case of the power signal, then the average power of the power signal, is given by

$$P_{av} \;=\; \frac{Energy/cycle}{period\ T_o}$$

, where the energy/cycle is given by

$$Energy/cycle \;=\; \int_{-T_0/2}^{T_0/2} |x_p(t)|^2\,dt$$

, and the average power will be

$$P_{av} \;=\; \frac{1}{T_o} \int_{-T_0/2}^{T_0/2} |x_p(t)|^2\,dt \qquad\qquad (1.9)$$

, where T_o is the periodic time of the periodic signal $x_p(t)$. In the frequency domain, the periodic signal can be represented by applying Fourier series and in a limiting sense Fourier transform.

On the other hand, the energy signal (unperiodic and deterministic) has periodic time T_o equals infinity, has finite energy and zero average power. Since the unperiodic signal $x(t)$ is a special case of the energy signal, then the energy content of the energy signal is given by

$$E = \int_{-\infty}^{\infty} |x(t)|^2\,dt \qquad\qquad (1.10)$$

In the frequency domain, the unperiodic signal can be represented by applying Fourier transform.

Problem 1.2
Evaluate the energy contents and the average power of the rectangular signals, Fig.1.2.

Solution
The signal x(t), Fig.1.2a, is an energy signal (unperiodic). The instantaneous power $p_{in}(t)$ is equal A^2 in the duration T and zero elsewhere. The energy content of x(t), Eq.(1.10), is given by

$$E = \int_{-T/2}^{T/2} A^2 \, dt$$

$$= A^2 \, T \qquad\qquad \text{joule} \qquad\qquad\qquad (1.11)$$

, and the average power of x(t) equals zero because it has periodic time T_o equals infinity. While the signal $x_p(t)$, Fig.1.2b, is a power signal (periodic), has finite periodic time T_o and the average power of $x_p(t)$, Eq.(1.9), is given by

$$P_{av} = \frac{1}{T_o} \int_{-T/2}^{T/2} A^2 \, dt$$

$$= A^2 \, \frac{T}{T_o} \qquad\qquad \text{watt} \qquad\qquad\qquad (1.12)$$

, and the energy content of $x_p(t)$ equals infinity. The ratio T/T_o, is defined by the duty cycle of the rectangular periodic signal, this ratio is very important in the design of the communication systems.

Problem 1.3
Evaluate the energy contents and the average power of the triangle signals, Fig.1.3.

Solution
The signal x(t), Fig.1.3a, is an energy signal (unperiodic). The total energy, Eq.(1.10), is given by

$$E = \int_{-T}^{0} \left[A + \frac{A}{T} t \right]^2 dt + \int_{0}^{T} \left[A - \frac{A}{T} t \right]^2 dt$$

$$= \frac{2}{3} A^2 T \qquad\qquad \text{joule}$$

, and the average power of x(t) equals zero because it has periodic time T_o equals infinity. While the signal $x_p(t)$, Fig.1.3b, is a power signal (periodic), has finite periodic time T_o and the average power of $x_p(t)$, Eq.(1.9), is given by

$$P_{av} = \frac{1}{T_o} \left\{ \int_{-T}^{0} \left[A + \frac{A}{T} t \right]^2 dt + \int_{0}^{T} \left[A - \frac{A}{T} t \right]^2 \right\} dt$$

$$= \frac{2}{3} A^2 \frac{T}{T_o} \qquad\qquad \text{watt}$$

, and the energy content of the periodic signal $x_p(t)$ equals infinity.

Problem 1.4
Evaluate the energy contents of the Radio Frequency RF signals, Fig.1.8.

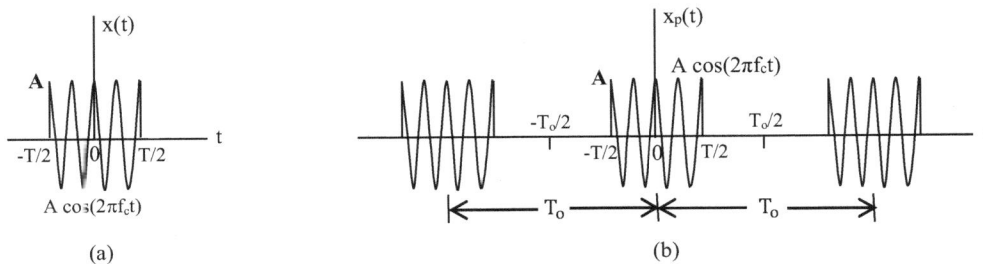

Fig.1.8. a) Unperiodic RF signal, b) Periodic RF signal.

Solution
The signal x(t), Fig.1.8a, is an energy signal (unperiodic). The instantaneous power $p_{in}(t)$ equals $A^2 \cos^2(2\pi f_c t)$ in the duration T and zero elsewhere. The energy content of x(t), Eq.(1.10), will be

$$E = \int_{-T/2}^{T/2} A^2 \cos^2(2\pi f_c t)\, dt$$

, making use of the formula "$2\cos^2(x) = 1 + \cos(2x)$", E yields

$$E = \tfrac{1}{2} \int_{-T/2}^{T/2} A^2 [1 + \cos(2\pi 2 f_c t)]\, dt$$

$$= \tfrac{1}{2} A^2 T \qquad \text{joule} \qquad (1.13)$$

, and the average power of x(t) equals zero because it has periodic time T_o equals infinity. While the signal $x_p(t)$, Fig.1.8b, is a power signal (periodic), has finite periodic time T_o and the average power of $x_p(t)$, Eq.(1.9), is given by

$$P_{av} = \frac{1}{T_o} \int_{-T/2}^{T/2} A^2 \cos^2(2\pi f_c t)\, dt$$

$$= \tfrac{1}{2} A^2 \frac{T}{T_o} \qquad \text{watt} \qquad (1.14)$$

, and the energy content of $x_p(t)$ equals infinity. The ratio T/T_o is defined by the duty cycle of the RF periodic signal which is very important in the design of the communication systems.

Problem 1.5
Evaluate the energy contents and the average power of the single sided decaying and rising exponential pulses $x_1(t) = e^{-t} u(t)$, and $x_2(t) = e^t u(-t)$, Fig.1.9.

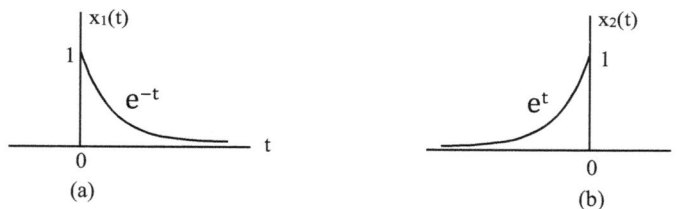

Fig.1.9. a) Single sided decaying pulse, b) Single sided rising pulse.

Solution
The signals $x_1(t)$ and $x_2(t)$ are energy signals (unperiodic). The instantaneous power $p_{in}(t)$ equal $|e^{-t}|^2$, and $|e^t|^2$ respectively in the duration of each sided. The energy contents of $x_1(t)$, and $x_2(t)$, Eq.(1.10), are given by

$$E_1 = \int_0^\infty |e^{-t}|^2\, dt = \int_0^\infty e^{-2t}\, dt = \frac{1}{2} \qquad \text{joule} \qquad (1.15)$$

, and

$$E_2 = \int_{-\infty}^0 |e^t|^2\, dt = \int_{-\infty}^0 e^{2t}\, dt = \frac{1}{2} \qquad \text{joule} \qquad (1.16)$$

, the average power of $x_1(t)$ and $x_2(t)$ equal zero because each of them have periodic time T_o equals infinity.

1.4.4. Causal and Non-Causal Signals

Causal signals are the signals that are zero for all negative time, Fig.1.10a, such as the unit step function u(t), Fig.1.4a, while the anti-causal signals are the signals that are zero for all positive time, Fig.1.10c, such as the negation unit step function u(–t), Fig.4.13. On the other hand, the non-causal signals are the signals that have nonzero values in both positive and negative time, Fig.1.10b. Shifting the signals along the time axis, change the nature of the signals from causal to non-causal to anti-causal signals.

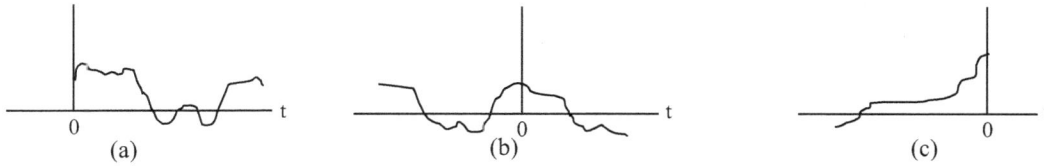

Fig.1.10. a) Causal signal, b) Non-causal signal, c) Anti-causal signal.

1.4.5. Analog and Digital Signals

An analog signal is a continuous function of time, Fig.1.11a, having continuous amplitude. It arises when a physical waveform such as:

i. Acoustic wave (sound) where the transducer (microphone) converts the sound pressure variations into corresponding electrical continuous signal.

ii. Light wave where the photo-electric cell converts the light intensity variations into corresponding electrical continuous signal.

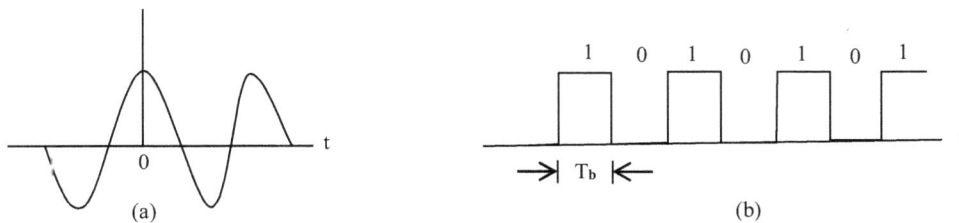

Fig.1.11. a) Analog signal, b) Digital signal.

On the other hand, the digital signal is a discrete time signal and is defined only at discrete times (bits), and usually uniformly spaced, Fig.1.11b. If the bit duration is T_b seconds, the data rate equals $1/T_b$ bits/sec.

In Digital Signal Processing DSP, the analog signal may be converted into digital signal, Fig.1.12. Consider a continuous band-limited signal m(t) is sampled one every T_s, where T_s is the sampling period. The sampling frequency f_s (f_s directly proportional to $1/T_s$), must be higher than or at least twice the maximum frequency of the input continuous band-limited signal m(t), in order to, the original signal m(t) can be reconstructed in the receiver with vanishingly small distortion (sampling theorem), (chapter X). The discrete time sampled signal $m_s(t)$ is distributed as sequences of samples with varying amplitude. The AND gate is a sampler when the continuous signal m(t) is fed to one input of the AND gate and the other input is fed by pulses having sampling frequency f_s. The output of the AND gate is then the discrete time sampling signal $m_s(t)$ which are passed through a pulse-shaping network to give them flat tops. These discrete samples are quantized and then coded (binary or octal or hexadecimal), to be in the form of digital signal, Fig.1.12.

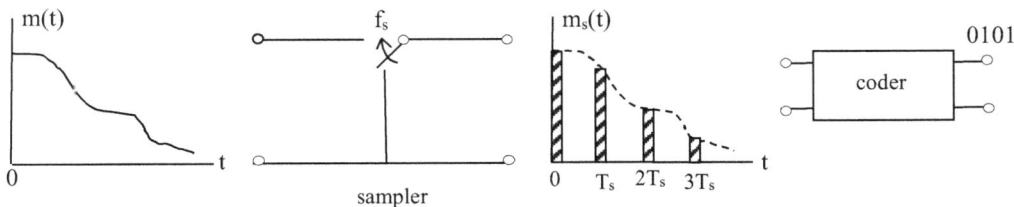

Fig.1.12. Analog to digital conversion.

1.4.6. Low-Pass and Band-Pass Signals

The information m(t) is a low-pass signal (baseband) has frequency components from or near dc up to some finite value usually less than a few MHz. This low-pass signal is a band-limited signal and is defined by its maximum frequency components W Hz. An example, the audio spectral content extends from 16 Hz up to 20 KHz, while the video spectral content extends from zero Hz up to about 7 MHz in color television, and the voice channel bandwidth extends from 300 Hz to 3.4 KHz. These signals contain very low frequency components which can not propagate very long distances. The transmitter prepares the baseband signal m(t) into suitable form to be transmitted in the channel, by performing the modulation processing for efficient transmission. Efficient transmission requires a shift of the range of the baseband low frequency components into other high frequency range of band-pass signal s(t) suitable for transmission in the channel, Fig.1.1. The band-pass signal s(t) is a modulated signal contains certain band of high frequency components around or adjacent to the carrier frequency f_c (f_c is much higher than W Hz) and being transmitted in the channel and propagates with very high speed (about the speed of light 3×10^8 mt/sec) and can be transmitted over a much longer distance. Fig.1.13. shows M(f) and S(f), are the representation of m(t) with maximum frequency W Hz and s(t) of bandwidth 2W Hz in the frequency domain, where the carrier wave is represented by a cosine function $\cos(2\pi f_c t)$, and the modulated wave is the product of the baseband m(t) by the carrier wave $\cos(2\pi f_c t)$.

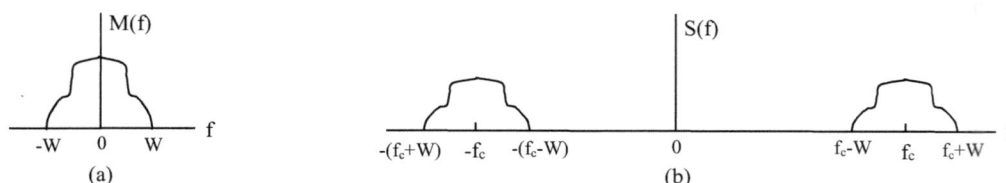

Fig.1.13. a) Low-pass signal, b) Band-pass signal.

Generally, the analog and digital signals are processed in the communication systems and may be described in the time and frequency domains. Fourier analyses play the role of resolution these signals into frequency spectrum, and provide a means of analyzing and designing frequency selective filters for the separation of signals on the basis of their frequency content. Fourier analyses are named after the renowned French mathematician Jean Baptiste Joseph Fourier (1768–1830). Fourier analyses are transform domain analyses that engineers find very useful because many of the properties and results of the transform have a physical interpretation in both the time domain and the transform domain. Fourier transform methods provide a unifying mathematical approach to the study of a variety of physical phenomena which also apply to problems in communication theory, control, information processing, and other linear analysis. Waveforms, either continuous-time or sampled, and their corresponding spectra arise in a variety of applications not only as electrical signals but also as optical or acoustical waveform.

Problems

1. A video signal channel being used to transmit digital data. The channel capacity achievable is 29.9 M bits/sec. The signal to noise power ratio of the video channel is 30 dB. Evaluate the bandwidth of the video signal channel.

2. A good voice reproduction of: telephone speech, wideband speech, and wideband audio, via PCM-encoded digital data, imply channel capacities achievable of 64 Kbits/sec, 224 Kbits/sec, and 768 Kbits/sec respectively. The voice signals occupy a bandwidth of 3100 Hz for telephone speech, 6.94 KHz for wideband speech, and 19.99 KHz for wideband audio. Evaluate the signal to noise power ratio in dB for the telephone speech, wideband speech, and wideband audio.

3. A teleprinter signal channel being used to transmit digital data. The bandwidth of this telegraph channel is 300 Hz. The signal to noise power ratio of the telegraph channel is 3 dB. Evaluate the channel capacity achievable ?

4. A channel of bandwidth 2 KHz and signal to noise power ratio of 24 dB. Calculate:
    i. The maximum capacity achievable of the channel.
    ii. The maximum capacity achievable of the channel when the channel bandwidth is halved, while the transmitted signal power remains constant.
    iii. The maximum capacity achievable of the channel when the channel bandwidth is doubled, while the transmitted signal power remains constant.

5. A TV picture signal channel being used to transmit digital data. The bandwidth of the channel is 4.5 MHz. The signal to noise power ratio of the video channel is 35 dB. Evaluate the capacity of the channel achievable ?

6. Evaluate the energy content and average power of the following exponential signals
   i. The single sided decaying exponential function $x_1(t) = e^{-at} u(t)$.
   ii. The single sided rising exponential function $x_2(t) = e^{at} u(-t)$.
   iii. The double sided exponential function $x_3(t) = e^{-|at|}$.

   , where the "a" is positive real constant $(a \geq 0)$.

7. Evaluate the energy content of the energy signals, Fig.1.14a,b,c,d.

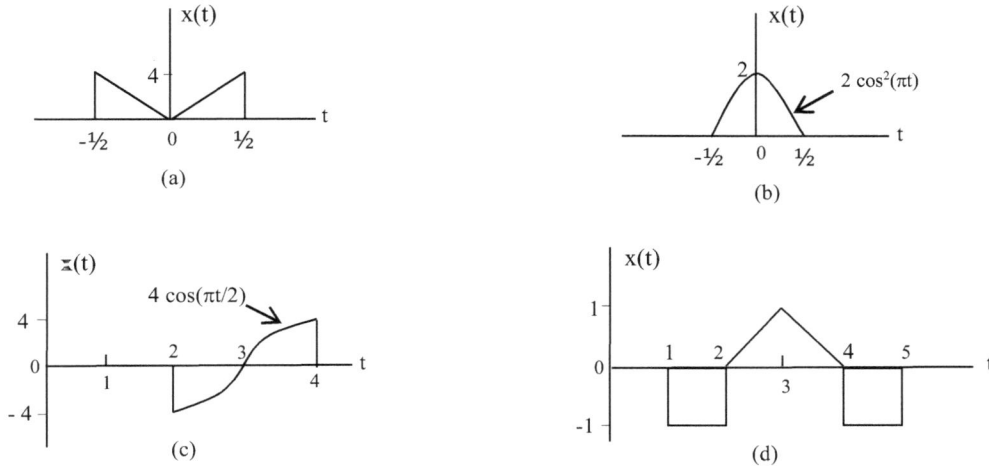

Fig.1.14. Energy signals.

8. Evaluate the following integrals

   i. $\int_{-\infty}^{\infty} (2 + t)^{-1} \, \delta(t) \, dt$

   ii. $\int_{-\infty}^{\infty} \sin(\pi t) \, e^{-j2\pi ft} \, \delta(t - \tfrac{1}{2}) \, dt$

   iii. $\int_{-\infty}^{\infty} (t^3 + 4) \, \delta(1 - \tfrac{1}{2} t) \, dt$

   iv. $\int_{-\infty}^{\infty} \cos(9t) \, \delta(t - 5) \, u(t) \, dt$

   v. $\int_{-\infty}^{\infty} e^{-t/4} \, \delta(t + 4) \, u(t - 2) \, dt$

   vi. $\int_{-\infty}^{\infty} e^{-t/3} \, \delta(9 - 3t) \, u(t + 4) \, dt$

   vii. $\int_{-\infty}^{\infty} (1 + 4t + 8t^2) \, \delta(2t + 4) \, dt$

   viii. $\int_{-\infty}^{\infty} e^{-\cos(t)} \, \sin(5t/2) \, \delta(2t - 3\pi) \, dt$

9. Evaluate the average power of the power signals, Fig.1.15a,b,c,d,e,f.

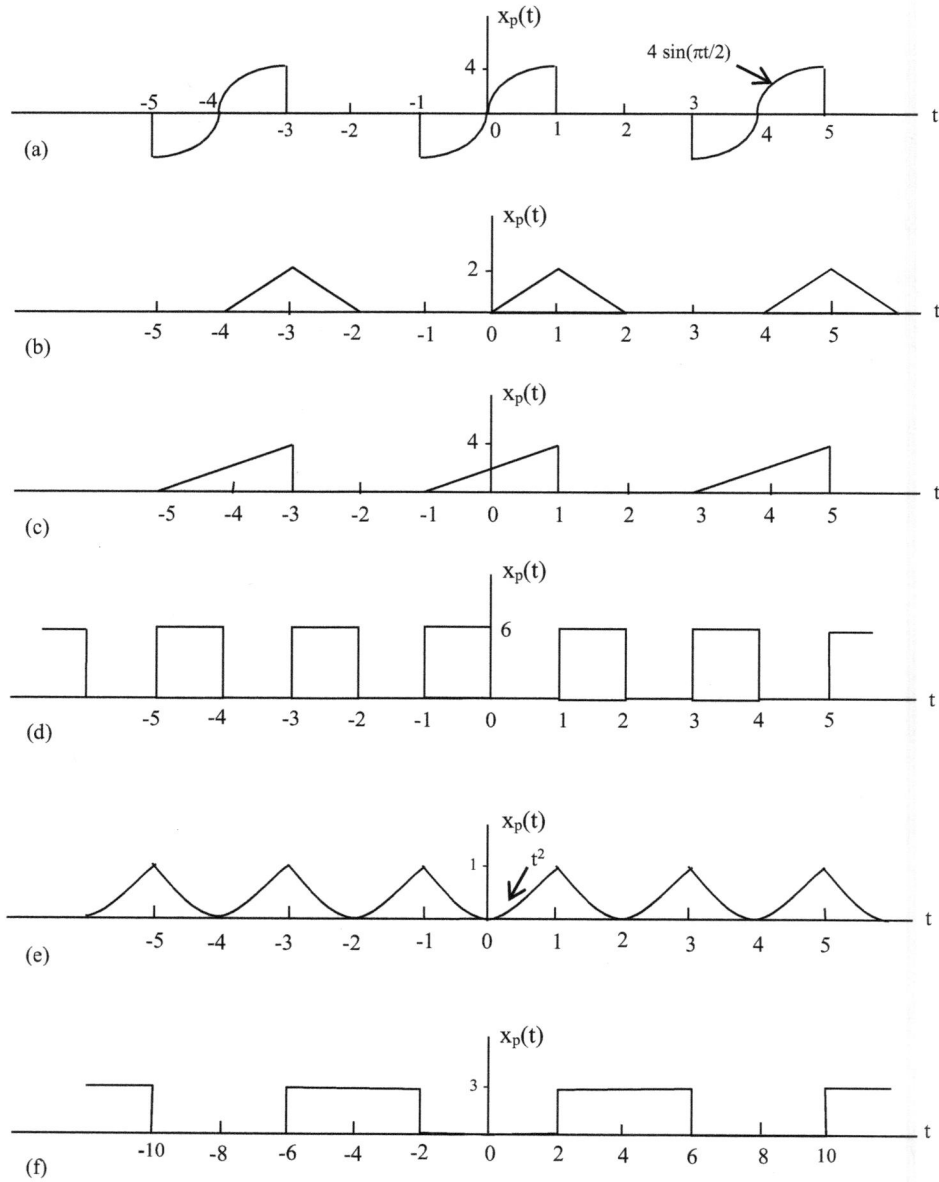

Fig.1.15. Power signals.

Send Orders for Reprints to reprints@benthamscience.net

CHAPTER 2

Fourier Series and Power Spectra

Abstract: Fourier analyses involve the resolution of signals into sinusoidal components to describe the spectral contents of the signals in the frequency domain. There are two major techniques of Fourier analyses namely: Fourier series and Fourier transform. Fourier series analysis resolves the periodic signal (power signal) into an infinite sum of sinusoidal wave components, while Fourier transform analysis performs a similar role in the analysis of the unperiodic signal (energy signal) which are more general use in the signal processing (chapter III). Fourier transform can also be used, in a limiting sense, to represent the periodic signals as an infinite sum of sinusoidal wave components (chapter V). An example, the frequency content of the direct current dc signal, is the zero frequency component only, while the frequency content of the signal x(t), Fig.2.1, are two components of frequencies: the zero frequency due to the dc component and the frequency f Hz due to the alternating current ac component imposed on the direct current. Then, as long as the amplitude of a signal varies, the signal contains certain frequency components.

Keywords: Fourier series analysis; Power spectra; Parseval's power theorem; Power spectral density.

Fourier series analysis is performed to obtain the discrete spectrum of the periodic signal $x_p(t)$ to describe its frequency components content. The periodic signal $x_p(t)$ of periodic time T_o is a special case of the power signal, having finite average power and infinite energy. There are two different ways of Fourier series analysis are applied to resolve the periodic signal namely: the real coefficients method and the complex exponential coefficient method. Both methods lead to the same spectral content for the same periodic signal $x_p(t)$ [4].

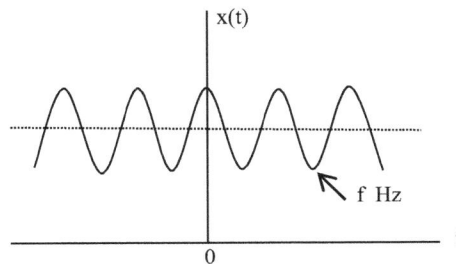

Fig.2.1. The frequency content of x(t) are zero and f Hz.

2.1. Fourier Series Real Coefficients

For a given periodic signal $x_p(t)$ of periodic time T_o, Fourier series expansion resolves the signal into an infinite sum of sinusoidal basic functions. Since T_o is the periodic time of the signal $x_p(t)$, then $1/T_o$ represents the fundamental frequency of the signal and n/T_o represent its harmonic frequencies, where n = 2, 3, 4, … . Then the periodic signal $x_p(t)$ may be expressed by the following formula

$$x_p(t) = \sum_{n=-\infty}^{\infty} a_n \cos(2\pi \frac{n}{T_o} t) + \sum_{n=-\infty}^{\infty} b_n \sin(2\pi \frac{n}{T_o} t) \qquad (2.1)$$

, where a_n and b_n are the real coefficients of the Fourier series expansion and represent the unknown amplitudes of the sinusoidal basic functions $\cos[2\pi(n/T_o)t]$ and $\sin[2\pi(n/T_o)t]$ which mainly depend on the shape of the periodic signal $x_p(t)$, and n = 0, ±1, ±2, ±3, ±4, … …

For the real coefficient a_n, multiply both sides of Eq.(2.1) by $\cos[2\pi(n/T_o)t]$ and then integrate both sides over the periodic time T_o, a_n yields

$$a_n = \frac{1}{T_o} \int_{-T_o/2}^{T_o/2} x_p(t) \cos(2\pi \frac{n}{T_o} t) \, dt \qquad (2.2)$$

, and for the real coefficient b_n, multiply both sides of Eq.(2.1) by $\sin[2\pi(n/T_o)t]$ and then integrate both sides over the periodic time T_o, b_n yields

$$b_n = \frac{1}{T_o} \int_{-T_o/2}^{T_o/2} x_p(t) \sin(2\pi \frac{n}{T_o} t) \, dt \qquad (2.3)$$

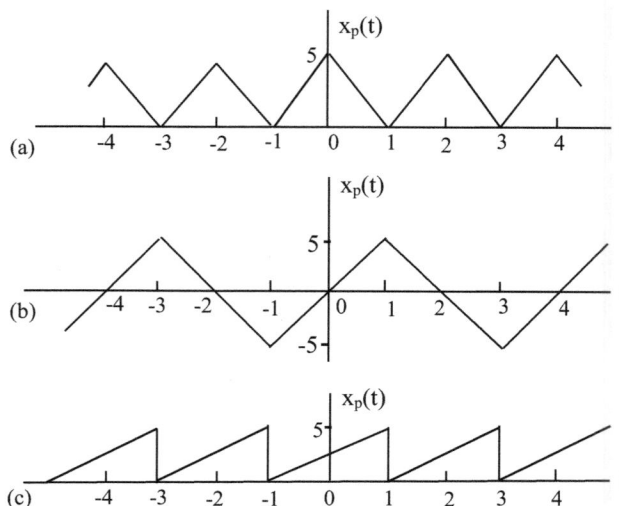

Fig.2.2.
a) Even periodic function
b) Odd periodic function
c) Neither even nor odd
   periodic function.

If the periodic signal $x_p(t)$ is even function, Fig.2.2a, the symmetry is being around the vertical axis, then $x_p(t) = x_p(-t)$, applying this condition for Eq.(2.1), yields b_n equals zero, and then the even periodic function $x_p(t)$ will contain the basic function of cosine terms only, and is given by

$$x_p(t) = \sum_{n=-\infty}^{\infty} a_n \cos(2\pi \frac{n}{T_o} t) \qquad \text{for even functions} \qquad (2.4)$$

, also if the periodic signal $x_p(t)$ is odd function, Fig.2.2b, the symmetry is being around the origin, then $x_p(t) = - x_p(-t)$, applying this condition for Eq.(2.1), yields a_n equals zero, and then the odd periodic function $x_p(t)$ will contain the basic function of sine terms only, and is given by

$$x_p(t) = \sum_{n=-\infty}^{\infty} b_n \sin(2\pi \frac{n}{T_o} t) \qquad \text{for odd functions} \qquad (2.5)$$

, but if the periodic signal $x_p(t)$ is neither even function nor odd function, Fig.2.2c, the expression of $x_p(t)$ will contain the basic functions cosine and sine terms, Eq.(2.1). Then the Fourier series real coefficients method resolve the periodic signal $x_p(t)$ by determining its real sets a_n and b_n, Equations.(2.2) and (2,3). Given the sets of the values a_n and b_n, then the original periodic signal $x_p(t)$ will be reconstructed, Eq.(2.1).

2.2. Orthogonality

The real basic functions $\cos[2\pi(n/T_o)t]$ and $\sin[2\pi(n/T_o)t]$, Eq.(2.1), form an orthogonal set in the periodic time T_o, so this kind of Fourier series analysis is defined by Quadrature Fourier series. Checking the orthogonality is performed by the following integrations

$$\int_{-T_o/2}^{T_o/2} \cos(2\pi \frac{n}{T_o} t) \sin(2\pi \frac{m}{T_o} t)\, dt = 0 \qquad \text{for } n = m, n \neq m$$

,

$$\int_{-T_o/2}^{T_o/2} \cos(2\pi \frac{n}{T_o} t) \cos(2\pi \frac{m}{T_o} t)\, dt = 0 \qquad \text{for } n \neq m$$

, and

$$\int_{-T_o/2}^{T_o/2} \sin(2\pi \frac{n}{T_o} t) \sin(2\pi \frac{m}{T_o} t)\, dt = 0 \qquad \text{for } n \neq m$$

, the proof of these three integrals is done, making use of the following formulas, $(x > y)$.

$$2 \cos(x) \cos(y) = \cos(x - y) + \cos(x + y)$$
,
$$2 \sin(x) \sin(y) = \cos(x - y) - \cos(x + y)$$
,
$$2 \sin(x) \cos(y) = \sin(x + y) + \sin(x - y)$$
, and
$$2 \cos(x) \sin(y) = \sin(x + y) - \sin(x - y)$$

Generally, for a two real functions $x_i(t)$ and $x_j(t)$, are said to be orthogonal with respect to each other over a certain interval $a < t < b$, if they satisfy the following orthogonality condition

$$\int_a^b x_i(t) \; x_j(t) \; dt = 0 \qquad\qquad \text{for } i \neq j \qquad\qquad (2.6)$$

, and for a two complex functions $x_i(t)$ and $x_j(t)$. The orthogonality condition yields

$$\int_a^b x_i(t) \; x_j^*(t) \; dt = 0 \qquad\qquad \text{for } i \neq j \qquad\qquad (2.7)$$

, or
$$\int_a^b x_i^*(t) \; x_j(t) \; dt = 0 \qquad\qquad \text{for } i \neq j$$

, where $x_i^*(t)$ and $x_j^*(t)$ are the complex conjugate of the functions $x_i(t)$ and $x_j(t)$ respectively. The zero result implies that these functions are "independent", or in "disagreement" in the interval $a < t < b$. If the result is not zero, then they are not orthogonal, and consequently, the two functions have some "dependence" or "alikeness" to each other [5].

Problem 2.1
Given a set of three functions defined by: $x_1(t) = 1$, $x_2(t) = t$, and $x_3(t) = 1.5 \, t^2 - 0.5$. Show that these functions are mutually orthogonal over the interval $-1 < t < 1$.

Solution
Checking the orthogonality of the functions $x_1(t)$ and $x_2(t)$, Eq.(2.6), yields

$$\int_{-1}^1 t \, dt = 0$$

, also checking the orthogonality of the functions $x_2(t)$ and $x_3(t)$, Eq.(2.6), yields

$$\int_{-1}^1 t \, (1.5 \, t^2 - 0.5) \, dt = 0$$

, and checking the orthogonality of the functions $x_1(t)$ and $x_3(t)$, Eq.(2.6), yields

$$\int_{-1}^1 (1.5 \, t^2 - 0.5) \; dt = 0$$

, then the three functions $x_1(t)$, $x_2(t)$, and $x_3(t)$ are mutually orthogonal over the interval $-1 < t < 1$.

Problem 2.2
Given a set of three functions $x_1(t)$, $x_2(t)$, and $x_3(t)$, Fig.2.3 Show that these functions are mutually orthogonal for all t.

Solution
Checking the orthogonality of the functions $x_1(t)$ and $x_2(t)$, Eq.(2.6) yields

$$\int_0^{T/2} A \, A \, dt + \int_{T/2}^T A(-A) \, dt = 0$$

, also checking the orthogonality of the functions $x_1(t)$ and $x_3(t)$, Eq.(2.6) yields

$$\int_0^{T/4} A^2 \, dt + \int_{T/4}^{T/2} (-A^2) \, dt + \int_{T/2}^{3T/4} A^2 \, dt + \int_{3T/4}^T (-A^2) \, dt = 0$$

, and checking the orthogonality of the functions $x_2(t)$ and $x_3(t)$, Eq.(2.6) yields

$$\int_0^{T/4} A^2 \, dt + \int_{T/4}^{T/2} (-A^2) \, dt + \int_{T/2}^{3T/4} (-A^2) \, dt + \int_{3T/4}^T (-A)^2 \, dt = 0$$

, then the three functions $x_1(t)$, $x_2(t)$, and $x_3(t)$ are mutually orthogonal for all t.

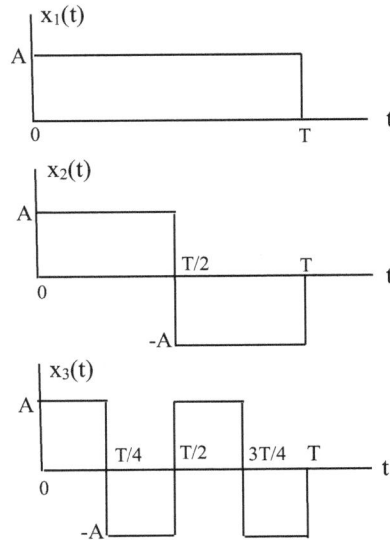

Fig.2.3. Set of three mutually orthogonal functions $x_1(t)$, $x_2(t)$, and $x_3(t)$.

Problem 2.3

Find the spectral analysis of the rectangular periodic signal, Fig.2.4. Check your results by graphical representation.

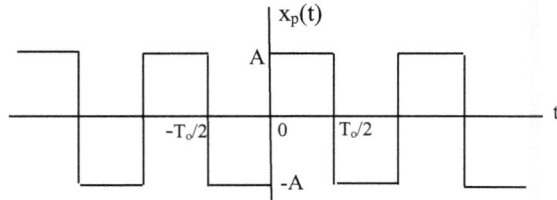

Fig.2.4. Odd rectangular periodic signal.

Solution

Since the signal, Fig.2.4, is odd function, Eq.(2.5), then the real coefficient b_n, Eq.(2.3) yields

$$b_n = \frac{1}{T_o} \int_{-T_o/2}^{0} (-A) \sin(2\pi \frac{n}{T_o} t) \, dt + \frac{1}{T_o} \int_{0}^{T_o/2} A \sin(2\pi \frac{n}{T_o} t) \, dt$$

, integrate and express the exponential terms by cosine function, the real coefficient b_n yields

$$b_n = \frac{A}{n\pi} [1 - \cos(\pi n)]$$

, the periodic signal $x_p(t)$, Eq.(2.5), yields

$$x_p(t) = 2 \sum_{n=1}^{\infty} \frac{A}{n\pi} [1 - \cos(\pi n)] \sin(2\pi \frac{n}{T_o} t) \tag{2.8}$$

, or equivalently

$$x_p(t) = \frac{4A}{\pi} \sin(2\pi \frac{1}{T_o} t) + \frac{4A}{3\pi} \sin(2\pi \frac{3}{T_o} t) + \frac{4A}{5\pi} \sin(2\pi \frac{5}{T_o} t) + \cdots$$

, the graphical representation of the fundamental and the harmonic frequencies, reconstructs the original periodic signal $x_p(t)$, Fig.2.5, where the frequency content of $x_p(t)$ are the odd harmonics $(1/T_o)$ Hz, $(3/T_o)$ Hz, $(5/T_o)$ Hz, ... , of amplitudes $(4A/\pi)$, $(4A/3\pi)$, $(4A/5\pi)$, ... , respectively. When increasing the harmonic terms n, the graphical summation approaches to the original rectangular signal $x_p(t)$, Fig.2.4.

A numerical values of the amplitude A is 10 volt and the period T_o is 1 ms, $x_p(t)$ is expressed by

$$x_p(t) = \frac{40}{\pi} \sin(2\pi 10^3 t) + \frac{40}{3\pi} \sin(2\pi \, 3 \times 10^3 t) + \frac{40}{5\pi} \sin(2\pi \, 5 \times 10^3 t) + \cdots \qquad \text{volt}$$

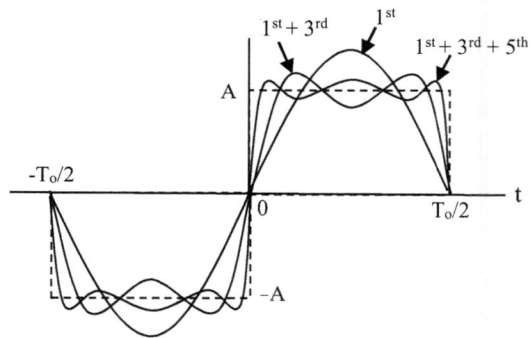

Fig.2.5.
Graphical successive representation
of odd rectangular periodic signal.

2.3. Fourier Series Complex Exponential Coefficient

The Fourier series real coefficients analysis represents the periodic signal $x_p(t)$ of periodic time T_o, Eq.(2.1), which can also be expressed by

$$x_p(t) = a_o + 2 \sum_{n=1}^{\infty} a_n \cos(2\pi \frac{n}{T_o} t) + 2 \sum_{n=1}^{\infty} b_n \sin(2\pi \frac{n}{T_o} t) \tag{2.9}$$

, where a_n and b_n are the Fourier series real coefficients, Equations (2.2) and (2.3), and the coefficient a_o denotes the average value of the periodic signal $x_p(t)$, is given by

$$a_o = \frac{1}{T_o} \int_{-T_o/2}^{T_o/2} x_p(t) \, dt \tag{2.10}$$

, rewrite $x_p(t)$, Eq.(2.9) in exponential form, yields

$$x_p(t) = a_0 + \sum_{n=1}^{\infty} a_n \left\{ e^{j2\pi \frac{n}{T_o} t} + e^{-j2\pi \frac{n}{T_o} t} \right\} - j \sum_{n=1}^{\infty} b_n \left\{ e^{j2\pi \frac{n}{T_o} t} - e^{-j2\pi \frac{n}{T_o} t} \right\}$$

, or equivalently

$$x_p(t) = a_0 + \sum_{n=1}^{\infty} (a_n - jb_n) e^{j2\pi \frac{n}{T_o} t} + \sum_{n=1}^{\infty} (a_n + jb_n) e^{-j2\pi \frac{n}{T_o} t} \tag{2.11}$$

, where the functions $e^{j2\pi \frac{n}{T_o} t}$ and $e^{-j2\pi \frac{n}{T_o} t}$ are the complex basic functions. The first summation of Eq.(2.11) denotes the terms for positive values of n, while the second summation denotes the terms for negative values of n. Let C_n denotes a complex coefficient related to the coefficients a_o, a_n and b_n by

$$
\begin{array}{lll}
C_n = a_n - jb_n & \text{for} & \text{n positive integer} \\
C_n = a_o & \text{for} & \text{n} = 0 \\
C_n = a_n + jb_n & \text{for} & \text{n negative integer}
\end{array}
$$

, and

, then $x_p(t)$ may be simplified and can be written in the following form

$$x_p(t) = \sum_{n=-\infty}^{\infty} C_n e^{j2\pi \frac{n}{T_o} t} \tag{2.12}$$

, where C_n is the complex coefficient of the Fourier series expansion and represent the unknown amplitudes of the exponential terms $e^{j2\pi \frac{n}{T_o} t}$, and $e^{-j2\pi \frac{n}{T_o} t}$ which mainly depend on the shape of the periodic signal $x_p(t)$.

For the complex coefficient C_n, multiply both sides of Eq.(2.12) by $e^{-j2\pi \frac{n}{T_o} t}$, and then integrate both sides over a complete periodic time T_o, C_n yields

$$C_n = \frac{1}{T_o} \int_{-T_o/2}^{T_o/2} x_p(t) \, e^{-j2\pi \frac{n}{T_o} t} \, dt \tag{2.13}$$

, the complex coefficient C_n is real value when the periodic signal $x_p(t)$ is even function, is imaginary value when $x_p(t)$ is odd function, and is complex value if $x_p(t)$ is neither even nor odd function. Then the complex coefficient Fourier series method resolve the periodic signal $x_p(t)$ by determining its complex set of C_n, Eq.(2.13). Given the set of the value C_n, then the original periodic signal $x_p(t)$ will be reconstructed, Eq.(2.12), and then the real value of $x_p(t)$ is taken into consideration. The magnitude and phase of the complex coefficient C_n in terms of the real coefficients a_n and b_n, are given by

$$|C_n| = \sqrt{(a_n)^2 + (b_n)^2}$$

, and
$$
\begin{array}{lll}
\arg[C_n] = \tan^{-1}[-b_n/a_n] & \text{for} & \text{n positive integer} \\
\quad\quad\quad = \tan^{-1}[b_n/a_n] & \text{for} & \text{n negative integer}
\end{array}
$$

The basic complex functions $e^{j2\pi \frac{n}{T_o} t}$ and $e^{-j2\pi \frac{n}{T_o} t}$, Eq.(2.11), form an orthogonal set of functions in the periodic time T_o, where

$$\int_{-T_o/2}^{T_o/2} e^{j2\pi \frac{n}{T_o} t} \, e^{-j2\pi \frac{m}{T_o} t} \, dt = 0 \qquad\qquad \text{for} \quad n \neq m$$

$$\int_{-T_o/2}^{T_o/2} e^{j2\pi \frac{n}{T_o} t} \, e^{j2\pi \frac{m}{T_o} t} \, dt = 0 \qquad\qquad \text{for} \quad n \neq m$$

$$\int_{-T_o/2}^{T_o/2} e^{-j2\pi \frac{n}{T_o} t} \, e^{-j2\pi \frac{m}{T_o} t} \, dt = 0 \qquad\qquad \text{for} \quad n \neq m$$

, these three simple integrals show that the basic functions $e^{j2\pi \frac{n}{T_o} t}$ and $e^{-j2\pi \frac{n}{T_o} t}$ form an orthogonal set of functions over the periodic time T_o.

Problem 2.4

Find the spectral analysis of the periodic signal $x_p(t)$, Fig.2.4, using complex coefficient Fourier series method.

Solution

The complex coefficient C_n, Eq.(2.13), is given by

$$C_n = \frac{1}{T_o} \int_{-T_o/2}^{0} (-A)\, e^{-j2\pi \frac{n}{T_o}t}\, dt \;+\; \frac{1}{T_o} \int_{0}^{T_o/2} A\, e^{-j2\pi \frac{n}{T_o}t}\, dt$$

, integrate, and express the exponential terms by cosine function, C_n yields

$$C_n = \frac{A}{jn\pi}[1 - \cos(n\pi)]$$

, C_n is imaginary value because the signal $x_p(t)$ is odd function. Given the set of the value C_n, therefore the original periodic signal $x_p(t)$ will be reconstructed, Eq.(2.12), $x_p(t)$ yields

$$x_p(t) = \sum_{n=-\infty}^{\infty} \frac{A}{jn\pi}[1 - \cos(n\pi)]\, e^{j2\pi \frac{n}{T_o}t}$$

, and the real value of $x_p(t)$ yields

$$[x_p(t)]_{real} = 2\sum_{n=1}^{\infty} \frac{A}{n\pi}[1 - \cos(n\pi)]\, \sin(2\pi \frac{n}{T_o}t) \tag{2.8}$$

, $[x_p(t)]_{real}$ contains sine terms only because $x_p(t)$ is odd function, and is the same spectral content of problem 2.3, where the real coefficients method.

2.4. Spectrum of Periodic Signals (discrete spectrum)

The representation of the periodic signal $x_p(t)$ by Fourier series analysis is equivalent to resolution $x_p(t)$ into its various harmonic components. The periodic signal $x_p(t)$ of periodic time T_o has components of frequencies equals n/T_o, $n = \pm 1, \pm 2, \pm 3, \dots$. If the fundamental frequency is defined by $f_o = 1/T_o$, the equivalent frequency domain $X(n/T_o)$ consists of discrete components of frequencies $\pm f_o, \pm 2f_o, \pm 3f_o, \dots$

The periodic signal $x_p(t)$ and its equivalent frequency domain $X(n/T_o)$ are separate aspects and are dependent of each other, and are related by Fourier series analysis. $x_p(t)$ is periodic in the time domain but in the frequency domain $X(n/T_o)$ is not periodic, except the function $X(n/T_o)$ represents the discrete frequency components content of $x_p(t)$. The waveform of periodic signal $x_p(t)$ can be displayed by oscilloscopes while its equivalent frequency domain $X(n/T_o)$ is displayed by frequency analyzers [6].

For a given periodic signal $x_p(t)$ of period T_o, the discrete spectrum $X(n/T_o)$ is given by

$$X(\frac{n}{T_o}) = \int_{-T_o/2}^{T_o/2} x_p(t)\, e^{-j2\pi \frac{n}{T_o}t}\, dt \tag{2.14}$$

, and the complex coefficient C_n will be

$$C_n = \frac{1}{T_o}\, X(\frac{n}{T_o}) \tag{2.15}$$

, since the periodic signal $x_p(t)$ may be real or imaginary or complex values, then its equivalent discrete spectrum $X(n/T_o)$ will also be real or imaginary or complex value, and is represented by

$$X(\frac{n}{T_o}) = \left|X(\frac{n}{T_o})\right| e^{j\arg\left[X(\frac{n}{T_o})\right]}$$

, where $|X(n/T_o)|$ and $\arg[X(n/T_o)]$, are the discrete amplitude spectrum and the discrete phase spectrum. Both these discrete amplitude and phase have non-zero values only for discrete frequencies that are integer (both positive and negative) multiples of the fundamental frequency [7].

For a real valued periodic signal $x_p(t)$, and its discrete spectrum definition $X(n/T_o)$, Eq,(2.14), then

$$X(-\frac{n}{T_o}) = X^*(\frac{n}{T_o})$$

, where $X^*(n/T_o)$ is the complex conjugate of $X(n/T_o)$, and therefore

$$\left| X(-\frac{n}{T_o}) \right| = \left| X(\frac{n}{T_o}) \right| \qquad , \text{ and } \qquad \arg\left[X(-\frac{n}{T_o}) \right] = -\arg\left[X(\frac{n}{T_o}) \right]$$

, hence the discrete amplitude spectrum is an even function, and the discrete phase spectrum is an odd function, Fig.2.6. Therefore, given the set of the discrete spectrum $X(n/T_o)$, Eq.(2.14), the original periodic signal $x_p(t)$ will be reconstructed by the following formula

$$x_p(t) = \frac{1}{T_o} \sum_{n=-\infty}^{\infty} X(\frac{n}{T_o}) e^{j2\pi \frac{n}{T_o} t} \tag{2.16}$$

, Equations (2.16) and (2.14) of Fourier series analysis relates the periodic signal $x_p(t)$ and its equivalent frequency domain $X(n/T_o)$ which are separate aspects and are dependent of each other.

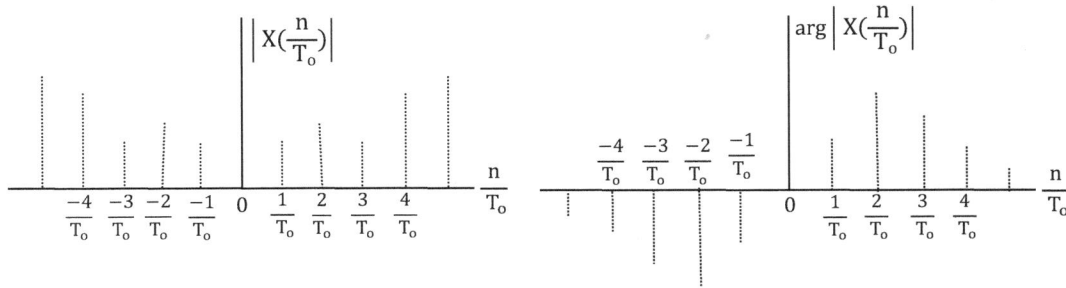

Fig.2.6. Amplitude and phase discrete spectrums for a real valued periodic signal.

2.5. Sinc Function

The sinc function, Fig.2.7, plays an important role in communication theory, is defined by

$$\text{sinc}(t) = \frac{\sin(\pi t)}{\pi t} \tag{2.17}$$

, the amplitude of the sinc function decreases because of the numerator $\sin(\pi t)$ lies between ± 1, while the denominator "πt" increases. Also its origin value is unity, because

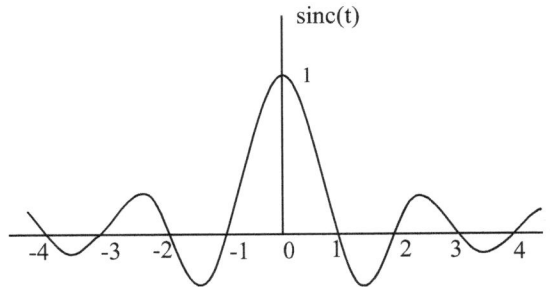

Fig.2.7. Sinc(t) function.

$$\lim_{t \to o} \frac{\sin(\pi t)}{\pi t} = 1$$

, the unity value of the function sinc(t) at the origin, is also obtained from the following trigonometric series

$$\sin(\pi t) = (\pi t) - \frac{1}{3!} (\pi t)^3 + \frac{1}{5!} (\pi t)^5 - \frac{1}{7!} (\pi t)^7 + \cdots$$

, then

$$\text{sinc}(t) = 1 - \frac{1}{3!} (\pi t)^2 + \frac{1}{5!} (\pi t)^4 - \frac{1}{7!} (\pi t)^6 + \cdots$$

, the zero crossing of the sinc function sinc(t) occurs when the term $\sin(\pi t)$ equals zero, at $\pi t = \pm n\pi$, then the points of zero crossing occurs at the points $t = \pm n$, where $n = 1, 2, 3, \ldots$. The sinc function is an even function and compromises of one main lob and other side lobes. The time period of the main lob is twice the time period of one side lob, Fig.2.7. The function sinc(t) is non-causal function because it extends from $-\infty$ to ∞.

Problem 2.5

Find the spectral analysis of the periodic rectangular signal, Fig.1.2b, in the period T_o, using the complex exponential Fourier series method.

Solution

The equivalent discrete spectrum $X(n/T_o)$, Eq.(2.14), is given by

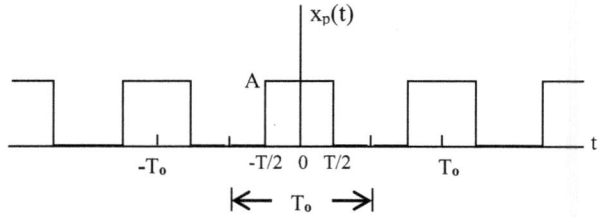

Fig.1.2b. Periodic rectangular signal.

$$X(\frac{n}{T_o}) = \int_{-T/2}^{T/2} A\, e^{-j2\pi\frac{n}{T_o}t}\, dt$$

, integrate, and express the exponential terms by sine function, $X(n/T_o)$ yields

$$X(\frac{n}{T_o}) = AT\; \frac{\sin(\pi\frac{n}{T_o}T)}{\pi\frac{n}{T_o}T}$$

, in terms of sinc function, $X(n/T_o)$ yields

$$X(\frac{n}{T_o}) = AT \operatorname{sinc}(\frac{n}{T_o}T)$$

This discrete spectrum $X(n/T_o)$ is the equivalent frequency domain of the periodic signal $x_p(t)$, and is real function. Fig.2.8. shows this discrete spectrum $X(n/T_o)$, function of the frequencies n/T_o, with discrete spacing of $1/T_o$, and with origin value equals AT, where A is the pulse amplitude and T is the pulse duration. The points of zero crossing of this sinc function occurs where the term $\sin[\pi(n/T_o)T]$ equals zero where

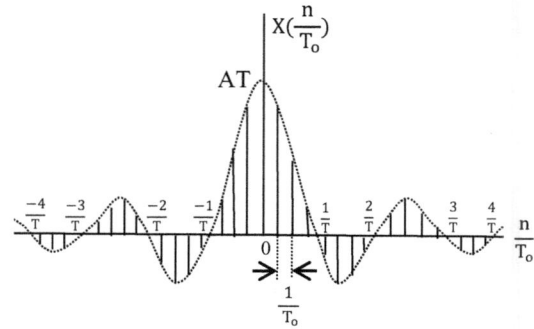

Fig.2.8. Sinc discrete spectrum of rectangular periodic signal.

$$\pi\frac{n}{T_o}T = m\pi \quad, \qquad m = \pm1, \pm2, \pm3, \dots \dots$$

, then the crossing points of $X(n/T_o)$ occurs at the discrete frequencies $(n/T_o) = (m/T)$.

Given the set of the value $X(n/T_o)$, therefore the original periodic signal $x_p(t)$ will be reconstructed, Eq.(2.16), $x_p(t)$ yields

$$x_p(t) = \sum_{n=-\infty}^{\infty} A\frac{T}{T_o}\operatorname{sinc}(\frac{n}{T_o}T)\left[\cos(2\pi\frac{n}{T_o}t) + j\sin(2\pi\frac{n}{T_o}t)\right]$$

, and the real value of $x_p(t)$ yields

$$[x_p(t)]_{real} = \sum_{n=-\infty}^{\infty} A\frac{T}{T_o}\operatorname{sinc}(n\frac{T}{T_o})\cos(2\pi\frac{n}{T_o}t)$$

, or equivalently

$$[x_p(t)]_{real} = A\frac{T}{T_o} + \sum_{n=1}^{\infty} 2A\frac{T}{T_o}\operatorname{sinc}(n\frac{T}{T_o})\cos(2\pi\frac{n}{T_o}t) \qquad (2.18)$$

, $[x_p(t)]_{real}$ contains cosine terms only because $x_p(t)$ is even function, Eq.(2.4), and the term AT/T_o is the average value a_o of the periodic signal $x_p(t)$, Eq.(2.10). The ratio T/T_o is defined by the duty cycle of the rectangular periodic signal.

Problem 2.6

Find the spectral analysis of the periodic Radio Frequency RF pulse, Fig.1.8b, in the periodic time T_o, using the complex exponential Fourier series method.

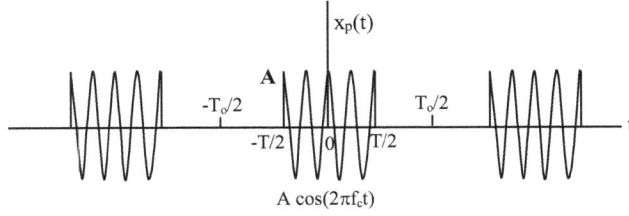

Fig.1.8b. Periodic RF signal.

Solution

The equivalent discrete spectrum $X(n/T_o)$, Eq.(2.14), is given by

$$X(\frac{n}{T_o}) = \int_{-T/2}^{T/2} A \cos(2\pi f_c t)\, e^{-j2\pi \frac{n}{T_o} t}\, dt$$

, express the cosine function by the exponential terms, $X(n/T_o)$ yields

$$X(\frac{n}{T_o}) = \int_{-T/2}^{T/2} \tfrac{1}{2}\, A \left\{ e^{j2\pi(f_c - \frac{n}{T_o})t} + e^{-j2\pi(f_c + \frac{n}{T_o})t} \right\} dt$$

, integrate, and in terms of sinc functions, $X(n/T_o)$ yields

$$X(\frac{n}{T_o}) = \tfrac{1}{2} AT\, \mathrm{sinc}\left[(f_c - \frac{n}{T_o})T\right] + \tfrac{1}{2} AT\, \mathrm{sinc}\left[(f_c + \frac{n}{T_o})T\right]$$

This discrete spectrum $X(n/T_o)$ is the equivalent frequency domain of the periodic RF pulse, and is real because $x_p(t)$ is even function. Fig.2.9 shows $X(n/T_o)$, function of the frequencies n/T_o, with discrete time spacing of $1/T_o$, and the two sinc functions are centered at the carrier frequency $\pm f_c$.

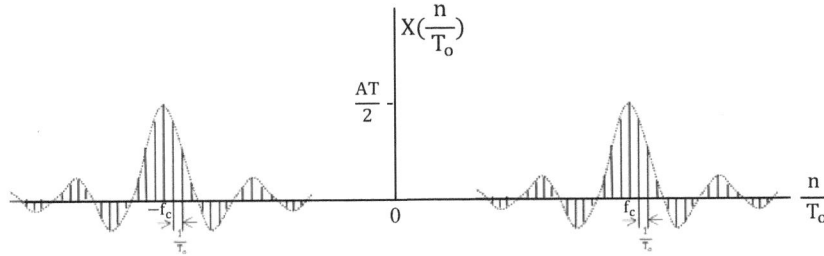

Fig.2.9. Two sinc discrete spectrum of periodic RF signal.

Given the set of the value $X(n/T_o)$, therefore the original periodic signal $x_p(t)$ will be reconstructed, Eq.(2.16), $x_p(t)$ yields

$$x_p(t) = \sum_{n=-\infty}^{\infty} \tfrac{1}{2} A\, \frac{T}{T_o} \left\{ \mathrm{sinc}\left[(f_c - \frac{n}{T_o})T\right] + \mathrm{sinc}\left[(f_c + \frac{n}{T_o})T\right] \right\} \left[\cos(2\pi \frac{n}{T_o} t) + j \sin(2\pi \frac{n}{T_o} t)\right]$$

, the real value of $x_p(t)$ yields

$$[x_p(t)]_{real} = \sum_{n=-\infty}^{\infty} \tfrac{1}{2} A\, \frac{T}{T_o} \left\{ \mathrm{sinc}\left[(f_c - \frac{n}{T_o})T\right] + \mathrm{sinc}\left[(f_c + \frac{n}{T_o})T\right] \right\} \cos(2\pi \frac{n}{T_o} t)$$

, or equivalently

$$[x_p(t)]_{real} = A\, \frac{T}{T_o}\, \mathrm{sinc}(f_c T) + \sum_{n=1}^{\infty} A\, \frac{T}{T_o} \left\{ \mathrm{sinc}\left[(f_c - \frac{n}{T_o})T\right] + \mathrm{sinc}\left[(f_c + \frac{n}{T_o})T\right] \right\} \cos(2\pi \frac{n}{T_o} t)$$

, $[x_p(t)]_{real}$ contains cosine terms only because $x_p(t)$ is even function, Eq.(2.4). The average value of $x_p(t)$ is $a_o = A\,(T/T_o)\,\mathrm{sinc}(f_c T)$, Eq.(2.10). A numerical values of the amplitude A is 10 volt, the periodic time T_o is 2 ms, the pulse duration T is 1 ms, where the duty cycle is half and the carrier frequency f_c is 1 GHz, and the average value a_o equals $5\,\mathrm{sinc}(10^6)$ equals zero, $x_p(t)$ is expressed by

$$[x_p(t)]_{real} = \sum_{n=1}^{\infty} 5 \left\{ \mathrm{sinc}\left[(10^6 - \frac{n}{2})\right] + \mathrm{sinc}\left[(10^6 + \frac{n}{2})\right] \right\} \cos(1000\, n\pi t) \qquad \text{volt}$$

Problem 2.7

Find the spectral analysis of the periodic triangle signal , Fig.1.3b , in the periodic time T_o , using the complex exponential Fourier series method.

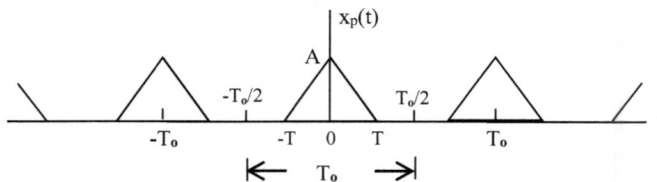

Solution

The equivalent discrete spectrum $X(n/T_o)$ of the periodic triangle signal, Fig.1.3b, Eq.(2.14), will be

Fig.1.3b. Periodic triangle signal.

$$X(\frac{n}{T_o}) = \int_{-T/2}^{0} (A + \frac{A}{T} t) e^{-j2\pi \frac{n}{T_o}t} dt + \int_{0}^{T/2} (A - \frac{A}{T} t) e^{-j2\pi \frac{n}{T_o}t} dt$$

, using integration by parts and express the exponential terms by sine functions, making use of the formula "$2 \sin^2(x) = 1 - \cos(2x)$", and in terms of sinc function, $X(n/T_o)$ yields

$$X(\frac{n}{T_o}) = AT \ sinc^2(\frac{n}{T_o}T)$$

This discrete spectrum $X(n/T_o)$ is the equivalent frequency domain of the periodic signal $x_p(t)$, and is real value because the signal $x_p(t)$ is even function. Fig.2.10. shows the discrete spectrum, function of the frequencies n/T_o , with discrete spacing of $1/T_o$, and within origin value equals AT. The zero crossing points of the sinc squared function occurs at the points m/T, and m = ±1, ±2, ±3, ...

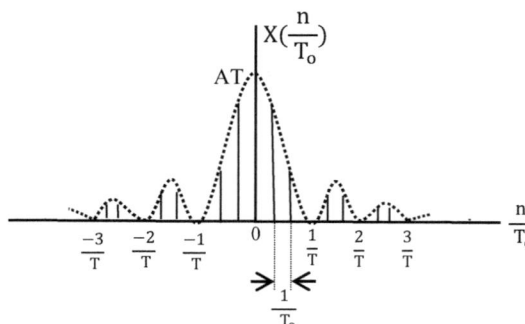

Fig.2.10. Sinc squared discrete spectrum of periodic triangle signal.

Given the set of the value $X(n/T_o)$, therefore the original periodic signal $x_p(t)$ will be reconstructed, Eq.(2.16), $x_p(t)$ yields

$$x_p(t) = \sum_{n=-\infty}^{\infty} A \frac{T}{T_o} \ sinc^2(\frac{n}{T_o}T) \left[\cos(2\pi \frac{n}{T_o}t) + j \sin(2\pi \frac{n}{T_o}t) \right]$$

, the real value of $x_p(t)$ yields

$$[x_p(t)]_{real} = A \frac{T}{T_o} + \sum_{n=1}^{\infty} 2A \frac{T}{T_o} \ sinc^2(n \frac{T}{T_o}) \cos(2\pi \frac{n}{T_o}t) \qquad (2.19)$$

, $[x_p(t)]_{real}$ contains cosine terms only because $x_p(t)$ is even function, Eq.(2.4), and the term AT/T_o is the average value a_o of the periodic signal $x_p(t)$, Eq.(2.10).

Problem 2.8

Find the spectral analysis of the periodic sinusoidal signal, Fig.2.11, using the complex exponential Fourier series method.

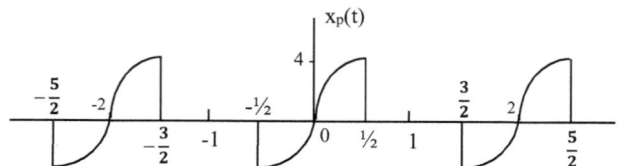

Fig.2.11. $x_p(t) = 4 \sin(\pi t)$ for $-\frac{1}{2} \leq t \leq \frac{1}{2}$
             $= 0$ for the remainder of the period.

Solution

The equivalent discrete spectrum $X(n/T_o)$ of the sinusoidal signal $x_p(t)$, Fig.2.11, Eq.(2.14), is given by

$$X(\frac{n}{2}) = \int_{-1/2}^{1/2} 4\sin(\pi t)\, e^{-j\pi n t}\, dt$$

, express the sine function by the exponential terms, $X(n/2)$ yields

$$X(\frac{n}{2}) = \frac{2}{j}\left\{ \int_{-1/2}^{1/2} e^{-j\pi(n-1)t}\, dt - \int_{-1/2}^{1/2} e^{-j\pi(n+1)t}\, dt \right\}$$

, integrate, express the exponential terms by sine functions, in terms of sinc functions, $X(n/2)$ yields

$$X(\frac{n}{2}) = -2j\left\{ \text{sinc}\left[\frac{1}{2}(n-1)\right] - \text{sinc}\left[\frac{1}{2}(n+1)\right] \right\}$$

This discrete spectrum $X(n/2)$ is the equivalent frequency domain of the periodic signal $x_p(t)$, and is imaginary value because the signal $x_p(t)$ is odd function. Fig.2.12 shows the discrete spectrum, function of the frequencies $f_n = n/2$, with discrete frequency spacing Δf equals 1/2.

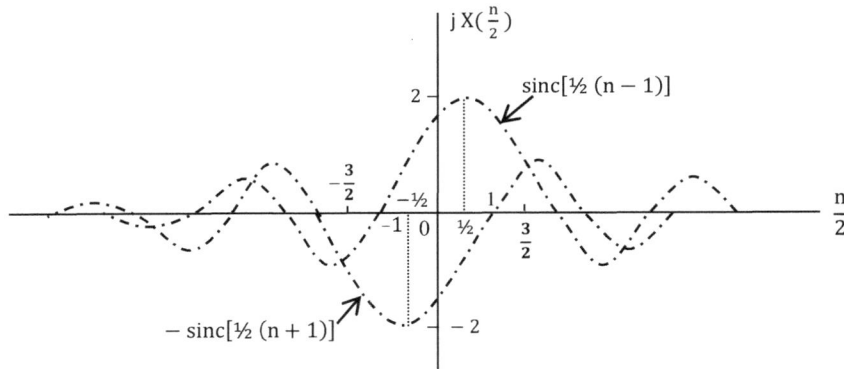

Fig.2.12. Discrete spectrum of periodic discrete sine function, Fig.2.11.

Given the set of the value $X(n/2)$, therefore the original periodic signal $x_p(t)$ will be reconstructed, Eq.(2.16), $x_p(t)$ yields

$$x_p(t) = \sum_{n=-\infty}^{\infty} -j\left\{ \text{sinc}\left[\frac{1}{2}(n-1)\right] - \text{sinc}\left[\frac{1}{2}(n+1)\right] \right\}\left[\cos(\pi n t) + j\sin(\pi n t)\right]$$

, the real value of $x_p(t)$ yields

$$[x_p(t)]_{real} = \sum_{n=1}^{\infty}\left\{ \text{sinc}\left[\frac{1}{2}(n-1)\right] - \text{sinc}\left[\frac{1}{2}(n+1)\right] \right\}\sin(\pi n t) \qquad (2.20)$$

, $[x_p(t)]_{real}$ contains sine terms only because $x_p(t)$ is odd function, Eq.(2.5).

Problem 2.9

Find the spectral analysis of the periodic signal, Fig.2.13, using the complex coefficient Fourier series method.

Solution

The equivalent discrete spectrum $X(n/T_o)$ of the signal $x_p(t)$, Fig.2.13 , Eq.(2.14), is given by

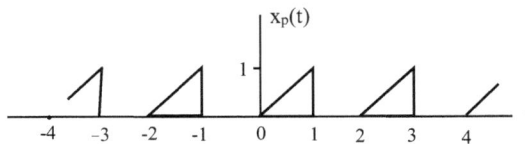

Fig.2.13. Periodic signal.

$$X(\frac{n}{2}) = \int_0^1 t \, e^{-j\pi nt} \, dt$$

, making use of integration by parts, and express the exponential terms by sine function. In terms of sinc and sinc squared functions, $X(n/2)$ yields

$$X(\frac{n}{2}) = \sum_{n=-\infty}^{\infty} \left\{ \left[sinc(n) - \tfrac{1}{2} sinc^2(\frac{n}{2}) \right] + j \left[\tfrac{1}{2}\pi n \, sinc^2(\frac{n}{2}) \right] \right\}$$

This discrete spectrum $X(n/2)$ is the equivalent frequency domain of the periodic signal $x_p(t)$, and is complex value because the signal $x_p(t)$ is neither even nor odd function. Given the set of the value $X(n/2)$, therefore the original periodic signal $x_p(t)$ will be reconstructed, Eq.(2.16), $x_p(t)$ yields

$$x_p(t) = \sum_{n=-\infty}^{\infty} \left\{ \left[sinc(n) - \tfrac{1}{2} sinc^2(\frac{n}{2}) \right] + j \left[\tfrac{1}{2}\pi n \, sinc^2(\frac{n}{2}) \right] \right\} \left[cos(\pi nt) + j \, sin(\pi nt) \right]$$

, then the real value of $x_p(t)$ yields

$$[x_p(t)]_{real} = \sum_{n=-\infty}^{\infty} \left[sinc(n) - \tfrac{1}{2} sinc^2(\frac{n}{2}) \right] cos(\pi nt) - \sum_{n=-\infty}^{\infty} \tfrac{1}{2}\pi n \, sinc^2(\frac{n}{2}) \, sin(\pi nt)$$

, $[x_p(t)]_{real}$ contain sine and cosine terms because $x_p(t)$ is neither even nor odd function, Eq.(2.1).

2.6. Parseval's Power Theorem

Parseval's power theorem facilitates the evaluation of the average power of the periodic signal, using the equivalent discrete frequency domain, Eq.(2.14). Consider a complex periodic signal $x_p(t)$ of periodic time T_o, its equivalent discrete frequency domain is $X(n/T_o)$, where $n = 0, \pm1, \pm2, \pm3, \dots$. A definition of the normalized average power of a complex periodic signal $x_p(t)$, it is the average power dissipated by a voltage $x_p(t)$ applied across a one ohm resistor, or by a current $x_p(t)$ passing through a one ohm resistor. The average power P_{av} is given by

$$P_{av} = \frac{1}{T_o} \int_{-T_o/2}^{T_o/2} \left| x_p(t) \right|^2 dt \qquad (1.9)$$

, in terms of the complex conjugate function of $x_p(t)$, P_{av} yields

$$P_{av} = \frac{1}{T_o} \int_{-T_o/2}^{T_o/2} x_p(t) \, x_p^*(t) \, dt$$

, in terms of the complex exponential Fourier series of $x_p(t)$, Eq.(2.16), P_{av} yields

$$P_{av} = \frac{1}{T_o} \int_{-T_o/2}^{T_o/2} \left\{ \frac{1}{T_o} \sum_{n=-\infty}^{\infty} X(\frac{n}{T_o}) \, e^{j2\pi \frac{n}{T_o}t} \right\} x_p^*(t) \, dt$$

, or equivalently

$$P_{av} = \frac{1}{T_o^2} \int_{-T_o/2}^{T_o/2} \left\{ \sum_{n=-\infty}^{\infty} X(\frac{n}{T_o}) \, e^{j2\pi \frac{n}{T_o}t} \right\} x_p^*(t) \, dt$$

, where $X(n/T_o)$ is the equivalent discrete spectrum of $x_p(t)$, Eq.(2.14). Interchange the order of summation and integration, P_{av} yields

$$P_{av} = \frac{1}{T_o^2} \sum_{n=-\infty}^{\infty} X(\frac{n}{T_o}) \left\{ \int_{-T_o/2}^{T_o/2} x_p^*(t) \, e^{j2\pi \frac{n}{T_o}t} \, dt \right\}$$

, in terms of $X^*(n/T_o)$, P_{av} yields

$$P_{av} = \frac{1}{T_o^2} \sum_{n=-\infty}^{\infty} X(\frac{n}{T_o}) \, X^*(\frac{n}{T_o})$$

, or equivalently

$$P_{av} = \frac{1}{T_o^2} \sum_{n=-\infty}^{\infty} \left| X(\frac{n}{T_o}) \right|^2 \qquad (2.21)$$

, in terms of the complex exponential Fourier coefficient C_n, Eq.(2.15), P_{av} is also expressed by

$$P_{av} = \sum_{n=-\infty}^{\infty} |C_n|^2 \qquad (2.22)$$

Parseval`s power theorem states that the average power of a periodic signal $x_p(t)$ equals the sum of the squared amplitudes of all the harmonic components $X(n/T_o)$ divided by the periodic time T_o, Eq.(2.21).

2.7. Power Spectral Density

The Power Spectral Density PSD $\Omega(f)$ is a very useful term to evaluate the average power of the periodic signal $x_p(t)$ using the frequency domain. The average power P_{av} in terms of the power spectral density plotted as a function of frequency, Fig.2.14, P_{av} is given by

$$P_{av} = \int_{-\infty}^{\infty} \Omega(f)\, df \qquad (2.23)$$

, where $\Omega(f)$ is defined by the Power Spectral Density PSD. In terms of rewriting Eq.(1.5) in the frequency domain, the Dirac delta function $\delta(f)$ at the harmonic components n/T_o, $\Omega(f)$ can be written in the form

$$\Omega(f) = \frac{1}{T_o^2} \sum_{n=-\infty}^{\infty} \left| X(\frac{n}{T_o}) \right|^2 \delta(f - \frac{n}{T_o}) \qquad (2.24)$$

, also in terms of the complex exponential Fourier coefficient C_n, Eq.(2.15), $\Omega(f)$ is also given by

$$\Omega(f) = \sum_{n=-\infty}^{\infty} |C_n|^2 \, \delta(f - \frac{n}{T_o}) \qquad (2.25)$$

, then the power spectral density of a periodic signal is a discrete function of frequency where it consists of a succession of weighted delta functions by the value $|C_n|^2$, Fig.2.14.

Also, the Power Spectral Density PSD $\Omega(f)$ facilitates the evaluation of the autocorrelation function of the periodic signal $x_p(t)$, where they constitute a Fourier transform pair (chapter VI).

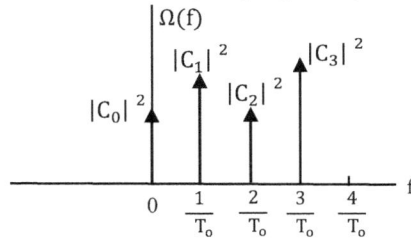

Fig.2.14. Power spectral density of periodic signal.

Problem 2.10

Evaluate the power spectral density and the average power of the following steady cosine periodic signal $x_p(t)$ of frequency $1/T_o$, using the frequency domain

$$x_p(t) = A \cos(2\pi \frac{1}{T_o} t)$$

Solution

The periodic signal $x_p(t)$ is a cosine function extends from $-\infty$ to $+\infty$ with constant frequency $1/T_o$ (periodic time T_o). The discrete spectrum $X(n/T_o)$ is the equivalent frequency domain of the periodic signal $x_p(t)$, Eq.(2.14), is given by

$$X(\frac{n}{T_o}) = \int_{-T_o/2}^{T_o/2} A \cos(2\pi \frac{1}{T_o} t) \, e^{-j2\pi \frac{n}{T_o} t} \, dt$$

, express the cosine function by the exponential terms, integrate, then express the exponential terms by sine functions, and in terms of sinc functions, $X(n/T_o)$ yields

$$X(\frac{n}{T_o}) = \tfrac{1}{2} A T_o \, \text{sinc}(n - 1) + \tfrac{1}{2} A T_o \, \text{sinc}(n + 1)$$

Then the power spectral density, Eq.(2.24), $\Omega(f)$ is given by

$$\Omega(f) = \frac{1}{T_o^2} \sum_{n=-\infty}^{\infty} |\,\tfrac{1}{2}\,AT_o\,\text{sinc}(n-1) + \tfrac{1}{2}\,AT_o\,\text{sinc}(n+1)|^2\,\delta(f - \frac{n}{T_o})$$

, since $x_p(t)$ contains single frequency only $(1/T_o)$, then $n = \pm 1$. Hence $\text{sinc}(0) = 1$, and $\text{sinc}(\pm 2) = 0$, $\Omega(f)$ yields

$$\Omega(f) = \tfrac{1}{4}\,A^2\,\delta(f - \frac{1}{T_o}) + \tfrac{1}{4}\,A^2\,\delta(f + \frac{1}{T_o})$$

, the power spectral density consists of two delta functions at the discrete frequency $\pm 1/T_o$, Fig.2.15, the average power P_{av}, Eq. (2.23), yields $P_{av} = \tfrac{1}{2}\,A^2$ watt.

This average power P_{av}, can also be obtained, using the time domain, Eq. (1.9), yields

$$P_{av} = \frac{1}{T_o} \int_{-T_o/2}^{T_o/2} A^2\,\cos^2(2\pi\frac{1}{T_o}t)\,dt$$

, making use of the formula "$2\cos^2(x) = 1 + \cos(2x)$", and integrates, P_{av} equals $\tfrac{1}{2}\,A^2$ watt.

Fig.2.15. Power spectral density of the steady cosine periodic signal (problem 2.10).

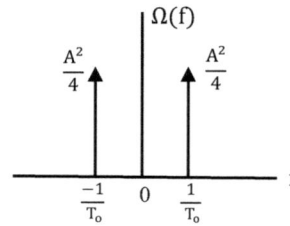

Problem 2.11
Evaluate the power spectral density and the average power of the following cosine squared periodic signal $x_p(t)$ of frequency $1/T_o$, using the frequency domain

$$x_p(t) = A\cos^2(2\pi\frac{1}{T_o}t)$$

Solution
The periodic signal $x_p(t)$ can be expressed by

$$x_p(t) = \tfrac{1}{2}\,A + \tfrac{1}{2}\,A\cos(2\pi\frac{2}{T_o}t)$$

, then the periodic signal $x_p(t)$ is composed of a cosine function, extends from $-\infty$ to $+\infty$ with constant frequency $2/T_o$ and another component of frequency equals zero. The discrete spectrum $X(n/T_o)$, is the equivalent frequency domain of the periodic signal $x_p(t)$, Eq.(2.14), is given by

$$X(\frac{n}{T_o}) = \int_{-T_o/2}^{T_o/2} \left\{\tfrac{1}{2}\,A + \tfrac{1}{2}\,A\cos(2\pi\frac{2}{T_o}t)\right\} e^{-j2\pi\frac{n}{T_o}t}\,dt$$

, express the cosine function by the exponential terms, integrate, then express the exponential terms by sine functions, and in terms of sinc functions, $X(n/T_o)$ yields

$$X(\frac{n}{T_o}) = \tfrac{1}{2}\,AT_o\,\text{sinc}(n) + \tfrac{1}{4}\,AT_o\,\text{sinc}(n-2) + \tfrac{1}{4}\,AT_o\,\text{sinc}(n+2)$$

, then the power spectral density, Eq.(2.24), $\Omega(f)$ is given by

$$\Omega(f) = \frac{1}{T_o^2} \sum_{n=-\infty}^{\infty} |\tfrac{1}{2}\,AT_o\,\text{sinc}(n) + \tfrac{1}{4}\,AT_o\,\text{sinc}(n-2) + \tfrac{1}{4}\,AT_o\,\text{sinc}(n+2)|^2\,\delta(f - \frac{n}{T_o})$$

, since $x_p(t)$ contains two frequency components (n/T_o) at $n = \pm 2$, and zero frequency at $n = 0$. Hence $\text{sinc}(0) = 1$, $\text{sinc}(\pm 2) = 0$, and $\text{sinc}(\pm 4) = 0$, $\Omega(f)$ yields

$$\Omega(f) = \frac{1}{4} A^2 \delta(f) + \frac{1}{16} A^2 \delta(f - \frac{2}{T_o}) + \frac{1}{16} A^2 \delta(f + \frac{2}{T_o})$$

Fig.2.16. Power spectral density of cosine squared periodic signal (problem 2.11).

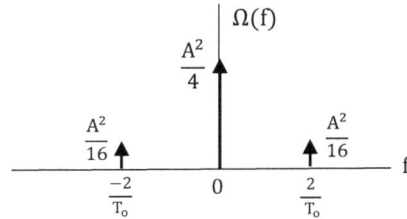

, then the power spectral density consists of three delta functions at the discrete frequencies $\pm 2/T_o$, and at the origin, Fig.2.16, the average power P_{av}, Eq. (2.23), yields

$$P_{av} = \frac{3}{8} A^2 \qquad \text{watt} \qquad\qquad (2.26)$$

This average power P_{av}, can also be obtained, using the time domain, Eq. (1.9), yields

$$P_{av} = \frac{1}{T_o} \int_{-T_o/2}^{T_o/2} A^2 \cos^4(2\pi \frac{1}{T_o} t)\ dt$$

, making use of the formula, "$\cos^4(x) = \frac{3}{8} + \frac{1}{2} \cos(2x) + \frac{1}{8} \cos(4x)$, integrate, P_{av} equals $\frac{3}{8} A^2$ watt, Eq.(2.26).

Problems

1. Given the set of three real functions defined by

$$x_1(t) = \cos(\pi t) \quad , \quad x_2(t) = \sin(2\pi t) \quad , \quad x_3(t) = \cos(3\pi t)$$

Show that these functions are mutually orthogonal over the interval $-1 < t < 1$.

2. Given the set of three complex functions defined by

$$x_1(t) = e^{j\pi t} \quad , \quad x_2(t) = e^{-j2\pi t} \quad , \quad x_3(t) = e^{j3\pi t}$$

Show that these functions are mutually orthogonal over the interval $-2 < t < 2$.

3. Check the orthogonality of the shown functions $x_1(t)$ and $x_2(t)$, Fig.2.17, for all t.

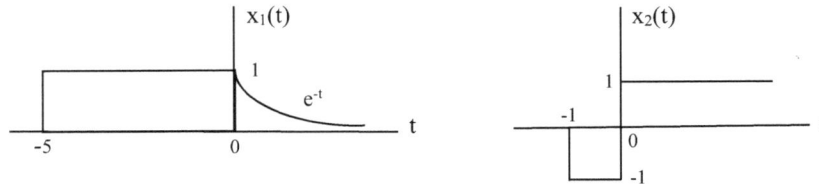

Fig.2.17. Two functions $x_1(t)$ and $x_2(t)$.

4. Check the mutually orthogonality for all t, of a set of three functions $x_1(t)$, $x_2(t)$, and $x_3(t)$, Fig.2.18.

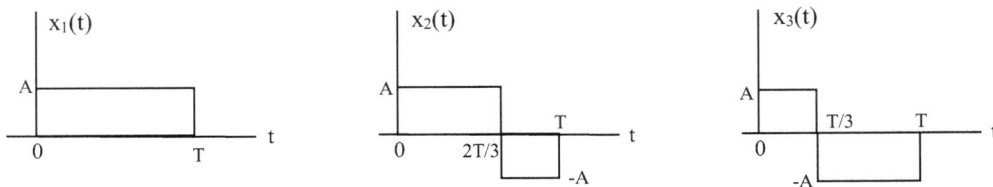

Fig.2.18. Set of three functions $x_1(t)$, $x_2(t)$, and $x_3(t)$.

5. Evaluate the power spectral density and the average power of the following cosine quadric periodic signal $x_p(t)$ of frequency $1/T_o$, using the frequency domain, is given by

$$x_p(t) = A \cos^4(2\pi \frac{1}{T_o} t)$$

6. Prove that the rectangular periodic signals $x_{p1}(t)$, and $x_{p2}(t)$, Fig.2.19, can be expressed by

$$x_{p1}(t) = \frac{1}{2} + \frac{2}{\pi} \sum_{n=1}^{\infty} \frac{(-1)^{n-1}}{2n-1} \cos\left[2\pi \frac{1}{T_o}(2n-1)t\right]$$

, and
$$x_{p2}(t) = \frac{4}{\pi} \sum_{n=1}^{\infty} \frac{(-1)^{n-1}}{2n-1} \cos\left[2\pi \frac{1}{T_o}(2n-1)t\right]$$

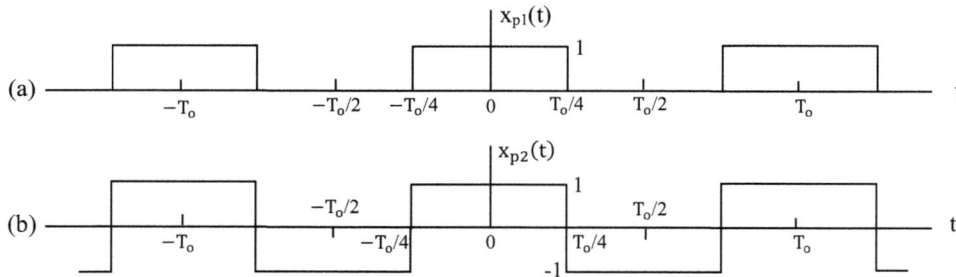

Fig.2.19. Two periodic functions $x_{p1}(t)$ and $x_{p2}(t)$.

7. Find the spectral analysis of the periodic signals, Fig.2.20a,b,c, using the real coefficients method and the complex exponential method of Fourier series analysis.

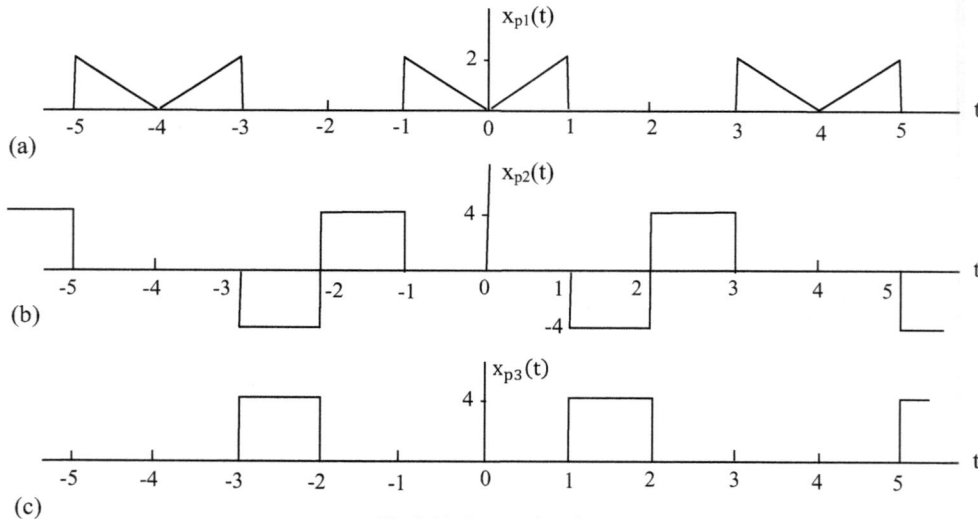

Fig.2.20. Power signals.

8. For the periodic rectangular signal $x_p(t)$, Fig.1.2b. Prove that the duty cycle (T/T_o) of $x_p(t)$ is given by

$$\frac{T}{T_o} = 1 / \sum_{n=-\infty}^{\infty} \text{sinc}^2(\frac{n}{T_o} T)$$

, where n/T_o are the harmonic frequencies, T is the width of the rectangular signal and T_o is the periodic time of the signal $x_p(t)$.

Send Orders for Reprints to reprints@benthamscience.net

CHAPTER 3

Fourier Transform and Energy Spectra

Abstract: In the time domain, the signal of constant amplitude does not contain frequency components while the signal of variable amplitude do contain frequency components. Fourier transform (Fourier integrals) is used for evaluating the frequency contents of the unperiodic signals and in a limiting sense for the periodic signals. Fourier transform provides a means of analyzing and designing frequency selective filters for the separation of signals on the basis of their frequency contents. This chapter is involved for performing the Fourier transform to obtain the continuous frequency spectrum of a given unperiodic signal x(t) such as energy signals and deterministic signals which has infinite periodic time T_o, finite energy, and zero average power. Fourier transform maps the signal x(t) from the time domain to its frequency domain X(f) in terms of the exponential functions. The waveform x(t) and its Fourier transform X(f) are separate aspects, dependent of each other and are related by Fourier transform analysis. Also in a limiting sense, Fourier transform represents the periodic signals such as power signals and random signals as an infinite sum of discrete sinusoidal wave components (chapter V).
Keywords: Fourier transform; Energy spectra; Rayleigh`s Energy theorem; Energy spectral density.

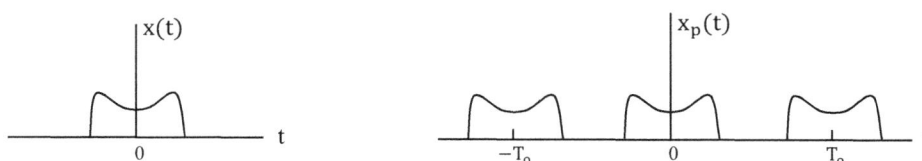

Fig.3.1. In the limit, as the period T_o approaches to infinity, $x_p(t)$ approaches to x(t).

Since the unperiodic signal x(t) defines one cycle of the periodic signal $x_p(t)$, Fig.3.1, so x(t) is considered the generating function of $x_p(t)$ in the periodic time T_o. Therefore, in the limit, as the periodic time T_o of the periodic signal become infinitely large and approaches to infinity, the periodic signal $x_p(t)$ approaches to the energy signal x(t), and x(t) can be obtained from

$$x(t) = \lim_{T_o \to \infty} x_p(t) \tag{1.4}$$

, the periodic signal $x_p(t)$, in terms of the complex exponential Fourier series analysis, is given by

$$x_p(t) = \frac{1}{T_o} \sum_{n=\infty}^{\infty} X(\frac{n}{T_o}) \, e^{j2\pi \frac{n}{T_o} t} \tag{2.16}$$

, where $X(n/T_o)$ is the equivalent discrete frequency domain of $x_p(t)$, Fig.2.6, is given by

$$X(\frac{n}{T_o}) = \int_{-T_o/2}^{T_o/2} x_p(t) \, e^{-j2\pi \frac{n}{T_o} t} \, dt \tag{2.14}$$

, the incremental spacing between two successive discrete frequencies is $\Delta f = 1/T_o$, and the harmonic frequencies $f_n = n/T_o$, where T_o is the periodic time, $n = 0, \pm 1, \pm 2, \pm 3, \ldots$. Then $X(f_n)$ and $x_p(t)$ yield

$$X(f_n) = \int_{-T_o/2}^{T_o/2} x_p(t) \, e^{j2\pi f_n t} \, dt$$

$$, \quad x_p(t) = \sum_{n=\infty}^{\infty} X(f_n) \, e^{j2\pi f_n t} \, \Delta f \quad , \quad \Delta f = \frac{1}{T_o} \quad , \quad n = 0, \pm 1, \pm 2, \pm 3, \ldots\ldots$$

, and the unperiodic signal x(t), Eq.(1.4) yields

$$x(t) = \lim_{T_o \to \infty} \sum_{n=\infty}^{\infty} X(f_n) \, e^{j2\pi f_n t} \, \Delta f$$

, as the periodic time T_o approaches infinity, Δf approaches zero, and the discrete frequency f_n approaches the continuous frequency variable f, and at any frequency f, the exponential function $e^{j2\pi ft}$ is weighted by the value "X(f) df", which is the contribution of X(f) in an infinitesimal interval df centered at the frequency f. Then x(t) and X(f) may be expressed by

$$x(t) = \int_{-\infty}^{\infty} X(f)\, e^{j2\pi ft}\, df \qquad (3.1)$$

, and
$$X(f) = \int_{-\infty}^{\infty} x(t)\, e^{-j2\pi ft}\, dt \qquad (3.2)$$

, x(t) and X(f) are said to constitute a Fourier transform pair and each of them is called the mate of the other, and each of them can be obtained in terms of the other, Equations (3.1) and (3.2). The waveform of the unperiodic signal x(t) can be displayed by the oscilloscopes while its Fourier transform X(f) is displayed by the frequency analysers. The Fourier transform exists if the integral of Eq.(3.1), is absolutely integrable, that is the signal x(t) must be unperiodic signal (energy signal) which has zero average power and finite energy (finite area), according to the following Dirichlet's conditions [1]

$$\int_{-\infty}^{\infty} |x(t)|\, dt < \infty \qquad (3.3a)$$

, and
$$\int_{-\infty}^{\infty} |x(t)|^2\, dt < \infty \qquad (3.3b)$$

On the other hand, there are some functions that are not absolutely integrable and do not satisfy Dirichlet's conditions and have Fourier transforms in the limit such as the Dirac delta function (unit impulse) $\delta(t)$, the unit step function u(t), and the signum function sgn(t), Fig.1.4, (chapter IV).

The general spectrum form of the Fourier transform X(f), is a complex function for either real or complex values of x(t), because Eq.(3.2) using Euler's formula, is given by

$$X(f) = \int_{-\infty}^{\infty} x(t)\, [\cos(2\pi ft) - j\sin(2\pi ft)]\, dt$$

, let x(t) is a complex function, where $x(t) = [x(t)]_{real} + j\,[x(t)]_{imag}$, then the real and imaginary parts of X(f) are given by

$$[X(f)]_{real} = \int_{-\infty}^{\infty} \{\, [x(t)]_{real}\cos(2\pi ft) + [x(t)]_{imag}\sin(2\pi ft)\,\}\, dt$$

, and
$$[X(f)]_{imag} = \int_{-\infty}^{\infty} \{[x(t)]_{imag}\cos(2\pi ft) - [x(t)]_{real}\sin(2\pi ft)\,\}\, dt$$

The symbol $X(f) = F[x(t)]$ indicates the Fourier transform operation, Eq.(3.2), while the symbol $x(t) = F^{-1}[X(f)]$ indicates the inverse Fourier transform operation, Eq.(3.1), (Fourier integrals). Another most convenient symbol of the Fourier transform pair, is more frequently used is

$$x(t) \rightleftharpoons X(f)$$

3.1. Spectrum of Unperiodic Signals (Continuous Spectrum)

The Fourier transform of the unperiodic signal x(t), of finite energy and zero average power, has a continuous spectrum due to Eq.(1.4), and expressed as a continuous sum of exponential functions with frequencies in the interval $-\infty$ to $+\infty$, and the amplitude of the frequency spectrum will be continuous function, Eq.(3.2). The Fourier transform X(f) is the equivalent frequency domain of the unperiodic signal x(t), is real value when the signal x(t) is even function, imaginary value when the signal x(t) is odd function, and is complex value if the unperiodic signal x(t) is neither even nor odd function. In terms of the continuous amplitude spectrum |X(f)| and the continuous phase spectrum arg[X(f)], the Fourier transform X(f) is represented by

$$X(f) = |X(f)|\, e^{j\arg[X(f)]}$$

For a real valued unperiodic signal x(t), and the definition of its spectrum X(f), Eq.(3.2), then

$$X(f) = X^*(-f)$$

, where $X^*(f)$ is the complex conjugate of X(f), therefore

$$|X(-f)| = |X(f)| \qquad \text{, and} \qquad \arg[X(-f)] = -\arg[X(f)]$$

, hence the continuous amplitude spectrum |X(f)| is even function, and the continuous phase spectrum arg[X(f)] is odd function.

3.2. Spectra of Some Important Integrable Signals

In communication systems, there are some famous integrable energy signals, are considered very vital pulses, and play an important role in the signal processing. These are: the rectangular pulse and its sinc spectrum, the triangle pulse and its sinc squared spectrum, the Gaussian pulse and its Gaussian spectrum, the Radio Frequency RF pulse and its two sinc spectrum, and also the decaying and rising single sided exponential pulses and their complex spectrums. It is necessary to obtain their Fourier transforms because these pulses are considered the basic functions for obtaining the frequency spectrums of any other complex problems in communication systems, to know their physical interpretation in both time domain and frequency domain, almost by inspection [8].

3.2.1. Rectangular Pulse and Sinc Spectra

Consider an energy signal, is a rectangular pulse (gate function) x(t), Fig.3.2a, centered at the origin, having amplitude A and duration T, x(t) may be expressed by

$$x(t) = A \, rect(\frac{t}{T})$$

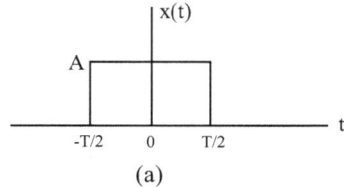

, the Fourier transform X(f), Eq.(3.2) is given by

$$X(f) = \int_{-T/2}^{T/2} A \, e^{-j2\pi ft} \, dt = AT \, \frac{\sin(\pi fT)}{\pi fT}$$

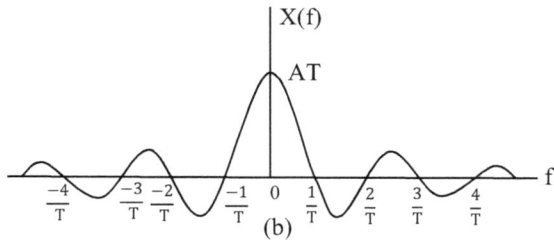

, in terms of sinc function, Eq.(2.17), X(f) yields

$$X(f) = AT \, sinc(fT)$$

, then the Fourier transform pair of this rectangular pulse yields

$$A \, rect(\frac{t}{T}) \; \rightleftharpoons \; AT \, sinc(fT) \qquad (3.4)$$

Fig.3.2. a) Rectangular pulse
b) Sinc spectrum.

, then an even rectangular pulse in the time domain of duration T and amplitude A, Fig.3.2a, is mapped into continuous real sinc function in the frequency domain extends from $-\infty$ to $+\infty$, Fig.3.2b. shows the continuous spectrum X(f), having origin value AT, and the zero crossing of the sinc spectrum occur at $f = \pm 1/T, \pm 2/T, \pm 3/T, \dots$.

Numerical examples

$$3 \, rect(\frac{t}{10}) \; \rightleftharpoons \; 30 \, sinc(10f)$$

, and

$$6 \, rect(3t) \; \rightleftharpoons \; 2 \, sinc(\frac{f}{3})$$

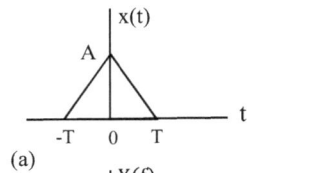

3.2.2. Triangle Pulse and Sinc Squared Spectra

Consider an energy signal, is a symmetric triangle pulse x(t) centered at the origin, Fig.3.3a, x(t) may be expressed by

$$x(t) = A \, tri(\frac{t}{T})$$

, the Fourier transform X(f), Eq.(3.2) is given by

$$X(f) = \int_{-T}^{0} (A + \frac{A}{T}t) \, e^{-j2\pi ft} \, dt + \int_{0}^{T} (A - \frac{A}{T}t) \, e^{-j2\pi ft} \, dt$$

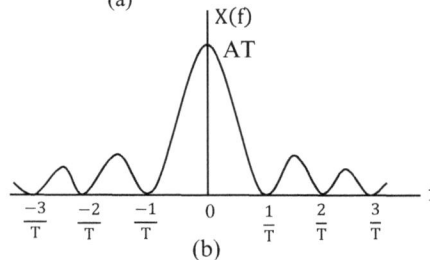

Fig.3.3. a) Triangle pulse
b) Sinc squared spectrum.

, using integration by parts technique, and express the exponential terms by sine functions, X(f) will be

$$X(f) = AT \frac{\sin^2(\pi fT)}{(\pi fT)^2}$$

, in terms of sinc function, X(f) yields

$$X(f) = AT \, sinc^2(fT)$$

, then an even triangle pulse in the time domain of base 2T and origin value A, Fig.3.3a, is mapped into continuous real sinc squared function in the frequency domain, Fig.3.3b. shows the continuous spectrum X(f), having origin value AT, and the zero crossing of the sinc squared function occur at $f = \pm 1/T$, $\pm 2/T, \pm 3/T, \dots$. The Fourier transform pair of this triangle pulse yields

$$A \, \mathrm{tri}(\frac{t}{T}) \; \rightleftharpoons \; AT \, \mathrm{sinc}^2(fT) \qquad\qquad (3.5)$$

Numerical examples

$$5 \, \mathrm{tri}(\frac{t}{2}) \; \rightleftharpoons \; 10 \, \mathrm{sinc}^2(2f)$$

, and
$$8 \, \mathrm{tri}(4t) \; \rightleftharpoons \; 2 \, \mathrm{sinc}^2(\frac{f}{4})$$

3.2.3. Gaussian pulse and Gaussian Spectra

Consider an energy signal, is a Gaussian pulse x(t) centered at the origin, Fig.3.4, x(t) is expressed by

$$x(t) = e^{-\pi t^2}$$

, the Fourier transform X(f), Eq.(3.2) is given by

$$X(f) = \int_{-\infty}^{\infty} e^{-\pi f^2} \, e^{-j2\pi ft} \, dt = \int_{-\infty}^{\infty} e^{-\pi(t^2 + j2ft)} \, dt$$

$$= \int_{-\infty}^{\infty} e^{-\pi(t + jf)^2} \, e^{-\pi f^2} \, dt$$

$$= e^{-\pi f^2} \int_{-\infty}^{\infty} e^{-\pi(t + jf)^2} \, dt$$

, let $x^2 = \pi(t + jf)^2$, then $x = \sqrt{\pi}\,(t + jf)$, and $dx = \sqrt{\pi}\, dt$, and as $t \rightarrow -\infty$ then $x \rightarrow -\infty$, and as $t \rightarrow \infty$ then $x \rightarrow \infty$, then the Fourier transform X(f) yields

$$X(f) = \frac{1}{\sqrt{\pi}} \, e^{-\pi f^2} \int_{-\infty}^{\infty} e^{-x^2} \, dx = e^{-\pi f^2} \qquad (3.6)$$

, where the area under the exponential e^{-x^2} equals $\sqrt{\pi}$, Eq.(3.48), and the Fourier transform pair of this Gaussian pulse yields

$$e^{-\pi t^2} \; \rightleftharpoons \; e^{-\pi f^2} \qquad\qquad (3.7)$$

, then the Fourier transform of the Gaussian pulse is also Gaussian pulse spectrum, Fig.3.4.

(a)

(b)

Fig.3.4. a) Gaussian pulse
b) Gaussian spectrum.

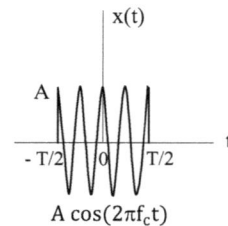

3.2.4. Radio Frequency Pulse and Two Sinc Spectra

Consider an energy signal, is a Radio Frequency RF pulse x(t), Fig.1.8a, of pulse duration T, pulse amplitude A, centered at the origin, and carrier frequency f_c, x(t) is expressed by

$$x(t) = A \, \mathrm{rect}(\frac{t}{T}) \, \cos(2\pi f_c t)$$

, the Fourier transform, Eq.(3.2), X(f) is given by

$$X(f) = \int_{T/2}^{T/2} A \cos(2\pi f_c t) \, e^{-j2\pi ft} \, dt$$

, express the cosine function by the exponential terms, X(f) yields

$$X(f) = \tfrac{1}{2} A \int_{-T/2}^{T/2} \left[e^{j2\pi f_c t} + e^{-j2\pi f_c t} \right] e^{-j2\pi ft} \, dt$$

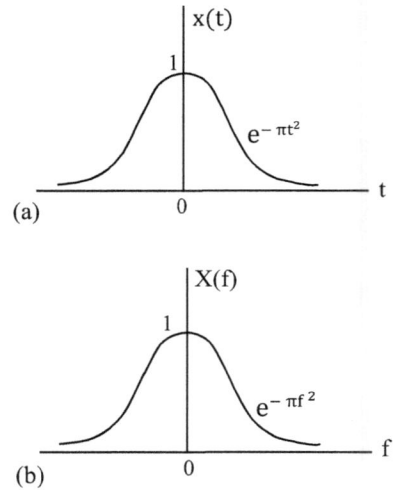

$A \cos(2\pi f_c t)$

Fig.1.8a. Radio frequency pulse.

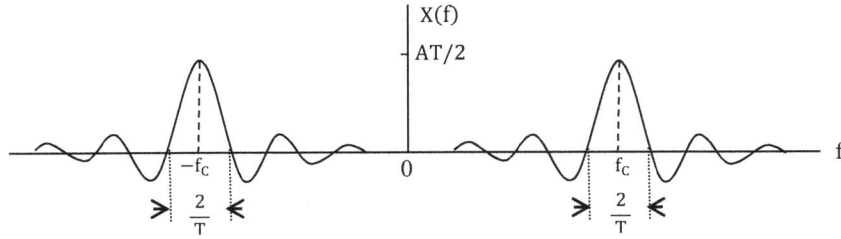

Fig.3.5. Two continuous sinc spectrum at the radio frequency $\pm f_c$.

, integrate, express the exponential terms by sine functions, and in terms of sinc functions, $X(f)$ yields

$$X(f) = \frac{1}{2} AT \text{ sinc}[T(f - f_c)] + \frac{1}{2} AT \text{ sinc}[T(f + f_c)] \qquad (3.8)$$

, then the Radio Frequency pulse in the time domain is mapped into two continuous real sinc functions in the frequency domain centered at the frequency carrying the pulse $\pm f_c$, Fig.3.5, and the Fourier transform pair yields

$$A \text{ rect}(\frac{t}{T}) \cos(2\pi f_c t) \quad \rightleftharpoons \quad \frac{1}{2} AT \text{ sinc}[T(f - f_c)] + \frac{1}{2} AT \text{ sinc}[T(f + f_c)]$$

, if the carrier frequency f_c equals zero, the RF pulse $x(t)$ becomes the rectangular pulse and its sinc spectra, Eq.(3.4).

Numerical example
A RF pulse has pulse duration T equals 500 micro second, carrier frequency equals one Giga Hz, and pulse amplitude A equals 100 volt, the Fourier transform pair yields

$$100 \text{ rect}(\frac{t}{0.0005}) \cos(2\pi 10^9 t) \quad \rightleftharpoons \quad 0.025 \text{ sinc}[0.0005(f - 10^9)] + 0.025 \text{ sinc}[0.0005(f + 10^9)]$$

, where in the time domain, the amplitude of the signal in volt and the time t in second, also in the frequency domain, the amplitude of the signal in volt and the frequency f in Hz.

3.2.5. Decaying and Rising Single Sided Exponential Pulses
Consider the energy signals, are the decaying and rising single sided exponential pulses, Fig.1.9, are expressed by $x_1(t) = e^{-t} u(t)$, and $x_2(t) = e^{t} u(-t)$, respectively. The Fourier transforms $X_1(f)$ and $X_2(f)$, Eq.(3.2), are given by

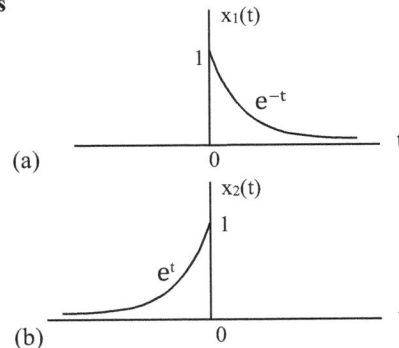

Fig.1.9. a) Single sided decaying signal, b) Single sided rising signal.

$$X_1(f) = \int_0^\infty e^{-t} e^{-j2\pi ft} dt = \frac{1}{1 + j2\pi f}$$

$$, \text{and} \quad X_2(f) = \int_{-\infty}^0 e^{t} e^{-j2\pi ft} dt = \frac{1}{1 - j2\pi f}$$

, in terms of the magnitude and phase continuous spectrums, Fig.3.6, $X_1(f)$ and $X_2(f)$ are expressed by

$$X_1(f) = \frac{1}{\sqrt{1 + (2\pi f)^2}} e^{-j \tan^{-1}(2\pi f)} \quad , \text{and} \quad X_2(f) = \frac{1}{\sqrt{1 + (2\pi f)^2}} e^{-j \tan^{-1}(-2\pi f)}$$

, then the decaying and rising single sided exponential pulses $e^{-t} u(t)$ and $e^{t} u(-t)$, are mapped into complex functions in the frequency domain, where the amplitude spectrums are even functions and the phase spectrums are odd functions. The Fourier transform pairs yield

$$e^{-t} u(t) \quad \rightleftharpoons \quad \frac{1}{1 + j2\pi f} \qquad (3.9)$$

$$, \text{and} \qquad e^{t} u(-t) \quad \rightleftharpoons \quad \frac{1}{1 - j2\pi f} \qquad (3.10)$$

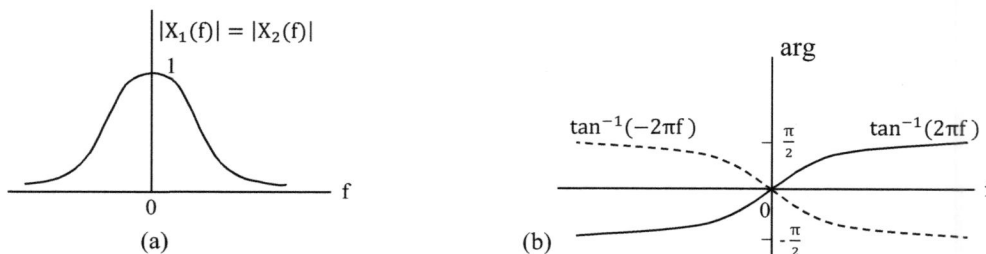

Fig.3.6. a) Magnitude spectrum of the single sided decaying and rising pulses.
b) Phase spectrum of the single sided decaying and rising pulses.

Problem 3.1
Prove that the inverse Fourier transform of the following complex spectrum

$$X(f) = \frac{1}{1 + j2\pi f}$$

, is the decaying single sided exponential pulse $x(t) = e^{-t}\, u(t)$, using Fourier integral, Eq.(3.1).
Solution
The inverse Fourier transform $x(t)$, Eq.(3.1), is given by

$$x(t) = \int_{-\infty}^{\infty} \frac{1}{1 + j2\pi f}\, e^{j2\pi ft}\, df$$

, let $w = 2\pi f$, then $dw = 2\pi\, df$, and as $f \rightarrow -\infty$ then $w \rightarrow -\infty$, and as $f \rightarrow \infty$ then $w \rightarrow \infty$, and the inverse Fourier transform $x(t)$ yields

$$x(t) = \frac{1}{2\pi} \int_{-\infty}^{\infty} \frac{1}{1 + jw}\, e^{jwt}\, dw$$

, multiplying the numerator and denominator by $(1 - jw)$, $x(t)$ yields

$$x(t) = \frac{1}{2\pi} \int_{-\infty}^{\infty} \frac{1}{(1 + jw)} \frac{(1 - jw)}{(1 - jw)}\, e^{jwt}\, dw$$

$$= \frac{1}{2\pi} \int_{-\infty}^{\infty} \frac{e^{jwt}}{1 + w^2}\, dw - \frac{1}{2\pi} \int_{-\infty}^{\infty} jw \frac{e^{jwt}}{1 + w^2}\, dw \qquad (3.11)$$

, according to the following closed contour integration (Cauchy`s Residue theorem)

$$\int_{c} f(w)\, e^{jwt}\, dw = 2\pi j \sum_{k=1}^{\infty} \text{Residue}\,[\,f(w), e^{jwt}, k] \qquad \text{for } t > 0$$

$$= 2\pi j \lim_{w \to k} (w - k)\, f(w)\, e^{jwt}$$

, where k is the number of poles inside the contour c which are poles at $w = \pm j$, $x(t)$, Eq.(3.11) yields

$$x(t) = \frac{1}{2\pi} \left[2\pi j \lim_{w \to j}(w - j) \frac{e^{jwt}}{(w - j)(w + j)} \right] - \frac{1}{2\pi} \left[2\pi j \lim_{w \to j}(w - j) \frac{jw\, e^{jwt}}{(w - j)(w + j)} \right]$$

$$= \frac{1}{2\pi} \left[2\pi j \frac{e^{-t}}{2j} \right] - \frac{1}{2\pi} \left[2\pi j \frac{(-e^{-t})}{2j} \right] \qquad \text{for } t > 0$$

, or equivalently

$$x(t) = e^{-t}\, u(t)$$

, then the following integral will be valued

$$\int_{-\infty}^{\infty} \frac{1}{1 + j2\pi f} \, e^{j2\pi ft} \, df \;=\; e^{-t} \, u(t)$$

, using the same procedures, for the rising single sided exponential pulse, Eq.(3.10), another integral will also be valued, given by

$$\int_{-\infty}^{\infty} \frac{1}{1 - j2\pi f} \, e^{j2\pi ft} \, df \;=\; e^{t} \, u(-t)$$

3.3. Fourier Transform Properties

Mapping the signals from the time domain into the frequency domain and vice versa, is considered the most important step in the design of any communication system, to determine which frequency band the subsystems operate and which frequency band the communication system itself serves in the different ways of our daily lives such as the audio frequency band for telephones, the medium and high frequency bands for radios, the very and ultra-high frequency bands for televisions, and the super and extremely high frequency bands in the senses for ships on the seas, aircraft in flights, and satellite in space.

Many Fourier transform problems need long time solutions, so there are very useful properties of Fourier transform which facilitate the solution of these problems and may be evaluate the Fourier transform almost by inspection. It is necessary to study these properties and how these properties allow the easy solution of many problems that would otherwise be quite difficult [4].

3.3.1. Linearity Property
Let the following Fourier Transform pairs

$$x_1(t) \rightleftharpoons X_1(f)$$

, and
$$x_2(t) \rightleftharpoons X_2(f)$$

, the linearity property states that, the Fourier transform of the sum of two functions, is the sum of the Fourier transform of the two functions, then the linearity property Fourier transform pair is

$$a_1\, x_1(t) + a_2\, x_2(t) \;\rightleftharpoons\; a_1\, X_1(f) + a_2\, X_2(f)$$

, where a_1 and a_2 are constants.

Proof:
The Fourier transform operation, Eq.(3.2) is given by

$$F[a_1\, x_1(t) + a_2\, x_2(t)] = \int_{-\infty}^{\infty} [a_1 x_1(t) + a_2\, x_2(t)] \; e^{-j2\pi ft} \, dt$$

$$= a_1 \int_{-\infty}^{\infty} x_1(t)\, e^{-j2\pi ft} \, dt \; + a_2 \int_{-\infty}^{\infty} x_2(t)\, e^{-j2\pi ft} \, dt$$

$$= a_1\, X_1(f) + a_2\, X_2(f)$$

Problem 3.2
Find the Fourier transform of the energy signals, $x_1(t)$ and $x_2(t)$, Fig.3.7, using linearity property.

Solution
The signals $x_1(t)$ and $x_2(t)$ are the algebraic summation of a rectangular and a triangle signals, and may be expressed by

$$x_1(t) \;=\; 5 \, \mathrm{rect}(\frac{t}{6}) \;+\; 3 \, \mathrm{tri}(\frac{t}{3})$$

, and

$$x_2(t) \;=\; 5 \, \mathrm{rect}(\frac{t}{6}) \;-\; 3 \, \mathrm{tri}(\frac{t}{3})$$

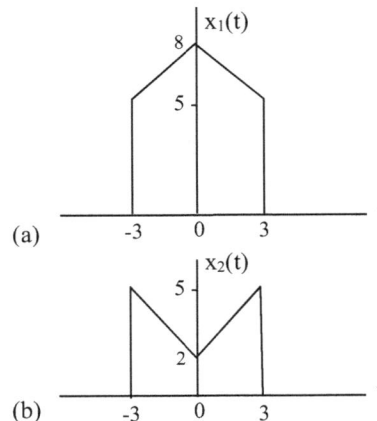

Fig.3.7. $x_1(t)$ and $x_2(t)$ are the algebraic summation of rectangular and triangle pulses.

, due to the linearity property, and the Fourier transforms of the rectangular and triangle pulses, Equations (3.4) and (3.5), the Fourier transforms $X_1(f)$ and $X_2(f)$ yield

$$X_1(f) = 30 \, \text{sinc}(6f) + 9 \, \text{sinc}^2(3f) \tag{3.12a}$$

, and
$$X_2(f) = 30 \, \text{sinc}(6f) - 9 \, \text{sinc}^2(3f) \tag{3.12b}$$

, these Fourier transforms can also be obtained, using Fourier integral, Eq.(3.2), $X_1(f)$ and $X_2(f)$ yield

$$X_1(f) = \int_{-3}^{0} (8 + t) \, e^{-j2\pi ft} \, dt + \int_{0}^{3} (8 - t) \, e^{-j2\pi ft} \, dt$$

, and

$$X_2(f) = \int_{-3}^{0} (2 - t) \, e^{-j2\pi ft} \, dt + \int_{0}^{3} (2 + t) \, e^{-j2\pi ft} \, dt$$

, making use of integration by parts technique, express the exponential terms by sine functions, the same spectrums, Equations (3.12), will be obtained, and the Fourier transform pairs will be

$$5 \, \text{rect}(\frac{t}{6}) + 3 \, \text{tri}(\frac{t}{3}) \;\rightleftharpoons\; 30 \, \text{sinc}(6f) + 9 \, \text{sinc}^2(3f)$$

$$5 \, \text{rect}(\frac{t}{6}) - 3 \, \text{tri}(\frac{t}{3}) \;\rightleftharpoons\; 30 \, \text{sinc}(6f) - 9 \, \text{sinc}^2(3f)$$

(a)

Problem 3.3
Find the Fourier transform of the double sided exponential pulse x(t), Fig.3.8a, using linearity property, where x(t) is given by

$$x(t) = e^{-|t|}$$

Solution
The pulse $x(t) = e^{-|t|}$, is the algebraic summation of the decaying and rising single sided exponential pulses $e^{-t} u(t)$ and $e^{t} u(-t)$, Fig.1.9, then x(t) is given by

$$e^{-|t|} = e^{-t} u(t) + e^{t} u(-t)$$

(b)

Fig.3.8. a) Double sided exponential pulse.
b) The Fourier transform of x(t) and has zero phase angle.

, due to the linearity property, and the Fourier transforms of the decaying and rising single sided exponential pulses, Equations (3.9) and (3.10), the Fourier transform X(f), Fig.3.8b, yields

$$X(f) = \frac{1}{1 + j2\pi f} + \frac{1}{1 - j2\pi f} = \frac{2}{1 + (2\pi f)^2} \tag{3.13}$$

, this Fourier transform can also be obtained, using Fourier integral, Eq.(3.2), X(f) yields

$$X(f) = \int_{-\infty}^{0} e^{t} e^{-j2\pi ft} \, dt + \int_{0}^{\infty} e^{-t} e^{-j2\pi ft} \, dt$$

, the same spectrum, Eq.(3.13), will be obtained, and the Fourier transform pair will be

$$e^{-|t|} \;\rightleftharpoons\; \frac{2}{1 + (2\pi f)^2} \tag{3.14}$$

Problem 3.4
Given two energy signals $x_1(t)$ and $x_2(t)$, Fig.3.9, are given by

$$x_1(t) = 2 \, \text{rect}(\frac{t}{8}) \quad , \text{and} \quad x_2(t) = 3 \, \text{tri}(\frac{t}{6})$$

Find the Fourier transform of the energy signal x(t), where x(t) is the product of $x_1(t)$ by $x_2(t)$, using the linearity property.

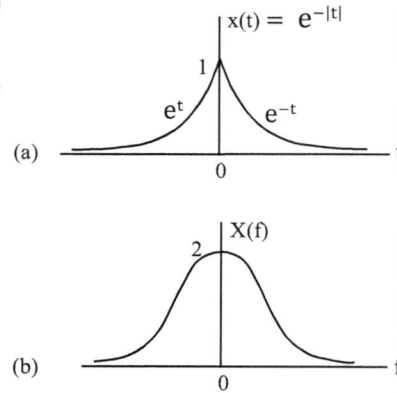

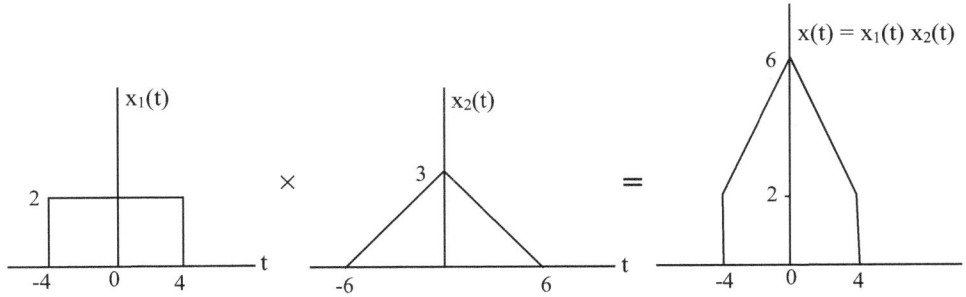

Fig.3.9. The graphical multiplication of rectangular and triangle pulses.

Solution

Since the signal x(t) is the product of $x_1(t)$ and $x_2(t)$, x(t) may expressed by

$$x(t) = 2 \, rect\left(\frac{t}{8}\right) \times 3 \, tri\left(\frac{t}{6}\right)$$

, Fig.3.9. shows the graphical multiplication of $x_1(t)$ by $x_2(t)$, and the signal x(t) is expressed by

$$x(t) = 2 \, rect\left(\frac{t}{8}\right) + 4 \, tri\left(\frac{t}{4}\right)$$

, due to the Fourier transform of the rectangular and triangle pulses, Equations (3.4) and (3.5). The Fourier transform X(f) will be

$$X(f) = 16 \, sinc(8f) + 16 \, sinc^2(4f) \qquad (3.15)$$

, this Fourier transform can also be obtained, using Fourier integral, Eq.(3.2), X(f) yields

$$X(f) = \int_{-4}^{0} (6+t) \, e^{-j2\pi ft} \, dt + \int_{0}^{4} (6-t) \, e^{-j2\pi ft} \, dt$$

, making use of the integration by parts technique, express the exponential terms by sine functions, the same spectrum, Eq.(3.15), will be obtained, and the Fourier transform pair will be

$$2 \, rect\left(\frac{t}{8}\right) \times 3 \, tri\left(\frac{t}{6}\right) \rightleftharpoons 16 \, sinc(8f) + 16 \, sinc^2(4f)$$

Problem 3.5

Find the Fourier transform of the energy signals, $x_1(t)$ and $x_2(t)$, Fig.3.10, hence, evaluate the Fourier transform of the function $x(t) = x_1(t) + x_2(t)$, using linearity property.

Solution

The Fourier transforms of the signals, $x_1(t)$ and $x_2(t)$, Eq.(3.2), are given by

$$X_1(f) = \int_{0}^{2} 7 \, e^{-j2\pi ft} \, dt$$

, and

$$X_2(f) = \int_{-2}^{0} 7 \, e^{-j2\pi ft} \, dt$$

, integrate, express the exponential terms by sine functions, and in terms of sinc function, $X_1(f)$ and $X_2(f)$ yield

$$X_1(f) = 14 \, sinc(2f) \, e^{-j2\pi f}$$

, and $X_2(f) = 14 \, sinc(2f) \, e^{j2\pi f}$

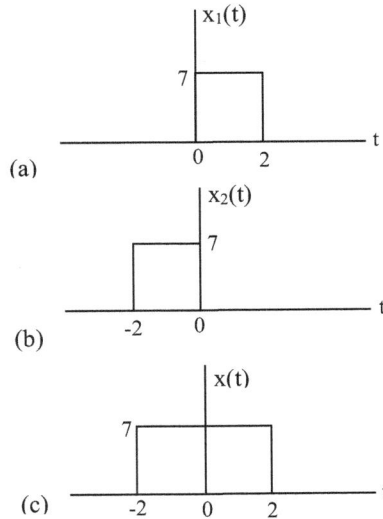

(a)

(b)

(c)

Fig.3.10. The summation of two shifted pulses, a) $x_1(t)$, b) $x_2(t)$ c) $x(t) = x_1(t) + x_2(t)$.

, due to the linearity property, the Fourier transform X(f) of the signal $x(t) = x_1(t) + x_2(t)$, will be

$$X(f) = 14 \, sinc(2f) \, e^{-j2\pi f} + 14 \, sinc(2f) \, e^{j2\pi f}$$

, express the exponential terms by cosine function and using the formula " $\sin(2x) = 2 \sin(x) \cos(x)$ ", and in terms of sinc function, $X(f)$ yields

$$X(f) = 28 \text{ sinc}(4f)$$

, which is the same Fourier transform of the signal $x(t)$, Fig.3.10c, is given by

$$x(t) = 7 \text{ rect}(\frac{t}{4})$$

3.3.2. Duality Property
Let the following Fourier transform pair

$$x(t) \rightleftharpoons X(f)$$

, the duality property states that, since the rectangular pulse $x(t)$ is mapped into a sinc function $X(f)$, then the rectangular continuous spectrum $x(f)$ is inversed mapped into a sinc function $X(-t)$, then the duality property Fourier transform pair is

$$X(-t) \rightleftharpoons x(f)$$

Proof:
Since the Fourier transform of the function $x(t)$, is given by

$$X(f) = \int_{-\infty}^{\infty} x(t) \ e^{-j2\pi ft} dt \qquad (3.2)$$

, put $(-f)$ instead of f, in both sides, Eq.(3.2) yields

$$X(-f) = \int_{-\infty}^{\infty} x(t) \ e^{j2\pi ft} dt \qquad (3.16)$$

, next, interchange f into t, in both sides, Eq.(3.16) yields

$$X(-t) = \int_{-\infty}^{\infty} x(f) \ e^{j2\pi ft} df \qquad (3.17)$$

, comparing Eq.(3.17) and the inverse Fourier transform operation, Eq.(3.1), then $X(-t)$ and $x(f)$ constitute a Fourier transform pair.

The following four examples illustrate how to apply the duality property:
i. For the decaying and rising single sided exponential pulses, Fig.1.9, Equations (3.9) and (3.10), and applying duality property, the inverse Fourier transform of the functions $X_1(f) = e^{-f} U(f)$, $X_2(f) = e^f U(-f)$, Fig.3.11, are deduced directly by inspection, and the Fourier transform pairs are

$$\frac{1}{1 - j2\pi t} \rightleftharpoons e^{-f} U(f) \qquad (3.18)$$

$$\frac{1}{1 + j2\pi t} \rightleftharpoons e^f U(-f) \qquad (3.19)$$

, these inverse Fourier transforms can also be obtained , using Fourier integral, Eq.(3.1), $x_1(t)$ and $x_2(t)$ yield

$$x_1(t) = \int_0^{\infty} e^{-f} e^{j2\pi ft} df = \frac{1}{1 - j2\pi t}$$

, and

$$x_2(t) = \int_{-\infty}^0 e^f e^{j2\pi ft} df = \frac{1}{1 + j2\pi t}$$

, where the same time domains, Equations (3.18) and (3.19).

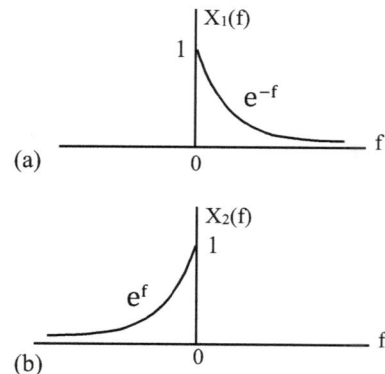

Fig.3.11. a) Single sided decaying spectrum b) Single sided rising spectrum.

ii. For the double sided exponential pulse $x(t) = e^{-|t|}$, Fig.3.8, Eq.(3.14), due to duality property, the inverse Fourier transform of the function $X(f) = e^{-|f|}$, Fig.3.12, is deduced by inspection, and the Fourier transform pair is

$$\frac{2}{1 + (2\pi t)^2} \rightleftharpoons e^{-|f|} \qquad (3.20)$$

, this inverse Fourier transform can also be obtained , using Fourier integral, Eq.(3.1), $x(t)$ yields

$$x(t) = \int_{-\infty}^{0} e^{f} e^{j2\pi ft} df + \int_{0}^{\infty} e^{-f} e^{j2\pi ft} df$$

, which leads to the same time domain, Eq.(3.20).

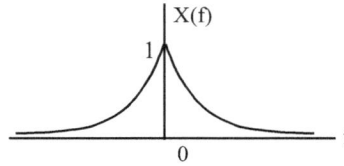

Fig.3.12. Double sided exponential spectrum.

iii. For the rectangular pulse, Eq. (3.4), due to the duality property, the inverse Fourier transform of the rectangular $X(f) = A \, rect(f/W)$, Fig.3.13a, is deduced directly by inspection, and the Fourier transform pair is

$$AW \, sinc(Wt) \rightleftharpoons A \, rect(\frac{f}{w}) \qquad (3.21)$$

, this inverse Fourier transform can also be obtained , using Fourier integral, Eq.(3.1), $x(t)$ yields

$$x(t) = \int_{-W/2}^{W/2} A \, e^{j2\pi ft} df$$

, which leads to the same time domain, Eq.(3.21).

iv. For the triangle pulse, Eq.(3.5), due to duality property, the inverse Fourier transform of the triangle $X(f) = A \, tri(f/W)$, Fig.3.14a, is deduced directly by inspection, and the Fourier transform pair is

$$AW \, sinc^2(Wt) \rightleftharpoons A \, tri(\frac{f}{w}) \qquad (3.22)$$

, this inverse Fourier transform can also be obtained, using Fourier integral, Eq.(3.1), $x(t)$ yields

cv

$$x(t) = \int_{-W}^{0} (A + \frac{A}{w}f) \, e^{j2\pi ft} df + \int_{0}^{w} (A - \frac{A}{w}f) \, e^{j2\pi ft} df$$

, which leads to the same time domain, Eq.(3.22).

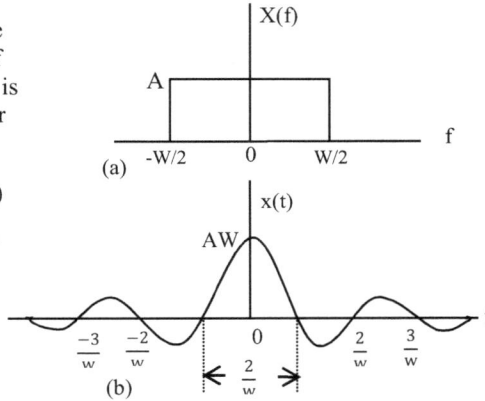

Fig.3.13. a) Rectangular spectrum
b) Sinc waveform.

3.3.3. Scaling Property
Let the following Fourier transform pair

$$x(t) \rightleftharpoons X(f)$$

, the scaling property Fourier transform pair is

$$x(at) \rightleftharpoons \frac{1}{|a|} X(\frac{f}{a})$$

, where the scaler "a" is positive real constant ($a \geq 0$).

Proof:
The Fourier transform operation of the function $x(at)$, Eq.(3.2), is given by

$$F[x(at)] = \int_{-\infty}^{\infty} x(at) \, e^{-j2\pi ft} dt$$

, let $\tau = at$, then $d\tau = a \, dt$, as $t \rightarrow -\infty$ then $\tau \rightarrow -\infty$, and as $t \rightarrow \infty$ then $\tau \rightarrow \infty$. The Fourier transform yields

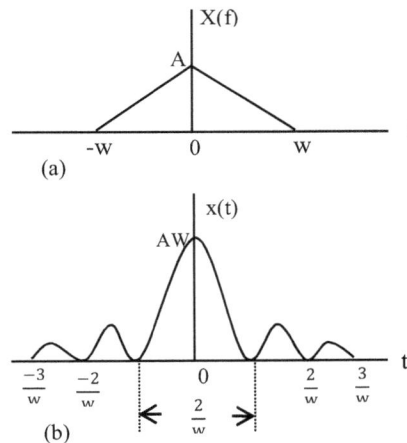

Fig.3.14. a) Triangle spectrum
b) Sinc squared waveform.

$$F[x(\tau)] = \frac{1}{a} \int_{-\infty}^{\infty} x(\tau)\, e^{-j2\pi\tau f/a}\, d\tau = \frac{1}{a} X\left(\frac{f}{a}\right)$$

, the scaling property shows that the expansion of a signal in the time scale by a certain scalar "a", has the effect of causing compression in the frequency domain by the same factor "a", with decrease of the amplitude by $1/|a|$, and vice versa, because if a function is expanded in the time scale, it varies slowly, and hence its frequency components are lowered. The factor "a" is positive real constant $(a \ge 0)$ because for $a < 0$, the limits of integration are interchanged.

Numerical examples

Applying the scaling property for Eq. (3.9), $a = 2$, then $e^{-2t} u(t) \rightleftharpoons \dfrac{1}{2} \dfrac{1}{(1 + j\pi f)}$

, applying the scaling property for Eq. (3.10), $a = 2\pi$, then $e^{2\pi t} u(-t) \rightleftharpoons \dfrac{1}{2\pi} \dfrac{1}{(1 - jf)}$

, and applying the scaling property for Eq. (3.18), $a = \dfrac{1}{2\pi}$, then $\dfrac{1}{1 - jt} \rightleftharpoons 2\pi\, e^{-2\pi f} U(f)$

Problem 3.6

Consider the energy pulse $x(t) = A\, \text{rect}(t/T)$.
Study the spectrum $X(f)$ of $x(t)$ when:

   i. The duration T is doubled
  ii. The duration T is halved.

Solution

Since the Fourier transform of the signal $x(t)$, Eq.(3.4), is $X(f) = AT\, \text{sinc}(fT)$, Fig.3.2b , let the pulse duration is 2T, the Fourier transform pair, Eq.(3.4) yields

$$A\, \text{rect}\left(\frac{t}{2T}\right) \rightleftharpoons 2AT\, \text{sinc}(2fT)$$

, and let the pulse duration is T/2, the Fourier transform pair, Eq.(3.4) yields

$$A\, \text{rect}\left(\frac{t}{T/2}\right) \rightleftharpoons \tfrac{1}{2}\, AT\, \text{sinc}(fT/2)$$

, both cases show that the expansion of a signal in the time scale by a certain scalar "a", has the effect of causing compression in the frequency domain by the same factor "a", with decrease of the amplitude by the $1/|a|$, and vice versa, Figures 3.15a,b.

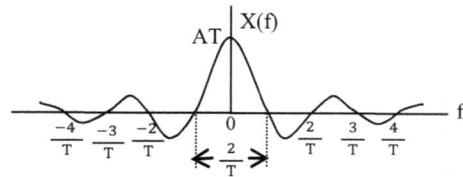

Fig.3.2b. Sinc spectrum. for pulse duration T.

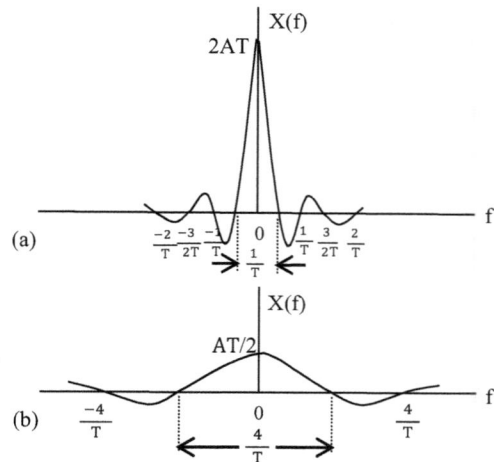

Fig.3.15. a) Sinc spectrum for pulse duration 2T, b) Sinc spectrum for pulse duration T/2.

Problem 3.7

Find the inverse Fourier transform of the following spectrum, using Fourier transform properties

$$X(f) = \frac{1}{(1 - j2\pi f)(2 + j2\pi f)}$$

Solution

Using partial fraction technique, $X(f)$ yields

$$X(f) = \frac{\tfrac{1}{3}}{1 - j2\pi f} + \frac{\tfrac{1}{3}}{2(1 + j\pi f)}$$

, making use of the linearity and scaling properties, and Equations (3.9) and (3.10), the inverse Fourier transform is given by

$$x(t) = \frac{1}{3} e^{t} u(-t) + \frac{1}{3} e^{-2t} u(t)$$

Problem 3.8
Based on the following Gaussian pulse Fourier transform pair

$$e^{-\pi t^2} \rightleftharpoons e^{-\pi f^2}$$

Find the Fourier transform of the Gaussian pulse $x(t) = e^{-t^2}$, using Fourier transform properties.

Solution

The Gaussian pulse $x(t) = e^{-t^2}$, can be expressed by $x(t) = e^{-\pi(t/\sqrt{\pi})^2}$, and due to the scaling property, its Fourier transform pair will be

$$e^{-\pi(t/\sqrt{\pi})^2} \rightleftharpoons \sqrt{\pi}\, e^{-\pi(\sqrt{\pi}f)^2}$$

, or equivalently

$$e^{-t^2} \rightleftharpoons \sqrt{\pi}\, e^{-\pi^2 f^2}$$

, this Fourier transform can also be obtained, using Fourier integral, Eq.(3.1), $X(f)$ yields

$$X(f) = \int_{-\infty}^{\infty} e^{-t^2}\, e^{j2\pi ft}\, dt$$

, using the same procedure of the Gaussian pulse and Gaussian spectrum, Eq.(3.7), $X(f) = \sqrt{\pi}\, e^{-\pi^2 f^2}$

3.3.4. Shifting Property
Let the following Fourier transform pair

$$x(t) \rightleftharpoons X(f)$$

, the shifting property Fourier transform pair is

$$x(t - t_o) \rightleftharpoons X(f)\, e^{-j2\pi ft_o}$$

, also

$$x(t + t_o) \rightleftharpoons X(f)\, e^{j2\pi ft_o}$$

, where t_o is the time shifting of $x(t)$, Fig.3.16.

Proof:
The Fourier transform operation, Eq.(3.2), of the signal $x(t - t_o)$ is given by

$$F[x(t - t_o)] = \int_{-\infty}^{\infty} x(t - t_o)\, e^{-j2\pi ft}\, dt$$

, let $\tau = t - t_o$, then $d\tau = dt$, as $t \rightarrow -\infty$ then $\tau \rightarrow -\infty$, and as $t \rightarrow \infty$ then $\tau \rightarrow \infty$. The Fourier transform operation yields

$$F[x(\tau)] = \int_{-\infty}^{\infty} x(\tau)\, e^{-j2\pi f(\tau + t_o)}\, d\tau = e^{-j2\pi ft_o} \int_{-\infty}^{\infty} x(\tau)\, e^{-j2\pi f\tau}\, d\tau$$

, or equivalently

$$F[x(t - t_o)] = X(f)\, e^{-j2\pi ft_o}$$

Fig.3.16. Shifted waveforms.

On the other hand, the inverse Fourier transform of a spectrum $X(f)$, shifted by a frequency shifting $\pm f_c$, can be deduced using the same procedure or directly by inspection using duality property, where the Fourier transform pairs will be

$$x(t)\, e^{j2\pi f_c t} \rightleftharpoons X(f - f_c)$$

, and

$$x(t)\, e^{-j2\pi f_c t} \rightleftharpoons X(f + f_c)$$

, then shifting the signal in the time domain by certain time shift t_o, has the effect of causing an angle $(2\pi ft_o)$ in the frequency domain in the same direction, while shifting the signal in the frequency domain by certain frequency shift f_c has the effect of causing an angle in the time domain $(2\pi f_c t)$ in the oppposite direction.

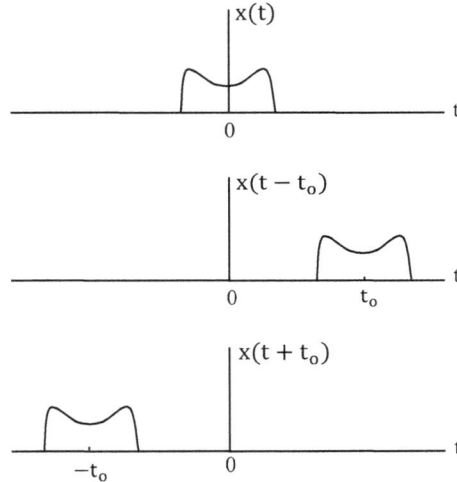

Numerical examples
Applying the shifting property for Eq.(3.9), then

$$e^{-(t-5)} u(t-5) \;\rightleftharpoons\; \frac{1}{1+j2\pi f} e^{-j2\pi f5}$$

, applying the shifting property for Eq.(3.4), then

$$5 \operatorname{rect}\left[\frac{t+3}{4}\right] \;\rightleftharpoons\; 20 \operatorname{sinc}(4f) \, e^{j2\pi f3}$$

, applying the shifting property for Eq.(3.5), then

$$6 \operatorname{tri}(\frac{t}{5}) \, e^{j2\pi 10^6 t} \;\rightleftharpoons\; 30 \operatorname{sinc}^2[5(f - 10^6)]$$

, and applying the shifting property for Eq.(3.10), then

$$e^t \, u(-t) \, e^{-j2\pi t2} \;\rightleftharpoons\; \frac{1}{1 - j2\pi(f+2)}$$

Problem 3.9
Find the Fourier transform of the energy signals, $x_1(t)$ and $x_2(t)$, Fig.3.10, using shifting property and hence evaluate the Fourier transform of the signal $x(t)$, where $x(t) = x_1(t) + x_2(t)$.

Solution
The energy signals $x_1(t)$ and $x_2(t)$ can be expressed by

$$x_1(t) = 7 \operatorname{rect}\left[\frac{t-1}{2}\right]$$

, and

$$x_2(t) = 7 \operatorname{rect}\left[\frac{t+1}{2}\right]$$

, due to shifting property, their Fourier transform pairs will be

$$7 \operatorname{rect}\left[\frac{t-1}{2}\right] \;\rightleftharpoons\; 14 \operatorname{sinc}(2f) \, e^{-j2\pi f}$$

$$7 \operatorname{rect}\left[\frac{t+1}{2}\right] \;\rightleftharpoons\; 14 \operatorname{sinc}(2f) \, e^{j2\pi f}$$

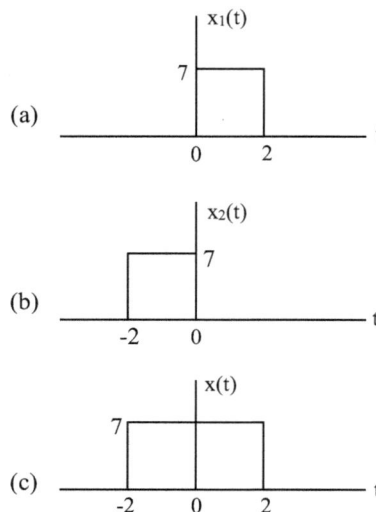

Fig.3.10. Summation of two shifted rectangular pulses , a) $x_1(t)$, b) $x_2(t)$ c) $x(t) = x_1(t) + x_2(t)$.

, for the signal $x(t) = x_1(t) + x_2(t)$, due to linearity property. The Fourier transform pair yields

$$7 \operatorname{rect}\left[\frac{t-1}{2}\right] + 7 \operatorname{rect}\left[\frac{t+1}{2}\right] \;\rightleftharpoons\; 14 \operatorname{sinc}(2f) \, e^{-j2\pi f} + 14 \operatorname{sinc}(2f) \, e^{j2\pi f}$$

, in the frequency domain, express the exponential terms by cosine function and making use of the formula "$\sin(2x) = 2 \sin(x) \cos(x)$", and the time domain is also expressed by $x(t) = 7 \operatorname{rect}(t/4)$, the Fourier transform pair yields

$$7 \operatorname{rect}(\frac{t}{4}) \;\rightleftharpoons\; 28 \operatorname{sinc}(4f)$$

, which is the same Fourier transform pair of problem 3.5, using Fourier integral operation.

Problem 3.10
Find the Fourier transform of the energy signal $x(t)$, Fig.3.17, using the shifting property .

Solution
The signal $x(t)$ is the algebraic summation of rectangular and triangle signals, and may be expressed by

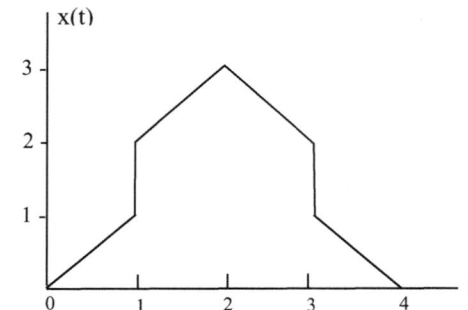

Fig.3.17. $x(t)$ is the algebraic summation of shifted rectangular and triangle pulses.

$$x(t) = rect\left[\frac{t-2}{2}\right] + 2\ tri\left[\frac{t-2}{2}\right]$$

, due to linearity property, and the Fourier transforms of the rectangular and triangle pulses, Equations (3.4) and (3.5), the Fourier transform X(f) yields

$$X(f) = 2\ sinc(2f)\ e^{-j2\pi f2} + 4\ sinc^2(2f)\ e^{-j2\pi f2} \qquad (3.23)$$

, this Fourier transforms can also be obtained, using Fourier integral, Eq.(3.2), X(f) yields

$$X(f) = \int_0^1 t\ e^{-j2\pi ft}\ dt + \int_1^2 (1+t)\ e^{-j2\pi ft}\ dt + \int_2^3 (4-t)\ e^{-j2\pi ft}\ dt + \int_3^4 (3-t)\ e^{-j2\pi ft}\ dt$$

, using integration by parts technique, express the exponential terms by sine functions, in terms of sinc functions, the same spectrum, Eq.(3.23), will be obtained, and the Fourier transform pairs will be

$$rect\left[\frac{t-2}{2}\right] + 2\ tri\left[\frac{t-2}{2}\right] \quad \rightleftharpoons \quad 2\ sinc(2f)\ e^{-j2\pi f2} + 4\ sinc^2(2f)\ e^{-j2\pi f2}$$

Problem 3.11
Find the Fourier transform of the Radio Frequency RF pulse x(t), Fig.1.8a, using shifting property.

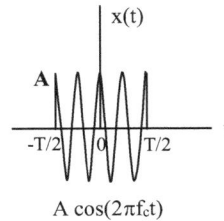

$$x(t) = A\ rect(\frac{t}{T})\ cos(2\pi f_c t)$$

Solution
Express the cosine function by the exponential terms, x(t) yields

A cos(2πf_ct)

Fig.1.8. a) Unperiodic RF pulse.

$$x(t) = \tfrac{1}{2}\ A\ rect(\frac{t}{T})\ e^{j2\pi f_c t} + \tfrac{1}{2}\ A\ rect(\frac{t}{T})\ e^{-j2\pi f_c t}$$

, due to linearity and shifting properties, the Fourier transform X(f) will be

$$X(f) = \tfrac{1}{2}\ AT\ sinc[T(f-f_c)] + \tfrac{1}{2}\ AT\ sinc[T(f+f_c)] \qquad (3.8)$$

, which is the same Fourier transform of the RF Pulse, Fig.3.5, using the Fourier integral, Eq.(3.2).

Problem 3.12
Find the Fourier transform of the energy signal x(t), Fig.3.18, using shifting property.

Solution
The signal x(t) is the algebraic summation of three signals, and can be expressed by

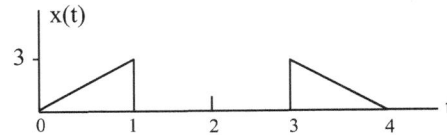

Fig.3.18. x(t) is the algebraic summation of shifted rectangular and triangle pulses.

$$x(t) = 6\ tri\left[\frac{t-2}{2}\right] - 3\ tri(t-2) - 3\ rect\left[\frac{t-2}{2}\right]$$

, the Fourier transform X(f) will be

$$X(f) = 12\ sinc^2(2f)\ e^{-j2\pi f2} - 3\ sinc^2(f)\ e^{-j2\pi f2} - 6\ sinc(2f)\ e^{-j2\pi f2}$$

, this Fourier transform can also be obtained, using the Fourier integral, Eq.(3.2), X(f) yields

$$X(f) = \int_0^1 3t\ e^{-j2\pi ft}\ dt + \int_3^4 (12-3t)\ e^{-j2\pi ft}\ dt \qquad (3.24)$$

, using the integration by parts technique, express the exponential terms by sine functions, and in terms of sinc and sinc squared functions, the same spectrum X(f), Eq.(3.24), will be obtained.

Problem 3.13
Find the Fourier transform of the following exponentially damped sinusoidal wave x(t), Fig.3.19a, using shifting property:

$$x(t) = e^{-t} \sin(2\pi f_c t) u(t)$$

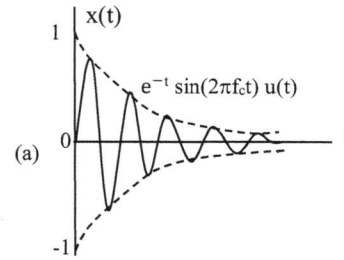
(a)

Solution
In terms of the exponential terms, the signal x(t) can be expressed by

$$x(t) = -\tfrac{1}{2} j\, e^{-t} [\, e^{j2\pi f_c t} - e^{-j2\pi f_c t} \,] u(t)$$

, based on the Fourier transform pair, Eq.(3.9), and using shifting property. The Fourier transform pairs of the two terms of x(t) are given by

$$e^{-t}\, e^{j2\pi f_c t}\, u(t) \;\rightleftharpoons\; \frac{1}{1 + j2\pi(f - f_c)}$$

, and

$$e^{-t}\, e^{-j2\pi f_c t}\, u(t) \;\rightleftharpoons\; \frac{1}{1 + j2\pi(f + f_c)}$$

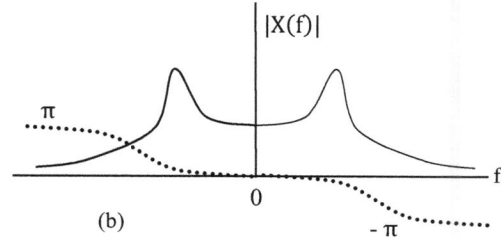
(b)

Fig.3.19. a) Exponentially damped sinusoidal wave x(t)
b) The Fourier transform X(f) of x(t).

, due to linearity property, the Fourier transform X(f), Fig.(3.19b), is given by

$$X(f) = -\frac{1}{2} j \left[\frac{1}{1 + j2\pi(f - f_c)} - \frac{1}{1 + j2\pi(f + f_c)} \right]$$

, or equivalently

$$X(f) = \frac{2\pi f_c}{(1 + j2\pi f)^2 + (2\pi f_c)^2}$$

3.3.5. Differentiation Property
Let the following Fourier transform pair

$$x(t) \rightleftharpoons X(f)$$

, the differentiation property Fourier transform pair is

$$\frac{dx(t)}{dt} \rightleftharpoons j2\pi f\, X(f) \qquad (3.25)$$

Proof:
Differentiate both sides of the inverse Fourier transform operation, Eq.(3.1), yields

$$\frac{dx(t)}{dt} = \frac{d}{dt}\left[\int_{-\infty}^{\infty} X(f)\, e^{j2\pi ft}\, df \right]$$

, changing the order of differentiation and integration, yields

$$\frac{dx(t)}{dt} = \int_{-\infty}^{\infty} X(f)\left[\frac{d}{dt} e^{j2\pi ft} \right] df$$

, or equivalently

$$\frac{dx(t)}{dt} = \int_{-\infty}^{\infty} [\, j2\pi f\, X(f)\,] e^{j2\pi ft}\, df \qquad (3.26)$$

, comparing Equations (3.26) and (3.1), then [dx(t)/dt] and [j2πf X(f)] constitute a Fourier transform pair. Then the differentiation of a function x(t) has the effect of multiplying its Fourier transform X(f) by (j2πf). On the other hand, due to duality property, the differentiation of a function in the frequency domain has the effect of multiplying its inverse Fourier transform by (−j2πt), and the Fourier transform pair will be

$$-j2\pi t\, x(t) \rightleftharpoons \frac{dX(f)}{df}$$

, then the differentiation property facilitates the evaluation of the Fourier transform of a new function equals [t x(t)], and the inverse Fourier transform of a new function equals [f X(f)], using the following Fourier transform pairs

$$t\ x(t) \rightleftharpoons \frac{1}{(-j2\pi)} \frac{dX(f)}{df} \tag{3.27a}$$

, and

$$\frac{1}{(j2\pi)} \frac{dx(t)}{dt} \rightleftharpoons f\ X(f) \tag{3.27b}$$

, repeating the differentiation operation, Equations (3.27), the Fourier transform pairs yield

$$t^n\ x(t) \rightleftharpoons \frac{1}{(-j2\pi)^n} \frac{d^n\ X(f)}{df^n} \tag{3.28a}$$

, and

$$\frac{1}{(j2\pi)^n} \frac{d^n\ x(t)}{dt^n} \rightleftharpoons f^n\ X(f) \tag{3.28b}$$

Problem 3.14
Find the Fourier transform of the signal x(t) = t rect(t/2), using differentiation property

Solution
The signal x(t) = t rect(t/2) can be represented by Fig.3.20, and based on the following Fourier transform pair

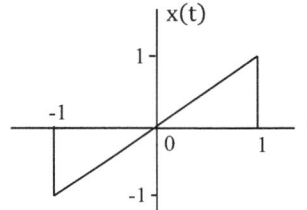

Fig.3.20. Energy signal x(t) = t rect(t/2).

$$rect(\frac{t}{2}) \rightleftharpoons 2\ sinc(2f)$$

, due to differentiation property in the frequency domain, Eq.(3.27a), the Fourier transform pair yields

$$t\ rect(\frac{t}{2}) \rightleftharpoons \frac{2}{(-j2\pi)} \frac{d\ sinc(2f)}{df}$$

, express the sinc function in terms of sine function, yields

$$t\ rect(\frac{t}{2}) \rightleftharpoons \frac{2}{(-j2\pi)} \frac{d}{df} \left[\frac{sin(2\pi f)}{2\pi f} \right]$$

, or

$$t\ rect(\frac{t}{2}) \rightleftharpoons j\ \frac{2\pi f\ cos(2\pi f) - sin(2\pi f)}{2\pi^2 f^2} \tag{3.29}$$

, the Fourier transform X(f) is imaginary value because x(t) is odd function, this Fourier transform can also be obtained, using Fourier integral, Eq.(3.2), X(f) yields

$$X(f) = \int_{-1}^{1} t\ e^{-j2\pi ft}\ dt$$

, using the integration by parts technique, the same spectrum X(f), Eq.(3.29), will be obtained.

Problem 3.15

Find the Fourier transform of the energy signal $x(t) = t\,e^{-t}\,u(t)$, Fig.3.21, using the differentiation property.

Solution

Based on the Fourier transform pair, Eq.(3.9), using differentiation property in the frequency domain, Eq.(3.27a), then

$$t\,e^{-t}\,u(t) \;\rightleftharpoons\; \frac{1}{(-j2\pi)}\,\frac{d}{df}\left[\frac{1}{1+j2\pi f}\right]$$

, or equivalently

$$t\,e^{-t}\,u(t) \;\rightleftharpoons\; \frac{1}{(1+j2\pi f)^2} \qquad (3.30)$$

, the Fourier transform $X(f)$ is complex value because $x(t)$ is neither even nor odd function. , this Fourier transform can also be obtained, using Fourier integral, Eq.(3.2) , $X(f)$ yields

$$X(f) = \int_0^\infty t\,e^{-t}\,e^{-j2\pi ft}\,dt$$

, using integration by parts technique, the same spectrum $X(f)$, Eq.(3.30), will be obtained.

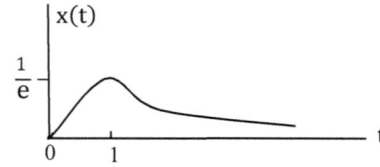

Fig.3.21. Energy signal $x(t) = t\,e^{-t}\,u(t)$.

Problem 3.16

Find the Fourier transform of the energy signal $x'(t)$, where $x'(t) = dx(t)/dt$, and $x(t)$ is shown in Fig.3.22a, using differentiation property.

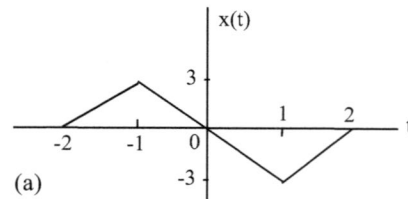

(a)

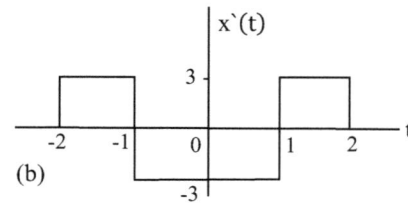

(b)

Fig.3.22. $x'(t)$ is the first derivative of $x(t)$.

Solution

The Fourier transform pair of the signal $x(t)$ is

$$3\,\mathrm{tri}(t+1) - 3\,\mathrm{tri}(t-1) \;\rightleftharpoons\; 3\,\mathrm{sinc}^2(f)\,e^{j2\pi f} - 3\,\mathrm{sinc}^2(f)\,e^{-j2\pi f}$$

, due to differentiation property in the time domain, Eq.(3.25), the Fourier transform pair yields

$$\frac{d}{dt}\left[3\,\mathrm{tri}(t+1) - 3\,\mathrm{tri}(t-1)\right] \;\rightleftharpoons\; j2\pi f\left[3\,\mathrm{sinc}^2(f)\,e^{j2\pi f} - 3\,\mathrm{sinc}^2(f)\,e^{-j2\pi f}\right]$$

, in the frequency domain, express the exponential terms by sine function, and the time domain equals the function $x'(t)$, the Fourier transform pair yields

$$x'(t) \;\rightleftharpoons\; -12\pi f\,\mathrm{sinc}^2(f)\,\sin(2\pi f) \qquad (3.31)$$

, another way to get this Fourier transform using the graphical differentiation of $x(t)$, Fig.3.22b, $x'(t)$ is expressed by

$$x'(t) = 3\,\mathrm{rect}(t+1.5) - 3\,\mathrm{rect}\left(\frac{t}{2}\right) + 3\,\mathrm{rect}(t-1.5)$$

, then the Fourier transform $X'(f)$ will be

$$X'(f) = 3\,\mathrm{sinc}(f)\,)\,e^{j\pi f3} - 6\,\mathrm{sinc}(2f) + 3\,\mathrm{sinc}(f)\,e^{-j\pi f3}$$

, express the exponential terms by cosine function, making use of the formulas "$\sin(2x) = 2\sin(x)\cos(x)$" and "$2\sin(x)\sin(y) = \cos(x-y) - \cos(x+y)$", and $x > y$, the same Fourier transform, Eq.(3.31), will be obtained.

3.3.6. Integration Property (of Zero Boundary Condition Functions)

Let the following Fourier transform pair

$$x(t) \rightleftharpoons X(f)$$

, and assuming that the boundary condition of the frequency domain $X(0)$ is zero, the integration property Fourier transform pair, Fig.3.23, is

$$\int_{-\infty}^{t} x(\tau)\, d\tau \rightleftharpoons \frac{X(f)}{j2\pi f} \qquad (3.32a)$$

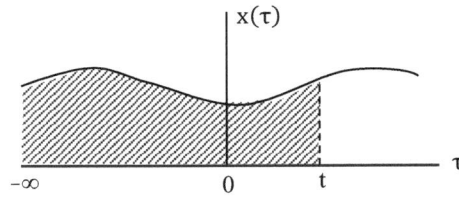

Fig.3.23. The integration function of the signal x(t).

Proof:

Integrate from $-\infty$ to t, the both sides of the inverse Fourier transform operation, Eq.(3.1), yields

$$\int_{-\infty}^{t} x(\tau)\, d\tau = \int_{-\infty}^{t} \left[\int_{-\infty}^{\infty} X(f)\, e^{j2\pi f \tau}\, df \right] d\tau$$

, changing the order of the integrations, yields

$$\int_{-\infty}^{t} x(\tau)\, d\tau = \int_{-\infty}^{\infty} X(f) \left[\int_{-\infty}^{\infty} X(f)\, e^{j2\pi f \tau}\, d\tau \right] df$$

$$= \int_{-\infty}^{\infty} \left[\frac{X(f)}{j2\pi f} \right] e^{j2\pi f \tau}\, df \qquad (3.32b)$$

, comparing Equations (3.32b) and (3.1), a Fourier transform pair of Eq.(3.32a) constitutes. Then the integration of a function x(t), has the effect of dividing its Fourier transform $X(f)$ by $(j2\pi f)$, assuming that the boundary condition of the frequency domain $X(0)$ is zero. On the other hand, due to duality property, the integration of a function in the frequency domain, has the effect of dividing its inverse Fourier transform by $(-j2\pi t)$ assuming that the boundary condition of the time domain $x(0)$ is zero, and the Fourier transform pair will be

$$\frac{x(t)}{-j2\pi t} \rightleftharpoons \int_{-\infty}^{f} X(\lambda)\, d\lambda \qquad (3.33)$$

, then integration property facilitates the evaluation of the Fourier transform of a new function equals $[x(t)/t]$, and the inverse Fourier transform of a new function equals $[X(f)/f]$, assuming that the boundary conditions of the time and frequency domains $x(0)$ and $X(0)$ are zero, using the following Fourier transform pairs

$$j2\pi \int_{-\infty}^{t} x(\tau)\, d\tau \rightleftharpoons \frac{X(f)}{f} \qquad (3.34)$$

, and

$$\frac{x(t)}{t} \rightleftharpoons - j2\pi \int_{-\infty}^{f} X(\lambda)\, d\lambda \qquad (3.35)$$

If the boundary conditions $x(0)$ and $X(0)$ are not zero, the Fourier transform pairs, Equations (3.32a) and (3.33) will be

$$\int_{-\infty}^{t} x(\tau)\, d\tau \rightleftharpoons \frac{X(f)}{j2\pi f} + \frac{X(0)\delta(f)}{2} \qquad (3.36)$$

, and

$$-\frac{x(t)}{j2\pi t} + \frac{x(0)\delta(t)}{2} \rightleftharpoons \int_{-\infty}^{f} X(\lambda)\, d\lambda \qquad (3.37)$$

, where $\delta(t)$ and $\delta(f)$ are the Dirac delta functions (unit impulse) in the time and frequency domains respectively. The Dirac delta function is nonintegrable function, do not satisfy Dirichlet's conditions, Eq.(3.3), but have Fourier transform in the limit (chapter IV).

Problem 3.17
Find the Fourier transform of the energy
signal $x_1(t)$, given by

$$x_1(t) = \int_{-\infty}^{t} x(\tau)\, d\tau$$

(a)

, where $x(t)$ is a doublet pulse, Fig.3.24a,
using integration property.

Solution
For the doublet pulse $x(t)$, Fig.3.24a, the
Fourier transform pair is

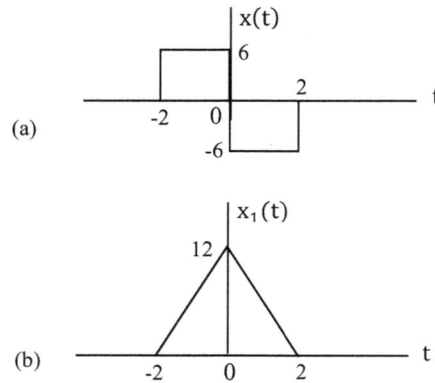

(b)

Fig.3.24. $x_1(t)$ is the integration of $x(t)$.

$$6\, \text{rect}\left[\frac{t+1}{2}\right] - 6\, \text{rect}\left[\frac{t-1}{2}\right] \quad \rightleftharpoons \quad \left\{\, 12\, \text{sinc}(2f)\, e^{j2\pi f} - 12\, \text{sinc}(2f)\, e^{-j2\pi f} \right\}$$

, due to integration property, the Fourier transform pair will be

$$\int_{-\infty}^{t} \left\{ 6\, \text{rect}\left[\frac{t+1}{2}\right] - 6\, \text{rect}\left[\frac{t-1}{2}\right] \right\} d\tau \quad \rightleftharpoons \quad \frac{12}{j2\pi f}\left\{ \text{sinc}(2f)\, e^{j2\pi f} - \text{sinc}(2f)\, e^{-j2\pi f} \right\}$$

, in the frequency domain, express the exponential terms by sine function, and the time domain is the
function $x_1(t)$, then in terms of sinc squared function, the Fourier transform pair yields

$$x_1(t) \quad \rightleftharpoons \quad 24\, \text{sinc}^2(2f) \qquad\qquad (3.38)$$

, this Fourier transform can also be obtained from the graphical integration, Fig.3.24b, then
$x_1(t) = 12\, \text{tri}(t/2)$, and the Fourier transform $X_1(f) = 24\, \text{sinc}^2(2f)$, is the same spectrum obtained, using
the integration property, Eq.(3.38).

Problem 3.18
Find the Fourier transform of the energy
signal $x_1(t)$, given by

$$x_1(t) = \int_{-\infty}^{t} x(\tau)\, d\tau$$

, where $x(t) = t\, e^{-|t|}$, Fig.3.25, using
integration property.

Fig.3.25. Energy signal $x(t) = t\, e^{-|t|}$.

Solution
Making use of the Fourier transform of the exponential double sided pulse $e^{-|t|}$, Eq.(3.14), and using
the differentiation property in the frequency domain, Eq.(3.27a), the Fourier transform pair of the
function $x(t) = t\, e^{-|t|}$, Fig.3.25, yields

$$t\, e^{-|t|} \quad \rightleftharpoons \quad \frac{1}{(-j2\pi)}\, \frac{d}{df}\left[\frac{2}{1+(2\pi f)^2}\right]$$

, or

$$t\, e^{-|t|} \quad \rightleftharpoons \quad -j\, \frac{8\pi f}{[1+(2\pi f)^2]^2}$$

, due to integration property, the Fourier transform pair will be

$$\int_{-\infty}^{t} \tau\, e^{-|\tau|}\, d\tau \quad \rightleftharpoons \quad \frac{1}{j2\pi f}\, \frac{-j8\pi f}{[1+(2\pi f)^2]^2}$$

, or equivalently

$$X(f) = -\frac{4}{[1+(2\pi f)^2]^2}$$

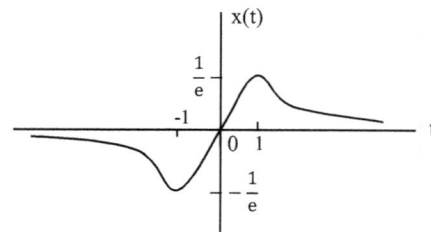

3.3.7. Convolution Property
Let the following Fourier transform pairs

$$x_1(t) \rightleftharpoons X_1(f)$$

, and

$$x_2(t) \rightleftharpoons X_2(f)$$

, the convolution property facilitates the evaluation of the Fourier transform of a function equals the product of $x_1(t)$ by $x_2(t)$. The Fourier transform pair convolution property in the frequency domain is

$$x_1(t) \; x_2(t) \rightleftharpoons X_1(f) \otimes X_2(f) \tag{3.39}$$

, where the symbol $\otimes$ denotes the convolution process, and the shorthand notation $X_1(f) \otimes X_2(f)$ is frequently used and called the convolution of $X_1(f)$ and $X_2(f)$, expressed in the frequency domain by the following convolution integral $X_{12}(f)$

$$X_{12}(f) = X_1(f) \otimes X_2(f) = F[x_1(t) \; x_2(t)] = \int_{-\infty}^{\infty} X_1(\lambda) \, X_2(f - \lambda) \, d\lambda \tag{3.40}$$

Proof:
The Fourier transform operation of the function $[x_1(t) \, x_2(t)]$ is given by

$$F[x_1(t) \; x_2(t)] = \int_{-\infty}^{\infty} x_1(t) \, x_2(t) \; e^{-j2\pi ft} \; dt$$

, in terms of the inverse Fourier transform operation of $x_2(t)$, the integration yields

$$F[x_1(t) \; x_2(t)] = \int_{-\infty}^{\infty} x_1(t) \left[\int_{-\infty}^{\infty} X_2(f) \; e^{j2\pi f^{\backprime} t} \; df^{\backprime} \right] e^{-j2\pi ft} \; dt$$

, or equivalently

$$F[x_1(t) \; x_2(t)] = \int_{-\infty}^{\infty} x_1(t) \left[\int_{-\infty}^{\infty} X_2(f) \, e^{-j2\pi(f - f^{\backprime})t} \; df^{\backprime} \right] \; dt$$

, let $\lambda = f - f^{\backprime}$, then $d\lambda = - \, df^{\backprime}$ as $f^{\backprime} \to -\infty$ then $\lambda \to \infty$, and as $f^{\backprime} \to \infty$ then $\lambda \to -\infty$, and interchange the order of integrations, the Fourier transform operation yields

$$F[x_1(t) \; x_2(t)] = \int_{-\infty}^{\infty} \left[\int_{-\infty}^{\infty} x_1(t) \; e^{-j2\pi\lambda t} \; dt \right] X_2(f - \lambda) \, d\lambda$$

, in terms of the Fourier transform of the function $x_1(t)$, Eq.(3.2), the convolution integral yields

$$F[x_1(t) \; x_2(t)] = \int_{-\infty}^{\infty} X_1(\lambda) \, X_2(f - \lambda) \, d\lambda$$

, this convolution integral in the frequency domain, is carried out over the frequency variable λ. The upper and lower limits of the integration can be deduced from the graphical multiplication of the fixed function $X_1(\lambda)$ and the moving function $X_2(f - \lambda)$ due to the frequency shift f, where f plays the role of a scanning parameter from $-\infty$ to ∞.

On the other hand, the convolution process in the time domain may be carried out using the same procedure or directly using duality property, where the inverse Fourier transform of a function equals the product of $X_1(f)$ by $X_2(f)$, and the Fourier transform pair will be

$$x_1(t) \otimes x_2(t) \rightleftharpoons X_1(f) \, X_2(f) \tag{3.41}$$

, in terms of the convolution integral, in the time domain, Eq.(3.41) yields

$$\int_{-\infty}^{\infty} x_1(\tau)\, x_2(t-\tau)\, d\tau \quad \rightleftharpoons \quad X_1(f)\, X_2(f)$$

, in the time domain the convolution integral of $x_1(t)$ and $x_2(t)$ is expressed by $x_{12}(t)$ and is given by

$$x_{12}(t) = x_1(t) \otimes x_2(t) = F^{-1}[X_1(f)\, X_2(f)] = \int_{-\infty}^{\infty} x_1(\tau)\, x_2(t-\tau)\, d\tau \qquad (3.42)$$

, this convolution integral in the time domain, is carried out over the time variable τ. The upper and lower limits of the integration can be deduced from the graphical multiplication of the fixed function $x_1(\tau)$ and the moving function $x_2(t-\tau)$ due to the time shift t, where t plays the role of a scanning parameter from $-\infty$ to ∞.

The convolution property concludes that the multiplication of two signals in any domain is mapped into the convolution of their individual Fourier transforms in the other domain [1].

The convolution process obeys the commutative law of algebra, where

$$x_1(t) \otimes x_2(t) = x_2(t) \otimes x_1(t) \qquad (3.43a)$$

, and
$$X_1(f) \otimes X_2(f) = X_2(f) \otimes X_1(f)$$

, the convolution process also obeys the distributive law of algebra, where

$$x_1(t) \otimes [x_2(t) + x_3(t)] = [x_1(t) \otimes x_2(t)] + [x_1(t) \otimes x_3(t)] \qquad (3.43b)$$

, and
$$X_1(f) \otimes [X_2(f) + X_3(f)] = [X_1(f) \otimes X_2(f)] + [X_1(f) \otimes X_3(f)]$$

, also the convolution process obeys the associative law of algebra, where

$$x_1(t) \otimes [x_2(t) \otimes x_3(t)] = x_1(t) \otimes x_2(t) \otimes x_3(t) \qquad (3.43c)$$

, and
$$X_1(f) \otimes [X_2(f) \otimes X_3(f)] = X_1(f) \otimes X_2(f) \otimes X_3(f)$$

The convolution property shows that there are two ways to solve the convolution problem, either in the same domain using the convolution integrals, Equations (3.40) or (3.42), or in the other domain by getting their individual transforms in the other domain and multiply and then return again to their first same domain by taking the inverse transform operation, using Equations (3.39) and (3.41).

The following examples are some important convolution processes of some famous signals such as the unit step function u(t), the rectangular pulse rect(t), the triangle pulse tri(t), the decaying and rising single sided exponential pulses, and the Gaussian pulse. It is necessary to evaluate the convolution processes of these pulses and each other, and other different known signals, to facilitate the evaluation of the Fourier transforms and also to obtain the continuous frequency spectrum of complex energy signals, to solve many applications in the communication systems [1].

u(t) ⊗ u(t)
The convolution function $x_{12}(t) = u(t) \otimes u(t)$, Fig.3.26 , using the time domain , Eq.(3.42), is given by:

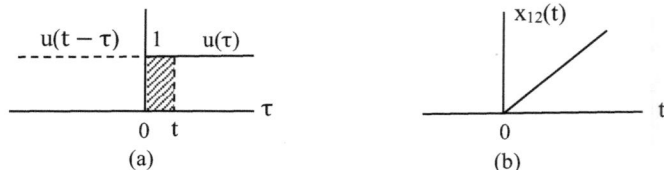

$$x_{12}(t) = \int_0^t d\tau = t\, u(t)$$

Fig.3.26. $x_{12}(t) = u(t) \otimes u(t)$.

rect(t) ⊗ rect(t)

The convolution function, using the time domain, $x_{12}(t) = \text{rect}(t) \otimes \text{rect}(t)$, Eq.(3.42), Fig.3.27a,b, is given by

$$x_{12}(t) = \int_{-\frac{1}{2}+t}^{\frac{1}{2}} d\tau \; = 1 - t \quad \text{for positive t}$$

$$, x_{12}(t) = \int_{-\frac{1}{2}}^{\frac{1}{2}+t} d\tau \; = 1 + t \quad \text{for negative t}$$

, another way, using the frequency domain, Eq.(3.41), the Fourier transform pair of $x_{12}(t)$ is given by

$$\text{rect}(t) \otimes \text{rect}(t) \; \rightleftharpoons \; \text{sinc}(f)\,\text{sinc}(f)$$

, taking the inverse Fourier transform , yields

$$\text{tri}(t) \; \rightleftharpoons \; \text{sinc}^2(f)$$

, then the convolution function $x_{12}(t)$, Fig.3.27c, is given by

$$x_{12}(t) \; = \; \text{rect}(t) \otimes \text{rect}(t) = \text{tri}(t)$$

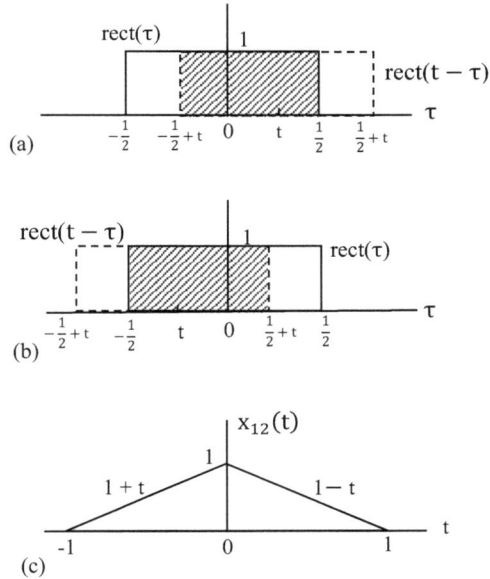

(a)

(b)

(c)

Fig.3.27. $x_{12}(t) = \text{rect}(t) \otimes \text{rect}(t)$, a) positive t. b) negative t, c) The convolution waveform.

rect(t) ⊗ u(t)

The convolution function
$x_{12}(t) = \text{rect}(t) \otimes u(t)$
, Fig.3.28a,b, using the time domain, Eq.(3.42), is given by

$$x_{12}(t) = \int_{-\frac{1}{2}}^{t} d\tau$$

$$= t + \tfrac{1}{2} \quad \text{for } -\tfrac{1}{2} \le t \le \tfrac{1}{2}$$

, and

$$x_{12}(t) = \int_{-\frac{1}{2}}^{\frac{1}{2}} d\tau = 1 \text{ for } t \ge \tfrac{1}{2}$$

, another way, Fig.3.28c,d, $x_{12}(t) = u(t) \otimes \text{rect}(t)$, is

$$x_{12}(t) = \int_{0}^{\frac{1}{2}+t} d\tau$$

$$= t + \tfrac{1}{2} \quad \text{for } -\tfrac{1}{2} \le t \le \tfrac{1}{2}$$

, and

$$x_{12}(t) = \int_{-\frac{1}{2}+t}^{\frac{1}{2}+t} d\tau = 1 \text{ for } \quad t \ge \tfrac{1}{2}$$

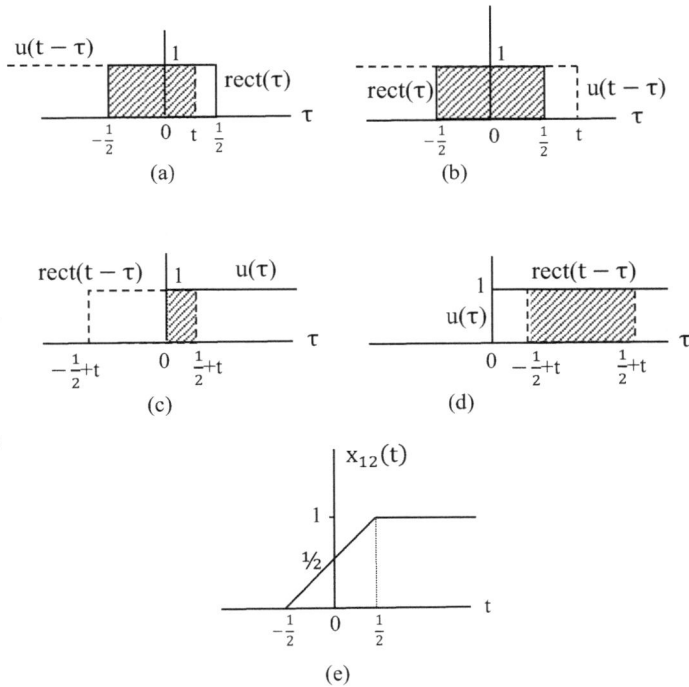

(a)

(b)

(c)

(d)

(e)

Fig.3.28. $x_{12}(t) = \text{rect}(t) \otimes u(t)$,
a) $-\tfrac{1}{2} \le t \le \tfrac{1}{2}$, b) $t \ge \tfrac{1}{2}$, c) $-\tfrac{1}{2} \le t \le \tfrac{1}{2}$ another way,
d) $t \ge \tfrac{1}{2}$ another way, e) The convolution waveform.

rect(t) ⊗ t u(t)

The convolution function $x_{12}(t) = \text{rect}(t) \otimes t\, u(t)$, Fig.3.29a,b, using the time domain, Eq.(3.42), where for the period $-\frac{1}{2} \leq t \leq \frac{1}{2}$, is given by

$$x_{12}(t) = \int_{-\frac{1}{2}}^{t} (t - \tau)\, d\tau$$

$$= \frac{1}{2}\, t^2 + \frac{1}{2}\, t + \frac{1}{8}$$

, and

$$x_{12}(t) = \int_{-\frac{1}{2}}^{\frac{1}{2}} (t - \tau)\, d\tau$$

$$= t \qquad \text{for } t \geq \frac{1}{2}$$

, another way, Fig.3.29c,d, $x_{12}(t) = t\, u(t) \otimes \text{rect}(t)$, where for the period $-\frac{1}{2} \leq t \leq \frac{1}{2}$, is

$$x_{12}(t) = \int_{0}^{\frac{1}{2}+t} \tau\ d\tau$$

$$= \frac{1}{2}\, t^2 + \frac{1}{2}\, t + \frac{1}{8}$$

, and

$$x_{12}(t) = \int_{-\frac{1}{2}+t}^{\frac{1}{2}+t} \tau\ d\tau$$

$$= t \qquad \text{for } t \geq \frac{1}{2}$$

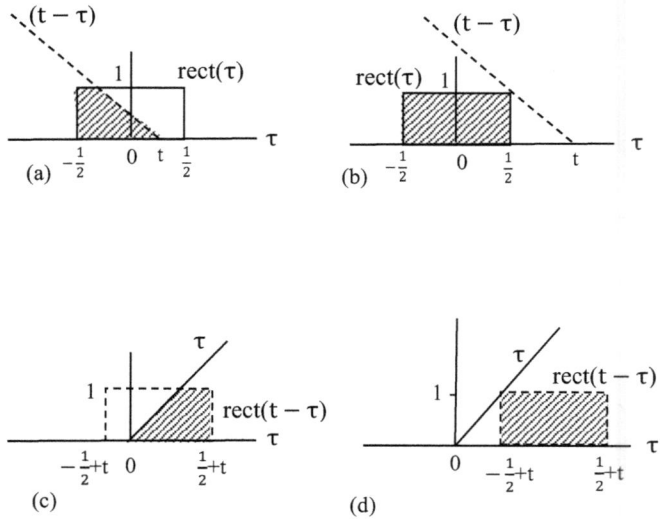

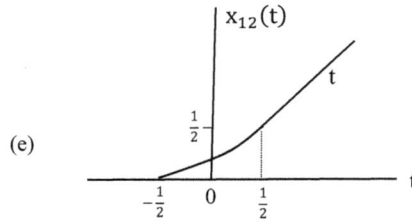

Fig.3.29. $x_{12}(t) = \text{rect}(t) \otimes t\, u(t)$,
a) $-\frac{1}{2} \leq t \leq \frac{1}{2}$, b) $t \geq \frac{1}{2}$,
c) $-\frac{1}{2} \leq t \leq \frac{1}{2}$ another way,
d) $t \geq \frac{1}{2}$ another way,
e) The convolution waveform.

tⁿ u(t) ⊗ u(t)

The convolution function

$$x_{12}(t) = t^n\, u(t) \otimes u(t)$$

, Fig.3.30a, using the time domain, Eq.(3.42), is given by

$$x_{12}(t) = \int_{0}^{t} \tau^n\ d\tau = \frac{1}{n+1}\, t^{n+1}\, u(t)$$

, another way $x_{12}(t) = u(t) \otimes t^n\, u(t)$, Fig.3.30b, is given by

$$x_{12}(t) = \int_{0}^{t} (t - \tau)^n\ d\tau = \frac{1}{n+1}\, t^{n+1}\, u(t) \qquad \text{, where } n = 1, 2, 3, 4, \ldots$$

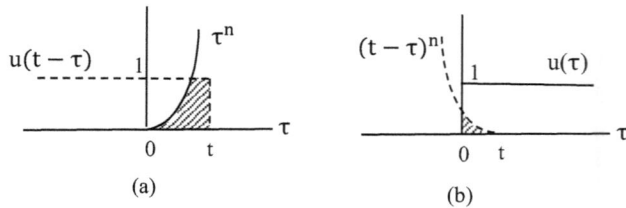

Fig.3.30. a) $x_{12}(t) = t^n\, u(t) \otimes u(t)$,
b) $x_{12}(t) = u(t) \otimes t^n\, u(t)$.

rect(t) ⊗ t²

The convolution function $x_{12}(t) = \text{rect}(t) \otimes t^2$, Fig.3.31a , using the time domain , Eq.(3.42), is given by

$$x_{12}(t) = \int_{-\frac{1}{2}}^{\frac{1}{2}} (t - \tau)^2 \, d\tau$$

$$= t^2 + (1/12)$$

, another way $x_{12}(t) = t^2 \otimes \text{rect}(t)$, Fig.3.31b, is given by

$$x_{12}(t) = \int_{-\frac{1}{2}+t}^{\frac{1}{2}+t} \tau^2 \, d\tau$$

$$= t^2 + (1/12)$$

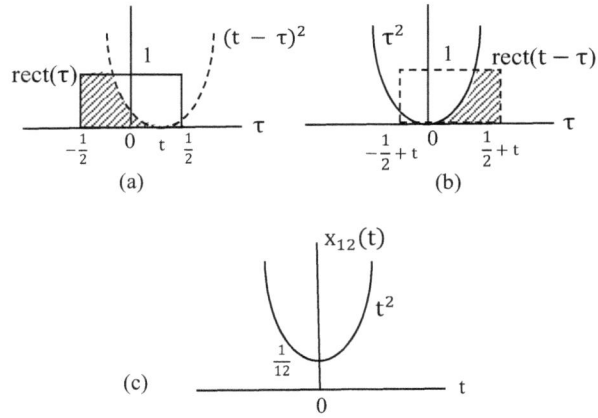

Fig.3.31. a) $x_{12}(t) = \text{rect}(t) \otimes t^2$, b) $x_{12}(t) = t^2 \otimes \text{rect}(t)$, c) The convolution waveform.

e⁻ᵗ u(t) ⊗ u(t)

The convolution function $x_{12}(t) = e^{-t} u(t) \otimes u(t)$, Fig.3.32a, using the time domain, Eq.(3.42), is given by

$$x_{12}(t) = \int_0^t e^{-\tau} \, d\tau$$

$$= [1 - e^{-t}] u(t)$$

, another way, Fig.3.32b, $x_{12}(t) = u(t) \otimes e^{-t} u(t)$, is given by

$$x_{12}(t) = \int_0^t e^{-(t-\tau)} d\tau$$

$$= [1 - e^{-t}] u(t)$$

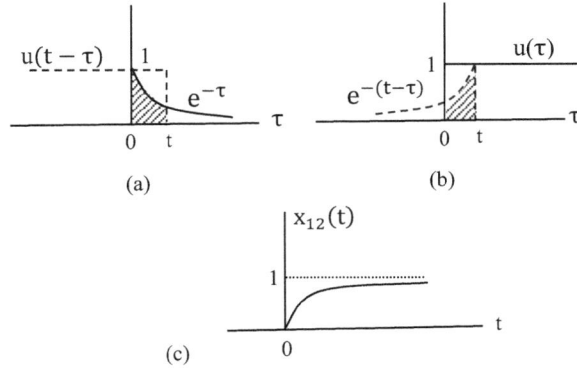

Fig.3.32. a) $x_{12}(t) = e^{-t} u(t) \otimes u(t)$, b) $x_{12}(t) = u(t) \otimes e^{-t} u(t)$ c) The convolution waveform.

e⁻ᵗ u(t) ⊗ t u(t)

The convolution function $x_{12}(t) = e^{-t} u(t) \otimes t\, u(t)$, Fig.3.33a, using the time domain, Eq.(3.42), is given by

$$x_{12}(t) = \int_0^t e^{-\tau} (t - \tau) \, d\tau$$

$$= t + e^{-t} - 1$$

, another way $x_{12}(t) = t\, u(t) \otimes e^{-t} u(t)$, Fig.3.33b, is given by

$$x_{12}(t) = \int_0^t \tau\, e^{-(t-\tau)} \, d\tau = t + e^{-t} - 1$$

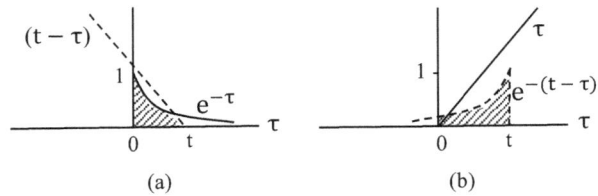

Fig.3.33. a) $x_{12}(t) = e^{-t} u(t) \otimes t\, u(t)$, b) $x_{12}(t) = t\, u(t) \otimes e^{-t} u(t)$.

$e^{-t} u(t) \otimes rect(t)$

The convolution function,

$$x_{12}(t) = e^{-t} u(t) \otimes rect(t)$$

, Fig.3.34a,b, using the time domain, Eq.(3.42), where for the period $-\frac{1}{2} \leq t \leq \frac{1}{2}$, is given by

$$x_{12}(t) = \int_{0}^{\frac{1}{2}+t} e^{-\tau} d\tau \approx 1 - 0.61 e^{-t}$$

, and for $t \geq \frac{1}{2}$, is given by

$$x_{12}(t) = \int_{-\frac{1}{2}+t}^{\frac{1}{2}+t} e^{-\tau} d\tau \approx 1.04 e^{-t}$$

, another way, Fig.3.34c,d, $x_{12}(t) = rect(t) \otimes e^{-t} u(t)$, where for the period $-\frac{1}{2} \leq t \leq \frac{1}{2}$, is given by

$$x_{12}(t) = \int_{-\frac{1}{2}}^{t} e^{-(t-\tau)} d\tau \approx 1 - 0.61 e^{-t}$$

, and for $t \geq \frac{1}{2}$, is given by

$$x_{12}(t) = \int_{-\frac{1}{2}}^{\frac{1}{2}} e^{-(t-\tau)} d\tau \approx 1.04 e^{-t}$$

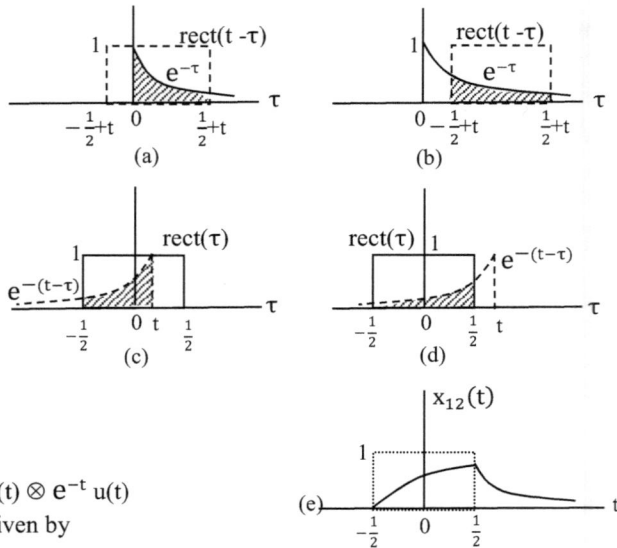

Fig.3.34. $x_{12}(t) = e^{-t} u(t) \otimes rect(t)$,
a) $-\frac{1}{2} \leq t \leq \frac{1}{2}$, b) $t \geq \frac{1}{2}$,
c) $-\frac{1}{2} \leq t \leq \frac{1}{2}$ another way,
d) $t \geq \frac{1}{2}$ another way,
e) The convolution waveform.

$tri(t) \otimes rect(t)$

The convolution function

$$x_{12}(t) = tri(t) \otimes rect(t)$$

, Fig.3.35, using the time domain, Eq.(3.42), for the period $-3/2 \leq t \leq -\frac{1}{2}$, $x_{12}(t)$ is given by

$$x_{12}(t) = \int_{-1}^{\frac{1}{2}+t} (1 + \tau) d\tau$$

$$= \frac{1}{2} t^2 + 1.5 t + (9/8)$$

, and for the period $-\frac{1}{2} \leq t \leq \frac{1}{2}$, $x_{12}(t)$ is given by

$$x_{12}(t) = \int_{-\frac{1}{2}+t}^{0} (1 + \tau) d\tau + \int_{0}^{\frac{1}{2}+t} (1 - \tau) d\tau$$

$$= (3/4) - t^2$$

, and for the period $\frac{1}{2} \leq t \leq 3/2$, $x_{12}(t)$ is given by

$$x_{12}(t) = \int_{-\frac{1}{2}+t}^{1} (1 - \tau) d\tau$$

$$= \frac{1}{2} t^2 - 1.5 t + (9/8)$$

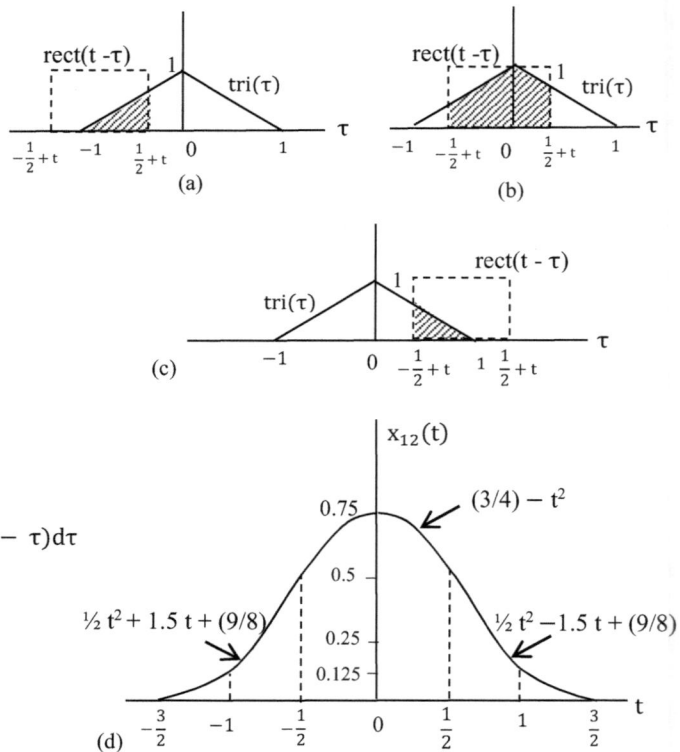

Fig.3.35. $x_{12}(t) = tri(t) \otimes rect(t)$, a) $-3/2 \leq t \leq -\frac{1}{2}$, b) $-\frac{1}{2} \leq t \leq \frac{1}{2}$,
c) $\frac{1}{2} \leq t \leq 3/2$, d) The convolution waveform.

$e^{-3t} u(t) \otimes e^{-t}$

The convolution function

$x_{12}(t) = e^{-3t} u(t) \otimes e^{-t}$

, Fig.3.36a, using the time domain, Eq.(3.42), is given by

$$x_{12}(t) = \int_{0}^{\infty} e^{-3\tau} e^{-(t-\tau)} \, d\tau$$

$$= \tfrac{1}{2} e^{-t}$$

, another way, Fig.3.36b,

$x_{12}(t) = e^{-t} \otimes e^{-3t} u(t)$

, is given by

$$x_{12}(t) = \int_{-\infty}^{t} e^{-\tau} e^{-3(t-\tau)} \, d\tau = \tfrac{1}{2} e^{-t}$$

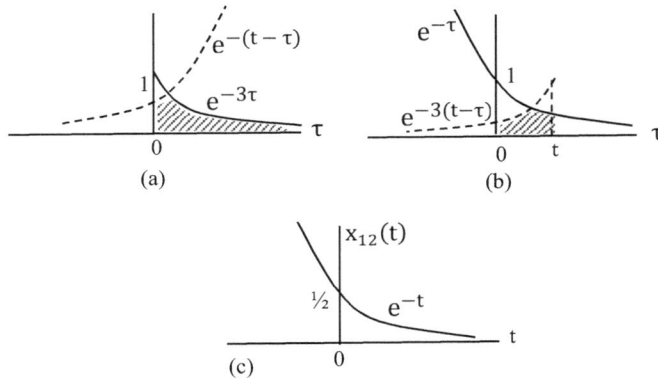

Fig.3.36. a) $x_{12}(t) = e^{-3t} u(t) \otimes e^{-t}$,
b) $x_{12}(t) = e^{-t} \otimes e^{-3t} u(t)$,
c) The convolution waveform.

$e^{-t} u(t) \otimes e^{-2t} u(t)$

The convolution function

$x_{12}(t) = e^{-t} u(t) \otimes e^{-2t} u(t)$

, Fig.3.37a, using the time domain, Eq.(3.42), $x_{12}(t)$, is given by

$$x_{12}(t) = \int_{0}^{t} e^{-\tau} e^{-2(t-\tau)} \, d\tau$$

$$= e^{-t} u(t) - e^{-2t} u(t)$$

$$= e^{-3t/2} \left[e^{\frac{1}{2}t} - e^{-\frac{1}{2}t} \right] u(t)$$

$$= 2 e^{-3t/2} \sinh(t/2) u(t)$$

, another way, Fig.3.37b,

$x_{12}(t) = e^{-2t} u(t) \otimes e^{-t} u(t)$

, is given by

$$x_{12}(t) = \int_{0}^{t} e^{-2\tau} e^{-(t-\tau)} \, d\tau$$

$$= 2 e^{-3t/2} \sinh(\frac{t}{2}) u(t)$$

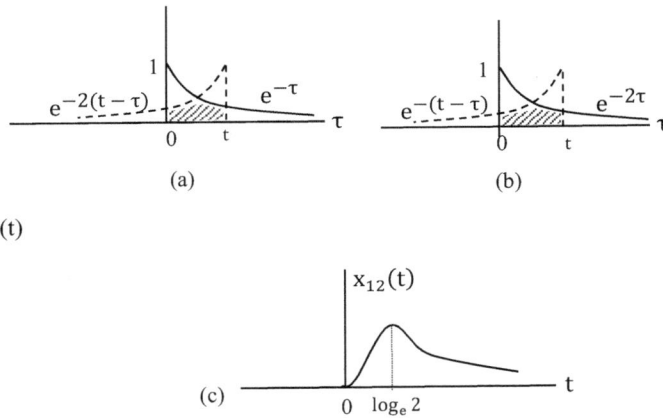

Fig.3.37. a) $x_{12}(t) = e^{-t} u(t) \otimes e^{-2t}$, b) $x_{12}(t) = e^{-2t} \otimes e^{-t} u(t)$,
c) The convolution waveform.

, this convolution function can also be obtained, using the frequency domain, Eq.(3.41), a Fourier transform pair is given by

$$e^{-t} u(t) \otimes e^{-2t} u(t) \quad \rightleftharpoons \quad \frac{1/2}{(1 + j2\pi f)(1 + j\pi f)}$$

, using partial fraction technique in the frequency domain, and taking the inverse Fourier transform using linearity and scaling properties, the Fourier transform pair yields

$$e^{-t} u(t) - e^{-2t} u(t) \quad \rightleftharpoons \quad \frac{1}{1 + j2\pi f} - \frac{1/2}{1 + j\pi f}$$

, then the time domain is the convolution function $x_{12}(t)$, and is given by

$$x_{12}(t) = e^{-t} u(t) - e^{-2t} u(t) = 2 e^{-3t/2} \sinh(\frac{t}{2}) u(t)$$

tri(t) ⊗ u(t)

The convolution function $x_{12}(t) = tri(t) \otimes u(t)$, Fig.3.38a,b,c, using the time domain, Eq.(3.42), where for the period $-1 \leq t \leq 0$, $x_{12}(t)$ is given by

$$x_{12}(t) = \int_{-1}^{t} (1 + \tau)\, d\tau = \tfrac{1}{2}\, t^2 + t + \tfrac{1}{2}$$

, for the period $0 \leq t \leq 1$, $x_{12}(t)$ is given by

$$x_{12}(t) = \int_{-1}^{0} (1 + \tau)\, d\tau + \int_{0}^{t} (1 - \tau)\, d\tau$$
$$= \tfrac{1}{2} + t - \tfrac{1}{2}\, t^2$$

, and for the period $t \geq 1$, $x_{12}(t)$ is given by

$$x_{12}(t) = \int_{-1}^{0} (1 + \tau)\, d\tau + \int_{0}^{1} (1 - \tau)\, d\tau = 1$$

, another way, $x_{12}(t) = u(t) \otimes tri(t)$, Fig.3.39d,e,f, where for the period $-1 \leq t \leq 0$, is given by

$$x_{12}(t) = \int_{0}^{1+t} [1 - (t - \tau)]\, d\tau = \tfrac{1}{2}\, t^2 + t + \tfrac{1}{2}$$

, for the period $0 \leq t \leq 1$, is given by

$$x_{12}(t) = \int_{0}^{t} [1 + (t - \tau)]\, d\tau + \int_{t}^{1+t} [1 - (t - \tau)]\, d\tau$$
$$= \tfrac{1}{2} + t - \tfrac{1}{2}\, t^2$$

, and for the period $t \geq 1$, is given by

$$x_{12}(t) = \int_{-1+t}^{t} [1 + (t - \tau)]\, d\tau + \int_{t}^{1+t} [1 - (t - \tau)\, d\tau = 1$$

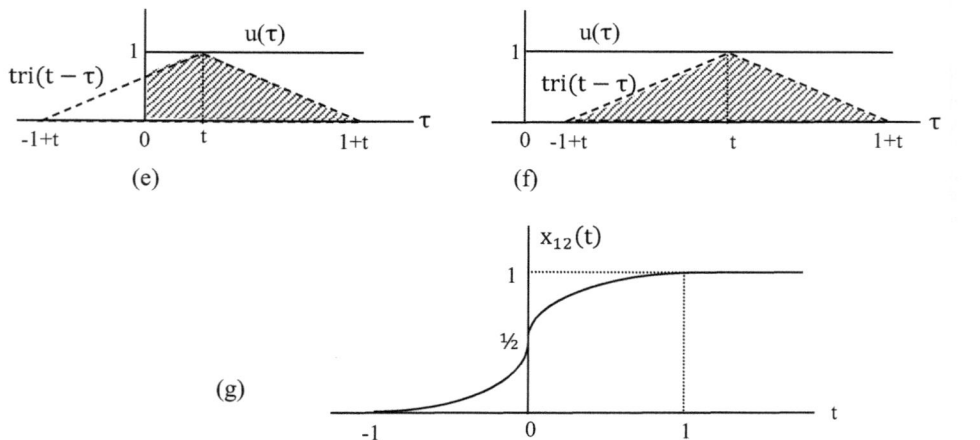

Fig.3.38. $x_{12}(t) = tri(t) \otimes u(t)$, a) $-1 \leq t \leq 0$, b) $0 \leq t \leq 1$, c) $t \geq 1$,
d) $-1 \leq t \leq 0$ another way, e) $0 \leq t \leq 1$ another way,
f) $t \geq 1$ another way, g) The convolution waveform.

$$e^{-\pi t^2} \otimes e^{-\pi t^2}$$

The convolution function $x_{12}(t) = e^{-\pi t^2} \otimes e^{-\pi t^2}$, using the frequency domain, Eq.(3.41), a Fourier transform pair is given by

$$e^{-\pi t^2} \otimes e^{-\pi t^2} \rightleftharpoons e^{-\pi f^2} \times e^{-\pi f^2}$$

, taking the inverse Fourier transform, and using scaling property, the Fourier transform pair yields

$$\frac{1}{\sqrt{2}} e^{-\pi t^2/2} \rightleftharpoons e^{-2\pi f^2}$$

, this inverse Fourier transform $x_{12}(t)$ of the function $e^{-2\pi f^2}$, can also be obtained, using the inverse Fourier transform operation, Eq.(3.1), yields

$$x_{12}(t) = F^{-1}[e^{-2\pi f^2}] = \int_{-\infty}^{\infty} e^{-2\pi f^2} e^{j2\pi ft} \, df$$

, using the same procedure of Gaussian pulse and Gaussian spectra, Eq.(3.7), $x_{12}(t)$ yields

$$x_{12}(t) = e^{-\pi t^2} \otimes e^{-\pi t^2} = \frac{1}{\sqrt{2}} e^{-\pi t^2/2}$$

, another way to obtain the convolution function $x_{12}(t) = e^{-\pi t^2} \otimes e^{-\pi t^2}$, using the time domain, Eq.(3.42), and using the same procedure of Gaussian pulse and Gaussian spectra, Eq.(3.7), $x_{12}(t)$ yields

$$x_{12}(t) = e^{-\pi t^2} \otimes e^{-\pi t^2} = \int_{-\infty}^{\infty} e^{-\pi \tau^2} e^{-\pi(t-\tau)^2} d\tau = \frac{1}{\sqrt{2}} e^{-\pi t^2/2}$$

Problem 3.19
Evaluate the following convolution functions

i. $X_{12}(f) = \text{rect}(f) \, e^{-j2\pi f} \otimes e^{-j2\pi f}$, using the frequency domain

ii. $x_{12}(t) = \dfrac{1}{1 - j2\pi t} \otimes \dfrac{1}{1 - j\pi t}$, using the frequency domain

iii. $x_{12}(t) = 2 \text{ sinc}(2t - 8) \otimes 14 \text{ sinc}(7t + 14)$, using the frequency domain

iv. $X_{12}(f) = 16 \text{ sinc}(8f) \otimes 18 \text{ sinc}^2(6f)$, using the time domain.

Solution

i. The convolution function $X_{12}(f) = \text{rect}(f) \, e^{-j2\pi f} \otimes e^{-j2\pi f}$, using the frequency domain, Eq.(3.40), is given by

$$X_{12}(f) = \int_{-\frac{1}{2}}^{\frac{1}{2}} e^{-j2\pi \lambda} e^{-j2\pi(f-\lambda)} d\lambda = e^{-j2\pi f} \int_{-\frac{1}{2}}^{\frac{1}{2}} d\lambda = e^{-j2\pi f}$$

ii. The convolution function

$$x_{12}(t) = \frac{1}{1 - j2\pi t} \otimes \frac{1}{1 - j\pi t}$$

, using the frequency domain, Eq.(3.41), a Fourier transform pair is given by

$$\frac{1}{1 - j2\pi t} \otimes \frac{1}{1 - j\pi t} \rightleftharpoons e^f u(-f) \times 2 e^{2f} u(-f)$$

, taking the inverse Fourier transform, using scaling property, a Fourier transform pair yields

$$\frac{\frac{2}{3}}{1 - j \frac{2}{3}\pi t} \rightleftharpoons 2 e^{3f} u(-f)$$

, then the convolution function $x_{12}(t)$ will be

$$x_{12}(t) = \frac{1}{1 - j2\pi t} \otimes \frac{1}{1 - j\pi t} = \frac{\frac{2}{3}}{1 - j \frac{2}{3}\pi t}$$

iii. The convolution function $x_{12}(t) = 2\,sinc(2t-8) \otimes 14\,sinc(7t+14)$, using the frequency domain, Eq.(3.41), a Fourier transform pair is given by

$$2\,sinc[(2(t-4)] \otimes 14\,sinc[7(t+2)] \rightleftharpoons rect(\frac{f}{2})\,e^{-j2\pi f4} \times 2\,rect(\frac{f}{7})\,e^{j2\pi f2}$$

, in the frequency domain, the graphical multiplication of the two rectangular functions rect(f/2) by rect(f/7), gives the rectangular function of the smallest frequency duration rect(f/2), and taking the inverse Fourier transform using shifting property, the Fourier transform pair yields

$$4\,sinc[2(t-2)] \rightleftharpoons 2\,rect(\frac{f}{2})\,e^{-j2\pi f\,2}$$

, then the convolution function $x_{12}(t)$ is given by

$$x_{12}(t) = 2\,sinc(2t-8) \otimes 14\,sinc(7t+14) = 4\,sinc(2t-4)$$

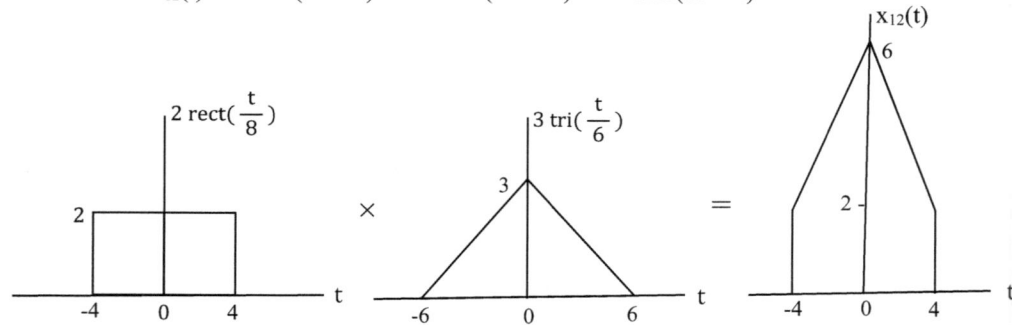

Fig.3.39. The graphical multiplication of rectangular and triangle pulses.

iv. The convolution function $X_{12}(f) = 16\,sinc(8f) \otimes 18\,sinc^2(6f)$, using the time domain, Eq.(3.39), a Fourier transform pair is given by

$$2\,rect(\frac{t}{8}) \times 3\,tri(\frac{t}{6}) \rightleftharpoons 16\,sinc(8f) \otimes 18\,sinc^2(6f)$$

, in the time domain, the graphical multiplication of the rectangular function rect(t/8) by the triangle function tri(t/6), gives another new function, is the algebraic summation of rectangular and triangle functions, Fig.3.39, and the time domain is given by

$$2\,rect(\frac{t}{8}) \times 3\,tri(\frac{t}{6}) = 2\,rect(\frac{t}{8}) + 4\,tri(\frac{t}{4})$$

, and the Fourier transform pair will be

$$2\,rect(\frac{t}{8}) + 4\,tri(\frac{t}{4}) \rightleftharpoons 16\,sinc(8f) + 16\,sinc^2(4f)$$

, then the convolution function $X_{12}(f)$ will be

$$X_{12}(f) = 16\,sinc(8f) \otimes 18\,sinc^2(4f) = 16\,sinc(8f) + 16\,sinc^2(4f)$$

Problem 3.20

Evaluate the convolution function of the two energy signals, $x_1(t)$ and $x_2(t)$, Fig.3.40, are given by

$$x_1(t) = e^{-t} \qquad 0 < t < T$$
$$,\quad x_2(t) = e^{t} \qquad 0 < t < T$$

Solution

The convolution function $x_{12}(t) = x_1(t) \otimes x_2(t)$, Fig.3.41a,b , using the time domain, Eq.(3.42) , is given by

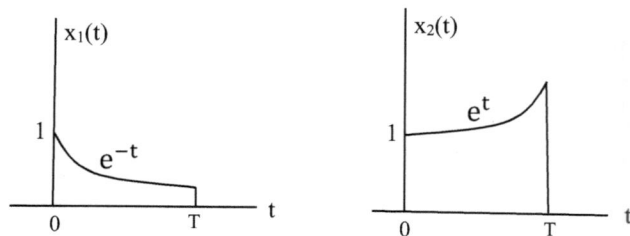

Fig.3.40. Two energy signals.

$$x_{12}(t) = \int_0^t e^{-\tau} e^{(t-\tau)} \, d\tau = \tfrac{1}{2}[e^t - e^{-t}] = \sinh(t) \qquad \text{for } 0 < t < T$$

, and

$$x_{12}(t) = \int_{t-T}^{T} e^{-\tau} e^{(t-\tau)} \, d\tau = \tfrac{1}{2}\left[e^{(2T-t)} - e^{-(2T-t)}\right] = \sinh(2T - t) \qquad \text{for } T < t < 2T$$

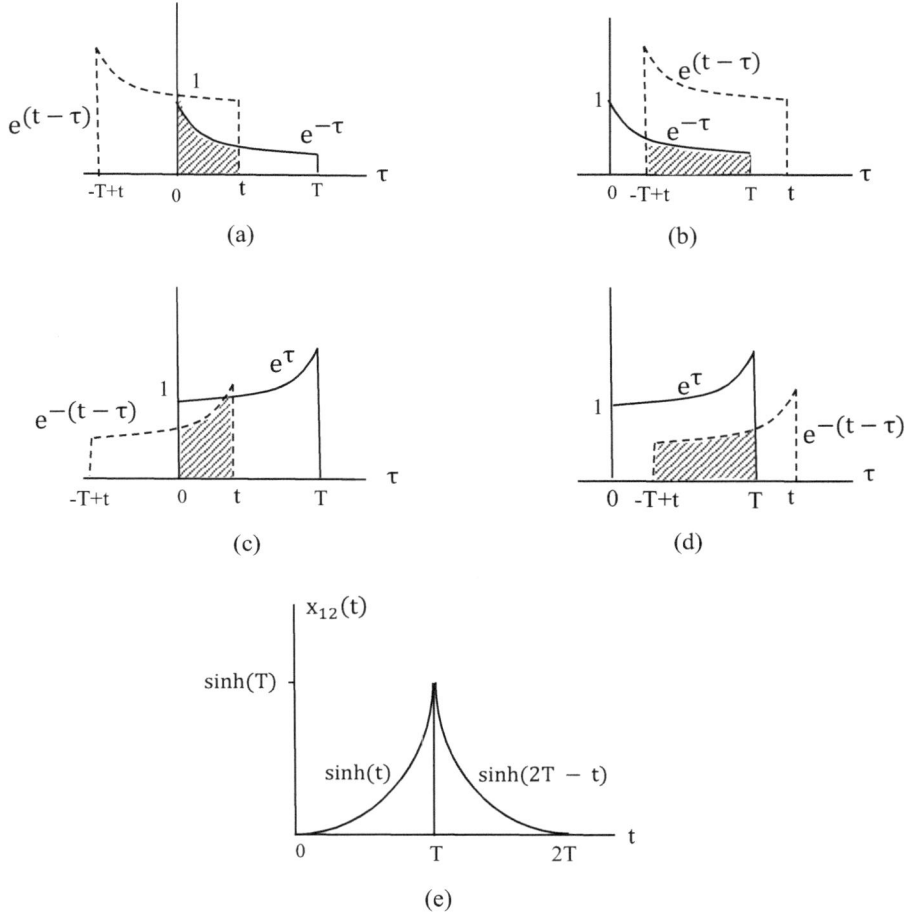

(a)   (b)

(c)   (d)

(e)

Fig.3.41. $x_{12}(t) = x_1(t) \otimes x_2(t)$, a) $0 \le t \le T$, b) $T \le t \le 2T$, c) $0 \le t \le T$ another way, d) $T \le t \le 2T$ another way , e) The convolution waveform.

, another way $x_{12}(t) = x_2(t) \otimes x_1(t)$, Fig.3.41c,d, using the time domain, Eq.(3.42), is given by

$$x_{12}(t) = \int_0^t e^{\tau} e^{-(t-\tau)} \, d\tau = \tfrac{1}{2}[e^t - e^{-t}] = \sinh(t) \qquad \text{for } 0 < t < T$$

, and

$$x_{12}(t) = \int_{-T+t}^{T} e^{\tau} e^{-(t-\tau)} \, d\tau = \tfrac{1}{2}\left[e^{(2T-t)} - e^{-(2T-t)}\right]$$

$$= \sinh(2T - t) \qquad \text{for } T < t < 2T$$

Problem 3.21
Find the Fourier transform of the following truncated sinc function, Fig.3.42, using the frequency domain

$$x(t) = A \, rect(\frac{t}{T}) \, sinc(Wt)$$

Fig.3.42. Truncated sinc function.

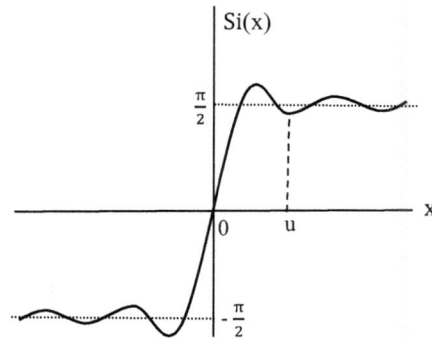

Solution
The Fourier transform pair of the truncated pulse x(t), is given by

$$A \, rect(\frac{t}{T}) \, sinc(Wt) \; \rightleftharpoons \; AT \, sinc(fT) \otimes \frac{1}{W} \, rect(\frac{f}{W})$$

, in the frequency domain, the convolution function $X_{12}(f)$, is given by

$$X_{12}(f) = \frac{AT}{W} \int_{-\infty}^{\infty} rect(\frac{\lambda}{W}) \, sinc[(f - \lambda)T] \, d\lambda = \frac{AT}{W} \int_{-W/2}^{W/2} \frac{\sin[\pi(f - \lambda)T]}{\pi(f - \lambda)T} \, d\lambda$$

, let $x = \pi(f - \lambda)T$, then $dx = -\pi T \, d\lambda$, as $\lambda \rightarrow -W/2$ then $x \rightarrow \pi(f + \frac{1}{2} W)T$, and as $\lambda \rightarrow W/2$ then $x \rightarrow \pi(f - \frac{1}{2} W)T$, the convolution operation $X_{12}(f)$ yields

$$X_{12}(f) = - \frac{A}{\pi W} \int_{X_1}^{X_2} \frac{\sin(x)}{x} \, dx$$

, in terms of two integrals, $X_{12}(f)$ yields

$$X_{12}(f) = \frac{A}{\pi W} \left\{ \int_{0}^{X_1} \frac{\sin(x)}{x} \, dx - \int_{0}^{X_2} \frac{\sin(x)}{x} \, dx \right\}$$

, where $x_1 = \pi(f + \frac{1}{2} W)T$, and $x_2 = \pi(f - \frac{1}{2} W)T$. In terms of sine integral function Si(u), $X_{12}(f)$ yields

Fig.3.43. The sine integral function $S_i(u)$.

$$X_{12}(f) = \frac{A}{\pi W} \{ Si[\pi(f + \frac{1}{2} W)T] - Si[\pi(f - \frac{1}{2} W)T] \} \qquad (3.44)$$

, where $\quad Si(u) = \int_{0}^{u} \frac{\sin(x)}{x} \, dx \qquad\qquad (3.45)$

, the sine integrals, Eq.(3.44), cannot be evaluated in a closed form. It is, however, extensively tabulated in standard tables under sine integral function Si(u), Fig.3.43.

Problem 3.22
Evaluate the convolution function of the two digits 4 and 7, in a binary form of three bits, using the time domain. Note that:

  1 is represented by positive pulse with amplitude A and duration T,
  0 is represented by negative pulse with amplitude A and duration T.

Solution
The convolution function $x_{12}(t) = 100 \otimes 111$, Fig.3.44, using the time domain, Eq.(3.42), is given by

$$x_{12}(t) = \int_{0}^{t} A^2 \, dt = A^2 \, t \qquad\qquad for \; 0 \leq t \leq T$$

$$, \qquad x_{12}(t) = \int_{0}^{T} A^2 \, dt - \int_{T}^{t} A^2 \, dt = 2 A^2 T - A^2 t \qquad for \; T \leq t \leq 4T$$

$$, and \qquad x_{12}(t) = - \int_{-3T+t}^{3T} A^2 \, dt = A^2 t - 6 A^2 T \qquad for \; 4T \leq t \leq 6T$$

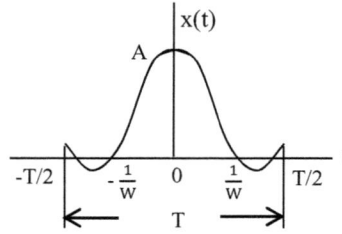

(a) $0 \leq t \leq T$ (b) $T \leq t \leq 4T$ (c) $4T \leq t \leq 6T$

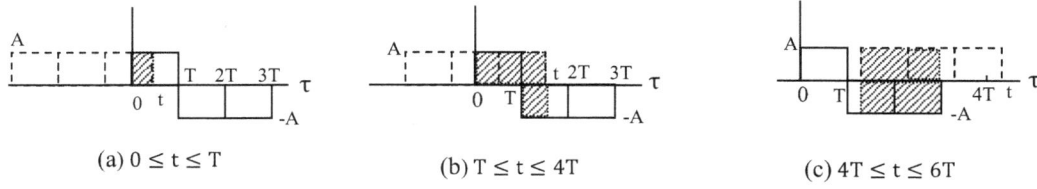

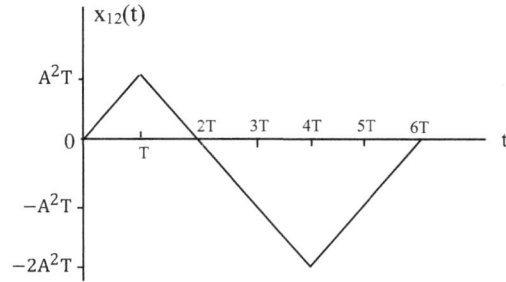

Fig.3.44. The convolution function $x_{12}(t) = 100 \otimes 111$.

, another way to get the convolution function $x_{12}(t) = 100 \otimes 111$, can also be deduced almost by inspection from table 3.1.

Table.3.1. The convolution function $x_{12}(t) = 100 \otimes 111$

$t = 0$		$x_{12}(t) = 0$
$t = T$		$x_{12}(t) = A^2 T$
$t = 2T$		$x_{12}(t) = 0$
$t = 3T$		$x_{12}(t) = -A^2 T$
$t = 4T$		$x_{12}(t) = -2 A^2 T$
$t = 5T$		$x_{12}(t) = -A^2 T$
$t = 6T$		$x_{12}(t) = 0$

3.8. Area property

Let the following Fourier transform pair

$$x(t) \rightleftharpoons X(f)$$

, where x(t) and X(f) are given in terms of each other, Equations (3.1) and (3.2), and their origin values are given by

$$x(0) = \int_{-\infty}^{\infty} X(f)\, df \qquad , and \qquad X(0) = \int_{-\infty}^{\infty} x(t)\, dt$$

, where $|x(0)|$ and $|X(0)|$ indicate the areas under the frequency domain X(f) and the time domain x(t) respectively, the magnitude is taken into consideration because the transforms may be real or imaginary or complex depends on the shape of the original signal. Then due to area property, the area of the signal in any domain equals the origin value of its transform in the other domain, and there is no relation between the area of the signal in the time domain and the area of its Fourier transform [1].

Numerical examples

i. The Fourier transform pairs of the decaying and rising single sided exponential pulses, Fig.1.9, Equations (3.9) and (3.10), and due to area property, where the area of the signal in any domain equals the origin value of its transform in the other domain, yield

$$\int_{-\infty}^{\infty} \frac{1}{1 + j2\pi f}\, df = [e^{-t} u(t)]_{t=0} = u(0) = \frac{1}{2}$$

, and

$$\int_{-\infty}^{\infty} \frac{1}{1 - j2\pi f}\, df = [e^{t} u(-t)]_{t=0} = u(0) = \frac{1}{2}$$

, where u(0) is the origin value of the unit step function u(t) and equals ½ , Fig.1.4a. Also

$$\int_{-\infty}^{\infty} e^{-t} u(t)\, dt = \left[\frac{1}{1 + j2\pi f}\right]_{f=0} = 1$$

, and

$$\int_{-\infty}^{\infty} e^{t} u(-t)\, dt = \left[\frac{1}{1 - j2\pi f}\right]_{f=0} = 1$$

ii. The Fourier transform pair of the Gaussian pulse, Fig.3.4, Eq.(3.7), and due to area property, where the area of the signal in any domain is the origin value of its transform in the other domain, yields

$$\int_{-\infty}^{\infty} e^{-\pi t^2}\, dt = \left[e^{-\pi f^2}\right]_{f=0} = 1 \qquad\qquad (3.46)$$

, and

$$\int_{-\infty}^{\infty} e^{-\pi f^2}\, df = \left[e^{-\pi t^2}\right]_{t=0} = 1 \qquad\qquad (3.47)$$

, Equations (3.46) and (3.47) are used to prove some important integrals, let $x^2 = \pi t^2$ in Eq.(3.46), then $x = \sqrt{\pi}\, t$, and $dx = \sqrt{\pi}\, dt$, and as $t \to -\infty$ then $x \to -\infty$, and as $t \to \infty$ then $x \to \infty$, Eq.(3.46) yields

$$\frac{1}{\sqrt{\pi}} \int_{-\infty}^{\infty} e^{-x^2}\, dx = 1$$

, then

$$\int_{-\infty}^{\infty} e^{-x^2}\, dx = \sqrt{\pi} \qquad\qquad (3.48)$$

, this integral, Eq.(3.48), is used before to prove that the Fourier transform of the Gaussian pulse is also Gaussian spectrum, Eq.(3.6). The integral, Eq.(3.48), can also be obtained using the traditional way, consider

$$I = \int_{-\infty}^{\infty} e^{-x^2} \, dx \quad , \text{and} \quad I = \int_{-\infty}^{\infty} e^{-y^2} \, dy$$

, then

$$I^2 = \int_{-\infty}^{\infty} \int_{-\infty}^{\infty} e^{-(x^2+y^2)} \, dx \, dy$$

, in terms of polar co-ordinates, $dx \, dy = r \, dr \, d\theta$
, and $r^2 = x^2 + y^2$, Fig.3.45, I^2 yields

$$I^2 = \int_{r=0}^{\infty} \int_{\theta=0}^{2\pi} e^{-r^2} \, r \, dr \, d\theta$$

$$= \pi \int_{r=0}^{\infty} e^{-r^2} \, dr^2 = \pi$$

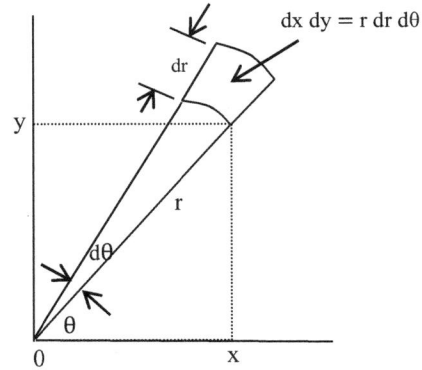

Fig.3.45. The polar co-ordinate, $dx \, dy = r \, dr \, d\theta$,
and $r^2 = x^2 + y^2$.

, then the integral I equals $\sqrt{\pi}$, Eq.(3.48).

Problem 3.23
Evaluate the area under the following sinc and sinc squared pulses, using Fourier transform properties.

      i. $x_1(t) = AW \, \text{sinc}(Wt)$ ii. $x_2(t) = AW \, \text{sinc}^2(Wt)$

Solution
The sinc and sinc squared pulses $x_1(t)$ and $x_2(t)$, extend from $-\infty$ to ∞, their Fourier transforms pair, Equations (3.21) and (3.22), and due to area property, the area of the signal in any domain equals the origin value of its transform in the other domain, the area under $x_1(t)$ and $x_2(t)$, yield

$$\int_{-\infty}^{\infty} AW \, \text{sinc}(Wt) \, dt = \left[A \, \text{rect}\left(\frac{f}{W}\right) \right]_{f=0} = A \qquad (3.49)$$

, and

$$\int_{-\infty}^{\infty} AW \, \text{sinc}^2(Wt) \, dt = \left[A \, \text{tri}\left(\frac{f}{W}\right) \right]_{f=0} = A \qquad (3.50)$$

, the area property of the sinc and sinc squared functions, is used to prove some important integrals, let $x = Wt$, then $dx = W \, dt$, as $t \to -\infty$ then $x \to -\infty$, and as $t \to \infty$ then $x \to \infty$, Equations (3.49) and (3.50) yield

$$\int_{-\infty}^{\infty} \text{sinc}(x) \, dx = 1 \qquad (3.51)$$

, and

$$\int_{-\infty}^{\infty} \text{sinc}^2(x) \, dx = 1 \qquad (3.52)$$

Other some useful integrals can be deduced from Equations (3.51) and (3.52), express sinc(x) and $\text{sinc}^2(x)$ in terms of sine functions, yield

$$\int_{-\infty}^{\infty} \frac{\sin(\pi x)}{\pi x} \, dx = 1 \qquad , \text{and} \qquad \int_{-\infty}^{\infty} \frac{\sin^2(\pi x)}{(\pi x)^2} \, dx = 1$$

, let $\lambda = \pi x$, then $d\lambda = \pi \, dx$, as $x \to -\infty$ then $\lambda \to -\infty$, and as $x \to \infty$ then $\lambda \to \infty$, the integrals yield

$$\frac{1}{\pi} \int_{-\infty}^{\infty} \frac{\sin(\lambda)}{\lambda} \, d\lambda = 1 \qquad , \text{and} \qquad \frac{1}{\pi} \int_{-\infty}^{\infty} \frac{\sin^2(\lambda)}{\lambda^2} \, d\lambda = 1$$

, or equivalently

$$\int_{-\infty}^{\infty} \frac{\sin(\lambda)}{\lambda} \, d\lambda = \pi \tag{3.53}$$

, and
$$\int_{-\infty}^{\infty} \frac{\sin^2(\lambda)}{\lambda^2} \, d\lambda = \pi \tag{3.54}$$

, also since the functions sinc(λ) and sinc^2(λ) are even functions, then

$$\int_{0}^{\infty} \frac{\sin(\lambda)}{\lambda} \, d\lambda = \int_{-\infty}^{0} \frac{\sin(\lambda)}{\lambda} \, d\lambda = \frac{\pi}{2} \tag{3.55}$$

, and
$$\int_{0}^{\infty} \frac{\sin^2(\lambda)}{\lambda^2} \, d\lambda = \int_{-\infty}^{0} \frac{\sin^2(\lambda)}{\lambda^2} \, d\lambda = \frac{\pi}{2} \tag{3.56}$$

Problem 3.24

Based on the Fourier transform pair of the Gaussian pulse, Eq.(3.7). Find the area under the function x(t), Fig.3.46, given by

$$x(t) = t^2 \, e^{-\pi t^2}$$

, using Fourier transform properties.

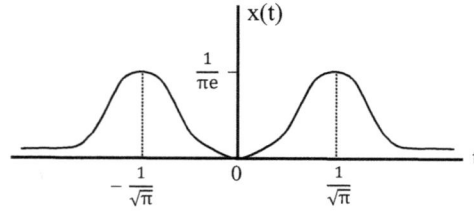

Fig.3.46. $x(t) = t^2 \, e^{-\pi t^2}$.

Solution

Due to differentiation property in the frequency domain, Eq.(3.28a), the Fourier transform pair, Eq.(3.7), yields

$$t^2 \, e^{-\pi t^2} \;\rightleftharpoons\; \frac{1}{(-j \, 2\pi)^2} \, \frac{d^2}{df^2} \, [e^{-\pi f^2}]$$

, double differentiating the Gaussian spectrum $e^{-\pi f^2}$, the frequency domain is given by

$$X(f) = \frac{1}{2\pi} \, [2\pi f^2 - 1] \, e^{-\pi f^2}$$

, then the area under the Gaussian pulse $x(t) = t^2 e^{-\pi t^2}$, equals the origin value of $X(f) = |X(0)| = 1/2\pi$.

3.3.9. Conjugate Property

Let the following Fourier transform pair

$$x(t) \;\rightleftharpoons\; X(f)$$

, the conjugate property Fourier transform pair is

$$x^*(t) \;\rightleftharpoons\; X^*(-f) \tag{3.57}$$

, where $x^*(t)$ is the complex conjugate of $x(t)$.

Proof:

Taking the complex conjugate of the both sides of the inverse Fourier transform operation, Eq.(3.1), yields

$$x^*(t) = \int_{-\infty}^{\infty} X^*(f) \, e^{-j2\pi ft} \, df \tag{3.58}$$

, next, replacing f with −f of both sides, Eq.(3.58), $x^*(t)$ yields

$$x^*(t) = - \int_{\infty}^{-\infty} X^*(-f) \ e^{j2\pi ft} \ df$$

$$= \int_{-\infty}^{\infty} X^*(-f) \ e^{j2\pi ft} \ df \qquad (3.59)$$

, comparing Equations (3.59) and (3.1), then $x^*(t)$ and $X^*(-f)$ constitute a Fourier transform pair.

There are two inportant applications of the conjugate property in communication theory, mainly, the evaluation of the energy content of the unperiodic signal, using the frequency domain (Rayleigh`s energy theorem), and the evaluation of the Fourier transform of the Real and Imaginary parts of a time function.

3.3.9.1. Rayleigh`s Energy Theorem
Rayleigh`s energy theorem facilitates the evaluation of the energy content E of the unperiodic signal x(t), using its equivalent continuous frequency transform X(f). A definition of the normalized energy of a complex signal x(t), it is the energy consumed by a voltage x(t) applied across a one ohm resistor or by a current x(t) passing through a one ohm resistor), E is given by

$$E = \int_{-\infty}^{\infty} |x(t)|^2 \ dt \qquad (1.10)$$

, in terms of the complex conjugate function, Eq.(1.10) yields

$$E = \int_{-\infty}^{\infty} x(t) \ x^*(t) \ dt$$

, due to the conjugate property, Eq.(3.57), the energy content E is given by

$$E = \int_{-\infty}^{\infty} x(t) \ [\int_{-\infty}^{\infty} X^*(-f) \ e^{j2\pi ft} \ df \ ] \ dt$$

, changing the order of the integrations, E yields

$$E = \int_{-\infty}^{\infty} X^*(-f) \ [\int_{-\infty}^{\infty} x(t) \ e^{j2\pi ft} \ dt \ ] \ df$$

$$= \int_{-\infty}^{\infty} X^*(-f) \ X(-f) \ df$$

, for a real valued function x(t), then $X(f) = X^*(-f)$, where $X^*(f)$ is the complex conjugate of X(f), and $|X(-f)| = |X(f)|$, then the energy content E is given by

$$E = \int_{-\infty}^{\infty} |X(f)|^2 \ dt \qquad joule \qquad (3.60)$$

, then $|X(f)|^2$ is defined by the Energy Spectral Density ESD $\Psi(f)$ in joule/Hz, and is given by

$$\Psi(f) = |X(f)|^2 \qquad joule/Hz \qquad (3.61)$$

3.9.2. Energy Spectral Density
The Energy Spectral Density ESD $\Psi(f)$ Eq.(3.61), is very useful term to evaluate the energy content of the unperiodic signals , where the total area under the energy spectral density curve plotted as a function of frequency, indicates the energy content E of the signal when it is developed across a one ohm resistor, Fig.3.47, E is given by

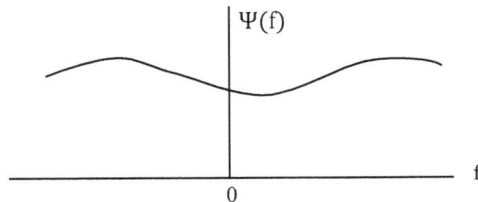

Fig.3.47. The Energy Spectral Density ESD $\Psi(f)$.

$$E = \int_{-\infty}^{\infty} \Psi(f) \; df \quad \text{joule} \tag{3.62}$$

, also the energy spectral density facilitates the evaluation of the autocorrelation function $R(\tau)$ of the energy signal $x(t)$, where they constitute a Fourier transform pair (chapter VI), hence the autocorrelation function plays an important part in the design of the communication systems.

Problem 3.25
Evaluate the energy content of the rectangular pulse $x(t) = A \; rect(t/T)$, using the frequency domain.
Solution
The Fourier transform of the rectangular pulse $x(t)$, is $X(f) = AT \; sinc(fT)$, Eq.(3.4), then the Energy Spectrum Density ESD $\Psi(f)$, Eq.(3.61), Fig.3.48, is given by

$$\Psi(f) = A^2 T^2 \; sinc^2(fT) \quad \text{joule/Hz}$$

, then due to the Rayleigh`s energy theorem
, Eq.(3.60), the energy content E is given by

$$E = \int_{-\infty}^{\infty} A^2 T^2 \; sinc^2(fT) \; df$$

, or equivalently

$$E = A^2 T^2 \int_{-\infty}^{\infty} \frac{sin^2(\pi fT)}{(\pi fT)^2} \; df$$

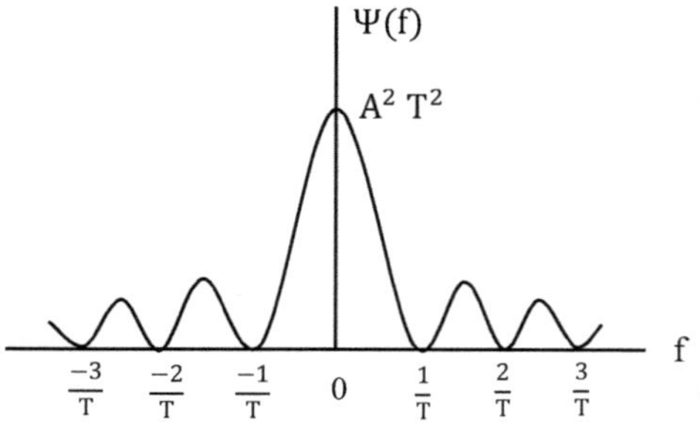

Fig.3.48. The Energy Spectral Density ESD $\Psi(f)$ is sinc squared function.

, let $\lambda = \pi fT$, then $d\lambda = \pi T \; df$, and as $f \to -\infty$ then $\lambda \to -\infty$, and as $f \to \infty$ then $\lambda \to \infty$, the energy content E yields

$$E = \frac{A^2 T}{\pi} \int_{-\infty}^{\infty} \frac{sin^2(\lambda)}{\lambda^2} \; d\lambda$$

, since the area under the integral equals π, Eq.(3.54), then the energy content E is given by

$$E = A^2 T \quad \text{joule} \tag{1.11}$$

, which is the same energy content obtained using the time domain, Eq.(1.11), Problem 1.2.

Problem 3.26
Evaluate the energy content of the Radio Frequency RF pulse

$$x(t) = A \; rect(t/T) \; cos(2\pi f_c t)$$
, Fig.1.8a, using the frequency domain.

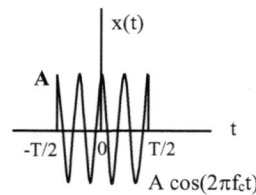

Fig.1.8a. Radio Frequency pulse.

Solution
The Fourier transform of the RF pulse $x(t)$, Eq.(3.8), $X(f)$ is given by

$$X(f) = \tfrac{1}{2} AT \; sinc[T(f - f_c)] + \tfrac{1}{2} AT \; sinc[T(f + f_c)] \tag{3.8}$$

, then the energy spectrum density $\Psi(f)$, Eq.(3.61), is given by

$$\Psi(f) = \left| \tfrac{1}{2} AT \; sinc[T(f - f_c)] + \tfrac{1}{2} AT \; sinc[T(f + f_c)] \right|^2 \quad \text{joule/Hz}$$

, due to the Rayleigh`s energy theorem, Eq.(3.60), the energy content E is given by

$$E = \int_{-\infty}^{\infty} |½\,AT\,\text{sinc}[T(f - f_c)] + ½\,AT\,\text{sinc}[T(f + f_c)]|^2 \, df$$

, or equivalently

$$E = \int_{-\infty}^{\infty} \left| ½\,AT\, \frac{\sin[\pi T(f - f_c)]}{\pi T(f - f_c)} + ½\,AT\, \frac{\sin[\pi T(f + f_c)]}{\pi T(f - f_c)} \right|^2 df$$

, let $\lambda_1 = \pi T(f - f_c)$, and $\lambda_2 = \pi T(f + f_c)$, then $d\lambda_1 = d\lambda_2 = d\lambda = \pi T\, df$, and as $f \to -\infty$ then $\lambda_1 \to -\infty$ and $\lambda_2 \to -\infty$, as $f \to \infty$ then $\lambda_1 \to \infty$ and $\lambda_2 \to \infty$, the energy E yields

$$E = \int_{-\infty}^{\infty} \frac{1}{\pi T} \left| ½\,AT\, \frac{\sin(\lambda_1)}{\lambda_1} + ½\,AT\, \frac{\sin(\lambda_2)}{\lambda_2} \right|^2 d\lambda$$

, since the area under the two sinc functions are the same, and the sinc function is even function, E yields

$$E = \frac{1}{\pi T} \int_{-\infty}^{\infty} 2\,(½\,AT)^2\, \frac{\sin^2(\lambda)}{\lambda^2} \, d\lambda$$

, or equivalently

$$E = \frac{A^2 T}{2\pi} \int_{-\infty}^{\infty} \frac{\sin^2(\lambda)}{\lambda^2} \, d\lambda$$

, since the area under the integral equals π, Eq.(3.54), then the energy E is given by

$$E = ½\,A^2\,T \qquad\qquad \text{joule} \qquad\qquad (1.13)$$

, which is the same energy content obtained using the time domain, Eq.(1.13), problem 1.4.

Problem 3.27

Evaluate the energy contents of the decaying and rising single sided exponential pulses , Fig.1.9,

$x_1(t) = e^{-t}\,u(t)$, and $x_2(t) = e^{t}\,u(-t)$, using the frequency domain.

Solution

The Fourier transform pairs of $x_1(t)$ and $x_2(t)$, are given by Equations (3.9) and (3.10), in terms of the magnitude and phase continuous spectrums, Fig.3.6, the energy spectrum densities, $\Psi_1(f)$ and $\Psi_2(f)$, Eq.(3.61), Fig.(3.49), are given by

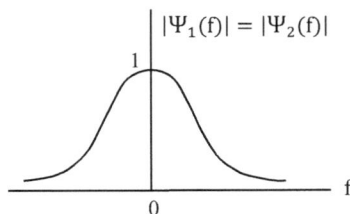

Fig. 3.49. The Energy Spectral Density ESD , Problem 3.27.

$$\Psi_1(f) = |X_1(f)|^2 = \frac{1}{1 + (2\pi f)^2} \qquad \text{joule/Hz}$$

$$\Psi_2(f) = |X_2(f)|^2 = \frac{1}{1 + (2\pi f)^2} \qquad \text{joule/Hz}$$

, then due to the Rayleigh`s energy theorem, Eq.(3.60), the energy contents E_1 and E_2 are given by

$$E_1 = \int_{-\infty}^{\infty} \frac{1}{1 + (2\pi f)^2} \, df \qquad , \qquad E_2 = \int_{-\infty}^{\infty} \frac{1}{1 + (2\pi f)^2} \, df$$

, let $\lambda = 2\pi f$, then $d\lambda = 2\pi\, df$, and as $f \to -\infty$ then $\lambda \to -\infty$, and as $f \to \infty$ then $\lambda \to \infty$, the energy contents E_1 and E_2 yield

$$E_1 = \frac{1}{2\pi} \int_{-\infty}^{\infty} \frac{1}{1+\lambda^2} \, d\lambda \qquad , \qquad E_2 = \frac{1}{2\pi} \int_{-\infty}^{\infty} \frac{1}{1+\lambda^2} \, d\lambda$$

, since $\Psi_1(f)$ and $\Psi_2(f)$ are even functions, E_1 and E_2 yield

$$E_1 = \frac{1}{\pi} \int_{0}^{\infty} \frac{1}{1+\lambda^2} \, d\lambda = \left[\frac{1}{\pi} \tan^{-1}(\lambda) \right]_{\lambda=0}^{\lambda=\infty} = \frac{1}{\pi} [\tfrac{1}{2}\pi - 0] = \frac{1}{2} \quad \text{joule} \quad (3.63a)$$

$$, \text{and} \quad E_2 = \frac{1}{\pi} \int_{0}^{\infty} \frac{1}{1+\lambda^2} \, d\lambda = \left[\frac{1}{\pi} \tan^{-1}(\lambda) \right]_{\lambda=0}^{\lambda=\infty} = \frac{1}{\pi} [\tfrac{1}{2}\pi - 0] = \frac{1}{2} \quad \text{joule} \quad (3.63b)$$

, which are the same energy contents obtained using the time domain, Equations (1.15) and (1.16), problem 1.5.

3.3.9.3. Fourier Transform of Real and Imaginary Parts of a Time Function
The conjugate property also facilitates the evaluation of the Fourier transform of the real and imaginary parts of a complex valued function x(t), which plays an important part in the analysis of the band-pass systems (chapter X), where the analogy of low-pass and band-pass systems. Consider x(t) is a complex valued function, it can be expressed by

$$x(t) = Re[x(t)] + j \, Im[x(t)] \qquad (3.64)$$

, where $Re[x(t)]$ and $Im[x(t)]$ denote the real and imaginary parts of the complex valued function x(t) respectively. The complex conjugate of x(t) is $x^*(t)$, is given by

$$x^*(t) = Re[x(t)] - j \, Im[x(t)] \qquad (3.65)$$

, adding and subtracting Equations (3.64) and (3.65), the real and the imaginary parts of x(t) can be expressed by

$$Re[x(t)] = \tfrac{1}{2}\{ x(t) + x^*(t) \} \qquad (3.66a)$$
$$Im[x(t)] = -\tfrac{1}{2}j \{ x(t) - x^*(t) \} \qquad (3.66b)$$

, then the Fourier transform of the real and the imaginary parts of the complex valued function x(t), the using conjugate property, are given by

$$F\{Re[x(t)]\} = \tfrac{1}{2} \{ X(f) + X^*(-f) \} \qquad (3.67a)$$
$$, \text{and} \quad F\{Im[x(t)]\} = -\tfrac{1}{2}j \{ X(f) - X^*(-f) \} \qquad (3.67b)$$

Problem 3.28
Find the Fourier transforms of the real and imaginary parts of the following complex time function, using conjugate property

$$x(t) = rect(t) + j \, tri(t)$$

Solution
The complex conjugate of x(t) is $x^*(t) = rect(t) - j \, tri(t)$, and their Fourier transforms are X(f) and $X^*(f)$ are given by

$$X(f) = sinc(f) + j \, sinc^2(f)$$
$$X^*(f) = sinc(f) - j \, sinc^2(f)$$

, since $sinc(f)$ and $sinc^2(f)$ are even functions, the Fourier transforms of the real and imaginary parts, Equations (3.67a) and (3.67b), are given by

$$F\{Re[x(t)]\} = \tfrac{1}{2} \{ [sinc(f) + j \, sinc^2(f)] + [sinc(f) - j \, sinc^2(f)] \}$$
$$, \text{and} \quad F\{Im[x(t)]\} = -\tfrac{1}{2}j \{ [sinc(f) + j \, sinc^2(f)] - [sinc(f) - j \, sinc^2(f)] \}$$

, or equivalently

$$F[rect(t)] = sinc(f)$$
$$, \text{and} \quad F[tri(t)] = sinc^2(f)$$

Problems
1. Find the Fourier transform of the shown energy signals, Fig.3.50a,b,c,d,e,f,g,h, using the linearity property.

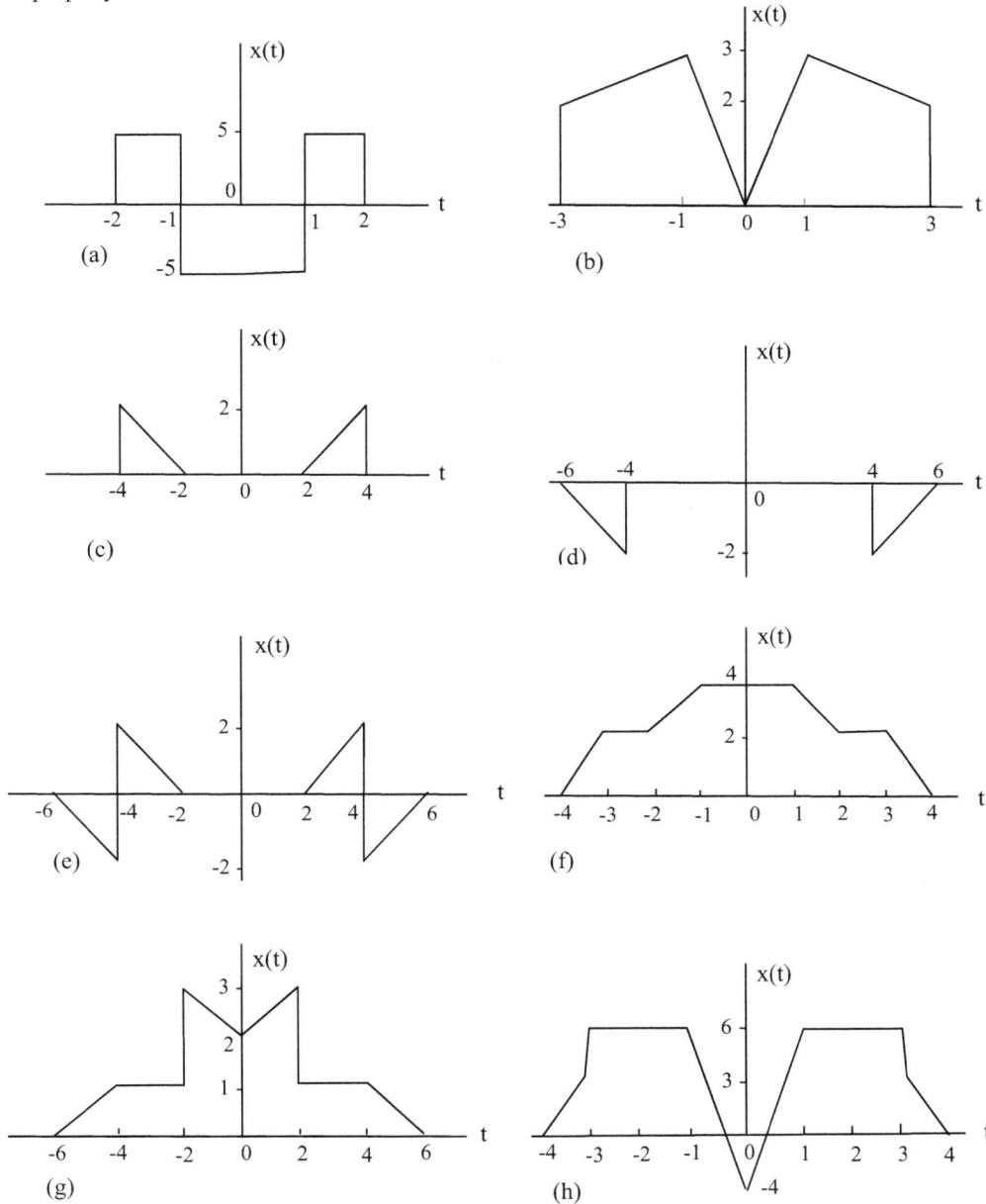

Fig.3.50. Energy signals.

2. Find the Fourier transform of the following energy signals:
   i. $x(t) = 5 \text{ rect}(t/6) \times \text{tri}(t/2)$ ii. $x(t) = 5 \text{ rect}(t/2) \times \text{tri}(t/6)$

3. Find the Fourier transform of the following energy signals, using the shifting property:
   i. $x(t) = 5 \text{ rect}(\frac{1}{2} t - 1)$ ii. $x(t) = 10 \text{ tri}(2t + \frac{1}{2})$
   iii. $x(t) = 12 \text{ sinc}(4t) \sin(4\pi t)$ iv. $x(t) = 8 \text{ rect}(t/4) \cos(2\pi 10^6 t)$
   v. $x(t) = e^{-2t} \text{ rect}[(t - 7)/4]$

4. Determine the complex valued constant k required for the following Fourier transform pairs

   i. $e^{j2\pi 10^6 t}\, x(t - 0.5) \;\rightleftharpoons\; k\,e^{-j\pi f}\, X(f - 10^6)$

   ii. $k\,e^{-j2000\pi t}\, x(t - 3) \;\rightleftharpoons\; e^{-j6\pi f}\, X(f + 1000)$

   , where X(f) is the Fourier transform of the energy signal x(t).

5. Find the Fourier transform of following energy signals, using the shifting and scaling properties:

   i. $x(t) = e^{-(t-2)}\, u(t)$ ii. $x(t) = e^{t}\, u(2 - t)$

   iii. $x(t) = e^{-(2t-4)}\, u(t)$ iv. $x(t) = e^{2t}\, u(4 - 2t)$

   v. $x(t) = e^{-3(t-2)}\, u(t - 5)$ vi. $x(t) = e^{-3(2-t)}\, u(5 - t)$

   vii. $x(t) = e^{-(t-2)}\, [u(t+5) - u(t-5)]$ viii. $x(t) = 10\, e^{-5t}\, \cos(2\pi 10^6 t)\, u(t)$

   , where u(t) is the unit step function in the time domain.

6. Find the Fourier transform of the energy signals, Fig.3.51a,b, using the linearity and shifting properties

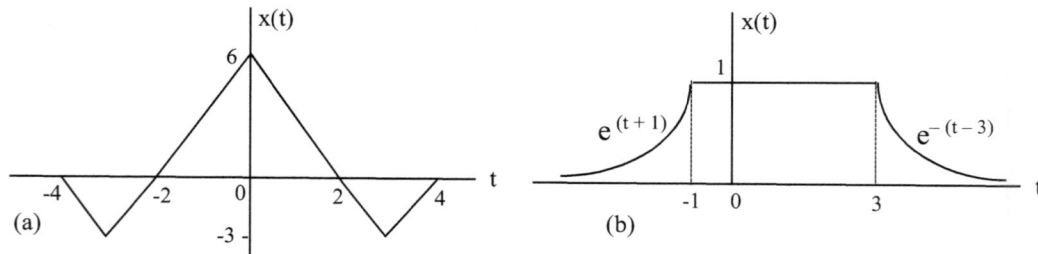

Fig.3.51. Energy signals.

7. Find the inverse Fourier transform of the following energy signals, using the shifting and scaling properties:

   i. $X(f) = e^{-(f-2)}\, U(f)$ ii. $X(f) = e^{f}\, U(2 - f)$

   iii. $X(f) = e^{-(2f-4)}\, U(f)$ iv. $X(f) = e^{2f}\, U(4 - 2f)$

   v. $X(f) = e^{-3(f-2)}\, U(f - 5)$ vi. $X(f) = e^{-3(2-f)}\, U(5 - f)$

   vii. $X(f) = e^{-(f-2)}\, [U(f+5) - U(f-5)]$ viii. $X(f) = 10\, e^{-5f}\, \sin(6\pi f)\, U(f)$

   , where U(f) is the unit step function in the frequency domain.

8. Find the Fourier transform of the following energy signals, using the scaling property

   i. $x(t) = \dfrac{1}{(1 + jt)(2 + jt)}$ ii. $x(t) = \dfrac{1}{t + j}$

   iii. $x(t) = \dfrac{1}{f + j2\pi t}$ iv. $x(t) = e^{-|t/3|}$

9. Find the inverse Fourier transform of the following spectrums, using the scaling property:

   i. $X(f) = \dfrac{1}{(2 + jf)(3 + jf)}$ ii. $X(f) = \dfrac{1}{f + j}$

   iii. $X(f) = \dfrac{1}{t + j2\pi f}$ iv. $X(f) = e^{-|5f|}$

   v. $X(f) = \dfrac{(1 - jw)(w^2 + 2)}{(w^4 + 3w^2 + 2)}$, where $w = 2\pi f$

10. Prove that the Fourier transform of the Gaussian pulse $x(t) = e^{-\pi t^2}$ is also Gaussian pulse $X(f) = e^{-\pi f^2}$, using the Fourier transform properties.

11. Find the Fourier transform of the following Gaussian pulses, using the scaling and shifting properties:

   i. $x(t) = e^{-5(t-3)^2}$ ii. $x(t) = e^{-9t^2}$

   iii. $x(t) = e^{-(t^2-2t)}$ iv. $x(t) = e^{-\pi t^2/9}\, e^{-j2\pi 10^6 t}$

   v. $x(t) = e^{-\pi(t-3)^2}$ vi. $x(t) = [1/\sqrt{2\pi}\,\sigma]\, e^{-t^2/2\sigma^2}$, σ^2 is the variance.

12. Find the inverse Fourier transform of the following Gaussian spectra, using the scaling and shifting properties:

   i. $X(f) = e^{-\pi f^2/4}$ ii. $X(f) = e^{-5f^2}$

   iii. $X(f) = e^{-(f-10^6)^2/3}$ iv. $X(f) = e^{-(f^2-2f)}$

13. Find the Fourier transform of the following energy signals, using the differentiation property:

   i. $x(t) = t\, e^{-4t}\, u(t)$ ii. $x(t) = (1-2t)\, e^{-3t}\, u(t)$

   iii. $x(t) = t\, e^{-(t-1)}\, u(t-1)$ iv. $x(t) = t\, e^{-(t-1)}\, u(t+2)$

   v. $x(t) = (1-t^2)\, e^{-2t}\, u(t)$ vi. $x(t) = (t-1)\, e^{-(t-1)}\, u(t-1)$

   vii. $x(t) = \dfrac{d}{dt}\left[e^{-4(t-1)}\, u(t+1) \right]$ viii. $x(t) = 2\,\text{rect}\left[\dfrac{t+2}{2}\right] \times 4\,\text{tri}(\dfrac{t}{4})$

   ix. $x(t) = t\,[u(t+2) - u(t-2)]$

14. Find the Fourier transform of the energy signals, Fig.3.52a,b,c,d, using the scaling, shifting, and differentiation properties

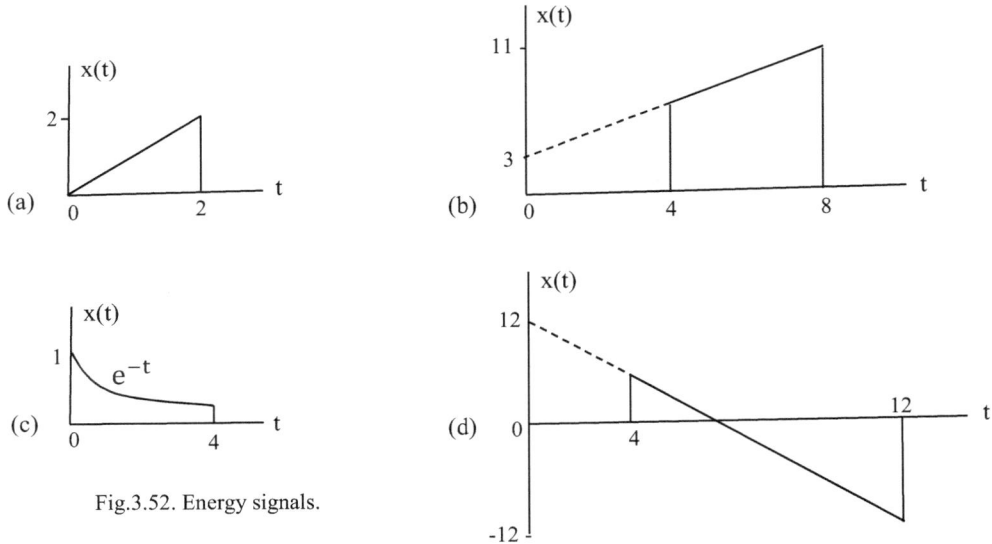

Fig.3.52. Energy signals.

15. Find the inverse Fourier transform of the following functions, using the differentiation property

   i. $X(f) = f\, e^{-\pi f^2}$ ii. $X(f) = \dfrac{d}{df}\left[e^{-\pi f^2} \right]$

   iii. $X(f) = (f-7)\, e^{-\pi(f-7)^2}$ iv. $X(f) = e^{-9(f-5)^2}$

   v. $X(f) = f\, \text{sinc}(f)$

16. Based on the following Fourier transform pair

$$e^{-t}u(t) \rightleftharpoons \frac{1}{1+j2\pi f}$$ (3.9)

Prove that

$$\frac{d^n}{df^n}\left[\frac{1}{1+j\,2\pi f}\right] = (-j2\pi)^n \frac{n!}{(1+j2\pi f)^{n+1}}$$

17. For the shown energy signal x(t), Fig.3.53. Find the Fourier transform of the following function, using the differentiation property

$$x'(t) = \frac{d}{dt}[x(t)]$$

Fig.3.53. Energy signal

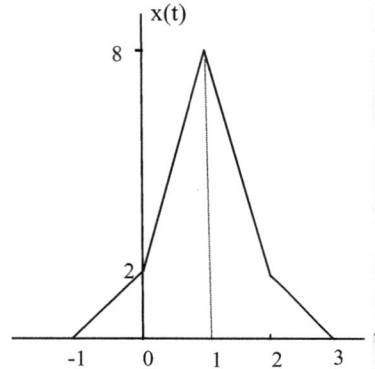

18. For the energy signal x(t), Fig.3.54. Find the Fourier transform of the following function, using the integration property

$$x_1(t) = \int_{-\infty}^{t} x(\tau)\,d\tau$$

Fig.3.54. Energy signal.

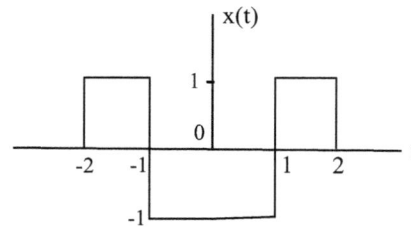

19. Prove that the Fourier transform of the energy signal $x^2(t)$, occupies spectrum equals double the spectrum of the band limited energy signal x(t), using the convolution property.

20. Evaluate the following convolution functions, using the time domain
    i. $x(t) = \text{rect}(t/6) \otimes u(t-2)$
    ii. $x(t) = \text{rect}(t/2) \otimes t^2$
    iii. $x(t) = \text{rect}(t) \otimes (t-1)$
    iv. $x(t) = e^{-t}u(t) \otimes \text{rect}[(t-1)/2]$
    v. $x(t) = \text{sinc}(t) \otimes \text{rect}(t-3)$
    vi. $x(t) = e^{-3t}u(t) \otimes \text{rect}[(t-5)/4]$
    vii. $x(t) = \text{tri}(t/2) \otimes \text{rect}[(t+1)/2]$
    viii. $x(t) = t^2 u(t) \otimes u(t-1)$
    ix. $x(t) = t\,u(t) \otimes t\,u(t)$
    x. $x(t) = u(t) \otimes u(t) \otimes u(t)$

21. Evaluate the following convolution functions, using the frequency domain
    i. $x(t) = e^{-\pi^2(t-3)^2} \otimes e^{-\pi^2 t^2}$
    ii. $x(t) = \text{rect}(t/8) \otimes e^{-j2\pi 10^6 t}$
    v. $x(t) = 3\,\text{sinc}(t-2) \otimes 8\,\text{sinc}(4t-12)$

22. Find the Fourier transform of the following convolution functions
    i. $x(t) = \frac{1}{1+t^2} \otimes e^{-\pi^2(t-3)^2}$
    ii. $x(t) = \frac{1}{1+(t-5)^2} \otimes e^{-(7t+35)}u(t+5)$

23. Evaluate the following convolution functions, using the time domain

    i. $X(f) = 10 \text{ sinc}(10f - 30) \otimes 20 \text{ sinc}(2f - 10)$

    ii. $X(f) = \quad 12 \text{ sinc}(8f) \otimes 30 \text{ sinc}^2(6f)$

    iii. $X(f) = \quad 10 \text{ sinc}(5f - 15) \otimes 20 \text{ sinc}(5f)$

24. Evaluate the following convolution functions, using the frequency domain

    i. $x(t) = \text{rect}(t/6) \otimes u(t - 2)$ ii. $X(f) = \text{rect}(f/10) \otimes \text{sinc}(4f)$

    iii. $X(f) = \dfrac{1}{1 + jf} \otimes \dfrac{1}{1 + j2f}$ iv. $X(f) = \quad e^{-j2\pi f} \otimes \text{rect}(f/2) \, e^{-j2\pi f}$

25. Evaluate the convolution function of the two energy signals $x_1(t)$ and $x_2(t)$, Fig.3.55 , using the time domain.

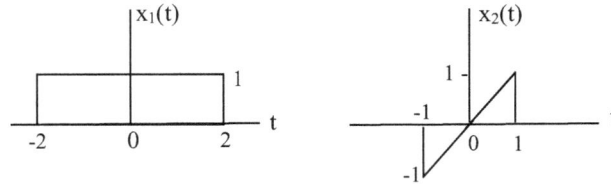

Fig.3.55. Energy signals.

26. Evaluate the convolution function of the two energy signals $x_1(t)$ and $x_2(t)$, Fig.3.56 , using the time domain.

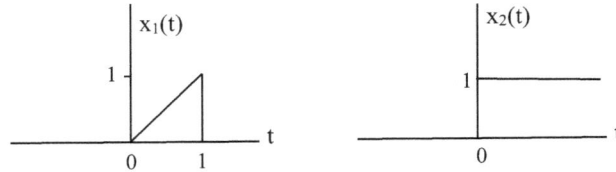

Fig.3.56. Energy signals.

27. Evaluate the convolution function of the two energy signals $x_1(t)$ and $x_2(t)$, Fig.3.57 , using the time domain.

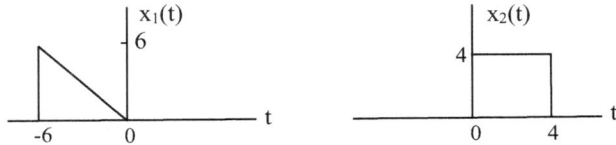

Fig.3.57. Energy signals.

28. Evaluate the convolution function of the two energy signals $x_1(t)$ and $x_2(t)$, Fig.3.58 , using the time domain.

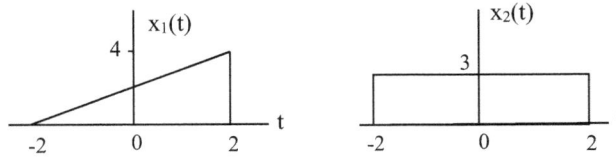

Fig.3.58. Energy signals.

29. Evaluate the convolution function of the two energy signals $x_1(t)$ and $x_2(t)$, Fig.3.59 , using the time domain.

Fig.3.59. Energy signals.

30. Evaluate the convolution function of the two energy signals $x_1(t)$ and $x_2(t)$, Fig.3.60, using the time domain.

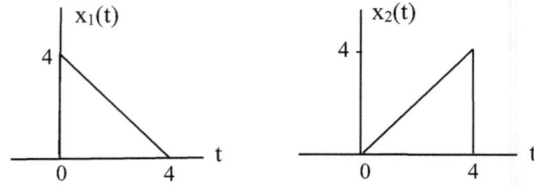

Fig.3.60. Energy signals.

31. Evaluate the convolution function of the two digits 2 and 5, in a binary of three bits, using the time domain. Note that :

        1 is represented by positive pulse with amplitude A and duration T,
        0 is represented by negative pulse with amplitude A and duration T.

32. Evaluate the following convolution functions, using the time domain.
    i. $x(t) = 1001 \otimes 0110$ ii. $x(t) = 1010 \otimes 1100$
  iii. $x(t) = 1000 \otimes 1001$
  Note that : 1 is represented by positive pulse with amplitude A and duration T,
                0 is represented by negative pulse with amplitude A and duration T.

33. Evaluate the area under the energy sinc function $x_1(t)$ and under the sinc squared spectrum $X_2(f)$, Fig.3.61a,b, using Fourier transform

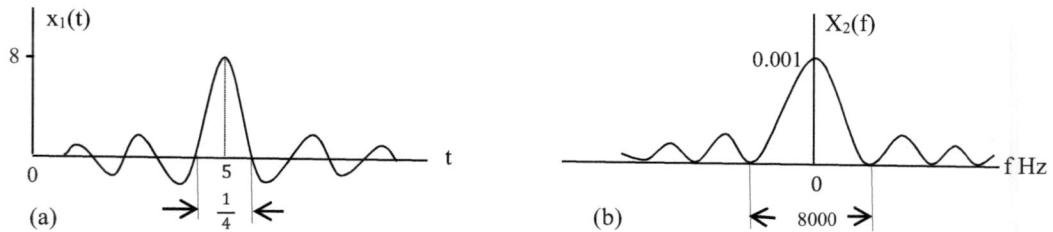

Fig.3.61. a) Sinc waveform, b) Sinc squared spectrum.

34. Find the area under the following energy signals, using the area property
    i. $x(t) = t\, e^{-2t}\, u(t)$ ii. $x(t) = 10\, \text{rect}(t/2) \times \text{tri}(t/10)$
  iii. $x(t) = (1 - 2t)\, e^{-3t}\, u(t)$ iv. $x(t) = e^{-|5t|}$
    v. $x(t) = e^{-(t-1)}\, u(t+1)$ vi. $x(t) = 5\, \text{sinc}[8(t-5)]$

  vii. $x(t) = \dfrac{d}{dt}\left[3\, \text{tri}(t - \dfrac{1}{2})\right]$

35. Find the area under the energy signals, Fig.3.63a,b, using the area property.

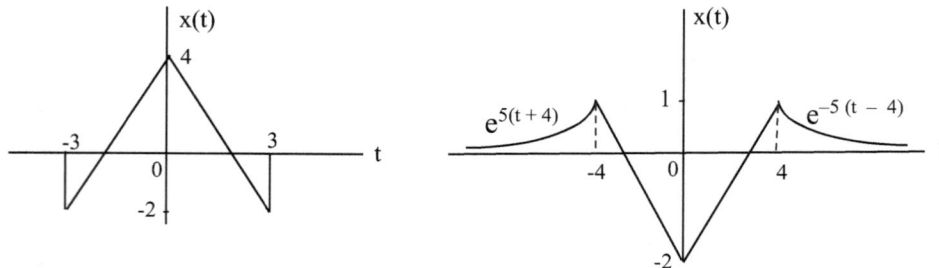

Fig.3.62. Energy signals.

36. Find the area under the following continuous spectra, using the area property

   i. $X(f) = 4 \operatorname{sinc}(2f)\, e^{-j2\pi f}$

   ii. $X(f) = 40 \operatorname{sinc}^2(4f)\, e^{-j4\pi f}$

   iii. $X(f) = 10 \operatorname{sinc}^2(8f + 4)$

   iv. $X(f) = 10 \operatorname{sinc}(5f - 15) \otimes 20 \operatorname{sinc}(4f)$

37. Evaluate the energy content of the following energy signal, using the frequency domain

$$x(t) = \frac{1}{1 + t^2}$$

38. Evaluate the energy contents of the following energy signals, using the frequency domain

   $x_1(t) = e^{-at}\, u(t)$, and $x_2(t) = e^{at}\, u(-t)$

   , where the scaler "a" is positive real constant $(a \geq 0)$.

39. Based on the Fourier transform of the Gaussian pulse $e^{-\pi t^2}$ is the Gaussian spectrum $e^{-\pi f^2}$, and using the Fourier transform properties. Prove that

$$\int_{-\infty}^{\infty} t^2\, e^{-t^2}\, dt = \sqrt{\pi}/2$$

40. Based on the Fourier transform of the triangle energy signal tri(t) is the sinc squared spectrum $\operatorname{sinc}^2(f)$, and using the Fourier transform properties. Prove that

$$\int_{-\infty}^{\infty} \operatorname{sinc}^4(t)\, dt = 2/3$$

41. A voltage v(t) has the triangle spectrum V(f), Fig.3.63, is applied to 4 KΩ resistor. Evaluate the energy content in the resistor.

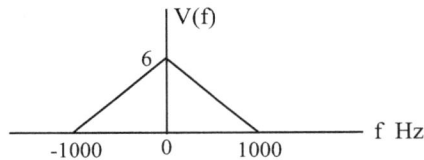

Fig.3.63. Triangle spectrum.

42. Evaluate the energy content of the following energy signals, using the frequency domain

   i. $x(t) = 4 \operatorname{sinc}(2t - 8) \otimes \operatorname{sinc}(5t + 15)$

   ii. $x(t) = t\, e^{-t}\, u(t)$

43. Find the Fourier transform of the energy signal x(t), Fig.3.64.

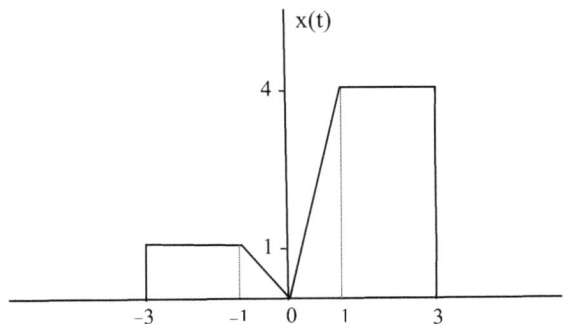

Fig.3.64. Energy signal x(t).

44. Prove that the Fourier transform of the trapezoidal energy signal x(t), Fig.3.65c, is the sum of the Fourier transforms of the two triangle energy signals $x_1(t)$ and $x_2(t)$, Fig.3.65a,b.

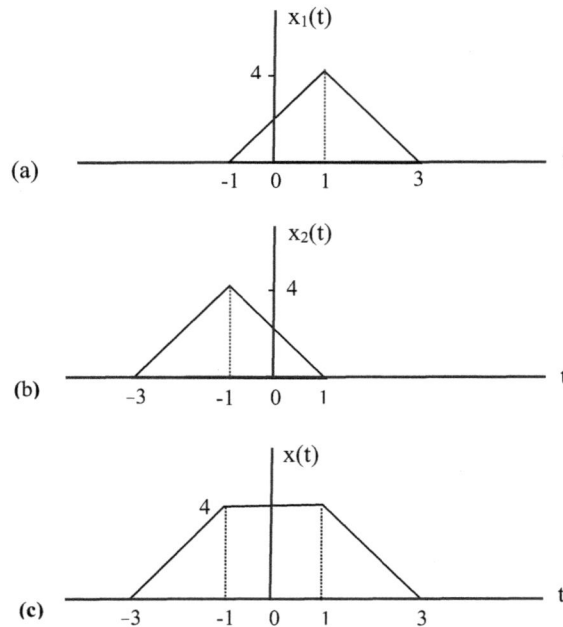

Fig.3.65. The energy signal x(t) is the sum of the energy signals $x_1(t)$ and $x_2(t)$.

45. For the following Gaussian monocycle pulse

$$x(t) = \frac{1}{\sqrt{2\pi}\ \sigma}\ e^{-(t-\mu)^2/2\sigma^2}$$

, where the constant parameters μ and σ are the mean and variance respectively of the statistical distribution of the Gaussian monocycle pulse. The first five derivatives of this Gaussian pulse, are the most widely used pulses in the Ultra Wide Band UWB communication systems. Find the first five derivatives of this Gaussian mono-cycle pulse and write their Fourier transforms.
Note: consider zero order Gaussian pulse, $\mu = 0$.

CHAPTER 4

Spectra of Nonintegrable Functions

Abstract: The main evidence of the Fourier transform concludes that when the integral, Eq.(3.2), is absolutely integrable (finite), the Fourier transform exists. More sufficient evidence for the existence of the Fourier transform is, since the magnitude of the exponential $e^{-j2\pi ft}$, equals unity, then the integral

$$\int_{-\infty}^{\infty} |x(t)| \, dt \; < \; \infty \qquad\qquad (3.3a)$$

, (Dirichlet`s conditions), Eq.(3.3a) must be finite (finite area), that is $x(t)$ must be unperiodic signal (energy signal) which has finite energy, zero average power, and infinite periodic time T_o. But, is the condition of the absolutely integrable is always necessary, the answer is No, because there are some special functions are not absolutely integrable (do not satisfy Dirichlet`s conditions) and have Fourier transforms in the limit such as the Dirac delta function (unit impulse) $\delta(t)$, the unit step function $u(t)$ and the signum function $sgn(t)$, Fig.1.4. It is necessary to evaluate the Fourier transform of these non-integrable functions to obtain their continuous frequency spectrum and study how these functions allow the easy solution of many communication problems. Also some of these non-integrable functions are related to each other by mathematical differentiation and integration formulas.

Keywords: Dirac delta function; Signum function; Unit step function; Error probability function.

4.1. Fourier Transform of Dirac Delta Function

Dirac Delta function (unit impulse) is non-integrable function and is an even function, and has a Fourier transform in the limit, and denoted by $\delta(t)$, Fig.1.4c. The delta function $\delta(t)$ is a pulse with duration equal zero, and given at the origin, its pulse amplitude goes to infinity with unit strength (unit area), and is defined by

$$\int_{-\infty}^{\infty} \delta(t) \, dt \; = \; 1 \qquad\qquad \text{at} \quad t = 0 \qquad\qquad (1.5)$$

, and $\qquad\qquad\qquad\qquad = 0 \qquad\qquad \text{at} \quad t \neq 0$

Now, to get the Fourier transform of the delta function $\delta(t)$, consider a unit area rectangular pulse with duration T and amplitude 1/T, Fig.1.6a, the Fourier transform pair will be

$$\frac{1}{T} \, rect(\frac{t}{T}) \quad \rightleftharpoons \quad sinc(fT)$$

, assume that the pulse duration T is a variable parameter, the delta function $\delta(t)$ is obtained by taking the limit T→0. The rectangular pulse then becomes infinitely narrow in duration and infinitely large in amplitude, remaining its area finite and fixed at unity. The delta function $\delta(t)$ is given by

$$\delta(t) = \lim_{T \to 0} \frac{1}{T} \, rect(\frac{t}{T}) \qquad\qquad (1.6)$$

, Fig.1.6a illustrates the sequence of such pulses as the parameter T varies, where the zero crossing of the spectrum $sinc(fT)$ will occur at infinitely, and the Fourier transform pair yields

$$\delta(t) \quad \rightleftharpoons \quad 1 \qquad\qquad (4.1)$$

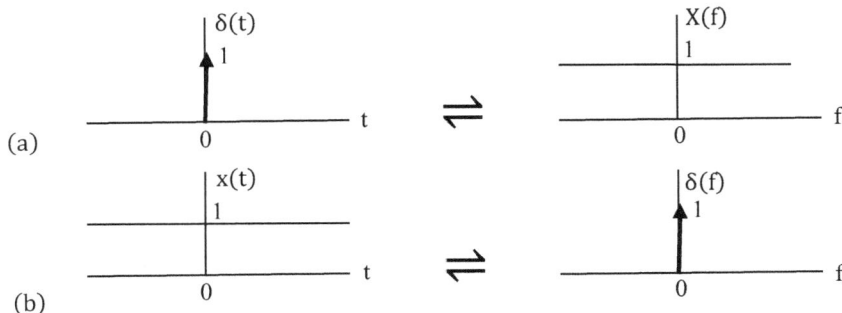

Fig.4.1. Fourier transform pairs of Dirac delta function.

, Fig.4.1a shows that the Fourier transform of the delta function $\delta(t)$, is a spectrum extends uniformly over the entire frequency interval from $-\infty$ to ∞, that is means, the delta function $\delta(t)$ contains infinite number of frequency components, and all with unity amplitude, then the inverse Fourier transform operation (Fourier integral), Eq.(3.1), and the delta function $\delta(t)$ yields

$$\delta(t) = \int_{-\infty}^{\infty} e^{j2\pi ft}\ df$$

, since the impulse signal $x(t) = \delta(t)$ is mapped into a spectrum $X(f) = 1$, and due to the duality property then the spectrum $X(f) = \delta(f)$ is inversed mapped into a waveform $x(t) = 1$, which contains only zero frequency (direct current), Fig.4.1b, and the duality property Fourier transform pair yields

$$1 \quad \rightleftharpoons \quad \delta(f) \qquad\qquad (4.2)$$

, and the Fourier transform operation of the direct current signal $x(t) = 1$, Eq.(3.2), $\delta(f)$ yields

$$\delta(f) = \int_{-\infty}^{\infty} e^{-j2\pi ft}\ dt \qquad\qquad (4.3)$$

, the definition of the delta function can also be obtained from the triangle function, the double sided exponential function, the sinc function, the sinc squared function, and the Gaussian function, Fig.4.2, respectively, by taking the following limits [4,5].

In terms of the triangle function, Fig.4.2a, Dirac delta function is given by

$$\delta(t) = \lim_{T\to 0} \frac{1}{T}\ tri(\frac{t}{T})$$

, in terms of the double sided exponential function, Fig.4.2b, Dirac delta function is given by

$$\delta(t) = \lim_{a\to 0} \frac{1}{a}\ e^{-|t/a|}$$

, in terms of the sinc function, Fig.4.2c, Dirac delta function is given by

$$\delta(t) = \lim_{W\to\infty} W\, sinc(Wt)$$

, also in terms of the sinc squared function, Dirac delta function is given by

$$\delta(t) = \lim_{W\to\infty} W\, sinc^2(Wt)$$

, and in terms of the Gaussian function, Fig.4.2d, Dirac delta function is given by

$$\delta(t) = \lim_{\tau\to 0} \frac{1}{\tau}\ e^{-(t/\tau)^2}$$

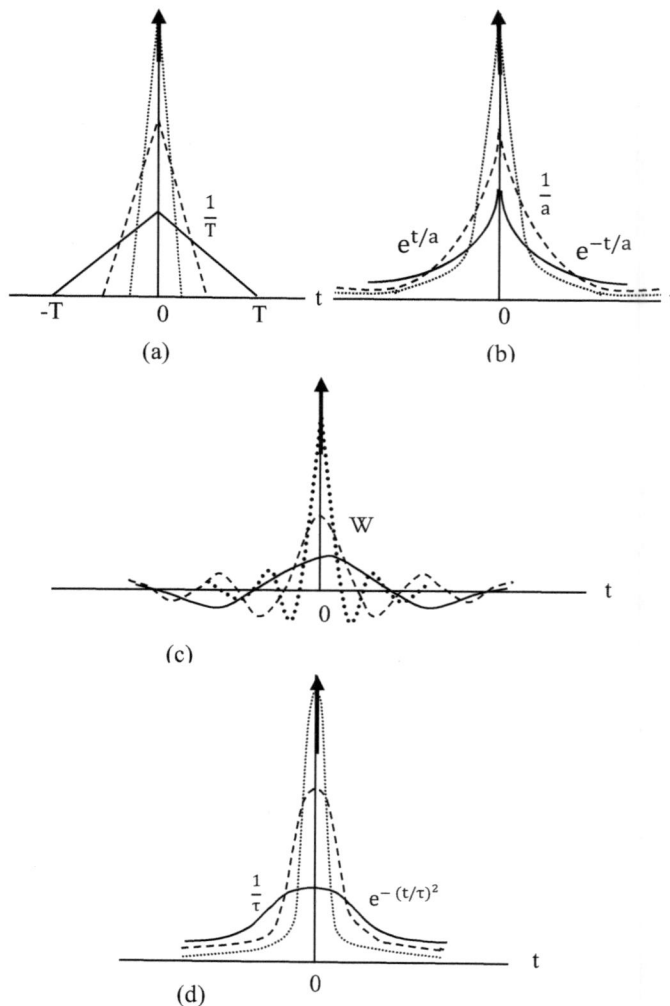

Fig.4.2. Derivation of Dirac delta function using:
a) Triangle pulse, b) Double sided pulse,
c) Sinc pulse, d) Gaussian pulse.

, studying the multiplication of an arbitrary function x(t) by the delta function $\delta(t)$, is given by

$$x(t)\, \delta(t) = x(0)\, \delta(t) \qquad \text{at} \quad t = 0 \qquad (1.7)$$
$$= 0 \qquad \text{otherwise}$$

, and

$$x(t)\, \delta(t - t_o) = x(t_o)\, \delta(t - t_o) \qquad \text{at} \quad t = t_o \qquad (1.8)$$
$$= 0 \qquad \text{otherwise}$$

, where t_o is the shifting time, $x(0)$ and $x(t_o)$ are the values of the arbitrary function x(t) at the origin and at the shifting time t_o respectively. The multiplication of a function x(t) by the delta function $\delta(t)$, Equations (1.5), (1.7), and (1.8), facilitate the evaluation of the following important integrals

$$\int_{-\infty}^{\infty} x(t)\, \delta(t)\, dt = x(0) \qquad (4.4a)$$

, and

$$\int_{-\infty}^{\infty} x(t)\, \delta(t - t_o)\, dt = x(t_o) \qquad (4.4b)$$

, the Fourier transform of the delta function $\delta(t)$ can also be obtained, Eq.(3.2), yields

$$F[\delta(t)] = \int_{-\infty}^{\infty} \delta(t)\, e^{-j2\pi ft}\, dt = \int_{-\infty}^{\infty} \delta(t)\, dt = 1$$

, replace τ instead of t, and t instead of t_o , in both sides of Eq.(4.4b), yields

$$\int_{-\infty}^{\infty} x(\tau)\, \delta(\tau - t)\, d\tau = x(t)$$

, since the delta function is an even function, the integration yields

$$\int_{-\infty}^{\infty} x(\tau)\, \delta(t - \tau)\, d\tau = x(t)$$

, this integration is the convolution integral, Eq.(3.42), of the arbitrary signal x(t) and the Dirac delta function $\delta(t)$, or equivalently

$$x(t) \otimes \delta(t) = x(t) \qquad (4.4c)$$

, another way to get the convolution of an arbitrary function x(t) and the delta function $\delta(t)$ using the frequency domain, by getting their individual Fourier transforms and then multiply and again return to the time domain by taking the inverse Fourier transform. Applying the same method to get the convolution of an arbitrary function x(t) and a shifted Dirac delta function $\delta(t - t_o)$ yields

$$x(t) \otimes \delta(t - t_o) = x(t - t_o) \qquad (4.5a)$$

, also

$$x(t - t_{o1}) \otimes \delta(t - t_{o2}) = x(t - t_{o1} - t_{o2}) \qquad (4.5b)$$

, referring to the definition of the Dirac delta function of constant area (unity), Eq.(1.6), where it is the Limit of T→0 for a rectangular pulse having duration T and amplitude 1/T, the time scaling of the Dirac delta function $\delta(at)$, where the "a" is positive real constant (a ≥ 0), is given by

$$\delta(at) = \frac{1}{|a|}\, \delta(t) \qquad (4.5c)$$

Dirac Delta function (unit impulse) is not a function in the usual sense, but it is very important concept or tool which allows the easy solution of many problems in communication theory that would otherwise quite difficult and in the electric circuits analysis [9].

Numerical examples

   i. $\cos(t)\, \delta(t - 3\pi) = \cos(3\pi)\, \delta(t - 3\pi) = -\,\delta(t - 3\pi)$
   ii. $u(t - 4)\, \delta(t) = 0$
   iii. $u(t + 4)\, \delta(t) = \delta(t)$
   iv. $u(t - 4) \otimes \delta(t) = u(t - 4)$
   v. $u(t - 4) \otimes \delta(t - 5) = u(t - 9)$
   vi. $\operatorname{sinc}(2t - \tfrac{1}{2}) \otimes \delta(2t - 1.5) = \operatorname{sinc}[2(t - \tfrac{1}{4})] \otimes \tfrac{1}{2}\,\delta(t - \tfrac{3}{4}) = \tfrac{1}{2}\operatorname{sinc}[2(t - 1)]$

   vii. $\displaystyle\int_{-\infty}^{\infty} \frac{\delta(2f - 1)}{1 + j2\pi f}\, e^{-j2\pi f}\, df = -\,\frac{\tfrac{1}{2}}{1 + j\pi}$

Problem 4.1

Evaluate the convolution function $x_{12}(t) = x_1(t) \otimes x_2(t)$, where $x_1(t)$ and $x_2(t)$ are shown in Fig.4.3.

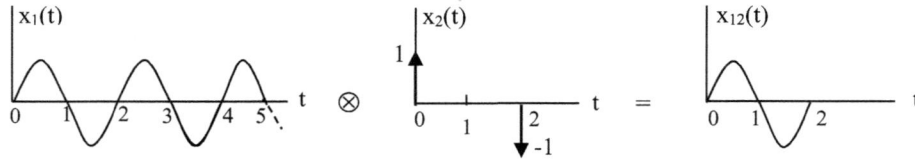

Fig.4.3. $x_{12}(t) = x_1(t) \otimes x_2(t)$.

Solution

The convolution function $x_{12}(t)$ can be expressed by

$$x_{12}(t) = \sin(\pi t)\, u(t) \otimes [\, \delta(t) - \delta(t-2)\,]$$

, or equivalently

$$x_{12}(t) = \sin(\pi t)\, u(t) \;-\; \sin[\pi(t-2)]\, u(t-2)$$

, then the convolution function $x_{12}(t)$, Fig.4.3c, yields

$$x_{12}(t) = \text{rect}\left[\frac{t-1}{2}\right] \sin(\pi t)$$

4.2. Fourier Transform of Exponential and Sinusoidal Functions

The Fourier transform of the exponential and sinusoidal functions is evaluated using the inverse Fourier transform of the shifted delta functions $\delta(f-f_c)$ and $\delta(f+f_c)$ which cause two angles $(j2\pi f_c t)$ and $(-j2\pi f_c t)$ respectively in the time domain and f_c is the frequency shifting. The Fourier transform pairs are given by

$$e^{j2\pi f_c t} \;\rightleftharpoons\; \delta(f-f_c)$$

, and

$$e^{-j2\pi f_c t} \;\rightleftharpoons\; \delta(f+f_c)$$

, the functions $\sin(2\pi f_c t)$ and $\cos(2\pi f_c t)$, Fig.4.4a,b, are the algebraic summation of the two exponential angles $e^{j2\pi f_c t}$ and $e^{-j2\pi f_c t}$, the Fourier transform pairs yield

$$\cos(2\pi f_c t) \;\rightleftharpoons\; \frac{1}{2}\,[\delta(f-f_c) + \delta(f+f_c)]$$

, and

$$\sin(2\pi f_c t) \;\rightleftharpoons\; \frac{1}{2j}\,[\delta(f-f_c) - \delta(f+f_c)]$$

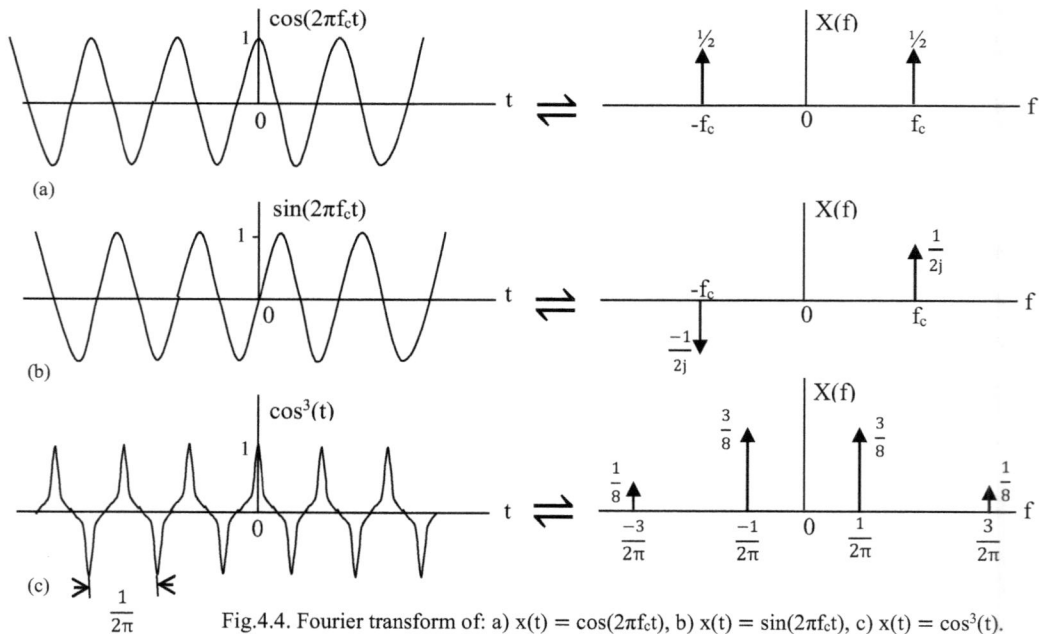

Fig.4.4. Fourier transform of: a) $x(t) = \cos(2\pi f_c t)$, b) $x(t) = \sin(2\pi f_c t)$, c) $x(t) = \cos^3(t)$.

The Fourier transform of the sinusoidal functions $\cos(2\pi f_c t)$ and $\sin(2\pi f_c t)$ can also be derived using Eq.(4.3), where the Fourier transform operations are given by

$$F[\cos(2\pi f_c t)] = \int_{-\infty}^{\infty} \frac{1}{2}[e^{j2\pi f_c t} + e^{-j2\pi f_c t}]\, e^{-j2\pi ft}\, dt = \frac{1}{2}[\delta(f - f_c) + \delta(f + f_c)]$$

, and

$$F[\sin(2\pi f_c t)] = \int_{-\infty}^{\infty} \frac{1}{2j}[e^{j2\pi f_c t} - e^{-j2\pi f_c t}]\, e^{-j2\pi ft}\, dt = \frac{1}{2j}[\delta(f - f_c) - \delta(f + f_c)]$$

Numerical example
The Fourier transform of the function $x(t) = \cos^3(t)$ can be obtained, where $x(t)$ is expressed by

$$x(t) = \frac{3}{4}\cos(t) + \frac{1}{4}\cos(3t)$$

, then the Fourier transform, Fig.4.4c, $X(f)$ is given by

$$X(f) = \frac{3}{8}\left[\delta(f - \frac{1}{2\pi}) + \delta(f + \frac{1}{2\pi})\right] + \frac{3}{8}\left[\delta(f - \frac{3}{2\pi}) + \delta(f + \frac{3}{2\pi})\right]$$

Problem 4.2
Find the Fourier transform of the band-pass signal $s(t) = m(t)\cos(2\pi f_c t)$, using Dirac delta function, where $m(t)$ is the baseband signal and has a spectrum $M(f)$ with maximum frequency W Hz, Fig.1.13a, and f_c much higher than W Hz.

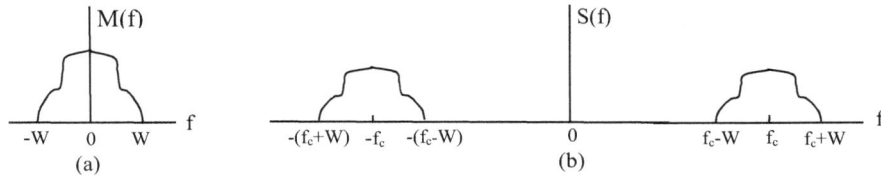

Fig.1.13. a) Low-pass signal, b) Band-pass signal.

Solution
The Fourier transform $S(f)$, Eq.(3.39), is given by

$$S(f) = M(f) \otimes \tfrac{1}{2}[\,\delta(f - f_c) + \delta(f + f_c)\,]$$

, or equivalently

$$S(f) = \tfrac{1}{2}[\,M(f - f_c) + M(f + f_c)\,]$$

, Fig.1.13b shows the spectrum $S(f)$ of the band-pass signal $s(t)$ and has bandwidth $2W$ Hz.

Problem 4.3
Find the Fourier transform of the Radio Frequency pulse $x(t) = A\,\mathrm{rect}(t/T)\cos(2\pi f_c t)$, Fig.1.8a, using the Dirac delta function.

Solution
The Fourier transform $X(f)$ of the RF pulse $x(t)$, Eq.(3.39), is given by

$$X(f) = AT\,\mathrm{sinc}(fT) \otimes \tfrac{1}{2}[\,\delta(f - f_c) + \delta(f + f_c)\,]$$

, or equivalently

$$X(f) = \tfrac{1}{2}AT\,\mathrm{sinc}[T(f - f_c)] + \tfrac{1}{2}AT\,\mathrm{sinc}[T(f + f_c)]$$

, which is the same Fourier transform of the RF pulse, using shifting property, problem 3.11, Eq.(3.8).

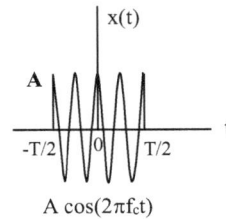

Fig.1.8a. Unperiodic RF signal.

(3.8)

Problem 4.4
Evaluate the following convolution function, using the frequency domain

$$x_{12}(t) = e^{-t}u(t) \otimes \cos(t)$$

Solution

The Fourier transform pair of the convolution function $x_{12}(t)$ is given by

$$e^{-t}u(t) \otimes \cos(t) \rightleftharpoons \frac{\frac{1}{2}}{1+j2\pi f}\left[\delta(f-\frac{1}{2\pi}) + \delta(f+\frac{1}{2\pi})\right]$$

, taking the inverse Fourier transform, yields

$$\frac{1}{2\sqrt{2}}\left[e^{jt}e^{-j\pi/4} + e^{-jt}e^{j\pi/4}\right] \rightleftharpoons \frac{\frac{1}{2}}{1+j}\delta(f-\frac{1}{2\pi}) + \frac{\frac{1}{2}}{1-j}\delta(f+\frac{1}{2\pi})$$

, or equivalently

$$\frac{1}{\sqrt{2}}\cos(t-45°) \rightleftharpoons \frac{\frac{1}{2}}{1+j}\delta(f-\frac{1}{2\pi}) + \frac{\frac{1}{2}}{1-j}\delta(f+\frac{1}{2\pi})$$

, then the convolution function $x_{12}(t)$ is given by

$$x_{12}(t) = e^{-t}u(t) \otimes \cos(t) = 0.707\cos(t-45°)$$

Problem 4.5

A current source $i(t) = 8\cos(4t)$, is applied at the shown inductive phase shift circuit, Fig.4.5. Find the current $i_o(t)$, passing in the inductance, using the Fourier transform.

Solution

The current $i_o(t)$ is given by

$$i_o(t) = i(t)\frac{4}{4+j2\pi f} \qquad (4.6)$$

, taking the Fourier transform of Eq.(4.6)
, $I_o(f)$ yields

$$I_o(f) = \frac{4}{4+j2\pi f}F[i(t)]$$

$$= \frac{16}{4+j2\pi f}[\delta(f-\frac{2}{\pi}) + \delta(f+\frac{2}{\pi})]$$

, or equivalently

$$I_o(f) = \frac{4}{1+j}\delta(f-\frac{2}{\pi}) + \frac{4}{1-j}\delta(f+\frac{2}{\pi}) \qquad (4.7)$$

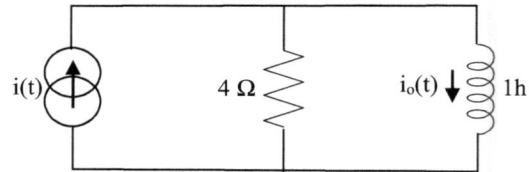

Fig.4.5. RL phase shift network.

, taking the inverse Fourier transform of Eq.(4.7), $i_o(t)$ yields

$$i_o(t) = \frac{4}{2\sqrt{2}}\left[e^{j4t}e^{-j\pi/4} + e^{-j4t}e^{j\pi/4}\right]$$

, or equivalently

$$i_o(t) = \sqrt{2}\cos(4t-45°)$$

Problem 4.6

A voltage source $v(t) = 2\sin(t)$, is applied at the shown capacitive phase shift circuit, Fig.4.6. Find the current $i(t)$ of the circuit, using the Fourier transform.

Solution

The current $i(t)$ is given by

$$i(t) = v(t)\frac{1}{4+\frac{1}{j2\pi f}} \qquad (4.8)$$

, or equivalently

$$i(t) = v(t)\frac{j2\pi f}{1+j8\pi f} \qquad (4.9)$$

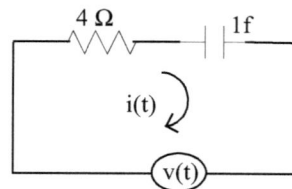

Fig.4.6. RC phase shift network.

, taking the Fourier transform of Eq.(4.9), $I(f)$ yields

$$I(f) = \frac{j2\pi f}{1 + j8\pi f} F[v(t)]$$

$$= \frac{j2\pi f}{1 + j8\pi f} \left\{ \frac{1}{j} \left[\delta(f - \frac{1}{2\pi}) - \delta(f + \frac{1}{2\pi}) \right] \right\}$$

, or equivalently

$$I(f) = \frac{1}{1 + j4} \delta(f - \frac{1}{2\pi}) + \frac{1}{1 - j4} \delta(f + \frac{1}{2\pi}) \qquad (4.10)$$

, taking the inverse Fourier transform of Eq.(4.10), i(t) yields

$$i(t) = \frac{4}{\sqrt{17}} \left[e^{jt} e^{-j\pi/2.37} + e^{-jt} e^{j\pi/2.37} \right]$$

, or equivalently

$$i(t) = 0.485 \cos(t - 76^\circ)$$

Problem 4.7
Evaluate the following convolution function
, using the time domain

$$X_{12}(f) = \text{rect}(f) e^{-j2\pi f} \otimes e^{-j2\pi f}$$

Solution
The Fourier transform pair of the convolution
$X_{12}(f)$, is given by

$$\text{sinc}(t - 1) \times \delta(t - 1) \rightleftharpoons \text{rect}(f) e^{-j2\pi f} \otimes e^{-j2\pi f}$$

, or equivalently

$$\delta(t - 1) \rightleftharpoons \int_{-\frac{1}{2}}^{\frac{1}{2}} e^{-j2\pi\lambda} e^{-j2\pi(f-\lambda)} d\lambda$$

, taking the Fourier transform, yields

$$\delta(t - 1) \rightleftharpoons e^{-j2\pi f}$$

, then the convolution function $X_{12}(f)$ is given by

$$X_{12}(f) = \text{rect}(f) e^{-j2\pi f} \otimes e^{-j2\pi f} = e^{-j2\pi f}$$

, which is the same Fourier transform, using the frequency domain, problem 3.19i.

4.3. Fourier Transform of Signum Function
The signum function is nonintegrable function
, denoted by sgn(t), and is odd function, and has a Fourier transform in the limit, Fig.1.4b, and is defined by

$$\text{sgn}(t) = 1 \qquad \text{for} \quad t > 0$$
$$= 0 \qquad \text{for} \quad t = 0$$
$$= -1 \qquad \text{for} \quad t < 0$$

, where the value of the signum function at the origin (point of discontinuity), is the mean of the values of the function, also $\text{sgn}(at) = \text{sgn}(\tau)$, where $\tau = at$, and the "a" is positive real constant $(a \geq 0)$.

Now, to get the Fourier transform of the signum function sgn(t), consider the energy signal x(t), Fig.4.7a, is given by

$$x(t) = e^{-at} u(t) - e^{at} u(-t)$$

, where the "a" is positive real constant $(a \geq 0)$
, the Fourier transform pair of x(t), is given by

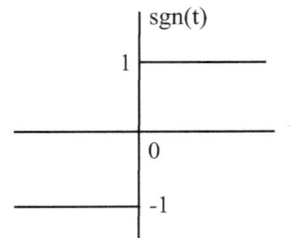

Fig.1.4b. Signum function.

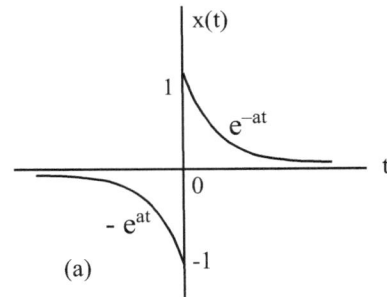

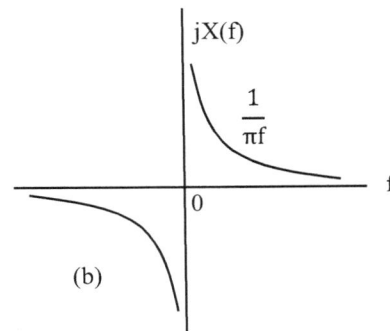

Fig.4.7. Fourier transform of x(t) = sgn(t).

$$e^{-at}\,u(t) \;-\; e^{at}\,u(-t) \;\rightleftharpoons\; \frac{1}{a + j2\pi f} - \frac{1}{a - j2\pi f}$$

, or equivalently

$$e^{-at}\,u(t) \;-\; e^{at}\,u(-t) \;\rightleftharpoons\; \frac{-j4\pi f}{a^2 + (2\pi f)^2} \qquad (4.11)$$

, assume that the positive real constant "a" ($a \geq 0$) is a variable parameter, the signum function sgn(t) is obtained by taking the limit a→0, and the energy signal x(t), Fig.4.7a, becomes a signum function sgn(t), Fig.1.4b, and is given by

$$sgn(t) = \lim_{a \to 0} \; [\, e^{-at}\,u(t) \;-\; e^{at}\,u(-t)\,]$$

, and the Fourier transform pair of the signum function sgn(t), Eq.(4.11), yields

$$sgn(t) \;\rightleftharpoons\; \frac{1}{j\pi f} \qquad (4.12)$$

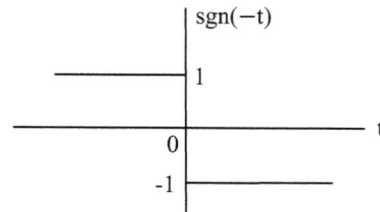

, then the signum function sgn(t) is represented in the frequency domain by an imaginary odd function (1/jπf), Fig.4.7b, then the inverse Fourier transform operation Eq.(3.1) of Eq.(4.12), is given by

$$\int_{-\infty}^{\infty} \frac{1}{j\pi f}\, e^{j2\pi ft}\, df \;=\; 1 \qquad\qquad \text{for } t > 0$$

$$= 0 \qquad\qquad \text{at } t = 0$$
$$= -1 \qquad\qquad \text{for } t < 0$$

, apply the duality property for Eq.(4.12), the Fourier transform pair will be

$$-\frac{1}{j\pi t} \;\rightleftharpoons\; sgn(f) \qquad (4.13)$$

, and the Fourier transform operation, Eq.(3.2), of Eq.(4.13) is given by

$$\int_{-\infty}^{\infty} (-\frac{1}{j\pi t})\, e^{-j2\pi ft}\, dt \;=\; 1 \qquad\qquad \text{for } f > 0$$

$$= 0 \qquad\qquad \text{at } f = 0$$
$$= -1 \qquad\qquad \text{for } f < 0$$

, Equations (4.13) and (4.12) facilitate the evaluation of the Fourier transform of the function $x(t) = 1/t$, and the inverse Fourier transform of the function $X(f) = 1/f$, and their Fourier transform pairs are given by

$$\frac{1}{t} \;\rightleftharpoons\; -j\pi\, sgn(f) \qquad (4.14)$$

, $$j\pi\, sgn(t) \;\rightleftharpoons\; \frac{1}{f} \qquad (4.15)$$

, on the other hand, the Fourier transform of the negation function sgn(−t), Fig.4.8, can also be obtained using the same procedure, and the Fourier transform pair yields

$$sgn(-t) \;\rightleftharpoons\; -\frac{1}{j\pi f}$$

Fig.4.8. Negation function x(t) = sgn(−t).

Numerical examples

i. $\int_{-\infty}^{\infty} \text{sinc}^2\left[\dfrac{t-2}{4}\right]\,\text{sgn}(t)\,\delta(\tfrac{1}{2}t-1)]\,dt = 2\int_{-\infty}^{\infty}\text{sgn}(2)\,\delta(t-2)\,dt = 2$

ii. $\int_{-\infty}^{\infty}\text{sgn}(t-1)\,\delta(t+10)\,dt = \int_{-\infty}^{\infty}(-1)\,\delta(t+10)\,dt = -1$

iii. The Fourier transform pair of the function $x(t) = \text{sgn}(t+2)$ is given by

$$\text{sgn}(t+2) \rightleftharpoons \dfrac{1}{j\pi f}\,e^{j2\pi f2}$$

iv. The Fourier transform pair of the function $X(f) = 1/(3+f)$, is given by

$$j\pi\,\text{sgn}(t)\,e^{-j2\pi t3} \rightleftharpoons \dfrac{1}{3+f}$$

Problem 4.8

A current source $i(t) = \text{sgn}(t)$, is applied at the shown inductive phase shift circuit, Fig.4.5. Find the current $i_o(t)$ passing in the inductance, using the Fourier transform.

Solution

The current $i_o(t)$ is given by

$$i_o(t) = i(t)\,\dfrac{4}{4+j2\pi f} \qquad (4.6)$$

, taking the Fourier transform of Eq.(4.6), $I_o(f)$ yields

$$I_o(f) = \dfrac{4}{4+j2\pi f}\,F[i(t)]$$
$$= \dfrac{4}{(4+j2\pi f)}\,\dfrac{1}{j2\pi f}$$

, using the partial fraction technique, $I_o(f)$ yields

$$I_o(f) = \dfrac{1}{j2\pi f} - \dfrac{1}{4+j2\pi f} \qquad (4.16)$$

, taking the inverse Fourier transform for Eq.(4.16), $i_o(t)$ yields

$$i_o(t) = \tfrac{1}{2}\,\text{sgn}(t) - e^{-4t}\,u(t)$$

, Fig.4.9 shows the current distribution $i_o(t)$.

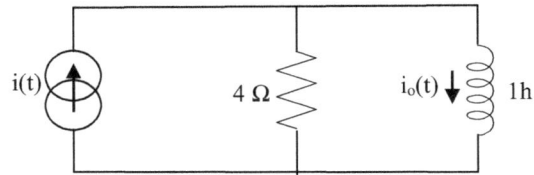

Fig.4.5. RL phase shift network.

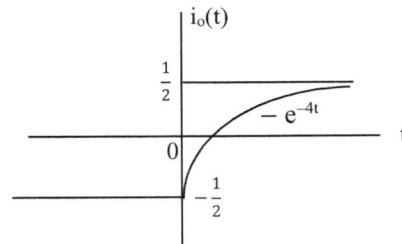

Fig.4.9. Current distribution in the inductance.

Problem 4.9

Evaluate the following convolution functions

i. $x(t) = e^{-2t}\,u(t) \otimes \text{sgn}(t)$
ii. $x(t) = \text{sgn}(t) \otimes \text{rect}(t)$

iii. $x(t) = \sin(2\pi10^6 t) \otimes \dfrac{1}{\pi t}$

Solution

i. The convolution function $x(t) = e^{-2t}\,u(t) \otimes \text{sgn}(t)$, Fig.4.10, using the time domain, Eq.(3.42), is given by

$$x(t) = \int_0^{\infty}(-1)\,e^{-2\tau}\,d\tau = -\dfrac{1}{2} \qquad \text{for negative t}$$

, and $\qquad x(t) = \int_0^{t} e^{-2\tau}\,d\tau - \int_t^{\infty} e^{-2\tau}\,d\tau = 1 - e^{-2t} \qquad \text{for positive t}$

, this convolution function can also be obtained, using the frequency domain , where its Fourier transform pair will be

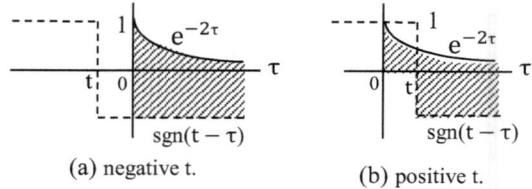

(a) negative t.   (b) positive t.

$$e^{-2t}\, u(t) \otimes sgn(t) \;\rightleftharpoons\; \frac{1}{(2 + j2\pi f)}\, \frac{1}{(j\pi f)}$$

, using the partial fraction technique, and taking the inverse Fourier transform yields

$$\tfrac{1}{2}\, sgn(t) - e^{-2t}\, u(t) \;\rightleftharpoons\; \frac{\tfrac{1}{2}}{j\pi f} - \frac{1}{2 + j2\pi f}$$

, then the convolution function $x_{12}(t)$ is given by

$$x(t) = e^{-2t}\, u(t) \otimes sgn(t)$$
$$= \tfrac{1}{2}\, sgn(t) - e^{-2t}\, u(t)$$

, which is the same convolution function $x_{12}(t)$, using the time domain, Fig.4.10.

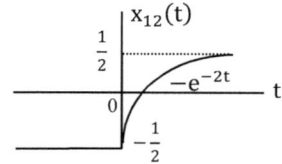

(c) The convolution waveform.

Fig.4.10. $x_{12}(t) = e^{-2t}\, u(t) \otimes sgn(t)$:
a) Negative t, b) Positive t,
c) The convolution waveform.

ii. The convolution function $x_{12}(t) = sgn(t) \otimes rect(t)$, Fig.4.11, using the time domain, Eq.(3.42), is given by

$$x_{12}(t) = \int_{t-\frac{1}{2}}^{t+\frac{1}{2}} (-1)\; d\tau = -1 \qquad\qquad for\ \ t \le -\tfrac{1}{2}$$

,
$$x_{12}(t) = \int_{t-\frac{1}{2}}^{0} (-1)\; d\tau + \int_{0}^{t+\frac{1}{2}} d\tau = 2t \qquad for\ \ -\tfrac{1}{2} \le t \le \tfrac{1}{2}$$

, and
$$x_{12}(t) = \int_{t-\frac{1}{2}}^{t+\frac{1}{2}} d\tau = 1 \qquad\qquad for\ \ t \ge \tfrac{1}{2}$$

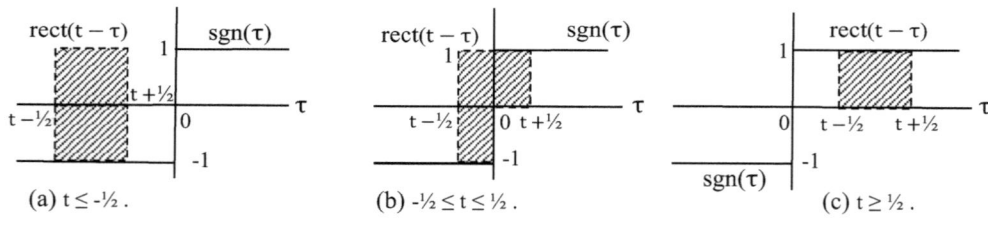

(a) $t \le -\frac{1}{2}$.   (b) $-\frac{1}{2} \le t \le \frac{1}{2}$.   (c) $t \ge \frac{1}{2}$.

Fig.4.11. $x_{12}(t) = sgn(t) \otimes rect(t)$:
a) $t \le -\frac{1}{2}$, b) $-\frac{1}{2} \le t \le \frac{1}{2}$,
c) $t \ge \frac{1}{2}$, d) The convolution waveform.

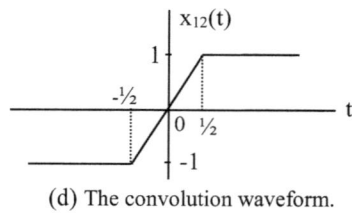

(d) The convolution waveform.

iii. The convolution function $x_{12}(t) = sin(2\pi 10^6 t) \otimes 1/\pi t$, using the frequency domain, is given by

$$sin(2\pi 10^6 t) \otimes \frac{1}{\pi t} \;\rightleftharpoons\; \frac{1}{2j}\, [\, \delta(f - 10^6) - \delta(f + 10^6)\,]\; [-j\, sgn(f)\,]$$

, in the frequency domain, multiply the signum function by the two delta functions, and taking the inverse Fourier transform, yields

$$-\cos(2\pi 10^6 t) \;\rightleftharpoons\; -\tfrac{1}{2}\, [\, \delta(f - 10^6) + \delta(f + 10^6)\,]$$

, then the convolution function $x_{12}(t)$ is given by

$$x_{12}(t) = sin(2\pi 10^6 t) \otimes \frac{1}{\pi t} = -\cos(2\pi 10^6 t)$$

4.4. Fourier Transform of Unit Step Function

The unit step function is a non-integrable function , has Fourier transform in the limit, and denoted by u(t). The unit step function is a causal function , Fig.1.4a, and is defined by

$$u(t) = 1 \qquad \text{for} \quad t > 0$$
$$= \tfrac{1}{2} \qquad \text{for} \quad t = 0$$
$$= 0 \qquad \text{for} \quad t < 0$$

, where the value of the step function at the origin (point of discontinuity), is the mean of the values of the function, also $u(at) = u(\tau)$, where $\tau = at$, and the "a" is positive real constant ($a \geq 0$). Since the signum function sgn(t) is a combination of two unit step functions by certain way, then

$$\text{sgn}(t) = 2\,u(t) - 1$$

, or equivalently

$$u(t) = \tfrac{1}{2} [\, \text{sgn}(t) + 1\,]$$

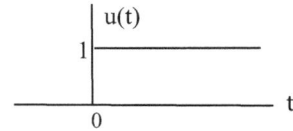

Fig.1.4a. Unit step function u(t).

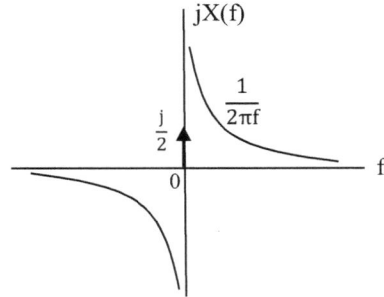

Fig.4.12. Fourier transform of unit step function u(t).

, then the Fourier transform of the unit step function u(t), is the same Fourier transform of the function " $\tfrac{1}{2}$ [sgn(t) + 1] ". In terms of the Fourier transform of sgn(t), Eq.(4.12), and using Eq.(4.2), the Fourier transform pair of the unit step function is given by

$$u(t) \quad \rightleftharpoons \quad \frac{1}{j2\pi f} + \frac{\delta(f)}{2} \qquad\qquad (4.17)$$

, then the unit step function u(t) is represented in the frequency domain by an imaginary odd function $(1/j2\pi f)$ plus a delta function $\delta(f)$ with strength equals half, Fig.4.12, then the inverse Fourier transform operation, Eq.(3.1), of Eq.(4.17), is given by

$$\int_{-\infty}^{\infty} [\frac{1}{j2\pi f} + \frac{\delta(f)}{2}]\, e^{j2\pi ft}\, df \quad = 1 \qquad\qquad \text{for} \ \ t > 0$$
$$= \tfrac{1}{2} \qquad\qquad \text{at} \quad t = 0$$
$$= 0 \qquad\qquad \text{for} \ \ t < 0$$

, applying the duality property for Eq.(4.17) , the Fourier transform pair will be

$$-\frac{1}{j2\pi t} + \frac{\delta(t)}{2} \quad \rightleftharpoons \quad U(f) \qquad (4.18)$$

On the other hand, the Fourier transform of the negation function u(−t), Fig.4.13 , can be obtained using the same procedure , and the Fourier transform pair yields

$$u(-t) \quad \rightleftharpoons \quad -\frac{1}{j2\pi f} + \frac{\delta(f)}{2}$$

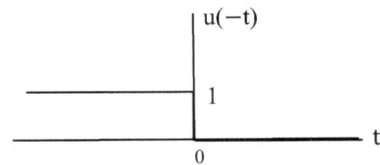

Fig.4.13. Negation unit step function u(−t).

, and the inverse Fourier transform operation, Eq.(3.1), of the negation function u(−t) is given by

$$\int_{-\infty}^{\infty} [-\frac{1}{j2\pi f} + \frac{\delta(f)}{2}]\, e^{j2\pi ft}\, df \quad = 0 \qquad\qquad \text{for} \ \ t > 0$$
$$= \tfrac{1}{2} \qquad\qquad \text{at} \quad t = 0$$
$$= 1 \qquad\qquad \text{for} \ \ t < 0$$

Numerical examples

i. $\int_{-\infty}^{\infty} [\delta(\tfrac{1}{2}t - 1)]\, u(t - 4)]\, dt = 2 \int_{-\infty}^{\infty} [\delta(t - 2)]\, u(t - 4)]\, dt = 0$

ii. $\int_{-\infty}^{\infty} t^2\, u(3 - t)\, u(t)\, dt = \int_{0}^{3} t^2\, dt = 9$

iii. $\int_{-\infty}^{\infty} u(t - 10)\, \delta(3t - 30)\, dt = \dfrac{1}{3} \int_{-\infty}^{\infty} u(t - 10)\, \delta(t - 10)\, dt = \dfrac{1}{3} \times \dfrac{1}{2} = \dfrac{1}{6}$

iv. $\int_{-\infty}^{\infty} \cos(\pi t)\, u(t - 2)\, \delta(t - 5)\, dt = -1$

v. The Fourier transform pair of the function x(t) = u(t + 2), is given by

$$u(t + 2) \rightleftharpoons \frac{1}{j2\pi f} e^{j2\pi f2} + \frac{\delta(f)}{2}$$

Problem 4.10
A voltage source v(t) = 20 u(t), is applied at the capacitive phase shift circuit, Fig.4.6. Find the current i(t) of the circuit, using the Fourier transform.
Solution
The current i(t) is given by

$$i(t) = v(t)\, \frac{1}{4 + \dfrac{1}{j2\pi f}} \qquad (4.8)$$

, or equivalently

$$i(t) = v(t)\, \frac{j2\pi f}{1 + j8\pi f} \qquad (4.9)$$

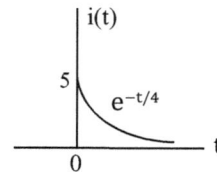
Fig.4.14. Current distribution in capacitance.

, taking the Fourier transform of Eq.(4.9), yields

$$I(f) = \frac{j2\pi f}{1 + j8\pi f}\, F[v(t)] = \frac{j2\pi f}{1 + j8\pi f}\, 20\, [\frac{1}{j2\pi f} + \frac{\delta(f)}{2}]$$

$$= \frac{20}{1 + j8\pi f} \qquad (4.19)$$

, taking the inverse Fourier transform for Eq.(4.19), the current i(t), Fig.4.14. yields

$$i(t) = 5\, e^{-t/4}\, u(t)$$

Problem 4.11
Evaluate the following convolution functions, using the frequency domain

i. $X(f) = \dfrac{1}{2f}\, U(f) \otimes U(f)$,   ii. $x(t) = \cos(2t) \otimes u(t)$

iii. $x(t) = u(t) \otimes e^{-2t}\, u(t)$,   iv. $x(t) = \delta(2t - 1) \otimes u(t + 2)$

Solution
i. The convolution function $X(f) = [(1/2f)\, U(f)] \otimes U(f)$
, Fig.4.15, using the convolution integral, Eq.(3.40)
, is given by

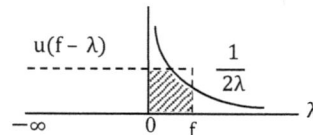
Fig.4.15. $X_{12}(f) = (1/2f)\, U(f) \otimes U(f).$

$$X(f) = \int_{-\infty}^{\infty} \frac{1}{2\lambda}\, U(\lambda)\, U(f - \lambda)\, d\lambda$$

$$= \int_{0}^{f} \frac{1}{2\lambda}\, d\lambda = [\frac{1}{2}\, \log(\lambda)]_{\lambda=0}^{\lambda=f} = \tfrac{1}{2}\, \log(f)$$

ii. The Fourier transform pair of the convolution function $x(t) = u(t) \otimes \cos(2t)$, is given by

$$u(t) \otimes \cos(2t) \;\rightleftharpoons\; \left[\frac{1}{j2\pi f} + \frac{\delta(f)}{2} \right] \frac{1}{2} [\delta(f - \frac{1}{\pi}) + \delta(f + \frac{1}{\pi})]$$

, or equivalently

$$u(t) \otimes \cos(2t) \;\rightleftharpoons\; -\frac{1}{4} j\, \delta(f - \frac{1}{\pi}) + \frac{1}{4} j\, \delta(f + \frac{1}{\pi})$$

, taking the inverse Fourier transform, the convolution function $x(t)$ yields

$$x(t) = u(t) \otimes \cos(2t) = -\frac{1}{2} \sin(2t)$$

iii. The Fourier transform pair of the convolution function $x(t) = e^{-2t} u(t) \otimes u(t)$, is given by

$$e^{-2t} u(t) \otimes u(t) \;\rightleftharpoons\; \frac{1}{2 + j2\pi f} \left[\frac{1}{j2\pi f} + \frac{\delta(f)}{2} \right]$$

, using the partial fraction technique, and taking the inverse Fourier transform, yields

$$\frac{1}{4} \operatorname{sgn}(t) - \frac{1}{2} e^{-2t} u(t) + \frac{1}{4} \;\rightleftharpoons\; \frac{\frac{1}{2}}{j2\pi f} - \frac{\frac{1}{2}}{2 + j2\pi f} + \frac{1}{4} \delta(f)$$

, then the convolution function $x(t)$ is given by

$$x(t) = e^{-2t} u(t) \otimes u(t) = \frac{1}{4} \operatorname{sgn}(t) - \frac{1}{2} e^{-2t} u(t) + \frac{1}{4}$$

iv. The Fourier transform pair of the convolution function $x(t) = \delta(2t - 1) \otimes u(t + 2)$, is given by

$$\delta(2t - 1) \otimes u(t + 2) \;\rightleftharpoons\; \frac{1}{2} e^{-j\pi f} \left[\frac{1}{j2\pi f} + \frac{\delta(f)}{2} \right] e^{j2\pi f2}$$

, taking the inverse Fourier transform, yields

$$\frac{1}{4} \operatorname{sgn}(t + \frac{3}{2}) + \frac{1}{4} \;\rightleftharpoons\; \frac{1}{j4\pi f} e^{j3\pi f} + \frac{\delta(f)}{4}$$

, then the convolution function $x(t)$ is given by

$$x(t) = \delta(2t - 1) \otimes u(t + 2) = \frac{1}{4} \operatorname{sgn}(t + \frac{3}{2}) + \frac{1}{4} = \frac{1}{2} u(t + \frac{3}{2})$$

4.5. Integration Property of Non-Zero Boundary Condition Functions

The integration property, Eq.(3.32a), evaluates the Fourier transform of the energy signal $x(t)$ where its Fourier transform $X(f)$ has zero boundary conditions $[X(0) = 0]$. Now, it is also important to derive the integration property in the case of the energy signals where its Fourier transform $X(f)$ have Non-zero boundary conditions $[X(0) \neq 0]$, Eq.(3.36), which facilitates the evaluation of the Fourier transform of some integrable and non-integrable signals that would otherwise be quite difficult. This study is considered very useful because it deduces the mathematical relation between the Dirac delta function $\delta(t)$ and the unit step function $u(t)$ [1].

Consider the convolution process $x(t) \otimes u(t)$, where $x(t)$ is an arbitrary function, using the frequency domain, the Fourier transform pair yields

$$x(t) \otimes u(t) \;\rightleftharpoons\; X(f) \left[\frac{1}{j2\pi f} + \frac{\delta(f)}{2} \right]$$

, or $$x(t) \otimes u(t) \;\rightleftharpoons\; \frac{X(f)}{j2\pi f} + \frac{X(0)\,\delta(f)}{2}$$

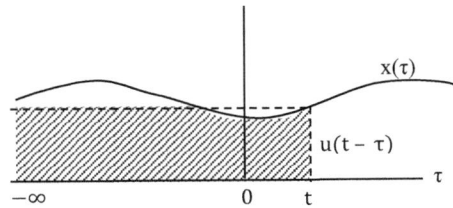

, and using the convolution integral in the time domain, Eq.3.42, Fig.4.16, the Fourier transform pair yields

Fig.4.16. The convolution function
$x_{12}(t) = x(t) \otimes u(t)$.

$$\int_{-\infty}^{t} x(\tau)\, d\tau \quad \rightleftharpoons \quad \frac{X(f)}{j2\pi f} + \frac{X(0)\delta(f)}{2} \qquad (3.36)$$

, comparing Eq.(3.36) and the integration property, Eq.(3.32a), they have the same time domain. The term $X(0)\,\delta(f)/2$ in the frequency domain, is the Non-zero boundary condition effect due to the integration of the energy signal in the time domain.

4.6. Relation between Dirac Delta Function and Unit Step Function

The relation between the unit step function u(t) and the delta function δ(t) can be deduced by assuming that x(t) is the Dirac delta function δ(t), the Fourier transform pair, Eq.(3.36), yields

$$\int_{-\infty}^{t} \delta(\tau)\, d\tau \quad \rightleftharpoons \quad \frac{1}{j2\pi f} + \frac{\delta(f)}{2} \qquad (4.20)$$

, comparing Equations (4.20) and (4.17), they have the same frequency domain, then their time domain should be the same, and a new expression of the unit step function u(t) is given by

$$u(t) \;=\; \int_{-\infty}^{t} \delta(\tau)\, d\tau \qquad (4.21a)$$

, then
$$\delta(t) \;=\; \frac{d}{dt}\, u(t) \qquad (4.21b)$$

, hence the unit step function u(t) is the integration of the Dirac delta function δ(t), and the delta function δ(t) is the differentiation of the unit step function u(t), Also, the differentiation of the negation step function u(−t) is given by

$$\frac{d}{dt}\, [u(-t)] \;=\; -\,\delta(t)$$

Numerical examples

i. $\displaystyle\int_{-\infty}^{t+5} \delta(\tau)\, d\tau \;=\; u(t+5)$ ii. $\displaystyle\int_{-\infty}^{5} \delta(\tau)\, d\tau \;=\; u(5) \;=\; 1$

iii. $\displaystyle\frac{d}{dt}\, [e^{-t} u(t)] \;=\; \delta(t) - e^{-t}\, u(t)$

Problem 4.12

Find the inverse Fourier transform of the following spectrums

i. $X(f) = \dfrac{1 + jf}{1 + j2\pi f}$ ii. $X(f) = \dfrac{1}{f\,(1 + jf)}$

Solution

i. The spectrum X(f) can be expressed by

$$X(f) = \frac{1}{1 + j2\pi f} + \frac{jf}{1 + j2\pi f}$$

, due to the differentiation property, Eq.(3.25), the inverse Fourier transform x(t) yields

$$x(t) \;=\; e^{-t}\, u(t) + \frac{1}{2\pi}\, \frac{d}{dt}\, [\,e^{-t} u(t)\,]$$

, or equivalently

$$x(t) \;=\; e^{-t}\, u(t) + \frac{1}{2\pi}\, [\,\delta(t) - e^{-t}\, u(t)\,]$$

ii. Making use of the partial fraction technique, the spectrum X(f) can be expressed by

$$X(f) = \frac{1}{f} - \frac{j}{1 + jf}$$

, the inverse Fourier transform x(t) yields

$$x(t) = j\pi \, sgn(t) - j2\pi \, e^{-2\pi t} \, u(t)$$

Problem 4.13
Prove the following Fourier transform pairs, using the delta function

i. $A \, rect(\frac{t}{T}) \rightleftharpoons AT \, sinc(fT)$

ii. $A \, tri(\frac{t}{T}) \rightleftharpoons AT \, sinc^2(fT)$

Solution

i. Differentiating the rectangular pulse $x(t) = A \, rect(t/T)$, Fig.4.17, x`(t) is expressed by

$$x`(t) = A \, \delta(t + \tfrac{1}{2} T) - A \, \delta(t - \tfrac{1}{2} T)$$

, taking the Fourier transform of x`(t), X`(f) is

$$X`(f) = A \, e^{j\pi fT} - A \, e^{-j\pi fT} \qquad (4.22)$$

, due to the integration property, Eq.(3.36), and check the boundary condition X`(0) of Eq.(4.22) which equals zero, X(f) yields

$$X(f) = \frac{1}{j2\pi f} [A \, e^{j\pi fT} - A \, e^{-j\pi fT}]$$

, express the exponential terms by sine function , the Fourier transform $X(f) = AT \, sinc(fT)$.

ii. Double differentiating the triangle pulse $x(t) = A \, tri(t/T)$, Fig.4.18, x``(t) is expressed by

$$x``(t) = \frac{A}{T} \, \delta(t + T) - \frac{2A}{T} \, \delta(t) + \frac{A}{T} \, \delta(t - T)$$

, the Fourier transform X``(f) of x``(t) will be

$$X``(f) = \frac{A}{T} \, e^{j2\pi f T} - \frac{2A}{T} + \frac{A}{T} \, e^{-j2\pi fT} \qquad (4.23)$$

, due to the integration property, Eq.(3.36), and check the boundary condition X``(0) of Eq.(4.23) which equals zero, X`(f) yields

$$X`(f) = \frac{A}{(j2\pi f)T} [e^{j2\pi fT} - 2 + e^{-j2\pi fT}] \qquad (4.24)$$

, also for X(f), check the boundary condition X`(0) of Eq.(4.24) which equals zero, X(f) is given by

$$X(f) = \frac{A}{(j2\pi f)^2 T} [e^{j2\pi fT} - 2 + e^{-j2\pi fT}]$$

, express the exponential terms by cosine function, making use of the formula "$2 \, sin^2(x) = 1 - cos(2x)$", and in terms of sinc function, the Fourier transform X(f) yields

$$X(f) = AT \, sinc^2(fT)$$

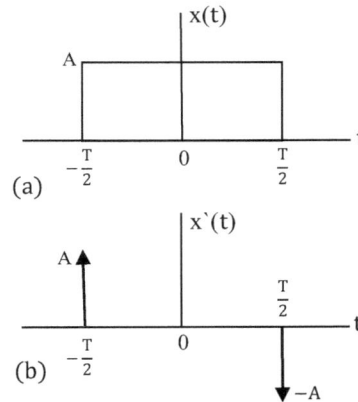

Fig.4.17. x(t) and its derivative x`(t).

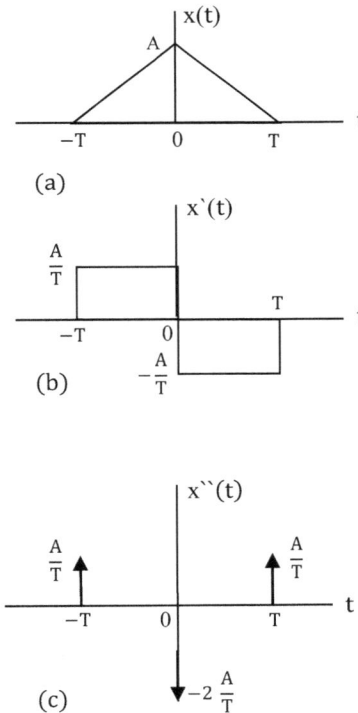

Fig.4.18. x(t) and its double derivatives x`(t) and x``(t).

Problem 4.14

Find the Fourier transform of the function x(t), Fig.4.19a, using the delta function.

Solution

Double differentiating the signal x(t), Fig.4.19a, x``(t) is expressed by

$$x``(t) = 3\,\delta(t+2) - 6\,\delta(t+1) + 6\,\delta(t-1) - 3\,\delta(t-2)$$

, taking the Fourier transform of x``(t), X``(f) will be

$$X``(f) = 3\,e^{j2\pi f2} - 6\,e^{j2\pi f} + 6\,e^{-j2\pi f} - 3\,e^{-j2\pi f2} \quad (4.25)$$

, due to integration property, Eq.(3.36), and check the boundary condition X``(0) of Eq.(4.25) which equals zero, X`(f) yields

$$X`(f) = \frac{3}{j2\pi f}\,[e^{j4\pi f} - 2\,e^{j2\pi f} + 2\,e^{-j2\pi f} - e^{-j4\pi f}] \quad (4.26)$$

, also for X(f), check the boundary condition X`(0) of Eq.(4.26) which equals zero, X(f) is given by

$$X(f) = \frac{3}{(j2\pi f)^2}\,[\,e^{j4\pi f} - 2\,e^{j2\pi f} + 2\,e^{-j2\pi f} - e^{-j4\pi f}\,]$$

, express the exponential terms by sine functions, and using the formulas "sin(2x) = 2 sin(x) cos(x), and "2 sin²(x) = 1 − cos(2x)", the Fourier transform is given by

$$X(f) = j\,6\,\text{sinc}^2(f)\,\sin(2\pi f)$$

, the Fourier transform X(f) is imaginary value because x(t) is odd function.

Problem 4.15

Find the Fourier transform of the energy signal x(t), Fig.4.20a, using the delta function.

Solution

Differentiating the signal x(t), Fig.4.20a, x`(t) is

$$x`(t) = 2\,\delta(t+1) + 3\,\text{rect}(t+\tfrac{1}{2}) - 3\,\text{rect}(t-\tfrac{1}{2}) - 2\,\delta(t-1)$$

, taking the Fourier transform of x`(t), X`(f) will be

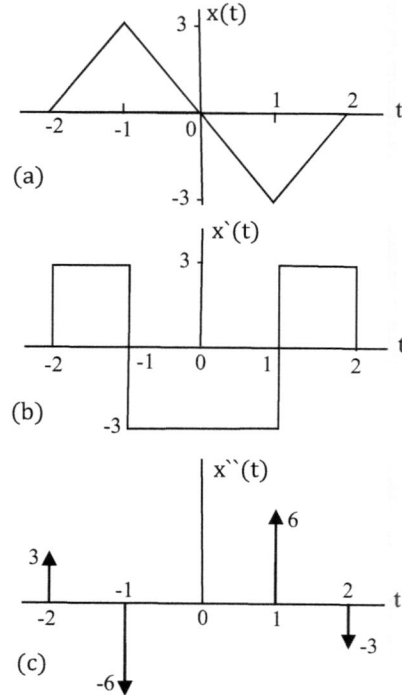

(a)

(b)

(c)

Fig.4.19. x(t) and its double derivatives x`(t) and x``(t).

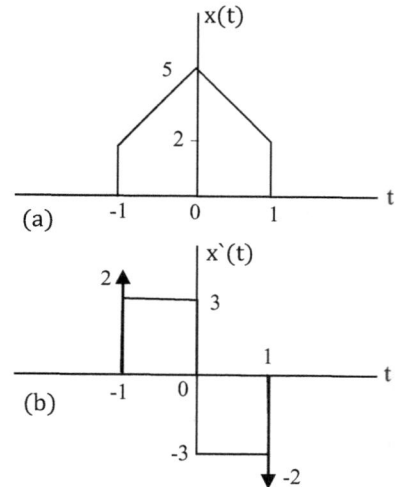

(a)

(b)

Fig.4.20. x(t) and its derivative x`(t).

$$X`(f) = 2\,e^{j2\pi f} + 3\,\text{sinc}(f)\,e^{j\pi f} - 3\,\text{sinc}(f)\,e^{-j\pi f} - 2\,e^{-j2\pi f} \quad (4.27)$$

, due to the integration property, Eq.(3.36), and check the boundary condition X`(0) of Eq.(4.27) which equals zero, X(f) yields

$$X(f) = \frac{1}{j2\pi f}\,[\,2\,e^{j2\pi f} + 3\,\text{sinc}(f)\,e^{j\pi f} - 3\,\text{sinc}(f)\,e^{-j\pi f} - 2\,e^{-j2\pi f}\,] \quad (4.28)$$

, express the exponential terms by sine functions, the Fourier transform is given by

$$X(f) = 3\,\text{sinc}^2(f) + 4\,\text{sinc}(2f)$$

Problem 4.16

Find the Fourier transforms of the functions $x_1(t)$ and $x_2(t)$, Fig.4.21a , using the delta function, and then Find the Fourier transform of the function $x(t)$, where $x(t) = x_1(t) + x_2(t)$

Solution

$x_1`(t)$ and $x_2`(t)$ are the differentiating functions of $x_1(t)$ and $x_2(t)$, Fig.4.21b , and are expressed by

$$x_1`(t) = 3 \, \text{rect}(t + \tfrac{1}{2}) - 3 \, \delta(t)$$

, and

$$x_2`(t) = 3 \, \delta(t) - 3 \, \text{rect}(t - \tfrac{1}{2})$$

, their Fourier transforms are given by

$$X_1`(f) = 3 \, \text{sinc}(f) \, e^{j\pi f} - 3 \qquad (4.29)$$

, and

$$X_2`(f) = 3 - 3 \, \text{sinc}(f) \, e^{-j\pi f} \qquad (4.30)$$

, due to the integration property , Eq.(3.36), and checking the boundary conditions $X_1`(0)$ and $X_2`(0)$ of Equations (4.29) and (4.30) which both equal zero $X_1(f)$ and $X_2(f)$ yield

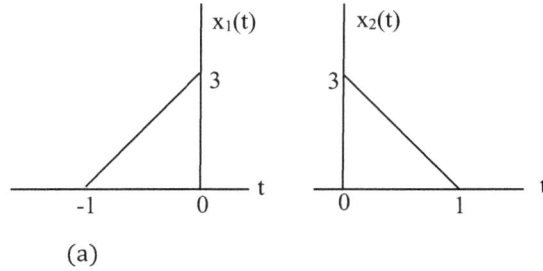

(a)

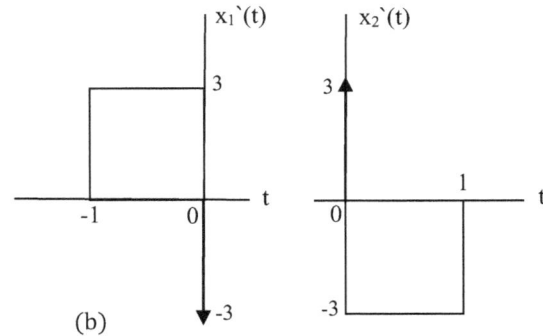

(b)

Fig.4.21. $x_1(t)$ and $x_2(t)$ and its derivatives $x_1`(t)$ and $x_2`(t)$.

$$X_1(f) = \frac{1}{j2\pi f} [\, 3 \, \text{sinc}(f) \, e^{j\pi f} - 3 \,]$$

, and

$$X_2(f) = \frac{1}{j2\pi f} [\, 3 - 3 \, \text{sinc}(f) \, e^{-j\pi f} \,]$$

, due to the linearity property, the Fourier transform of the function $x(t) = x_1(t) + x_2(t)$, is $X(f) = X_1(f) + X_2(f)$, and express the exponential terms by sine function, $X(f)$ yields

$$X(f) = 3 \, \text{sinc}^2(f)$$

Problem 4.17

Prove the following convolution operation, using the delta function

$$\cos(2t + \theta) \otimes \frac{1}{\pi t} = \sin(2t + \theta)$$

Solution

Due to the convolution property, the Fourier transform pair is given by

$$\cos(2t + \theta) \otimes \frac{1}{\pi t} \rightleftharpoons \frac{1}{2} \left[\delta(f - \frac{1}{\pi}) + \delta(f + \frac{1}{\pi}) \right] e^{j\pi f\theta} \, [-j \, \text{sgn}(f) \,]$$

, taking the inverse Fourier transform, the Fourier transform pair is given by

$$\sin(2t + \theta) \rightleftharpoons \frac{1}{2j} \left[\delta\left(f - \frac{1}{\pi}\right) - \delta(f + \frac{1}{\pi}) \right] e^{j\pi f\theta}$$

, then

$$\cos(2t + \theta) \otimes \frac{1}{\pi t} = \sin(2t + \theta)$$

Problem 4.18

Evaluate the following integral

$$x(t) = \int_{t+2}^{t+6} \delta(\tau)\, d\tau$$

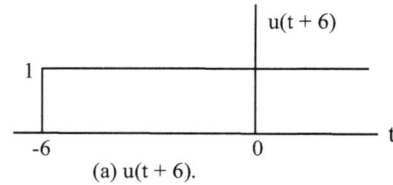
(a) u(t + 6).

Solution

The function x(t) can be expressed by

$$x(t) = \int_{t+2}^{-\infty} \delta(\tau)\, d\tau + \int_{-\infty}^{t+6} \delta(\tau)\, d\tau$$

, or equivalently

$$x(t) = -\int_{-\infty}^{t+2} \delta(\tau)\, d\tau + \int_{-\infty}^{t+6} \delta(\tau)\, d\tau$$

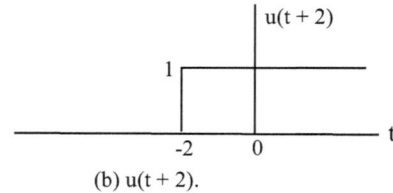
(b) u(t + 2).

, since the unit step function u(t) is the integration of the delta function $\delta(t)$, Eq.(4.21a), then x(t), Fig.4.22, yields

$$x(t) = -u(t+2) + u(t+6)$$

$$= rect\left[\frac{t+4}{4}\right]$$

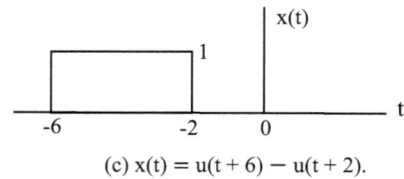
(c) x(t) = u(t + 6) − u(t + 2).

Fig.4.22. a) u(t + 6), b) u(t + 2),
c) x(t) = u(t + 6) − u(t + 2).

Problem 4.19

Find the Fourier transform of the following function x(t), using the delta function

$$x(t) = \frac{d}{dt}\left[e^{-t}\, u(t) \right]$$

Solution

Differentiate the function $e^{-t}\, u(t)$, x(t) yields

$$x(t) = \delta(t) - e^{-t}\, u(t)$$

, taking the Fourier transform, yields

$$X(f) = 1 - \frac{1}{1 + j2\pi f} = \frac{j2\pi f}{1 + j2\pi f}$$

, this Fourier transform X(f) can also be obtained, based on the Fourier transform pair, Eq.(3.9), and using the differentiation property, where the differentiation of a function x(t) has the effect of multiplying its Fourier transform X(f) by (j2πf).

Problem 4.20

Find the Fourier transform of the following energy signal, using the delta function

$$x(t) = \frac{t}{1 + jt}$$

Solution

Based on the following Fourier transform pair

$$\frac{1}{1 + jt} \quad \rightleftharpoons \quad 2\pi\, e^{2\pi f}\, u(-f)$$

, and due to the differentiation property in the frequency domain, Eq.(3.27a), the Fourier transform pair yields

$$\frac{t}{1 + jt} \quad \rightleftharpoons \quad \frac{1}{(-j2\pi)}\, \frac{d}{df}\left[2\pi\, e^{2\pi f}\, u(-f) \right]$$

, then

$$X(f) = j\left[2\pi\, e^{2\pi f}\, u(-f) - \delta(f) \right]$$

, the Fourier transform X(f) is imaginary value because x(t) is odd function.

Problem 4.21
Find the Fourier transform of the following
energy signal, using the delta function

$$x(t) = t \, \text{rect}(t)$$

Solution
Differentiating the signal x(t), Fig.4.23a, x`(t)
, Fig.4.23b, is expressed by

$$x`(t) = -\delta(t+1) + \text{rect}(\frac{t}{2}) - \delta(t-1)$$

, taking the Fourier transform of x`(t), X`(f) is

$$X`(f) = -e^{j2\pi f} + 2\,\text{sinc}(2f) - e^{-j2\pi f} \qquad (4.31)$$

, due to integration property, Eq.(3.36), and
check the boundary condition X`(0) of
Eq.(4.31) which equals zero, X(f) yields

$$X(f) = \frac{1}{j2\pi f}[-e^{j2\pi f} + 2\,\text{sinc}(2f) - e^{-j2\pi f}]$$

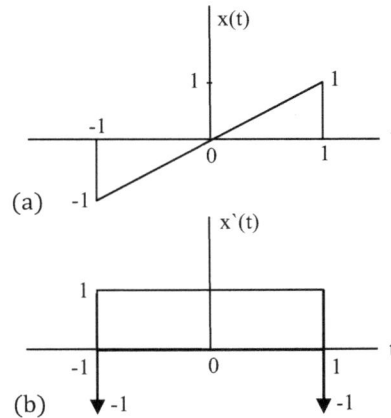

Fig.4.23. x(t) = t rect(t) and its derivative x`(t).

, express the exponential terms by cosine function and in terms of sine function, X(f) yields

$$X(f) = j\,\frac{2\pi f \cos(2\pi f) - \sin(2\pi f)}{2\,\pi^2\,f^2} \qquad (3.29)$$

, which is the same spectrum X(f) of problem 3.14, using the differentiation property.

4.7. Fourier Transform of Error Probability Function erf(t)
The integration property. Eq.(3.36), facilitates the definition and the evaluation of the Fourier transform
of the error probability function erf(t), Fig.4.24, which plays an important part in the probability
theory.

Consider the Gaussian pulse $x(t) = e^{-\pi t^2}$, Eq.(3.7) yields

$$\int_{-\infty}^{t} e^{-\pi\tau^2}d\tau \quad \rightleftharpoons \quad \frac{e^{-\pi f^2}}{j2\pi f} + \frac{\delta(f)}{2} \qquad (4.32)$$

, let $x^2 = 2\pi\tau^2$, then $x = \sqrt{2\pi}\,\tau$, and $dx = \sqrt{2\pi}\,d\tau$, as $\tau \to -\infty$, then $x \to -\infty$, and as $\tau \to t$ then $x \to \sqrt{2\pi}\,t$,
Eq.(4.32) yields

$$\frac{1}{\sqrt{2\pi}}\int_{-\infty}^{\sqrt{2\pi}\,t} e^{-x^2/2}\,dx \quad \rightleftharpoons \quad \frac{e^{-\pi f^2}}{j2\pi f} + \frac{\delta(f)}{2} \qquad (4.33)$$

, the time domain of Eq.(4.33) is defined by the error probability function erf($\sqrt{2\pi}$ t), and its
Fourier transform operation is given by

$$F\left[\text{erf}(\sqrt{2\pi}\,t)\right] = \frac{e^{-\pi f^2}}{j2\pi f} + \frac{\delta(f)}{2} \qquad (4.34)$$

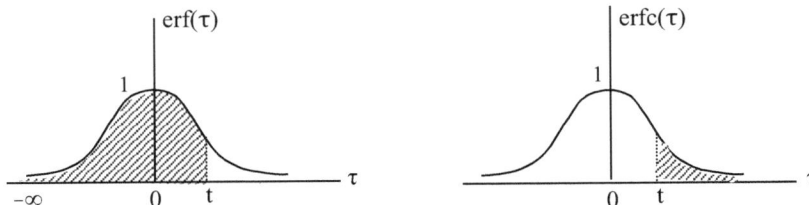

Fig.4.24. a) Error probability function erf(t), b) Error probability complementary function erfc(t).

, since the error probability function erf(t) and the error complementary function erfc(t) are given by

$$erf(t) = \frac{1}{\sqrt{2\pi}} \int_{-\infty}^{t} e^{-x^2/2} \, dx \qquad (4.35)$$

, and

$$erfc(t) = \frac{1}{\sqrt{2\pi}} \int_{t}^{\infty} e^{-x^2/2} \, dx \qquad (4.36)$$

, then making use of the scaling property, the Fourier transform of the error probability function erf(t), Eq.(4.34), yields

$$F[\,erf(t)] = \left[\frac{e^{-2(\pi f)^2}}{j2\pi f} + \frac{\sqrt{\pi}}{2} \, \delta(f) \right] \qquad (4.37)$$

, also since the error probability complementary function "erfc(t) = 1 − erf(t)", then the Fourier transform of the error probability complementary function erfc(t), using the linearity property, is given by

$$F[\,erfc(t)] = \left[- \frac{e^{-2(\pi f)^2}}{j2\pi f} + (1 - \frac{\sqrt{\pi}}{2}) \, \delta(f) \right]$$

, the integral erf(t), Eq.(4.35), cannot be evaluated in a closed form. It is however, extensively tabulated in standard tables (probability integral).

Problems

1. Evaluate the convolution function $x_{12}(t) = x_1(t) \otimes x_2(t)$, where $x_1(t)$ and $x_2(t)$ are shown in Fig.4.25.

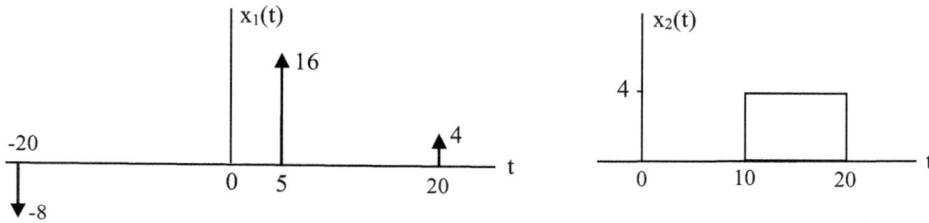

Fig.4.25. Two energy signals.

2. Evaluate the convolution function $x_{12}(t) = x_1(t) \otimes x_2(t)$, where $x_1(t)$ and $x_2(t)$ are shown in Fig.4.26.

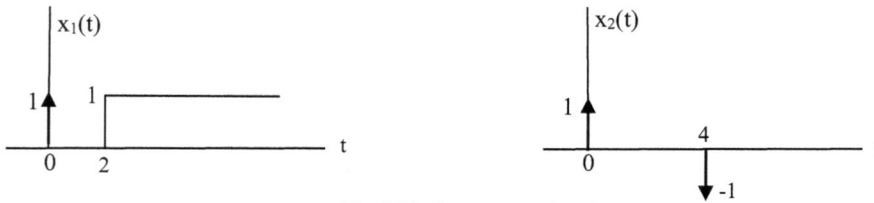

Fig.4.26. Two energy signals.

3. Evaluate the convolution function $x_{12}(t) = x_1(t) \otimes x_2(t)$, where $x_1(t)$ and $x_2(t)$ are shown in Fig.4.27.

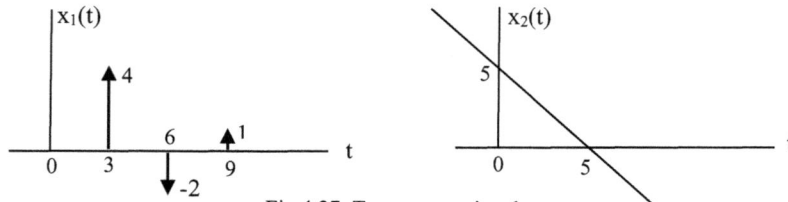

Fig.4.27. Two energy signals.

4. Evaluate the following convolution functions, using the frequency domain:

  i. $x(t) = \text{sgn}(t) \otimes \text{rect}(t/6)$ ii. $x(t) = \text{sgn}(t) \otimes \sin(3t)$

  iii. $x(t) = e^{-5t} u(t) \otimes \cos(5t)$ iv. $x(t) = u(t) \otimes e^{-(t-2)} u(t)$

  v. $x(t) = (1/\pi t) \otimes \cos(2t)$ vi. $x(t) = (1/\pi t) \otimes \cos(4t + \theta)$

  vii. $x(t) = u(t-1) \otimes \text{rect}(t/6)$ viii. $x(t) = u(-t) \otimes [\delta(t) - \delta(t-2)]$

  ix. $x(t) = \delta(2t-1) \otimes u(t+2)$ x. $x(t) = \text{sgn}(t) \otimes e^{-2t} u(t)$

  xi. $x(t) = \delta(t-1) \otimes [1/(1-jt)]$ xii. $x(t) = \delta(t-4) \otimes 1/[1 - j(t-5)]$

5. Evaluate the following convolution functions, using the frequency domain:

  i. $X(f) = \text{sgn}(f) \otimes \text{sgn}(f) e^{-j2\pi f}$ ii. $X(f) = e^{-j\pi f} \otimes \text{rect}(f+1) e^{-j2\pi f}$

6. Find the Fourier transform of the following energy signals, using the frequency domain:

  i. $x(t) = 1/(t+1)$ ii. $x(t) = t/(t-2)$

  iii. $x(t) = 1/(t^2 - 2t)$ iv. $x(t) = 1/(1+jt)$

  v. $x(t) = t/(1+jt)$ vi. $x(t) = t/(1+t^2)$

  vii. $x(t) = 1/(f+t)$ viii. $x(t) = t^2$

  ix. $x(t) = (1-jt)/(1+jt)$ x. $x(t) = t^2 - 2t$

  xi. $x(t) = \dfrac{d}{dt}[\text{sgn}(t+6)]$ xii. $x(t) = \sin(2\pi 10^3 t) \sin(2\pi 10^6 t)$

7. Find the Fourier transform of the energy signals, Fig.4.28a,b,c,d,e,f, using the delta function.

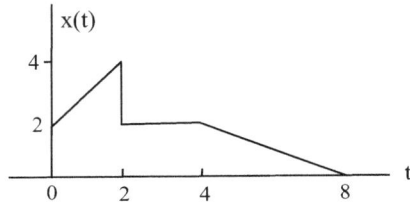
(a)

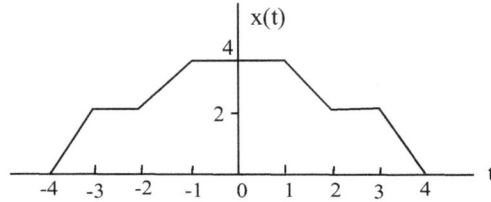
(b)

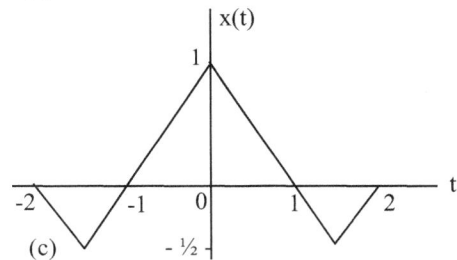
(c)

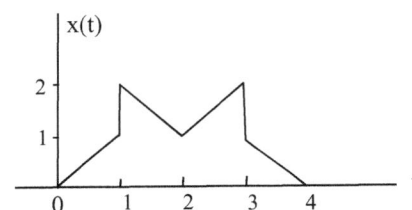
(d)

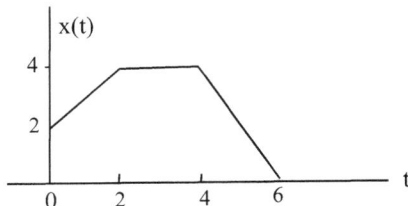
(e)

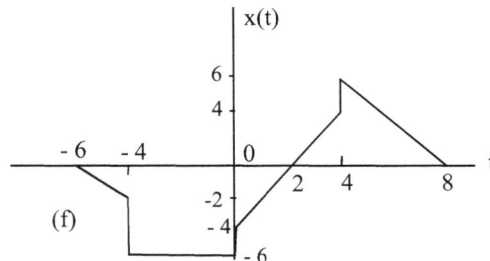
(f)

Fig.4.28. Energy signals.

8. Find the Fourier transform of the following integrals, using the Fourier transform properties:

  i. $x(t) = \displaystyle\int_{0}^{t} e^{-2\tau}\, d\tau$ ii. $x(t) = \displaystyle\int_{-\infty}^{5-t} \delta(\tau)\, d\tau$

  iii. $x(t) = \displaystyle\int_{-\infty}^{t} e^{-\pi(\tau-3)^2}\, d\tau$ iv. $x(t) = \displaystyle\int_{-\infty}^{t} \delta(\tau-3)\, d\tau$

9. A current source $i(t) = u(t)$, is applied at the shown inductive phase shift circuit, Fig.4.29. Find the current $i_o(t)$, passing in the resistor, using the Fourier transform.

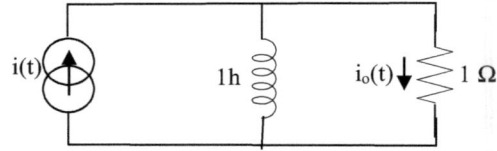

Fig.4.29. Inductive circuit.

10. A voltage source $v(t) = sgn(t)$, is applied at the shown capacitive phase shift circuit, Fig.4.30. Find the voltage across the capacitance, using the Fourier transform.

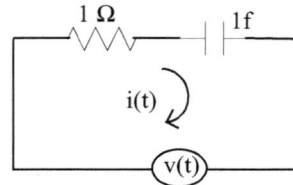

Fig.4.30. Capacitive circuit.

11. An AM DSB-SC band-pass signal $s(t) = m(t) \cos(2\pi 10^6 t)$. Find the Fourier transform $S(f)$ for the following baseband signals:

    i. $m(t) = 2 \sin(2\pi 10^3 t)$
    ii. $m(t) = 4 \sin^2(2\pi 10^3 t)$
    iii. $m(t) = 2 \sin(2000\pi t) + 2 \cos(4000\pi t)$

12. Prove that the Fourier transform of the signum function $sgn(t)$ is $1/(j\pi f)$, using the shown doublet pulse $x(t)$, Fig.4.31.

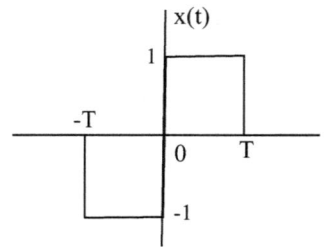

Fig.4.31. Doublet pulse.

13. Find the inverse Fourier transform of the following spectrums

    i. $X(f) = 1/(f + t)$ ii. $X(f) = 1/(jf + t)$
    iii. $X(f) = 1/(f + jt)$ iv. $X(f) = f^2 - 2f$
    v. $X(f) = 1/(f^2 - 2f)$ vi. $X(f) = 1/(f^2 + 5f + 6)$
    vii. $X(f) = f/(f + 2)$

14. Find the Fourier transforms of the real and imaginary parts of the following complex time functions, using Equations (3.67)

    i. $x(t) = 1 + j\dfrac{1}{t}$ ii. $x(t) = e^{jat}$

15. Sketch the following energy signal $x(t)$ and Find its Fourier transform, using the delta function

$$x(t) = 8 \, tri\left[\frac{t + 2}{2}\right] - 4 \, tri\left[\frac{t - 2}{2}\right]$$

16. A current source $i(t) = 50 \cos(t)$, is applied at the shown capacitive phase shift circuit, Fig.4.32. Find the current $i_o(t)$, passing in the condenser, using the Fourier transform.

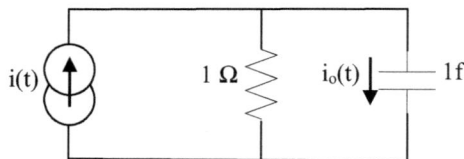

Fig.4.32. Capacitive circuit.

CHAPTER 5

Fourier Transform and Power Spectra

Abstract: Fourier series analysis is performed to obtain the discrete spectrum representation of a given periodic signal (power signal) $x_p(t)$ which has finite periodic time T_o, finite average power and infinite energy, to describe its frequency components content (n/T_o), where $n = 0, 1, 2, 3, 4, \ldots$, by either using the real coefficients method to obtain the real coefficients a_n and b_n, Equations (2.2) and (2.3), to construct $x_p(t)$, Eq.(2.1), or by using the complex coefficient method to obtain the complex coefficient C_n, Eq.(2.13) to construct the real value of $x_p(t)$, Eq.(2.12), (chapter II). While Fourier transform analysis is performed to obtain the continuous spectrum representation of a given unperiodic signal (energy signal) $x(t)$ which has infinite periodic time T_o, finite energy, and zero average power (chapter III). Fourier transform is also used in a limiting sense, to evaluate the frequency content of the periodic signal $x_p(t)$. Moreover, there are some special periodic functions cannot be solved using Fourier series analysis such as the periodic Dirac delta function, in this case, Fourier transform is the only way to evaluate its frequency content.

Keywords: Spectral Analysis; Fourier transform; Power spectra; Periodic signals.

Hence consider an energy signal $x(t)$, its Fourier transform is $X(f)$. If $x(t)$ is repeated every T_o seconds, then the periodic signal $x_p(t)$ is expressed by

$$x_p(t) = \sum_{m=-\infty}^{\infty} x(t - mT_o) \qquad \text{for all } t, \quad m = 0,1,2,3,\ldots \qquad (1.3)$$

, the Fourier transform pair of the periodic signal $x_p(t)$, is given by

$$\sum_{m=-\infty}^{\infty} x(t - mT_o) \quad \rightleftharpoons \quad \frac{1}{T_o} \sum_{m=-\infty}^{\infty} X(\frac{n}{T_o}) \, \delta(f - \frac{n}{T_o}) \qquad (5.1)$$

, where $X(n/T_o)$ is the Fourier transform $X(f)$ of the energy signal $x(t)$ at the frequencies $f = n/T_o$.

Proof:

Consider a periodic signal $x_p(t)$ of period T_o, $x_p(t)$ can be represented, in terms of the complex exponential Fourier series, by

$$x_p(t) = \sum_{n=-\infty}^{\infty} C_n \, e^{j2\pi \frac{n}{T_o} t} \qquad (2.12)$$

, where C_n is the complex exponential coefficient of the Fourier series expansion, is given by

$$C_n = \frac{1}{T_o} \int_{-T_0/2}^{T_0/2} x_p(t) \, e^{-j2\pi \frac{n}{T_o} t} \, dt \qquad (2.13)$$

, or $\qquad C_n = \frac{1}{T_o} X(\frac{n}{T_o}) \qquad (2.15)$

, then the periodic signal $x_p(t)$, Eq.(2.12) yields

$$x_p(t) = \frac{1}{T_o} \sum_{n=-\infty}^{\infty} X(\frac{n}{T_o}) \, e^{j2\pi \frac{n}{T_o} t} \qquad (2.16)$$

, where $X(n/T_o)$ is given by

$$X(\frac{n}{T_o}) = \int_{-T_0/2}^{T_0/2} x_p(t) \, e^{-j2\pi \frac{n}{T_o} t} \, dt \qquad (2.14)$$

, the discrete spectrum $X(n/T_o)$, Eq.(2.14) is considered the Fourier transform of the one cycle energy signal $x(t)$ from $-T_o/2$ to $T_o/2$, which generates the periodic signal $x_p(t)$, evaluated at the frequencies $f = n/T_o$, According to the properties of the Dirac delta function, the Fourier transform pair of the periodic function $x_p(t)$, Eq.(2.16), is given by

$$\frac{1}{T_o} \sum_{n=-\infty}^{\infty} X(\frac{n}{T_o}) \, e^{j2\pi \frac{n}{T_o} t} \quad \rightleftharpoons \quad \frac{1}{T_o} \sum_{n=-\infty}^{\infty} X(\frac{n}{T_o}) \, \delta(f - \frac{n}{T_o}) \qquad (5.2)$$

, the periodic signal $x_p(t)$ is the time domain of Eq.(5.2), then Eq.(1.3) yields

$$\sum_{m=-\infty}^{\infty} x(t - mT_0) = \frac{1}{T_0} \sum_{m=-\infty}^{\infty} X(\frac{n}{T_0}) e^{j2\pi \frac{n}{T_0} t} \tag{5.3}$$

, where $x(t)$ is the unperiodic signal and defines one cycle of the periodic signal $x_p(t)$, so $x(t)$ is considered the generating function of $x_p(t)$ in the periodic time T_0, then using Equations (5.2) and (5.3), the discrete spectrum of $x_p(t)$ is given from the following Fourier transform pair

$$\sum_{m=-\infty}^{\infty} x(t - mT_0) \rightleftharpoons \frac{1}{T_0} \sum_{n=-\infty}^{\infty} X(\frac{n}{T_0}) \delta(f - \frac{n}{T_0}) \tag{5.1}$$

, Eq.(5.1) is defined by Poisson's sum formula which shows that the Fourier transform of the periodic signal $x_p(t)$ of periodic time T_0, consists of delta functions occurring at the frequencies n/T_0, and each delta function is weighted by a factor equals $X(n/T_0)$ divided by T_0, where $X(n/T_0)$ is the Fourier transform of the generating function $x(t)$ at the discrete frequencies $f = n/T_0$. The Fourier transform pair, Eq.(5.1) can be written in the form

$$\sum_{m=-\infty}^{\infty} x(t - mT_0) \rightleftharpoons \frac{X(0) \delta(f)}{T_0} + \frac{1}{T_0} \sum_{n=1}^{\infty} X(\frac{n}{T_0}) \delta(f - \frac{n}{T_0})$$
$$+ \frac{1}{T_0} \sum_{n=-\infty}^{-1} X(\frac{n}{T_0}) \delta(f + \frac{n}{T_0}) \tag{5.4}$$

, where $X(0)$ is the value of $X(n/T_0)$ at $n = 0$. The inverse Fourier transform of Eq.(5.4) is the time domain of Eq.(5.1), will be in the form

$$x_p(t) = \sum_{n=-\infty}^{\infty} x(t - mT_0) = a_0 + 2 \sum_{n=1}^{\infty} a_n \cos(2\pi \frac{n}{T_0} t) + 2 \sum_{n=1}^{\infty} b_n \sin(2\pi \frac{n}{T_0} t) \tag{5.5}$$

, where the average value a_0 is given by

$$a_0 = \frac{1}{T_0} X(0)$$

, and $X(0)$, using Eq.(2.10), is given by

$$X(0) = \int_{-T_0/2}^{T_0/2} x_p(t) \, dt$$

, while a_n and b_n are determined from the strengths, signs, and the combination of the delta functions at the positive and negative discrete frequencies $\pm n/T_0$, where together are inversed Fourier transformed to the time domain to be cosine and sine functions with amplitudes equals $2 X(n/T_0)/T_0$ [1].

5.1. Spectral Analysis of Periodic Delta Function Using Fourier Transform

The periodic infinite sequence delta function, Fig.5.1, can be expressed by

$$x_p(t) = \sum_{m=-\infty}^{\infty} \delta(t - mT_0)$$

, where the generating function is $x(t) = \delta(t)$, and its Fourier transform is $X(f) = 1$, Eq.(4.1), then using the Fourier transform of the periodic signals, Eq.(5.1), the Fourier transform pair is given by

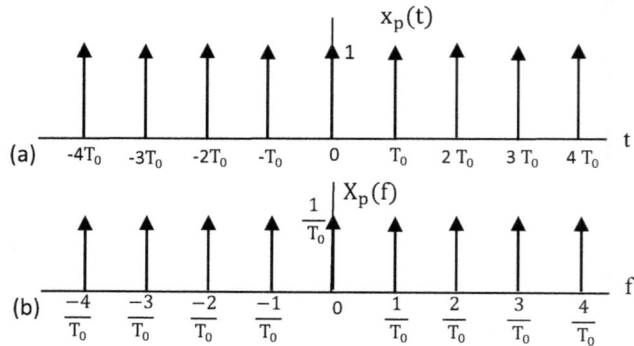

Fig.5.1. a) Periodic delta function,
b) Fourier transform of periodic delta function.

$$\sum_{m=-\infty}^{\infty} \delta(t - mT_0) \rightleftharpoons \frac{1}{T_0} \sum_{n=-\infty}^{\infty} \delta(f - \frac{n}{T_0}) \tag{5.6}$$

, then the Fourier transform of a periodic delta functions spaced by T_o seconds, is another set of periodic delta functions spaced by $1/T_o$ and weighted by the factor $1/T_o$. In other words, the sampling in one domain has the effect of the periodicity in the other domain. The Fourier transform pair, Eq.(5.6) can be written in the form

$$\sum_{m=-\infty}^{\infty} \delta(t - mT_o) \rightleftharpoons \frac{1}{T_o} \delta(f) + \frac{1}{T_o} \sum_{n=1}^{\infty} \delta(f - \frac{n}{T_o}) + \frac{1}{T_o} \sum_{n=-\infty}^{-1} \delta(f + \frac{n}{T_o}) \quad (5.7)$$

, the inverse Fourier transform of Eq.(5.7), is the time domain of Eq.(5.6), is given by

$$x_p(t) = \sum_{m=-\infty}^{\infty} \delta(t - mT_o) = \frac{1}{T_o} + \frac{2}{T_o} \sum_{n=1}^{\infty} \cos(2\pi \frac{n}{T_o} t) \quad (5.8)$$

, another way to obtain the spectral analysis of the periodic delta function, Eq.(5.8), by taking the inverse Fourier transform of Eq.(5.6), yields

$$[x_p(t)]_{real} = \sum_{m=-\infty}^{\infty} \delta(t - mT_o) = Re[\frac{1}{T_o} \sum_{m=-\infty}^{\infty} e^{j2\pi \frac{n}{T_o} t}] \quad (5.9)$$

, where Re[ ] is the real part of the time function $x_p(t)$, then

$$[x_p(t)]_{real} = \frac{1}{T_o} \sum_{m=-\infty}^{\infty} \cos(2\pi \frac{n}{T_o} t) = \frac{1}{T_o} + \frac{2}{T_o} \sum_{n=1}^{\infty} \cos(2\pi \frac{n}{T_o} t)$$

, which is the same spectral analysis, Eq.(5.8).

Problem 5.1
Find the spectral analysis of the periodic double polarity infinite sequence delta function , uniformly spaced, Fig.5.2, using the Fourier transform.

Solution
The double polarity infinite sequence delta function uniformly spaced, Fig.5.2, can be expressed by

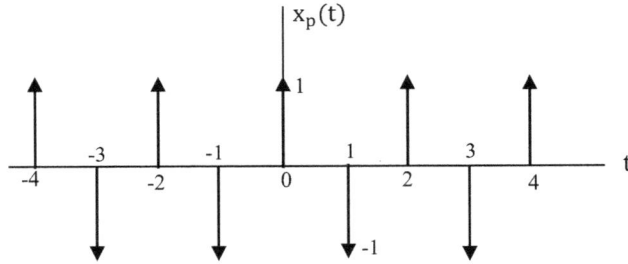

Fig.5.2. Periodic double polarity delta function.

$$x_p(t) = \sum_{m=-\infty}^{\infty} \delta(t - 2m) - \sum_{m=-\infty}^{\infty} \delta[(t - 1) - 2m]$$

, taking the Fourier transform of the periodic signal using Eq.(5.1), the Fourier transform pair will be

$$\sum_{m=-\infty}^{\infty} \delta(t - 2m) - \sum_{m=-\infty}^{\infty} \delta[(t - 1) - 2m] \rightleftharpoons \frac{1}{2} \sum_{n=-\infty}^{\infty} \delta(f - \frac{n}{2}) - \frac{1}{2} \sum_{n=-\infty}^{\infty} \delta(f - \frac{n}{2}) e^{-j\pi n}$$

, or equivalently

$$\sum_{m=-\infty}^{\infty} \delta(t - 2m) - \sum_{m=-\infty}^{\infty} \delta[(t - 1) - 2m] \rightleftharpoons \frac{1}{2} \sum_{n=1}^{\infty} \delta(f - \frac{n}{2}) + \frac{1}{2} \sum_{n=-\infty}^{-1} \delta(f + \frac{n}{2})$$

$$- \frac{1}{2} \sum_{n=1}^{\infty} \delta\left(f - \frac{n}{2}\right) e^{-j\pi n} - \frac{1}{2} \sum_{n=-\infty}^{-1} \delta(f + \frac{n}{2}) e^{j\pi n}$$

, and then taking the inverse Fourier transform yields

$$\sum_{m=-\infty}^{\infty} \delta(t - 2m) - \sum_{m=-\infty}^{\infty} \delta[(t - 1) - 2m] = \sum_{n=1}^{\infty} \cos(n\pi t) - \sum_{n=1}^{\infty} \cos(\pi n) \cos(n\pi t)$$

, then $x_p(t)$ is expressed by

$$x_p(t) = \sum_{n=1}^{\infty} [1 - (-1)^n] \cos(n\pi t)$$

5.2. Spectral Analysis of Periodic Rectangular Function Using Fourier Transform

The periodic rectangular signal, Fig.1.2b, can be expressed by

$$x_p(t) = \sum_{m=\infty}^{\infty} A \, rect\left[\frac{t - mT_o}{T}\right]$$

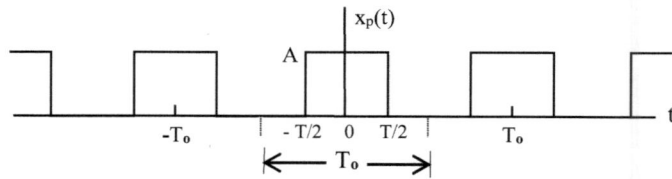
Fig.1.2b. Periodic rectangular signal.

, the generating function will be $x(t) = A \, rect(t/T)$, Fig.1.2a, its Fourier transform is given by

$$X(f) = AT \, sinc(fT) \qquad (3.4)$$

, then using the Fourier transform of the periodic signals, Eq.(5.1), the Fourier transform pair is given by

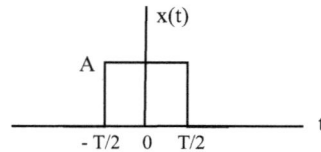
Fig.1.2a. Unperiodic rectangular signal.

$$\sum_{m=\infty}^{\infty} A \, rect\left[\frac{t - mT_o}{T}\right] \;\rightleftharpoons\; \frac{1}{T_o} \sum_{n=-\infty}^{\infty} AT \, sinc(\frac{nT}{T_o}) \, \delta(f - \frac{n}{T_o}) \qquad (5.10)$$

, the Fourier transform pair, Eq.(5.10), can be written in the form

$$\sum_{m=\infty}^{\infty} A \, rect\left[\frac{t - mT_o}{T}\right] \;\rightleftharpoons\; \frac{AT}{T_o}\delta(f) + \frac{AT}{T_o} \sum_{n=1}^{\infty} sinc(\frac{nT}{T_o}) \, \delta(f - \frac{n}{T_o})$$

$$+ \frac{AT}{T_o} \sum_{n=-\infty}^{-1} sinc(\frac{nT}{T_o}) \, \delta(f + \frac{n}{T_o}) \qquad (5.11)$$

, the inverse Fourier transform of Eq.(5.11) is the time domain of Eq.(5.10), is given by

$$\sum_{m=-\infty}^{\infty} A \, rect\left[\frac{t - mT_o}{T}\right] = \frac{AT}{T_o} + \sum_{n=1}^{\infty} \frac{2AT}{T_o} \, sinc(\frac{n}{T_o}T) \, cos(2\pi \frac{n}{T_o}t)$$

, or equivalently

$$x_p(t) = A \frac{T}{T_o} + \sum_{n=1}^{\infty} 2A \frac{T}{T_o} \, sinc(n\frac{T}{T_o}) \, cos(2\pi\frac{n}{T_o}t) \qquad (2.18)$$

, which is the same spectral analysis, problem 2.5, Eq.(2.18), using the complex exponential Fourier series method. The ratio T/T_o is defined by the duty cycle of the periodic signal $x_p(t)$ [1].

Problem 5.2

Find the spectral analysis of the shifted periodic rectangular signal, Fig.5.3, using the Fourier transform.

Solution

The shifted periodic rectangular signal, Fig.5.3, is expressed by

$$x_p(t) = \sum_{m=-\infty}^{\infty} A \, rect\left[\frac{t - t_o - mT_o}{T}\right]$$

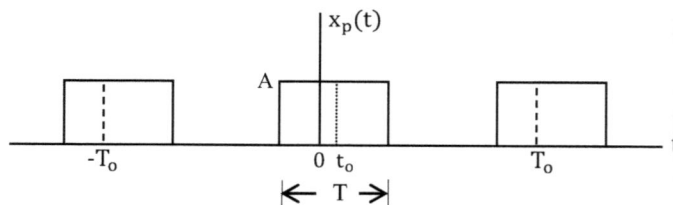
Fig.5.3. Shifted periodic rectangular signal.

, where t_o is the time shifting. The generating function will be $x(t) = A \, rect[(t - t_o)/T]$, its Fourier transform is $X(f) = AT \, sinc(fT) \, e^{-j2\pi f t_o}$, then using the Fourier transform of the periodic signals, Eq.(5.1), the Fourier transform pair will be

$$\sum_{m=\infty}^{\infty} A \, rect\left[\frac{t - t_o - mT_o}{T}\right] \;\rightleftharpoons\; \frac{1}{T_o} \sum_{n=\infty}^{\infty} AT \, sinc(n\frac{T}{T_o}) \, e^{-j2\pi \frac{n}{T_o} t_o} \, \delta(f - \frac{n}{T_o}) \qquad (5.12)$$

, the Fourier transform pair, Eq.(5.12), can be written in the form

$$\sum_{m=\infty}^{\infty} A \, rect\left[\frac{t - t_o - mT_o}{T}\right] \rightleftharpoons \frac{AT}{T_o} \, \delta(f) + \frac{AT}{T_o} \sum_{n=1}^{\infty} sinc\left(n\frac{T}{T_o}\right) e^{-j2\pi\frac{n}{T_o}t_o} \delta\left(f - \frac{n}{T_o}\right)$$

$$+ \frac{AT}{T_o} \sum_{n=-\infty}^{-1} sinc(n\frac{T}{T_o}) \, e^{-j2\pi\frac{n}{T_o}t_o} \, \delta(f + \frac{n}{T_o}) \quad (5.13)$$

, then taking the inverse Fourier transform, Eq.(5.13), yields

$$\sum_{m=-\infty}^{\infty} A \, rect\left[\frac{t - t_o - mT_o}{T}\right] = \frac{AT}{T_o} + \sum_{n=-\infty}^{\infty} \frac{2AT}{T_o} sinc(n\frac{T}{T_o}) \, e^{-j2\pi\frac{n}{T_o}t_o} \, cos(2\pi\frac{n}{T_o}t)$$

, or equivalently

$$x_p(t) = \frac{AT}{T_o} + \sum_{n=1}^{\infty} \frac{2AT}{T_o} sinc(n\frac{T}{T_o}) \, e^{-j2\pi\frac{n}{T_o}t_o} \, cos(2\pi\frac{n}{T_o}t)$$

Problem 5.3
Find the spectral analysis of the periodic rectangular signal , Fig.5.4a, using the Fourier transform.

Solution
The periodic rectangular signal, Fig.5.4b, is expressed by

$$x_p(t) + 4 = \sum_{m=-\infty}^{\infty} 2 \, rect\left[\frac{t - 4m}{2}\right]$$

, the generating function of the signal $[x_p(t) + 4]$ will be $x(t) = 2 \, rect(t/2)$, its Fourier transform is $X(f) = 4 \, sinc(2f)$, then using the Fourier transform of the periodic signals, Eq.(5.1), the Fourier transform pair will be

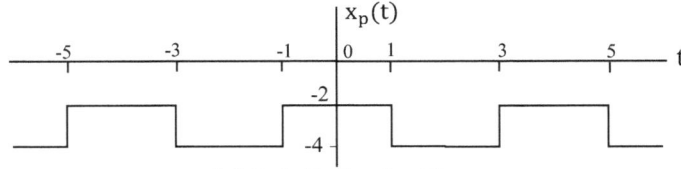

(a) Periodic signal $x_p(t)$.

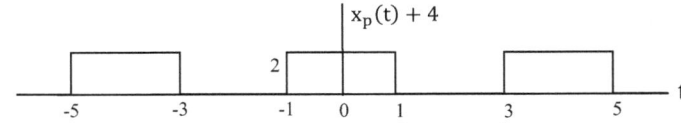

(b) Periodic signal $[x_p(t) + 4]$.

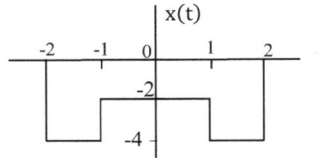

(c) Unperiodic signal $x(t)$.

Fig. 5.4. a) Periodic signal $x_p(t)$, b) Periodic signal $[x_p(t) + 4]$, c) Unperiodic signal $x(t)$.

$$\sum_{m=-\infty}^{\infty} 2 \, rect\left[\frac{t - 4m}{2}\right] \rightleftharpoons \frac{1}{4} \sum_{n=-\infty}^{\infty} 4 \, sinc(\frac{2n}{4}) \, \delta(f - \frac{n}{4})$$

, or equivalently

$$\sum_{m=-\infty}^{\infty} 2 \, rect\left[\frac{t - 4m}{2}\right] \rightleftharpoons \delta(f) + \sum_{n=1}^{\infty} sinc(\frac{n}{2}) \, \delta(f - \frac{n}{4}) + \sum_{n=-\infty}^{-1} sinc(\frac{n}{2}) \, \delta(f + \frac{n}{4}) \quad (5.14)$$

, then taking the inverse Fourier transform, Eq.(5.14) yields

$$1 + \sum_{n=1}^{\infty} 2 \, sinc(\frac{n}{2}) \, cos(2\pi\frac{n}{4}t) \rightleftharpoons \delta(f) + \sum_{n=1}^{\infty} sinc(\frac{n}{2}) \, \delta(f - \frac{n}{4}) + \sum_{n=-\infty}^{-1} sinc(\frac{n}{2}) \, \delta(f + \frac{n}{4})$$

, then the signal $[x_p(t) + 4]$ is given by

$$x_p(t) + 4 = 1 + \sum_{n=1}^{\infty} 2 \, sinc(\frac{n}{2}) \, cos(2\pi\frac{n}{4}t)$$

, or equivalently

$$x_p(t) = -3 + \sum_{n=1}^{\infty} 2 \, sinc(\frac{n}{2}) \, cos(2\pi\frac{n}{4}t) \quad (5.15)$$

, another way to obtain the spectral analysis $x_p(t)$, Fig.5.4a, consider its generating function $x(t)$, Fig.5.4c, and its Fourier transform, are given by

$$x(t) = -4 \; \text{rect}(\frac{t}{4}) + 2 \; \text{rect}(\frac{t}{2})$$

, and

$$X(f) = -16 \; \text{sinc}(4f) + 4 \; \text{sinc}(2f)$$

, then using the Fourier transform of the periodic signals, Eq.(5.1), the Fourier transform pair will be

$$\sum_{m=-\infty}^{\infty} \left\{ -4 \; \text{rect}\left[\frac{t-4m}{4}\right] + 2 \; \text{rect}\left[\frac{t-4m}{2}\right] \right\} \; \rightleftharpoons \; \frac{1}{4} \sum_{m=-\infty}^{\infty} \left[-16 \; \text{sinc}(n) + 4 \; \text{sinc}(\frac{n}{2}) \right] \delta(f - \frac{n}{4})$$

, in the frequency domain, at $n = 0$, the term will be "$-3 \; \delta(f)$", and then since $\text{sinc}(n)$ equals zero, $n = 1, 2, 3, \dots$, the Fourier transform pair yields

$$\sum_{m=-\infty}^{\infty} \left\{ -4 \; \text{rect}\left[\frac{t-4m}{4}\right] + 2 \; \text{rect}\left[\frac{t-4m}{2}\right] \right\} \; \rightleftharpoons \; -3 \; \delta(f) + \sum_{n=1}^{\infty} \text{sinc}(\frac{n}{2}) \; \delta(f - \frac{n}{4})$$

$$+ \sum_{n=-\infty}^{-1} \text{sinc}(\frac{n}{2}) \; \delta(f + \frac{n}{4}) \quad (5.16)$$

, taking the inverse Fourier transform for Eq.(5.16), the spectral analysis $x_p(t)$, Eq.(5.15) is obtained.

5.3. Spectral Analysis of Periodic Triangle Function Using Fourier Transform

The periodic triangle signal, Fig.1.3b, can be expressed by

$$x_p(t) = \sum_{m=\infty}^{\infty} A \; \text{tri}\left[\frac{t-mT_o}{T}\right]$$

, the generating function will be $x(t) = A \; \text{tri}(t/T)$, Fig.1.3a, its Fourier transform is given by

$$X(f) = AT \; \text{sinc}^2(fT)$$

, Eq.(3.5), then using the Fourier transform of the periodic signals, Eq.(5.1), the Fourier transform pair is given by

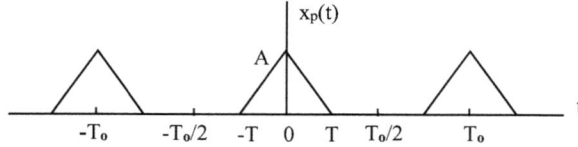

Fig.1.3b. Periodic triangle signal.

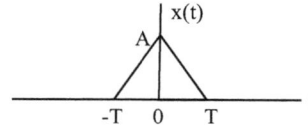

Fig.1.3a. Unperiodic triangle signal.

$$\sum_{m=-\infty}^{\infty} A \; \text{tri}\left[\frac{t-mT_o}{T}\right] \; \rightleftharpoons \; \frac{1}{T_o} \sum_{n=-\infty}^{\infty} AT \; \text{sinc}^2(\frac{nT}{T_o}) \; \delta(f - \frac{n}{T_o}) \quad (5.17)$$

, the Fourier transform pair, Eq.(5.17), can be written in the form

$$\sum_{m=-\infty}^{\infty} A \; \text{tri}\left[\frac{t-mT_o}{T}\right] \; \rightleftharpoons \; \frac{AT}{T_o} \delta(f) + \frac{AT}{T_o} \sum_{n=1}^{\infty} \text{sinc}^2(\frac{nT}{T_o}) \; \delta(f - \frac{n}{T_o})$$

$$+ \frac{AT}{T_o} \sum_{n=-\infty}^{-1} \text{sinc}^2(\frac{nT}{T_o}) \; \delta(f + \frac{n}{T_o}) \quad (5.18)$$

, the inverse Fourier transform of Eq.(5.18) is the time domain of Eq.(5.17), yields

$$\sum_{m=-\infty}^{\infty} A \; \text{tri}\left[\frac{t-mT_o}{T}\right] = \frac{AT}{T_o} + \sum_{n=1}^{\infty} \frac{2AT}{T_o} \; \text{sinc}^2(\frac{nT}{T_o}) \cos(2\pi \frac{n}{T_o} t)$$

, or equivalently

$$x_p(t) = A \frac{T}{T_o} + \sum_{n=1}^{\infty} 2A \frac{T}{T_o} \; \text{sinc}^2(\frac{nT}{T_o}) \cos(2\pi \frac{n}{T_o} t) \quad (2.19)$$

, which is the same spectral analysis, problem 2.7, Eq.(2.19), using the complex exponential Fourier series method [1].

Problem 5.4
Find the spectral analysis of the periodic sinusoidal signal $x_p(t)$, Fig.2.11, using the Fourier transform, $x_p(t)$ is given by

$$x_p(t) = 4\sin(\pi t) \qquad\qquad -\tfrac{1}{2} \le t \le \tfrac{1}{2}$$
$$= 0 \qquad\qquad \text{for the remainder of the period}$$

Solution

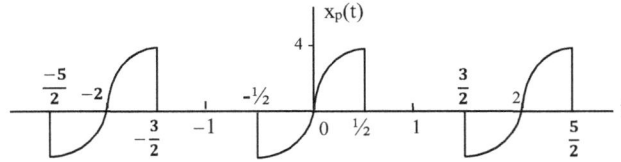

$$\text{Fig.2.11.} \quad x_p(t) = 4\sin(\pi t) \quad \text{for} \;\; -\tfrac{1}{2} \le t \le \tfrac{1}{2}$$
$$= 0 \qquad \text{for the remainder of the period.}$$

The periodic sinusoidal signal $x_p(t)$, Fig.2.11, can be expressed by

$$x_p(t) = \sum_{m=-\infty}^{\infty} 4\,\text{rect}(t - 2m)\,\sin[\pi(t - 2m)]$$

, the generating function $x(t)$ will be $x(t) = 4\,\text{rect}(t)\sin(\pi t)$, its Fourier transform is given by

$$X(f) = -2j\,[\delta(f - \tfrac{1}{2}) - \delta(f + \tfrac{1}{2})] \otimes \text{sinc}(f)$$

, or equivalently

$$X(f) = -2j\,[\text{sinc}(f - \tfrac{1}{2}) - \text{sinc}(f + \tfrac{1}{2})]$$

, then using the Fourier transform of the periodic signals, Eq.(5.1), the Fourier transform pair will be

$$\sum_{m=-\infty}^{\infty} 4\,\text{rect}(t - 2m)\,\sin[\pi(t-2m)] \;\rightleftharpoons\; \frac{1}{2}\sum_{m=-\infty}^{\infty} -2j\left[\text{sinc}\left(\frac{n}{2}-\frac{1}{2}\right) - \text{sinc}\left(\frac{n}{2}+\frac{1}{2}\right)\right]\delta(f - \frac{n}{2}) \quad (5.19)$$

, the Fourier transform pair, Eq.(5.19), can be written in the form

$$\sum_{m=-\infty}^{\infty} 4\,\text{rect}(t - 2m)\,\sin[\pi(t-2m)] \;\rightleftharpoons\; \frac{1}{j}\sum_{n=1}^{\infty}\left[\text{sinc}\left(\frac{n}{2}-\frac{1}{2}\right) - \text{sinc}\left(\frac{n}{2}+\frac{1}{2}\right)\right]\delta(f - \frac{n}{2})$$
$$-\frac{1}{j}\sum_{n=-\infty}^{-1}\left[\text{sinc}\left(\frac{n}{2}-\frac{1}{2}\right) - \text{sinc}\left(\frac{n}{2}+\frac{1}{2}\right)\right]\delta(f + \frac{n}{2})$$

, then the inverse Fourier transform is the time domain of Eq.(5.19), is given by

$$\sum_{m=-\infty}^{\infty} \text{rect}(t - 2m)\,\sin[\pi(t-2m)] = \sum_{n=1}^{\infty}\left[\text{sinc}\left(\frac{n}{2}-\frac{1}{2}\right) - \text{sinc}\left(\frac{n}{2}+\frac{1}{2}\right)\right]\sin(\pi nt)$$

, or equivalently

$$x_p(t) = \sum_{n=1}^{\infty}\left[\text{sinc}\left(\frac{n}{2}-\frac{1}{2}\right) - \text{sinc}\left(\frac{n}{2}+\frac{1}{2}\right)\right]\sin(\pi nt)$$

, which is the same spectral analysis of problem 2.8, Eq.(2.20), using the complex exponential Fourier series method.

Problem 5.5

Find the Fourier transform of the shown ladder signal x(t), Fig.5.5a.

Solution

The ladder signal x(t), Fig.5.5a. is composed of the periodic saw tooth signal $x_1(t)$, Fig.5.5b. plus the ramp signal $x_2(t)$, Fig.5.5c.

Then the ladder signal x(t) can be expressed by

$$x(t) = x_1(t) + x_2(t)$$

, or equivalently

$$x(t) = x_1(t) + \tfrac{1}{2} t$$

, due the linearity property, the Fourier transform of x(t) is the sum of the Fourier transforms of $x_1(t)$ and $x_2(t)$ and is given by

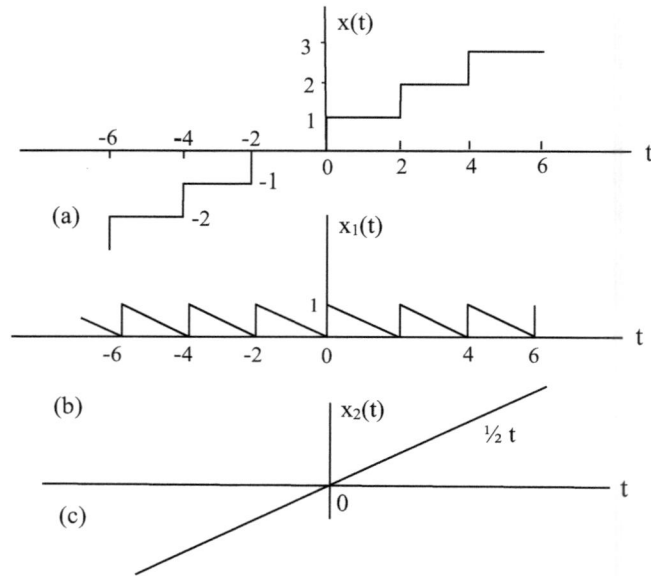

(a)

(b)

(c)

Fig.5.5. The ladder periodic signal x(t) is composed of $x_1(t)$ and $x_2(t)$.

$$F[x(t)] = F[x_1(t) + x_2(t)]$$
$$= F[x_1(t)] + F[\tfrac{1}{2} t] \tag{5.20}$$

, where the Fourier transform of the periodic signal $x_1(t)$, using Eq.(5.1) , is given by

$$F[x_1(t)] = \frac{1}{2} \sum_{m=-\infty}^{\infty} X_1(\tfrac{n}{2}) \; \delta(f - \tfrac{n}{2})$$

, and $X_1(n/2)$, using Eq.(2.14), is given by

$$X_1(\tfrac{n}{2}) = \int_{-1}^{0} (-\tfrac{1}{2} t) \, e^{-j\pi n t} \; dt + \int_{0}^{1} (1 - \tfrac{1}{2} t) \, e^{-j\pi t} \; dt$$

, using the integration by parts technique, integrate, express the exponential terms by sine function, and in terms of sinc function, $X_1(n/2)$ yields

$$X_1(\tfrac{n}{2}) = \operatorname{sinc}(n) + j \frac{1}{\pi n} [\operatorname{sinc}(n) - 1] \tag{5.21}$$

, also the Fourier transform of the function $x_2(t) = \tfrac{1}{2} t$, using Equations (4.2) and (3.27a), is given by

$$F[x_2(t)] = F[\tfrac{1}{2} t] = \frac{1}{2} \left(\frac{1}{-j2\pi} \right) \frac{d\,\delta(f)}{df} \tag{5.22}$$

Then the Fourier transform of x(t), using Equations (5.20), (5.21) and (5.22), is given by

$$X(f) = \sum_{n=-\infty}^{\infty} \frac{1}{2} \operatorname{sinc}(n) \; \delta(f - \tfrac{n}{2}) + j \left\{ \sum_{n=-\infty}^{\infty} \frac{1}{2\pi n} [\operatorname{sinc}(n) - 1] \; \delta(f - \tfrac{n}{2}) + \frac{1}{4\pi} \frac{d\,\delta(f)}{df} \right\}$$

, where $n = 0, \pm1, \pm2, \pm3 \ldots \ldots$. The Fourier transform X(f) is complex value because the ladder periodic signal x(t), Fig5.5a, is neither even nor odd function.

Problem 5.6
Find the spectral analysis of the periodic signal $x_p(t)$, Fig.5.6a, using the Fourier transform,

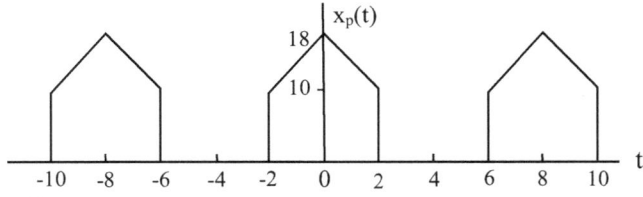

Solution
Consider the generating function x(t), Fig.5.6b, of the periodic signal $x_p(t)$, is given by

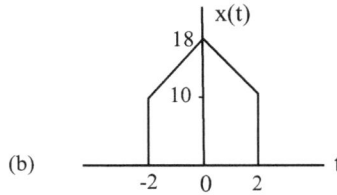

$$x(t) = 10 \ \text{rect}(\frac{t}{4}) + 8 \ \text{tri}(\frac{t}{2})$$

, the Fourier transform of the generating function is given by

Fig 5.6. a) Periodic signal $x_p(t)$, b) Unperiodic signal x(t).

$$X(f) = 40 \ \text{sinc}(4f) + 16 \ \text{sinc}^2(2f)$$

, then the periodic signal $x_p(t)$, Eq.(2.12), the complex coefficient C_n, Eq.(2.13), is given by

$$C_n = \frac{1}{8} \ X(f)\big|_{\text{at } f = n/8}$$

, or equivalently

$$C_n = 5 \ \text{sinc}(\frac{n}{2}) + 2 \ \text{sinc}^2(\frac{n}{4})$$

, then the periodic signal $x_p(t)$, Eq.(2.12), is given by

$$x_p(t) = \sum_{m=-\infty}^{\infty} \left[5 \ \text{sinc}(\frac{n}{2}) + 2 \ \text{sinc}^2(\frac{n}{4}) \right] \left[\cos(\pi \frac{n}{4} t) + j \sin(2\pi \frac{n}{8} t) \right]$$

, and the real value of $x_p(t)$ yields

$$\left[x_p(t) \right]_{\text{real}} = \sum_{m=-\infty}^{\infty} \left[5 \ \text{sinc}(\frac{n}{2}) + 2 \ \text{sinc}^2(\frac{n}{4}) \right] \cos(2\pi \frac{n}{8} t)$$

, or equivalently

$$\left[x_p(t) \right]_{\text{real}} = 7 + \sum_{m=1}^{\infty} \left[5 \ \text{sinc}(\frac{n}{2}) + 2 \ \text{sinc}^2(\frac{n}{4}) \right] \cos(2\pi \frac{n}{8} t)$$

Problems
1. Find the spectral analysis of the shown periodic signals, Fig.5.7a,b,c, using the Fourier transform.

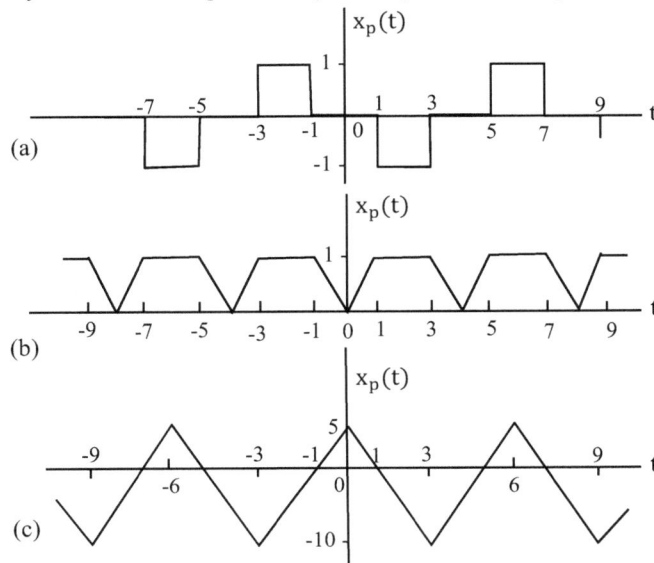

Fig.5.7. Periodic signals.

2. Find the spectral analysis of the shifted periodic triangle signal, Fig.5.8, using the Fourier transform.

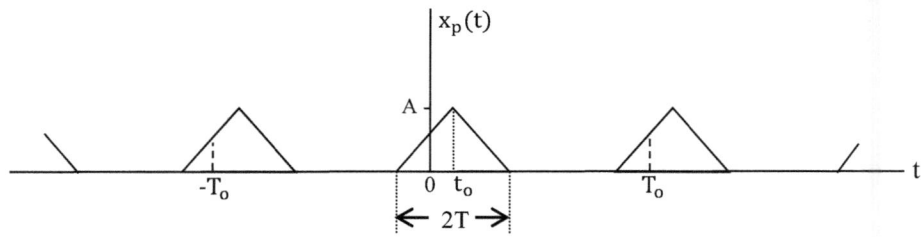

Fig.5.8. Shifted periodic triangle signal.

3. Based on the Fourier transform pair of the signal $x(t) = A\ rect(t/T)$ is $X(f) = AT\ sinc(fT)$, Eq.(3.4), and using the Fourier transform of the periodic signals. Prove the following Fourier transform pair

$$\sum_{n=-\infty}^{\infty} rect\left[\frac{t - mT}{T}\right] \rightleftharpoons \delta(f)$$

, where $\delta(f)$ is the Dirac delta function in the frequency domain.

4. Find the Fourier transform of the ladder signal $x(t)$, Fig.5.9.

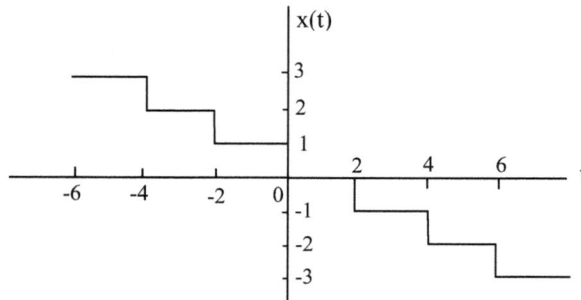

Fig.5.9. Ladder signal $x(t)$.

5. Find the spectral analysis of the periodic signals $x_{1p}(t)$ and $x_{2p}(t)$, Fig.5.10a,b, using the Fourier transform.

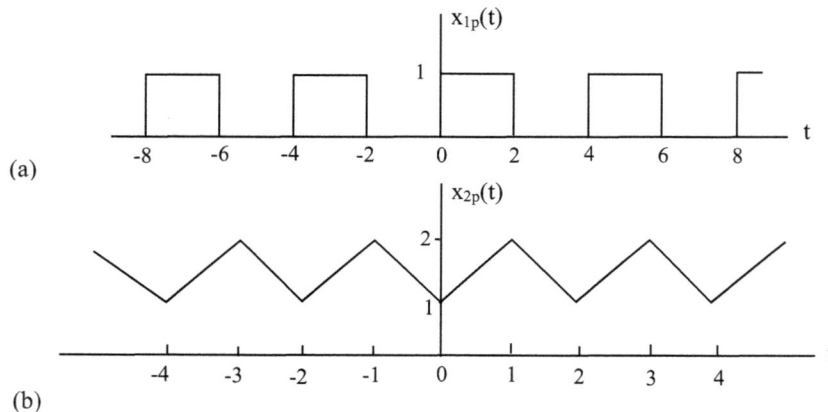

(a)

(b)

Fig.5.10. Two power signals.

6. Find the spectral analysis of the periodic Radio Frequency signal, Fig1.8b, using the Fourier transform.

CHAPTER 6

Correlation Function and Spectral Density

Abstract: Real random processes (stochastic processes) exhibit unpredictable changes with time and defy precise prediction. Practically, it is not possible to determine the probability distribution function of these real random processes. Then it is important to describe partially the mean, the correlation function and the covariance function of these real random processes which is defined by stationary processes or Ergodic processes. The collection of all possible recording of these processes, is called an ensemble. The ensemble correlation functions play an important role in the design of the correlation receiver and the matched filter receiver. The correlation function studies the statistically dependent and independent relation between two real random processes. The correlation function is independent of the time and depends only on the time difference (delay).

Keywords: Autocorrelation function; Cross-Correlation function; Energy signals; Power signals.

6.1. Spectral Density

Spectral Density is a very useful term to evaluate the energy content and the average power of the energy signals (unperiodic) and the power signals (periodic) respectively, using the frequency domain. Spectral density and correlation function constitute a Fourier transform pair, where the correlation function is a measure of the similarity or the coherence between a signal and the time delayed version of another signal delayed by a variable amount of time delay τ. Also, the correlation function checks the orthogonality of the signals which is considered very important in the design of the communication systems [1,4,6].

6.1.1. Energy Spectral Density

Energy Spectral Density ESD $\Psi(f)$ facilitates the evaluation of the energy content of the unperiodic signals, where the total area under the Energy Spectral Density ESD curve, plotted as a function of frequency, indicates the energy content of the signal , Fig.3.47. The energy spectral density is given by

$$\Psi(f) = |X(f)|^2 \qquad \text{joule/Hz} \qquad (3.61)$$

, where $X(f)$ is the Fourier transform of the energy signal $x(t)$. The energy content of the signal $x(t)$ normalized to one ohm, E is given by

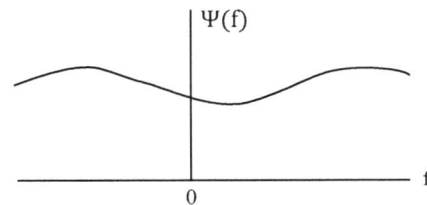

Fig.3.47. The Energy Spectral Density ESD.

$$E = \int_{-\infty}^{\infty} \Psi(f)\, df \qquad \text{joule} \qquad (3.62)$$

Problem 6.1

A current passing in resistor R ohm, is $i(t) = e^{-t} u(t)$. What is the percentage of the energy content in the resistor, can associated within the frequency band $0 \le f \le 1/\pi$ Hz.

Solution

The Fourier transform of the current passing in the resistor is given by

$$I(f) = \frac{1}{1 + j2\pi f}$$

, in terms of the magnitude and phase angle, I(f) yields

$$I(f) = \frac{1}{\sqrt{1 + (2\pi f)^2}} e^{-j\tan^{-1}(2\pi f)}$$

, and the Energy Spectrum Density ESD $\Psi(f)$, Eq.(3.61), is given by

$$\Psi(f) = |I(f)|^2 = \frac{1}{1 + (2\pi f)^2} \qquad \text{joule/Hz}$$

, then due to Rayleigh`s energy theorem, Eq.(3.62), the energy content normalized to one ohm resistor, and can be associated with the frequency band $-\infty \le f \le \infty$ Hz, E is given by

$$E = \int_{-\infty}^{\infty} \frac{1}{1 + (2\pi f)^2}\, df \qquad \text{joule}$$

, and the energy normalized to one ohm resistor and can be associated with the frequency band $0 \le f \le 1/\pi$ Hz, Fig.6.1, E` is given by

$$E` = \int_{-1/\pi}^{1/\pi} \frac{1}{1 + (2\pi f)^2} \, df \qquad \text{joule}$$

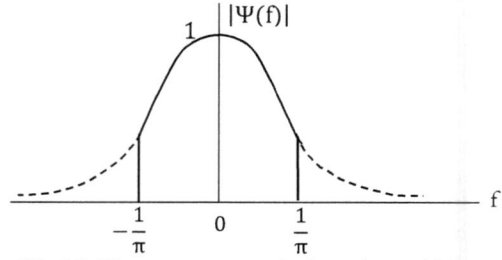

Fig.6.1. The energy content in the resistor within the frequency band $0 \le f \le 1/\pi$ Hz.

, let $\lambda = 2\pi f$, then $d\lambda = 2\pi \, df$, as $f \to -\infty$, then $\lambda \to -\infty$, and as $f \to \infty$ then $\lambda \to \infty$, for the energy E. Also, as $f \to -1/\pi$, then $\lambda \to -2$, as $f \to 1/\pi$, then $\lambda \to 2$, for the energy E`, then the energy contents E and E` yield

$$E = \frac{1}{2\pi} \int_{-\infty}^{\infty} \frac{1}{1 + \lambda^2} \, d\lambda \qquad \text{, and} \qquad E` = \frac{1}{2\pi} \int_{-2}^{2} \frac{1}{1 + \lambda^2} \, d\lambda$$

, since the function $1/(1 + \lambda^2)$ is an even function, the energy contents E and E` yield

$$E = \frac{1}{\pi} \int_{0}^{\infty} \frac{1}{1 + \lambda^2} \, d\lambda \;\; = \left[\frac{1}{\pi} \tan^{-1}(\lambda) \right]_{\lambda=0}^{\lambda=\infty} = \frac{1}{\pi} \left[\frac{\pi}{2} - 0 \right] = 0.5 \qquad \text{joule}$$

$$\text{, and} \quad E` = \frac{1}{\pi} \int_{0}^{2} \frac{1}{1 + \lambda^2} \, d\lambda \;\; = \left[\frac{1}{\pi} \tan^{-1}(\lambda) \right]_{\lambda=0}^{\lambda=2} = \frac{1}{\pi} \left[\frac{\pi}{2.84} - 0 \right] = 0.352 \quad \text{joule}$$

, then the percentage of the energy content in the resistor R, associated in the frequency band $0 \le f \le 1/\pi$ Hz, is o.352/0.5 equals 70%, Fig.6.1.

6.1.2. Power Spectral Density

Power Spectral Density PSD $\Omega(f)$ facilitates the evaluation of the average power of the periodic signal $x_p(t)$, where in terms of the Dirac delta function $\delta(f)$ at the harmonic components n/T_o, the power spectral density $\Omega(f)$ plotted as a function of frequency, Fig.2.14, is given by

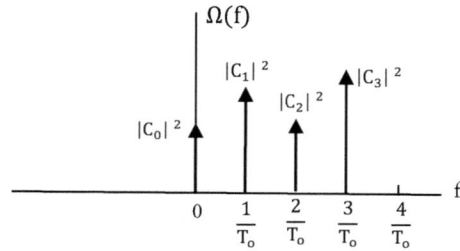

Fig.2.14. Power spectral density of periodic signal.

$$\Omega(f) = \frac{1}{T_o^2} \sum_{n=-\infty}^{\infty} \left| X(\frac{n}{T_o}) \right|^2 \delta(f - \frac{n}{T_o}) \quad (2.24)$$

, where $X(n/T_o)$ is the Fourier transform of the energy signal $x(t)$ which generates the periodic signal $x_p(t)$ at $f = n/T_o$. The average power normalized to one ohm of the periodic signal $x_p(t)$, is given by

$$P_{av} = \int_{-\infty}^{\infty} \Omega(f) \, df \qquad (2.23)$$

, also the average power is given by

$$P_{av} = \frac{1}{T_o^2} \sum_{n=-\infty}^{\infty} \left| X(\frac{n}{T_o}) \right|^2 \qquad (2.21)$$

, where $X(n/T_o)$ is given by

$$X(\frac{n}{T_o}) = \int_{-T_o/2}^{T_o/2} x_p(t) \, e^{-j2\pi \frac{n}{T_o} t} \, dt \qquad (2.14)$$

, where $x_p(t)$ is the periodic signal of periodic time T_o.

Problem 6.2

Evaluate the power spectral density and the average power of the following cosine cubed periodic signal, using the frequency domain

$$x_p(t) = A\cos^3(2\pi\frac{1}{T_o}t)$$

Solution

The periodic signal $x_p(t)$ is expressed by

$$x_p(t) = \frac{3}{4}A\cos(2\pi\frac{1}{T_o}t) + \frac{1}{4}A\cos(2\pi\frac{3}{T_o}t)$$

, then the periodic signal $x_p(t)$ is composed of two cosine functions, both extend from $-\infty$ to ∞, their discrete frequencies are $1/T_o$ and $3/T_o$. The discrete spectrum $X(n/T_o)$, is the equivalent frequency domain of the periodic signal $x_p(t)$, Eq.(2.14), is given by

(a)

(b)

Fig.6.2. a) The waveform $x(t) = A\cos^3[2\pi(n/T_o)t]$, b) The spectral density of $x(t)$.

$$X(\frac{n}{T_o}) = \int_{-T_o/2}^{T_o/2} [\frac{3}{4}A\cos(2\pi\frac{1}{T_o}t) + \frac{1}{4}A\cos(2\pi\frac{3}{T_o}t)]\ e^{-j2\pi\frac{n}{T_o}t}\ dt$$

, integrate, express the exponential terms by sine functions, and in terms of sinc functions, $X(n/T_o)$ yields

$$X(\frac{n}{T_o}) = \frac{3}{8}AT_o\ \mathrm{sinc}(n-1) + \frac{3}{8}AT_o\ \mathrm{sinc}(n+1) + \frac{1}{8}AT_o\ \mathrm{sinc}(n-3) + \frac{1}{8}AT_o\ \mathrm{sinc}(n+3)$$

, the power spectral density, Eq.(2.24), $\Omega(f)$ is given by

$$\Omega(f) = \sum_{n=-\infty}^{\infty} \left|\frac{3}{8}A\ \mathrm{sinc}(n-1) + \frac{3}{8}A\ \mathrm{sinc}(n+1) + \frac{1}{8}A\ \mathrm{sinc}(n-3) + \frac{1}{8}A\ \mathrm{sinc}(n+3)\right|^2 \delta(f-\frac{n}{T_o})$$

, since $x_p(t)$ contains two frequencies $(1/T_o)$ at $n=\pm 1$, and $(3/T_o)$ at $n=\pm 3$. Hence $\mathrm{sinc}(0) = 1$, and $\mathrm{sinc}(\pm n) = 0$, where $n = 1, 2, 3, 4, \dots$, $\Omega(f)$ yields

$$\Omega(f) = (\frac{3}{8}A)^2\ \delta(f-\frac{1}{T_o}) + (\frac{3}{8}A)^2\ \delta(f+\frac{1}{T_o}) + (\frac{1}{8}A)^2\ \delta(f-\frac{3}{T_o}) + (\frac{1}{8}A)^2\ \delta(f+\frac{3}{T_o})$$

, then the Power Spectral Density PSD consists of four delta functions at the discrete frequencies $\pm 1/T_o$, and $\pm 3/T_o$. Fig.6.2, and making use of Eq.(1.5), the average power P_{av}, Eq. (2.23), yields

$$P_{av} = \frac{10}{32}A^2 \qquad \text{watt}$$

, this average power P_{av}, an also be obtained using the time domain, Eq. (1.9), yields

$$P_{av} = \frac{1}{T_o} \int_{-T_o/2}^{T_o/2} A^2\cos^6(2\pi\frac{1}{T_o}t)\ dt$$

, making use of the formula "$\cos^3(x) = \frac{3}{4}\cos(x) + \frac{1}{4}\cos(3x)$", the average power P_{av} yields

$$P_{av} = \frac{1}{T_o} \int_{-T_o/2}^{T_o/2} \left[\frac{3}{4}A\cos(2\pi\frac{1}{T_o}t) + \frac{1}{4}A\cos(2\pi\frac{3}{T_o}t)\right]^2 dt$$

, or equivalently

$$P_{av} = \frac{1}{T_o} \int_{-T_o/2}^{T_o/2} \left\{(\frac{3}{4}A)^2\cos^2(2\pi\frac{1}{T_o}t) + \frac{3}{8}A^2\cos(2\pi\frac{1}{T_o}t)\cos(2\pi\frac{3}{T_o}t) + (\frac{1}{4}A)^2\cos^2(2\pi\frac{3}{T_o}t)\right\} dt$$

, making use of the formulas, "$2\cos^2(x) = 1 + \cos(2x)$", and "$2\cos(x)\cos(y) = \cos(x+y) + \cos(x-y)$", and integrate, P_{av} equals $10\ A^2/32$ watt.

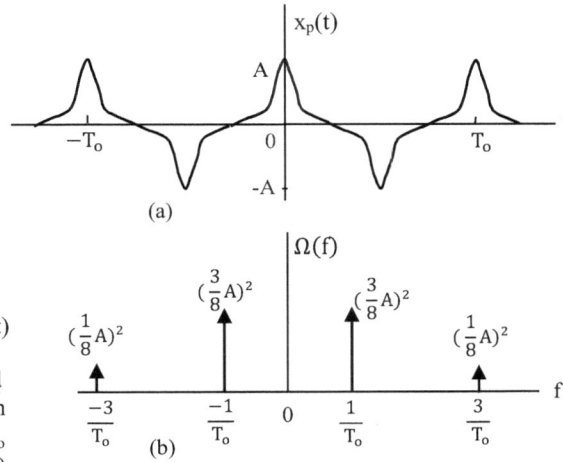

6.2. Correlation Function

Correlation function provides a measure of the similarity or the coherence between a signal and the time delayed version of another signal by a variable amount of time delay τ. Correlation function indicates the variation of this coherence between these two signals, plotted as a function of this variable time delay τ. The correlation function decreases when the delay τ increases and vice versa, then the maximum value of the correlation function occurs at the delay τ equals zero. On the other hand, in the frequency domain, where the correlation function and the spectral density constitute a Fourier transform pair, so it may be a measure, how much the spectrums of these two signals are overlapped or not, which depends on the variable amount of time delay τ. The overlap of the two spectrums decreases when the correlation function decreases and vice versa. For some larger value of delay τ, the correlation function decreases, say to the 0.707 times the maximum value of the correlation function, if the absolute value of τ at the 0.707 point is large, the spectrums overlap of these two signals will tend to have small spectral spread, while a small absolute value of τ corresponds to a broad spectral spread [1,3,5,6,7].

Correlation function is defined by autocorrelation function when the two signals are a given signal and its replica, or by the cross-correlation function when the two signals are different. The formula of the correlation function depends on whether the pair of signals being considered are energy signals (unperiodic) or power signals (periodic).

6.3. Autocorrelation Function

Autocorrelation function provides a measure of the similarity or the coherence between a given signal $x(t)$ and its replica delayed by a variable amount of time delay τ, $x(t-\tau)$. In the frequency domain, the autocorrelation function measures, how much the spectrums of these two spectrums $X(f)$ and $X(f)$ $e^{-j2\pi f \tau}$, are overlapped or not, which are very useful in the design of all communication systems, such as the transmitter and the receiver of the RADAR (Radio Amplification and Detection And Ranging) system, where a signal $x(t)$ is a carrier wave modulated by a rectangular function, is transmitted in the space, to detect a target. Many useful information about the target can be obtained from the delayed received signal $x(t-\tau)$, such as the speed and range of the target, Fig.6.3. This autocorrelation function depends on the variable delay τ, it decreases when τ increases and it increases when τ decreases. Autocorrelation function indicates the variation of this correlation, plotted as a function of this variable time delay τ. The formula of the autocorrelation function depends on whether the signals being considered, energy signals or power signals.

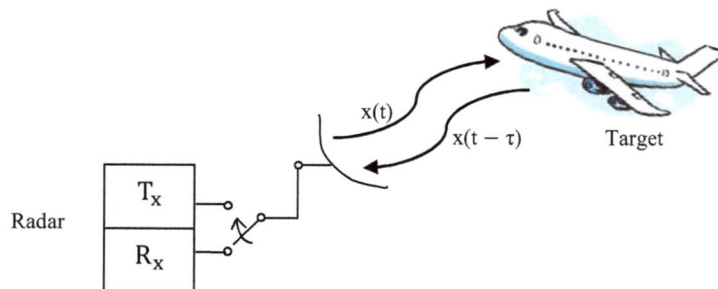

Fig.6.3. The autocorrelation function of a given signal $x(t)$ and its replica delayed by a variable amount of time delay τ, $x(t-\tau)$.

6.3.1. Autocorrelation Function of Energy Signals

Let x(t) is a complex energy signal (unperiodic). The autocorrelation function R(τ) of x(t) involves the integrating product of x(t) and its complex conjugate x*(t) delayed by variable amount of time delay τ, Fig.6.3, normalized to one ohm, R(τ) is given by

$$R(\tau) = \int_{-\infty}^{\infty} x(t)\, x^*(t - \tau)\, dt \qquad (6.1)$$

, where x*(t − τ) is the complex conjugate version of the signal x(t), and the time delay τ plays the role of a scanning parameter from −∞ to ∞, till the intersection of the signals x(t) and x*(t − τ) exists, then the limits of the integration are determined. If the signal x(t) is complex-valued then the autocorrelation function R(τ) is also complex value as well. Eq.(6.1) shows that the complex conjugate x*(t − τ) is delayed by τ with respect to the signal x(t), then the autocorrelation function R(τ) is also given by

$$R(\tau) = \int_{-\infty}^{\infty} x(t + \tau)\, x^*(t)\, dt \qquad (6.2)$$

, where the complex conjugate signal x*(t) is delayed by τ with respect to the signal x(t + τ). The autocorrelation function exhibits conjugate symmetry, where R(τ) = R*(−τ), Equations (6.1) and (6.2).

6.3.2. Fourier Transform of Autocorrelation Function for Energy Signals

Fourier transform of the autocorrelation function can be obtained by making use of the convolution property, where the autocorrelation function can also be expressed by

$$R(\tau) = x(\tau) \otimes x^*(-\tau) \qquad (6.3)$$

, or

$$R(\tau) = \int_{-\infty}^{\infty} x(t)\, x^*[-(\tau - t)]\, dt$$

, which is the same autocorrelation function, Eq.(6.1), then taking the Fourier transform for Eq.(6.3), a Fourier transform pair will be

$$x(\tau) \otimes x^*(-\tau) \rightleftharpoons X(f)\, X^*(f)$$

, or

$$R(\tau) \rightleftharpoons |X(f)|^2 \qquad (6.4)$$

, then the Fourier transform of the autocorrelation function R(τ) is the Energy Spectrum Density ESD, Ψ(f) of the signal x(t), where Ψ(f) = $|X(f)|^2$ joule/Hz, Eq.(3.61), or equivalently

$$\Psi(f) = \int_{-\infty}^{\infty} R(\tau)\, e^{-j2\pi f\tau}\, d\tau \qquad (6.5a)$$

, and

$$R(\tau) = \int_{-\infty}^{\infty} \Psi(f)\, e^{j2\pi f\tau}\, df \qquad (6.5b)$$

, Equations (6.5a,b) are known as the Wiener-Khintchine relations for the energy signals. The evaluation of the autocorrelation function R(τ) for an energy signal x(t) can be obtained by two ways, either in the time domain using Eq.(6.1) to get R(τ) directly for varying τ, or in the frequency domain by getting their individual Fourier transforms X(f) and the complex conjugate X*(f) and multiply, and then return again to the time domain by taking the inverse Fourier transform, Eq.(6.4) to obtain R(τ).

6.3.3. Evaluation of Energy Content in terms of Autocorrelation Function

The energy content E of the signal x(t) is the origin value of the autocorrelation function, R(0), Eq.(6.1) yields

$$R(0) = \int_{-\infty}^{\infty} x(t)\, x^*(t)\, dt$$

, or $$E = R(0) = \int_{-\infty}^{\infty} |x(t)|^2\, dt \qquad (6.6)$$

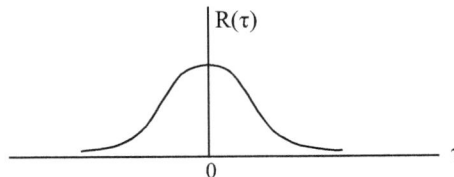

Fig.6.4. The autocorrelation function R(τ).

, which is the same energy content of the signal x(t), Eq.(1.10), and also is the maximum value of the autocorrelation function R(τ), occurs at the origin, Fig.6.4, where $|R(\tau)| \le R(0)$ for all τ.

Problem 6.3

For the doublet pulse x(t), Fig.6.5. Evaluate:

   i. The energy spectral density $\Psi(f)$ of x(t)

   ii. The autocorrelation function $R(\tau)$ of x(t)

   iii. The energy content of x(t) from (ii)

Solution

The doublet pulse, Fig.6.5, is expressed by

   $x(t) = rect(t + \frac{1}{2}) - rect(t - \frac{1}{2})$

, and its Fourier transform X(f) is given by

   $X(f) = sinc(f)\ e^{j\pi f} - sinc(f)\ e^{-j\pi f}$

, and the complex conjugate $X^*(f)$ will be

   $X^*(f) = sinc(f)\ e^{-j\pi f} - sinc(f)\ e^{j\pi f}$

, then the Energy Spectral Density will be

   $\Psi(f) = |X(f)|^2 = X(f)\ X^*(f)$

, or equivalently

   $\Psi(f) = sinc^2(f)\ [\ 2 - (e^{j2\pi f} + e^{-j2\pi f})\ ]$ (6.7)

, express the exponential terms by cosine function, $\Psi(f)$ yields

   $\Psi(f) = sinc^2(f)\ [\ 2 - 2\cos(2\pi f)\ ]$

, using the formula "$2\sin^2(x) = 1 - \cos(2x)$"
, $\Psi(f)$ yields

   $\Psi(f) = 4\ sinc^2(f)\ \sin^2(\pi f)$

, then the autocorrelation function $R(\tau)$ is obtained by taking the inverse Fourier transform of the energy spectral density , Eq.(6.7), $R(\tau)$ yields

$R(\tau) = 2\ tri(\tau) - tri(\tau + 1) - tri(\tau - 1)$

, Fig.6.6. shows the autocorrelation function $R(\tau)$. The energy content E equals the origin value of the autocorrelation function, is given by

   $E = R(0) = 2$ joule

The autocorrelation function $R(\tau)$ can also be obtained using the time domain, Eq.(6.1), where for negative τ, Fig.6.7a,b, $R(\tau)$ is given by

$$R(\tau) = \int_{-1}^{1+\tau} (-1)\,dt = -\tau - 2 \quad \text{for } -2 \le \tau \le -1$$

$$, R(\tau) = \int_{-1}^{\tau} dt - \int_{\tau}^{0} dt + \int_{0}^{1+\tau} dt$$

$$= 2 + 3\tau \qquad \text{for } -1 \le \tau \le 0$$

, for positive τ, Fig.6.7c,d, $R(\tau)$ is given by

$$R(\tau) = \int_{-1+\tau}^{0} dt - \int_{0}^{\tau} dt + \int_{\tau}^{1} dt$$

$$= 2 - 3\tau \qquad \text{for } 0 \le \tau \le 1$$

$$, R(\tau) = \int_{-1+\tau}^{1} (-1)\,dt = \tau - 2 \quad \text{for } 1 \le \tau \le 2$$

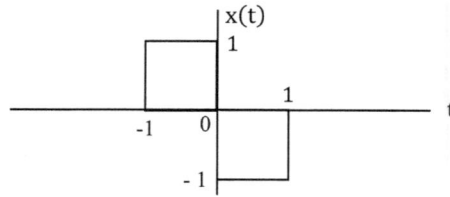

Fig.6.5. A doublet pulse.

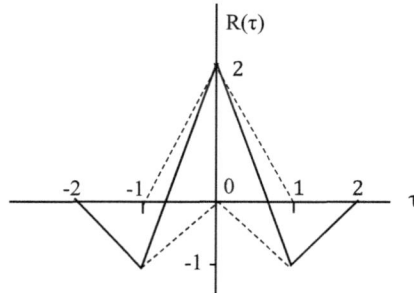

Fig.6.6. The autocorrelation function $R(\tau)$ of the doublet pulse, Fig.6.5.

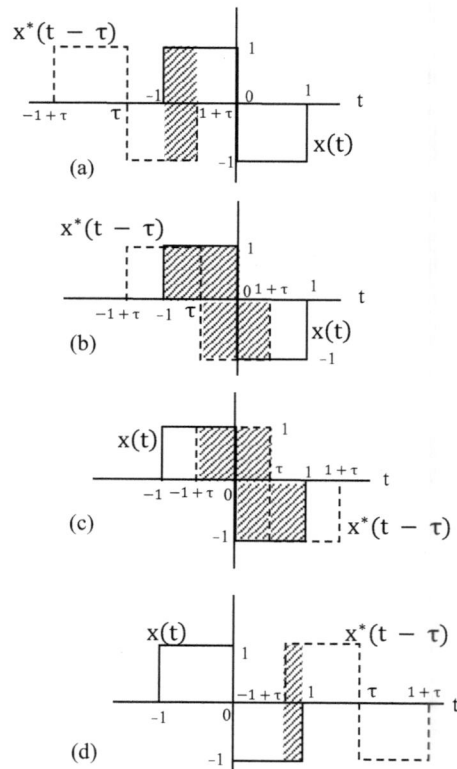

(a)

(b)

(c)

(d)

Fig.6.7.
The autocorrelation function $R(\tau)$ with varying the delay τ of the doublet pulse of Fig.6.5,
a) $-2 \le \tau \le -1$, b) $-1 \le \tau \le 0$,
c) $0 \le \tau \le 1$, d) $1 \le \tau \le 2$.

, Fig.6.6. shows the autocorrelation function $R(\tau)$ which co-incidences with the four previous separate equations of the four different time period regions.

Problem 6.4

For the energy signal $x(t) = A\ \text{rect}(t/T)$, Fig.1.2a. Evaluate:

i. The energy spectral density $\Psi(f)$ of $x(t)$.
ii. The autocorrelation function $R(\tau)$ of $x(t)$.
iii. The energy content of $x(t)$ from (ii).

Solution

The Fourier transform $X(f)$ of $x(t)$ is given by

$$X(f) = AT\ \text{sinc}(fT) \qquad (3.4)$$

, and the complex conjugate will be

$$X^*(f) = AT\ \text{sinc}(fT)$$

, the energy spectral density is given by

$$\Psi(f) = |X(f)|^2 = X(f)\,X^*(f)$$

, or equivalently, $\Psi(f)$, Fig.6.8, is given by

$$\Psi(f) = A^2 T^2\ \text{sinc}^2(fT) \qquad (6.8)$$

, the autocorrelation function $R(\tau)$ is obtained by taking the inverse Fourier transform of the energy spectral density, Eq (6.8), $R(\tau)$ is given by

$$R(\tau) = A^2 T\ \text{tri}\left(\frac{\tau}{T}\right) \qquad (6.9)$$

The energy content E is the origin value of the autocorrelation function, yields

$$E = R(0) = A^2 T \quad \text{joule} \qquad (1.11)$$

, which is the same energy content of problem 1.2 , Eq.(1.11), using the time domain, and problem 3.25, using the frequency domain.

The autocorrelation function $R(\tau)$ can also be obtained using the time domain, Eq.(6.1) , Fig.6.9a,b, $R(\tau)$ is given by

$$R(\tau) = \int_{-T/2}^{\tau+\frac{1}{2}T} A^2\ dt = A^2 T + A^2 \tau \quad \text{for negative } \tau$$

$$,R(\tau) = \int_{\tau-\frac{1}{2}T}^{T/2} A^2\ dt = A^2 T - A^2 \tau \quad \text{for positive } \tau$$

, Fig.6.8a. shows the autocorrelation function $R(\tau)$, which co-incidences with the two previous separate equations of the negative and positive τ.

Problem 6.5

For the Radio Frequency RF pulse, Fig.1.8a, $x(t) = A\ \text{rect}(t/T)\ \cos(2\pi f_c t)$, where the carrier frequency f_c is much higher than $1/T$, and T is the pulse duration. Evaluate:

i. The energy spectral density $\Psi(f)$ of $x(t)$.
ii. The autocorrelation function $R(\tau)$ of $x(t)$.
iii. The energy content of $x(t)$ from (ii).

Solution

The Fourier transform $X(f)$ of the RF pulse $x(t) = A\ \text{rect}(t/T)\ \cos(2\pi f_c t)$, Eq.(3.8), is given by

$$X(f) = \tfrac{1}{2}\,AT\ \text{sinc}[T(f - f_c)] + \tfrac{1}{2}\,AT\ \text{sinc}[T(f + f_c)] \qquad (3.8)$$

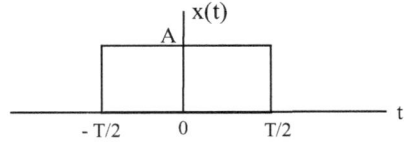

Fig.1.2a. Unperiodic rectangular signal.

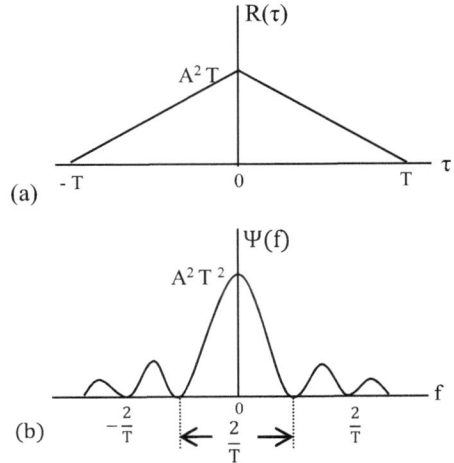

Fig.6.8. For the pulse $x(t) = A\ \text{rect}(t/T)$
a) The autocorrelation function $R(\tau)$,
b) The energy spectral density of the signal.

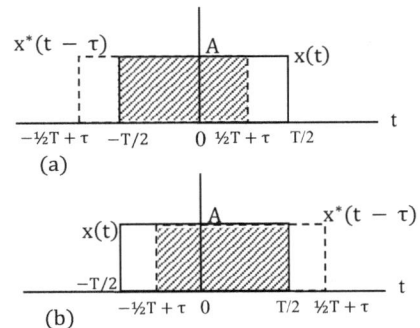

Fig.6.9. The autocorrelation function $R(\tau)$ with varying the delay τ of the signal $x(t) = A\ \text{rect}(t/T)$,
a) Negative τ, b) Positive τ.

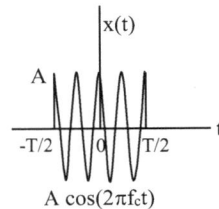

Fig.1.8a. Unperiodic RF pulse.

, the complex conjugate $X^*(f)$ will be

$$X^*(f) = \frac{1}{2} AT \text{ sinc}[T(f - f_c)] + \frac{1}{2} AT \text{ sinc}[T(f + f_c)]$$

, the energy spectral density is given by

$$\Psi(f) = |X(f)|^2 = X(f) X^*(f)$$

, or equivalently

$$\Psi(f) = \frac{1}{4} A^2 T^2 \text{ sinc}^2[T(f - f_c)] + \frac{1}{2} A^2 T^2 \text{ sinc}[T(f + f_c)] \text{ sinc}[T(f - f_c)] + \frac{1}{4} A^2 T^2 \text{ sinc}^2[T(f + f_c)]$$

, since the carrier frequency f_c is much higher than $1/T$, where T is the pulse duration, the energy spectral density $\Psi(f)$, Fig.6.10, yields

$$\Psi(f) = \frac{1}{4} A^2 T^2 \text{ sinc}^2[T(f - f_c)] + \frac{1}{4} A^2 T^2 \text{ sinc}^2[T(f + f_c)] \tag{6.10}$$

, the autocorrelation function is obtained by taking the inverse Fourier transform of the energy spectral density, Eq.(6.10), $R(\tau)$ is given by

$$R(\tau) = \frac{1}{4} A^2 T \text{ tri}(\frac{\tau}{T}) e^{j2\pi f_c t} + \frac{1}{4} A^2 T \text{ tri}(\frac{\tau}{T}) e^{-j2\pi f_c t}$$

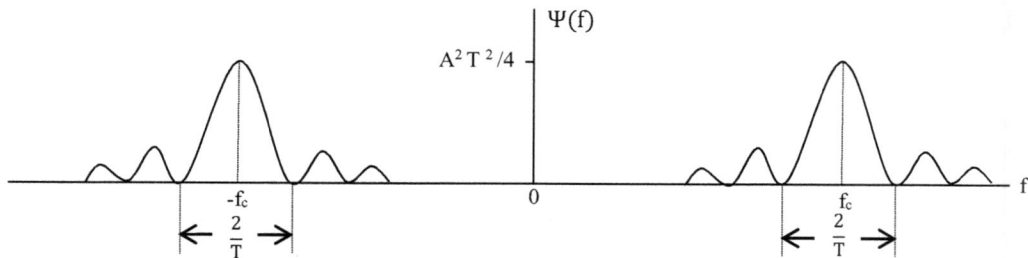

Fig.6.10. The spectral energy density of the Radio Frequency pulse $x(t) = A \text{ rect}(t/T) \cos(2\pi f_c t)$.

, express the exponential terms by cosine function, $R(\tau)$, Fig.6.11, yields

$$R(\tau) = \frac{1}{2} A^2 T \text{ tri}(\frac{\tau}{T}) \cos(2\pi f_c t) \tag{6.11}$$

The energy content E is the origin value of the autocorrelation function, yields $E = R(0) = \frac{1}{2} A^2 T$ joule, which is the same energy content of problem 1.4 , Eq.(1.13), using the time domain, and problem 3.26, using the frequency domain.

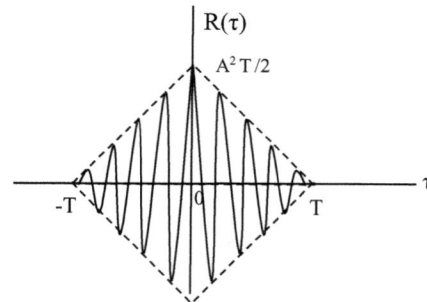

Fig.6.11. The autocorrelation function of the RF pulse, $x(t) = A \text{ rect}(t/T) \cos(2\pi f_c t)$.

Problem 6.6

Evaluate the autocorrelation function $R(\tau)$ and the energy content E, using the time domain, of the following energy signal, Fig.6.12a.

$$x(t) = t \text{ rect}(\frac{t}{2})$$

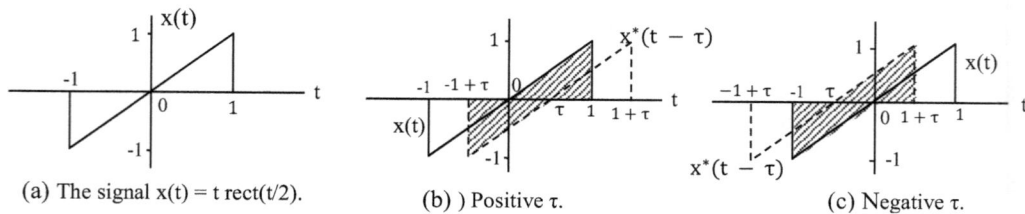

(a) The signal $x(t) = t \text{ rect}(t/2)$.

(b)) Positive τ.

(c) Negative τ.

Fig.6.12. a) The signal $x(t) = t \text{ rect}(t/2)$, b) Positive τ, c) Negative τ.

Solution

The autocorrelation function $R(\tau)$, Eq.(6.1), where for positive τ, Fig.6.12b, is given by

$$R(\tau) = \int_{-1+\tau}^{1} t\,(t-\tau)\,dt = \frac{\tau^3}{6} - \tau + \frac{2}{3}$$

, and for negative τ, Fig.6.12c, is given by

$$R(\tau) = \int_{-1}^{1+\tau} t\,(t-\tau)\,dt = -\frac{\tau^3}{6} + \tau + \frac{2}{3}$$

The energy content E equals the origin value of the autocorrelation function, yields $E = R(0) = 2/3$ joule , Fig.6.13. shows the autocorrelation function $R(\tau)$.

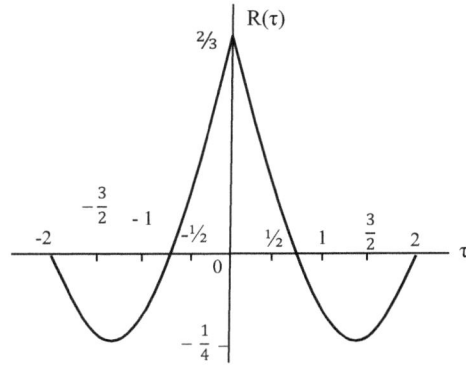

Fig.6.13. The autocorrelation function of the signal $x(t) = t\,\mathrm{rect}(t/2)$.

Problem 6.7

For the energy signal $x(t) = 12\,\mathrm{sinc}[6(t-3)]$. Evaluate:

i. The energy spectral density $\Psi(f)$ of $x(t)$.
ii. The autocorrelation function $R(\tau)$ of $x(t)$.
iii. The energy content of $x(t)$ from (ii).

Solution

The Fourier transform of the energy signal $x(t) = 12\,\mathrm{sinc}[6(t-3)]$, is given by

$$X(f) = 2\,\mathrm{rect}\left(\frac{f}{6}\right)\,e^{-j2\pi f\,3}$$

, and the energy spectral density, $\Psi(f)$ will be

$$\Psi(f) = |X(f)|^2 = 4\,\mathrm{rect}\left(\frac{f}{6}\right)$$

, the autocorrelation function $R(\tau)$ is obtained , by taking the inverse Fourier transform of the energy spectral density $\Psi(f)$, $R(\tau)$ yields

$$R(\tau) = 24\,\mathrm{sinc}(6\tau)$$

The energy content E equals the origin value of the autocorrelation function $R(\tau)$, yields $E = 24$ joule. Fig.6.14. shows the autocorrelation function $R(\tau)$ and the energy spectral density $\Psi(f)$.

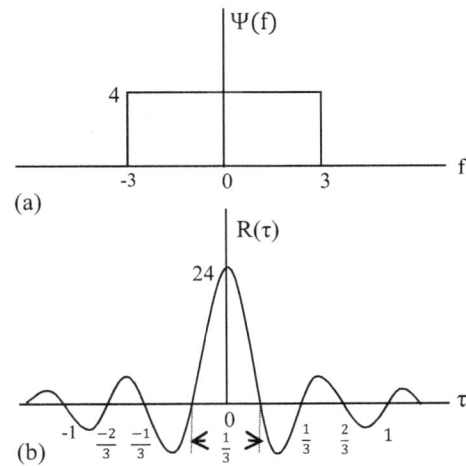

Problem 6.8

For the spectrum $X(f)$, Fig.6.15. Evaluate:

i. The energy spectral density $\Psi(f)$ of $X(f)$.
ii. The autocorrelation function $R(\tau)$ of $X(f)$.
iii The energy content of $X(f)$ from (ii).

Solution

The spectrum $X(f)$, Fig.6.15, is expressed by

$$X(f) = -2\,\mathrm{rect}(f+1) + 2\,\mathrm{rect}(f) - 2\,\mathrm{rect}(f-1)$$

(a)

(b)

Fig.6.14.
The autocorrelation function and the energy spectral density of the signal $x(t) = 12\,\mathrm{sinc}[6(t-3)]$.

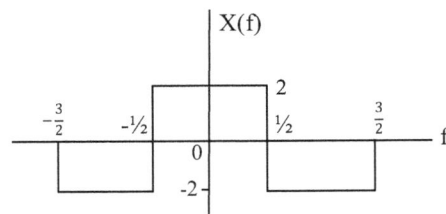

Fig.6.15. Spectrum of energy signal.

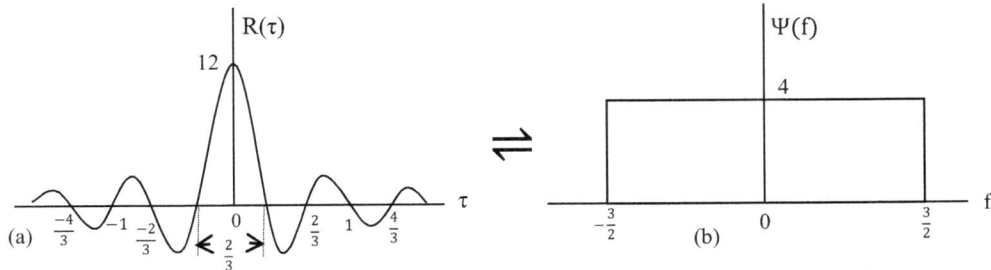

Fig.6.16. For the spectrum Fig.6.15, a) The autocorrelation function, b) The energy spectral density.

, then $|X(f)| = 2 \, \text{rect}(f/3)$, and the energy spectral density, Fig.6.16b, $\Psi(f)$ is given by

$$\Psi(f) = |X(f)|^2 = 4 \, \text{rect}(\frac{f}{3})$$

the autocorrelation function is obtained by taking the inverse Fourier transform of the energy spectral density $\Psi(f)$, $R(\tau)$ yields

$$R(\tau) = 12 \, \text{sinc}(3\tau)$$

The energy content equals the origin value of the autocorrelation function, yields $E = 12$ joule. Fig.6.16. shows the energy spectral density $\Psi(f)$ and the autocorrelation function $R(\tau)$.

Problem 6.9

For the single sided decaying energy signal $x(t) = e^{-t} u(t)$, Fig.1.9a. Evaluate:
 i. The energy spectral density $\Psi(f)$ of $x(t)$.
 ii. The autocorrelation function $R(\tau)$ of $x(t)$.
 iii. The energy content of $x(t)$ from (ii).

Solution

The Fourier transform of the energy signal $x(t) = e^{-t} u(t)$, is given by

$$X(f) = \frac{1}{1 + j2\pi f}$$

, the energy spectral density $\Psi(f)$ will be

$$\Psi(f) = |X(f)|^2 = \frac{1}{1 + (2\pi f)^2}$$

, the autocorrelation function $R(\tau)$ is obtained by taking the inverse Fourier transform of the energy spectral density $\Psi(f)$, $R(\tau)$ yields

$$R(\tau) = \tfrac{1}{2} \, e^{-|\tau|}$$

The energy content equals the origin value of the autocorrelation function, yields $E = R(0) = \tfrac{1}{2}$ joule, which is the same energy content of problem 1.5, Eq.(1.15), using the time domain and problem 3.27, Eq.(3.63a), using the frequency domain. Fig.6.17. shows the energy spectral density $\Psi(f)$ and the autocorrelation function $R(\tau)$.

Problem 6.10

For the energy signal $x(t) = 4 \, \text{rect}(t) \cos(\pi t)$, Fig.6.18a, using time domain. Evaluate:
 i. The autocorrelation function $R(\tau)$ of $x(t)$.
 ii. The energy spectral density $\Psi(f)$ of $x(t)$.
 iii. The energy content of $x(t)$ from (ii).

Solution

The autocorrelation function $R(\tau)$ can be obtained using the time domain, Eq.(6.1) , and using the formula
"$2 \cos(x) \cos(y) = \cos(x + y) + \cos(x - y)$",
then for negative τ, Fig.6.18b, $R(\tau)$ will be

$$R(\tau) = \int_{-\frac{1}{2}}^{\frac{1}{2}+\tau} 4 \cos(\pi t) \, 4 \cos[\pi(t - \tau)] \, dt$$

$$= [8 + \tau] \cos(\pi\tau)$$

, and for positive τ, Fig.6.18c, $R(\tau)$ is given by

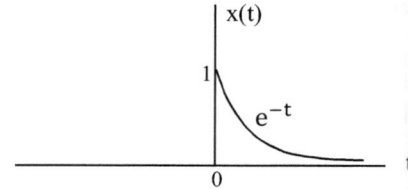

Fig.1.9a. Single sided decaying signal.

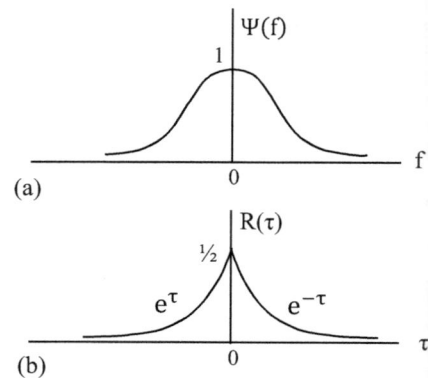

(a)

(b)

Fig.6.17. For the signal $x(t) = e^{-t} u(t)$, Fig.9a.
 a) The energy spectral density,
 b) The autocorrelation function.

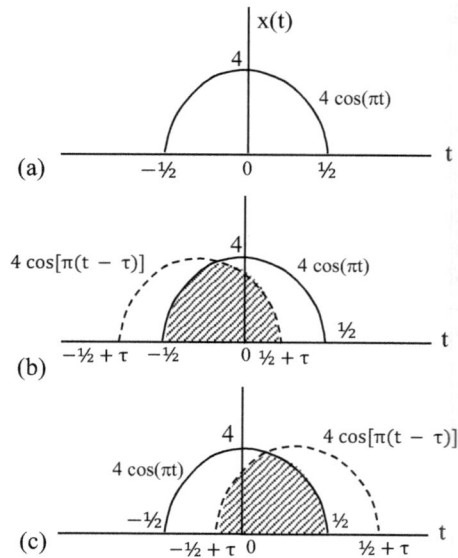

(a)

(b)

(c)

Fig.6.18. a) The signal $x(t) = 4 \, \text{rect}(t) \cos(\pi t)$,
 b) Negative τ, c) Positive τ.

$$R(\tau) = \int_{-\frac{1}{2}+\tau}^{\frac{1}{2}} 4\cos(\pi t)\, 4\cos[\pi(t-\tau)]\, dt = [8-\tau]\cos(\pi\tau)$$

The autocorrelation function $R(\tau)$ can also be obtained, using the frequency domain, Eq.(3.61), where the Fourier transform of the signal $x(t) = 4\,\text{rect}(t)\cos(\pi t)$, $X(f)$ is given by

$$X(f) = 2\,\text{sinc}(f-\tfrac{1}{2}) + 2\,\text{sinc}(f+\tfrac{1}{2})$$

, and the energy spectral density is

$$\Psi(f) = |X(f)|^2 = 4\,\text{sinc}^2(f-\tfrac{1}{2}) + 8\,\text{sinc}(f+\tfrac{1}{2})\,\text{sinc}(f-\tfrac{1}{2}) + 4\,\text{sinc}^2(f+\tfrac{1}{2})$$

, or equivalently

$$\Psi(f) = 4\,\text{sinc}^2(f-\tfrac{1}{2}) + 4\,\text{sinc}^2(f+\tfrac{1}{2})$$

, the autocorrelation function is obtained by taking the inverse Fourier transform of the energy spectral density, $R(\tau)$ will be

$$R(\tau) = 4\,\text{tri}(\tau)\,e^{j\pi\tau} + 4\,\text{tri}(\tau)\,e^{-j\pi\tau}$$

, express the exponential terms by cosine function, $R(\tau)$ yields

$$R(\tau) = 8\,\text{tri}(\tau)\cos(\pi\tau)$$

The energy content E equals the origin value of the autocorrelation function, $E = R(0) = 8$ joule. Fig.6.19a,b. show the energy spectral density $\Psi(f)$ and the autocorrelation function $R(\tau)$ which co-incidences with the two previous separate equations of the negative and positive τ.

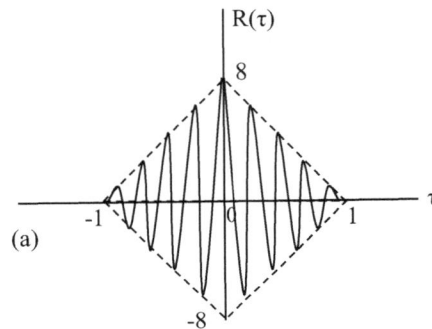

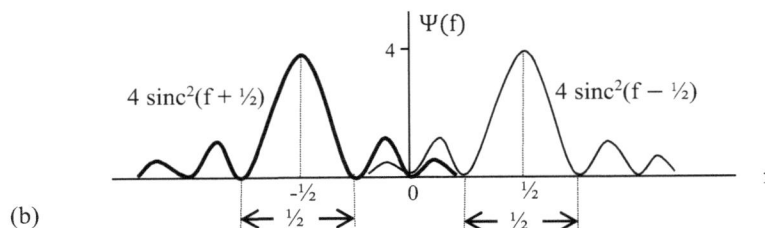

(a)

(b)

Fig.6.19. a) The autocorrelation function, b) The energy spectral density, for the pulse, Fig.6.18a.

Problem 6.11

Evaluate the autocorrelation function $R(\tau)$ and the energy content, using the time domain, of the pulse, Fig.6.20a, and given by

$$x(t) = e^{-at}\,[u(t) - u(t-T)]$$

(a) The signal x(t)

(b)) Negative τ.

(c) Positive τ.

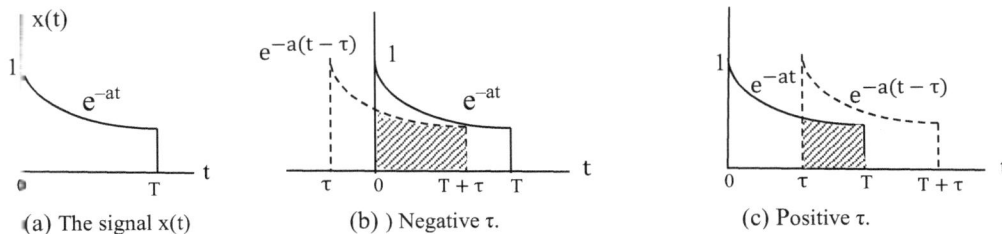

Fig.6.20. The pulse $x(t) = e^{-at}[u(t) - u(t-T)]$ and varying τ for the autocorrelation $R(\tau)$.

Solution
The energy signal x(t), Fig.6.20a, can be expressed by

$$x(t) = e^{-at} \; rect\left[\frac{t - \frac{1}{2}T}{T}\right]$$

, the autocorrelation function of x(t), Eq.(6.1), for negative τ, Fig.6.20b, $R(\tau)$ is given by

$$R(\tau) = \int_0^{T+\tau} e^{-at} \, e^{-a(t-\tau)} \, dt = \frac{e^{-aT}}{a} \; sinh[a(T + \tau)] \qquad -T \le \tau \le 0$$

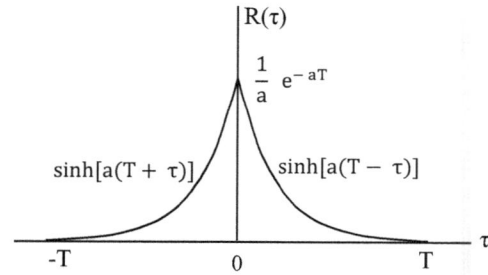

, for positive τ, Fig.6.20c, $R(\tau)$ is given by

$$R(\tau) = \int_\tau^T e^{-at} \, e^{-a(t-\tau)} \, dt = \frac{e^{-aT}}{a} \; sinh[a(T - \tau)] \qquad 0 \le \tau \le T$$

, then the autocorrelation function, Fig.6.21, can be expressed by

$$R(\tau) = \frac{e^{-aT}}{a} \; sinh[a(T - |\tau|)] \qquad -T \le \tau \le T \qquad (6.12)$$

, the energy content equals the origin value of the autocorrelation function $R(\tau)$, E is given by

$$E = R(0) = \frac{e^{-aT}}{a} \; sinh(aT) \qquad joule$$

, another way to get the energy content, using the time domain, Eq.(1.10), E is given by

$$E = \int_0^T e^{-at} \, e^{-at} \, dt = \frac{e^{-aT}}{a} \; sinh(aT) \qquad joule$$

Fig.6.21. The autocorrelation function $R(\tau)$.

6.3.4. Autocorrelation Function of Power Signals
Let $x_p(t)$ is a complex power signal (periodic) of periodic time T_o. The autocorrelation function $R_p(\tau)$ of $x_p(t)$ is expressed by the integrating product of $x_p(t)$ and its complex conjugate delayed by variable amount of time delay τ, normalized to one ohm, $R_p(\tau)$ is given by

$$R_p(\tau) = \frac{1}{T_o} \int_{-T_o/2}^{T_o/2} x_p(t) \, x_p^*(t - \tau) \, dt \qquad (6.13)$$

, where $x_p^*(t - \tau)$ is the complex conjugate version of the signal and τ is the time delay which plays the role of a scanning parameter from $-\infty$ to ∞, till the intersection of $x_p(t)$ and $x_p^*(t - \tau)$ exists, then the limits of the integration are determined. If the signal $x_p(t)$ is complex-valued then the autocorrelation function $R_p(\tau)$ is also complex value as well. Eq.(6.13) shows that the complex conjugate $x_p^*(t - \tau)$ is delayed by τ with respect to the signal $x_p(t)$, then the autocorrelation function $R_p(\tau)$ is also given by

$$R_p(\tau) = \frac{1}{T_o} \int_{-T_o/2}^{T_o/2} x_p(t + \tau) \, x_p^*(t) \, dt \qquad (6.14)$$

, where the complex conjugate of the signal $x_p^*(t)$ is also delayed by τ with respect to the signal $x_p(t + \tau)$. The autocorrelation function of the periodic signal $R_p(\tau)$ exhibits conjugate symmetry where $R_p(\tau) = R_p^*(-\tau)$, Equations (6.13) and (6.14).

6.3.5. Periodicity of Autocorrelation Function for Power Signals
Since each cycle of the periodic signal behaves as an independent energy signal, and has its own autocorrelation function $R(\tau)$, then the autocorrelation function of the periodic signal $R_p(\tau)$ is also periodic within the same periodic time T_o of the periodic signal itself. Consider a periodic signal $x_p(t)$ of periodic time T_o, it can be expressed by

$$x_p(t) = \sum_{m=-\infty}^{\infty} x(t - mT_o) \qquad \text{for all t,} \quad m = 0, 1, 2, 3, \dots \qquad (1.3)$$

, or equivalently

$$x_p(t - \tau) = \sum_{m=-\infty}^{\infty} x(t - \tau - mT_o) \qquad \text{for all t,} \quad m = 0, 1, 2, 3, \dots$$

, where x(t) is the generating function of the periodic signal $x_p(t)$. Then the autocorrelation function of the periodic signal $R_p(\tau)$, Eq.(6.13) yields

$$R_p(\tau) = \frac{1}{T_o} \int_{-T_o/2}^{T_o/2} x_p(t) \left\{ \sum_{m=-\infty}^{\infty} x^*(t - \tau - mT_o) \right\} dt$$

, interchange the order of summation and integration, $R_p(\tau)$ yields

$$R_p(\tau) = \frac{1}{T_o} \sum_{m=-\infty}^{\infty} \left\{ \int_{-T_o/2}^{T_o/2} x_p(t) \, x^*(t - \tau - mT_o) \right\} dt$$

, or equivalently

$$R_p(\tau) = \frac{1}{T_o} \sum_{m=-\infty}^{\infty} R(\tau - mT_o) \qquad , m = 0, 1, 2, 3, \dots \qquad (6.15)$$

, Eq.(6.15) shows that the autocorrelation function of the periodic signal $R_p(\tau)$ is also periodic within the same periodic time T_o as the periodic signal itself, and $R(\tau)$ is the autocorrelation function of the energy signal x(t), where x(t) is the generating function of the periodic signal $x_p(t)$.

6.3.6. Fourier Transform of Autocorrelation Function for Power Signals

Using the Fourier transform technique of the periodic signals, Eq.(5.1), and making use of Eq.(6.4), the Fourier transform pair of the autocorrelation function $R_p(\tau)$, Eq.(6.15), is given by

$$\frac{1}{T_o} \sum_{m=-\infty}^{\infty} R(\tau - mT_o) \rightleftharpoons \frac{1}{T_o^2} \sum_{n=-\infty}^{\infty} \left| X\left(\frac{n}{T_o}\right) \right|^2 \delta\left(f - \frac{n}{T_o}\right) \qquad (6.16)$$

, where $\left| X(n/T_o) \right|^2$ is the Fourier transform of the autocorrelation function (energy spectral density), Eq.(6.4), at the discrete frequencies n/T_o, and $X(n/T_o)$ is the Fourier transform of the generating function x(t) for the periodic signal $x_p(t)$ at the frequencies $f = n/T_o$. Equations (6.16) and Eq.(2.24) have the same frequency domain which is the Power Spectral Density PSD $\Omega(f)$, is given by

$$\Omega(f) = \frac{1}{T_o^2} \sum_{n=-\infty}^{\infty} \left| X\left(\frac{n}{T_o}\right) \right|^2 \delta\left(f - \frac{n}{T_o}\right) \qquad (2.24)$$

Then the autocorrelation function $R_p(\tau)$ of the periodic signal and the Power Spectral Density PSD $\Omega(f)$, constitute a Fourier transform pair, or equivalently

$$\Omega(f) = \int_{-\infty}^{\infty} R_p(\tau) \, e^{-j2\pi f \tau} \, d\tau \qquad (6.17a)$$

$$, \text{and} \quad R_p(\tau) = \int_{-\infty}^{\infty} \Omega(f) \, e^{j2\pi f \tau} \, df \qquad (6.17b)$$

, Equations (6.17a,b) are known as the Wiener-Khintchine relations for the power signals. The evaluation of the autocorrelation function $R_p(\tau)$ of the power signal $x_p(t)$ may be obtained by three ways, either in the time domain using Eq.(6.13) to get $R_p(\tau)$ directly for varying τ, or in the frequency domain by getting the power spectral density $\Omega(f)$ of the periodic signal $x_p(t)$, Eq.(2.24), and then return to the time domain by taking the inverse Fourier transform, Eq.(6.17b). The third way is carried out by getting the autocorrelation function $R(\tau)$ of the generating function x(t) as an energy signal, which generates the periodic signal $x_p(t)$ and then get the autocorrelation function $R_p(\tau)$, Eq.(6.15), of the periodic signal $x_p(t)$, in terms of the autocorrelation function $R(\tau)$ of the generating function x(t) [1].

6.3.7. Evaluation of Average Power in terms of Autocorrelation Function

The average power of the periodic signal $x_p(t)$ equals the origin value of the autocorrelation function $R_p(\tau)$, Eq.(6.13) yields

$$R_p(0) = \frac{1}{T_o} \int_{-T_0/2}^{T_0/2} x_p(t) \, x_p^*(t) \; dt$$

, or equivalently

$$P_{av} = R_p(0) = \frac{1}{T_0} \int_{-T_0/2}^{T_0/2} |x_p(t)|^2 \; dt \qquad\qquad (6.18)$$

, which is the same average power of the periodic signal $x_p(t)$, Eq.(1.9), and also the maximum value of the autocorrelation function $R_p(\tau)$ which occurs at the origin, where $|R_p(\tau)| \le R_p(0)$, for all τ [1].

Problem 6.12

For the energy pulse $x(t) = e^{-aT} [u(t) - u(t-T)]$, Fig.6.20a, problem 6.11. If $x(t)$ is repeated periodically every 2T seconds, Fig.6.22a. Find the autocorrelation function and the average power of the periodic signal $x_p(t)$.

Solution

The signal $x(t)$, Fig.6.20a, is considered the generating function and its autocorrelation function is given by Eq.(6.12), then the autocorrelation function $R_p(\tau)$ of the periodic signal $x_p(t)$, Fig.6.22b, Eq.(6.15) can be expressed by

$$R_p(\tau) = \frac{1}{2T} \frac{e^{-aT}}{a} \sum_{m=-\infty}^{\infty} \sinh\{a[(1 - 2m)T - |\tau|]\} \qquad , m = 0, 1, 2, 3, \dots$$

, the average power equals the origin value of the autocorrelation function $R_p(\tau)$, Eq.(6.18), P_{av} is given by

$$p_{av} = R_p(0) = \frac{e^{-aT}}{2aT} \sinh(aT) \qquad\qquad \text{watt}$$

, another way to obtain P_{av}, Eq.(1.9), P_{av} is given by

$$p_{av} = \frac{1}{2T} \int_0^T e^{-at} \, e^{-at} \, dt = \frac{e^{-aT}}{2aT} \sinh(aT) \qquad \text{watt}$$

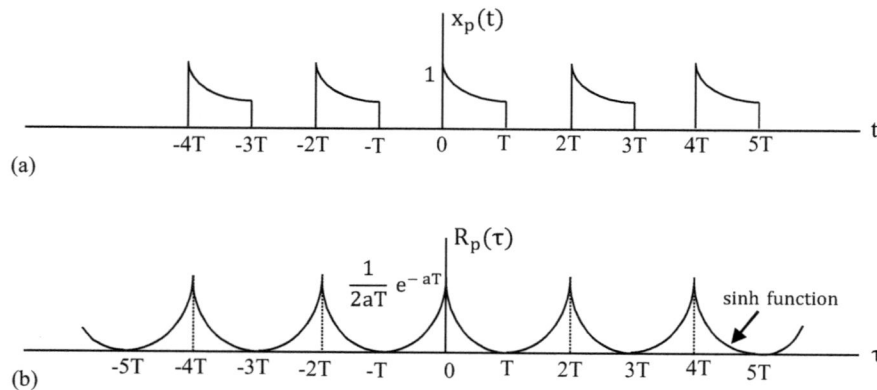

(a)

(b)

Fig.6.22. a) A periodic signal $x_p(t)$, b) The autocorrelation function $R_p(\tau)$.

Problem 6.13

For the rectangular Pulse $x(t) = A \, rect(t/T)$, Fig.1.2a, problem 6.4. If $x(t)$ is repeated periodically every T_o seconds, Fig.1.2b. Find the autocorrelation function and the average power of the periodic signal $x_p(t)$.

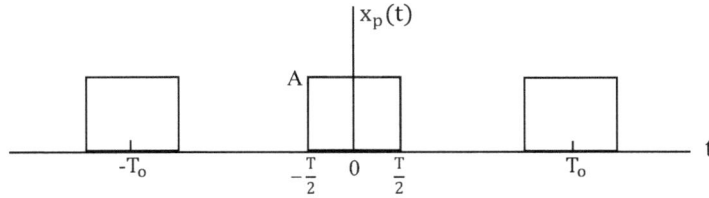

Fig.1.2b. A periodic signal $x_p(t)$.

Solution

The signal $x(t)$ is considered the generating function and its autocorrelation function is given by Eq.(6.9), then the autocorrelation function $R_p(\tau)$ of the periodic signal $x_p(t)$, Fig.6.23, Eq.(6.15) can be expressed by

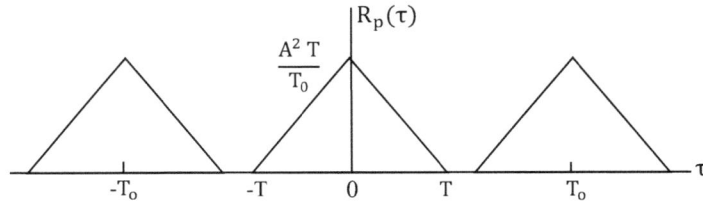

Fig.6.23. The autocorrelation function $R_p(\tau)$.

$$R_p(\tau) = \frac{1}{T_o} \sum_{m=-\infty}^{\infty} A^2 \, T \, tri[\frac{\tau - mT_o}{T}] \qquad , m = 0, 1, 2, 3, \dots$$

, the average power equals the origin value of the autocorrelation function $R_p(\tau)$, Eq.(6.18) yields $P_{av} = R_p(0) = A^2 T/T_o$ watt, which is the same average power, problem 1.2, Eq.(1.12), using the time domain.

Problem 6.14

For the RF pulse, $x(t) = A \, rect(t/T) \cos(2\pi f_c t)$, Fig.1.8a, problem 6.5. If $x(t)$ is repeated periodically every T_o seconds, Fig.1.8b. Find the autocorrelation function and the average power of the periodic signal $x_p(t)$.

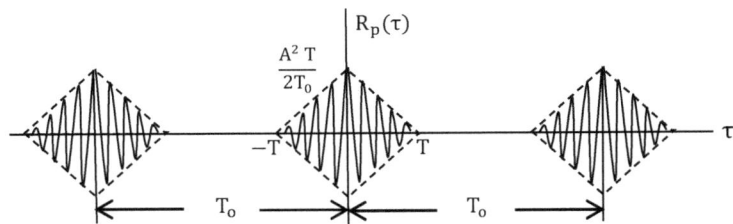

Fig.1.8b. Periodic RF pulse.

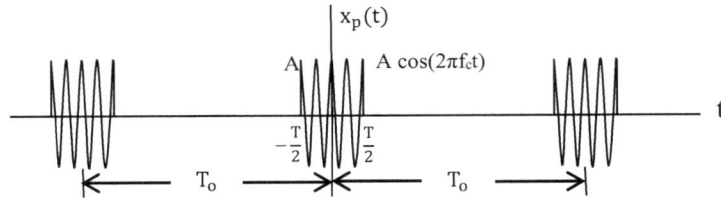

Fig.6.24. The autocorrelation function $R_p(\tau)$ of the Periodic RF pulse.

Solution

The signal $x(t)$ is considered the generating function and its autocorrelation function is given by Eq.(6.11), and the autocorrelation function $R_p(\tau)$ of the periodic signal $x_p(t)$, Fig.6.24, Eq.(6.15), can be expressed by

$$R_p(\tau) = \frac{1}{T_o} \sum_{m=-\infty}^{\infty} \frac{1}{2} A^2 \, T \, tri\left[\frac{\tau - mT_o}{T}\right] \cos[2\pi f_c (\tau - mT_o)] \qquad , m = 0, 1, 2, 3, \dots$$

, the average power equals the origin value of the auto-correlation function $R_p(\tau)$, Eq.(6.18) yields $P_{av} = \frac{1}{2} A^2 T/T_o$ watt, which is the same average power of problem 1.4, Eq.(1.14), using the time domain.

Problem 6.15

Find the autocorrelation function and the average power of the power signal , Fig.6.25a, using the time domain, and is given by

(a)

$$x_p(t) = A \cos(2\pi f_c t)$$

Solution

The autocorrelation function $R_p(\tau)$ can be obtained from the time domain, Eq.(6.13), $R_p(\tau)$ is given by

(b)

(c)

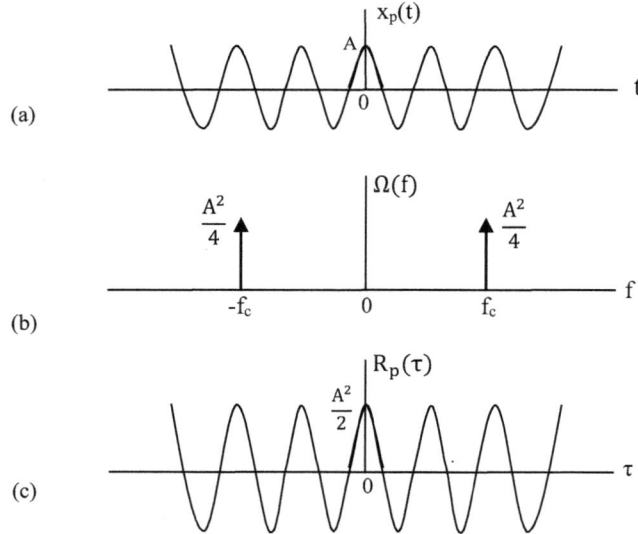

Fig.6.25. a) Periodic signal $x_p(t)$, b) The spectral density, c) The autocorrelation function.

$$R_p(\tau) = \frac{1}{T_o} A^2 \int_{-T_o/2}^{T_o/2} \cos(2\pi f_c t) \cos[2\pi f_c(t - \tau)] \, dt$$

, using the formula "$2 \cos(x) \cos(y) = \cos(x + y) + \cos(x - y)$", $R_p(\tau)$ yields

$$R_p(\tau) = \tfrac{1}{2} A^2 \cos(2\pi f_c \tau)$$

, the autocorrelation function $R_p(\tau)$ can also be obtained, using the frequency domain, where the Fourier transform of the signal $x_p(t) = A \cos(2\pi f_c t)$, $X(f)$ is given by

$$X(f) = \tfrac{1}{2} A \, \delta(f - f_c) + \tfrac{1}{2} A \, \delta(f + f_c)$$

, the spectral density $\Omega(f)$, Fig.6.25b, Eq.(3.61), is given by

$$\Omega(f) = |X(f)|^2 = \tfrac{1}{4} A^2 \, \delta(f - f_c) + \tfrac{1}{4} A^2 \, \delta(f + f_c)$$

, the autocorrelation function $R_p(\tau)$ is obtained by taking the inverse Fourier transform of the spectral density, $R_p(\tau)$ is given by

$$R_p(\tau) = \tfrac{1}{2} A^2 \cos(2\pi f_c \tau)$$

, the average power P_{av} equals the origin value of the autocorrelation function $R_p(\tau)$, P_{av} is given by

$$P_{av} = R_p(0) = \tfrac{1}{2} A^2 \qquad \text{watt}$$

, the average power can also be obtained, using the time domain, Eq.(1.9), where $f_c = 1/T_o$, P_{av} is given by

$$P_{av} = \frac{1}{T_o} \int_{-T_o/2}^{T_o/2} A^2 \cos^2(2\pi f_c t) \, dt = \tfrac{1}{2} A^2 \qquad \text{watt}$$

, Fig.6.25b,c. show the spectral density and the autocorrelation function respectively.

6.4. Cross-Correlation Function

Cross-correlation function is a measure of the similarity between a signal and the time delayed version of another signal. The cross-correlation function and the cross spectral density constitute Fourier transform pair. Cross-correlation function is a useful term for checking the orthogonality of two different signals. The formula of the cross-correlation function depends on whether the pair of signals being considered are energy signals or power signals [1,4].

6.4.1. Cross-Correlation Function of Energy Signals

Let $x_1(t)$ and $x_2(t)$ are two different complex energy signals (unperiodic). The cross-correlation function $R_{12}(\tau)$ of this pair of signals, involves the integrating product of $x_1(t)$ and the complex conjugate of the signal $x_2(t)$ delayed by variable amount of time delay τ, $R_{12}(\tau)$ is expressed by

$$R_{12}(\tau) = \int_{-\infty}^{\infty} x_1(t)\; x_2^*(t-\tau)\; dt \qquad (6.19)$$

, or
$$R_{12}(\tau) = \int_{-\infty}^{\infty} x_1(t+\tau)\; x_2^*(t)\; dt \qquad (6.20)$$

, and the cross-correlation function $R_{21}(\tau)$ is given by

$$R_{21}(\tau) = \int_{-\infty}^{\infty} x_2(t)\; x_1^*(t-\tau)\; dt \qquad (6.21)$$

, or
$$R_{21}(\tau) = \int_{-\infty}^{\infty} x_2(t+\tau)\; x_1^*(t)\; dt \qquad (6.22)$$

, the cross-correlation functions of the energy signals $R_{12}(\tau)$ and $R_{21}(\tau)$ show that $R_{12}(\tau) = R_{21}^*(-\tau)$, the cross-correlation process does not obey the commutative law of algebra, where $R_{12}(\tau) \neq R_{21}(\tau)$, unlike the convolution function principle, Eq.(3.43a).

6.4.2. Fourier Transform of Cross-Correlation Function for Energy signals

Fourier transform of the cross-correlation function is obtained by making use of the convolution function property, where the cross-correlation function can also be expressed by

$$R_{12}(\tau) = x_1(\tau) \otimes x_2^*(-\tau)$$

$$= \int_{-\infty}^{\infty} x_1(t)\; x_2^*[-(\tau-t)]\; dt \qquad (6.23)$$

, which is the same cross-correlation $R_{12}(\tau)$, Eq.(6.19). The Fourier transform of the cross-correlation function $R_{12}(\tau)$, Eq.(6.23), is the product of the Fourier transform of $x_1(\tau)$ multiplied by the Fourier transform of $x_2^*(-\tau)$, and is defined by the Cross Energy Spectral Density CESD $\Psi_{12}(f)$.

6.4.3. Cross Spectral Density of Energy Signals

Cross spectral density $\Psi_{12}(f)$ of the energy signals is the frequency domain of the cross correlation function $R_{12}(\tau)$, and can be obtained by taking the Fourier transform of Eq.(6.23), where the Fourier transform pair will be

$$x_1(\tau) \otimes x_2^*(-\tau) \quad \rightleftharpoons \quad X_1(f)\; X_2^*(f)$$

, then
$$R_{12}(\tau) \quad \rightleftharpoons \quad X_1(f)\; X_2^*(f) \qquad (6.24)$$

, in the frequency domain, the product $X_1(f)\, X_2^*(f)$ is the Cross Energy Spectral Density CESD $\Psi_{12}(f)$ of the two energy signals $x_1(t)$ and $x_2(t)$, and is given by

$$\Psi_{12}(f) = X_1(f)\; X_2^*(f) \qquad \text{joule/Hz} \qquad (6.25)$$

, or equivalently

$$\Psi_{12}(f) = \int_{-\infty}^{\infty} R_{12}(\tau)\; e^{-j2\pi f\tau}\; d\tau \qquad (6.26a)$$

, and
$$R_{12}(\tau) = \int_{-\infty}^{\infty} \Psi_{12}(f)\; e^{j2\pi f\tau}\; df \qquad (6.26b)$$

, also
$$R_{21}(\tau) \quad \rightleftharpoons \quad X_2(f)\; X_1^*(f)$$

, where
$$\Psi_{21}(f) = X_2(f)\; X_1^*(f) \qquad \text{joule/Hz}$$

The evaluation of the cross-correlation function $R_{12}(\tau)$ of the energy signals $x_1(t)$ and $x_2(t)$ may be carried out by two ways, either in the time domain using Eq.(6.19) to get $R_{12}(\tau)$ directly for varying τ, or in the frequency domain where the Fourier transforms $X_1(f)$ and $X_2^*(f)$ of the signals $x_1(t)$ and $x_2(t)$, Eq.(6.24), are obtained and then multiply and again return to the time domain by taking the inverse Fourier transform. Note that, if the two energy signals are identical, the cross-correlation function, Eq.(6.19) becomes the autocorrelation function, Eq.(6.1).

6.4.4. Orthogonal Energy Signals in terms of Cross-correlation Function
The origin value of the cross-correlation $R_{12}(\tau)$, Eq.(6.19), is given by

$$R_{12}(0) = \int_{-\infty}^{\infty} x_1(t) \, x_2^*(t) \, dt \qquad (6.27)$$

, if the origin value $R_{12}(0) = 0$, then the two complex energy signals $x_1(t)$ and $x_2(t)$ are orthogonal, Eq.(2.7).

Problem 6.16
For the two energy pulses $x_1(t)$ and $x_2(t)$, Fig.6.26a,b, given by

$$x_1(t) = 10 \text{ rect}(t)$$

, and $\quad x_2(t) = 5 \text{ rect}(t/2)$

i. Find the cross-correlation function $R_{12}(\tau)$.
ii. Find the cross energy spectral density $\Psi_{12}(f)$.
iii. Check the orthogonality of the two energy pulses $x_1(t)$ and $x_2(t)$.

Solution
The cross-correlation function $R_{12}(\tau)$ can be obtained, using the time domain, Eq.(6.19), Fig.6.27a,b,c. is given by

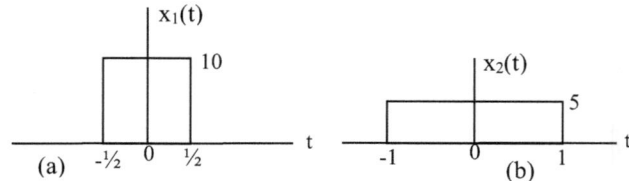
Fig.6.26. Two energy signals.

$$R_{12}(\tau) = \int_{-\frac{1}{2}}^{1+\tau} 10 \times 5 \; dt$$

$$= 50\,\tau + 75 \qquad \text{for} \quad -\frac{3}{2} \leq \tau \leq -\frac{1}{2}$$

$$, R_{12}(\tau) = \int_{-\frac{1}{2}}^{\frac{1}{2}} 10 \times 5 \; dt = 50 \quad \text{for} -\frac{1}{2} \leq \tau \leq \frac{1}{2}$$

$$, R_{12}(\tau) = \int_{-1+\tau}^{\frac{1}{2}} 10 \times 5 \; dt$$

$$= -50\,\tau + 75 \qquad \text{for} \quad \frac{1}{2} \leq \tau \leq 3/2$$

, the cross-correlation function $R_{12}(\tau)$ can also be obtained, using the frequency domain, where the energy signals $x_1(t) = 10 \text{ rect}(t)$, and $x_2(t) = 5 \text{ rect}(t/2)$, their Fourier transforms $X_1(f)$ and $X_2(f)$ are $X_1(f) = 10 \text{ sinc}(f)$, and $X_2(f) = 10 \text{ sinc}(2f)$, and the complex conjugate $X_2^*(f) = 10 \text{ sinc}(2f)$, then the Cross Energy Spectral Density CESD $\Psi_{12}(f)$ is given by

$$\Psi_{12}(f) = X_1(f) \; X_2^*(f) = 100 \text{ sinc}(f) \text{ sinc}(2f)$$

, express the sinc function $\text{sinc}(2f)$ in terms of sine function, the using the formula $\sin(2x) = 2 \sin(x) \cos(x)$, then express the cosine function by exponential terms, $\Psi_{12}(f)$ yields

$$\Psi_{12}(f) = 50 \text{ sinc}^2(f) \, [\, e^{-j\pi f} + e^{j\pi f} \,] \qquad (6.28)$$

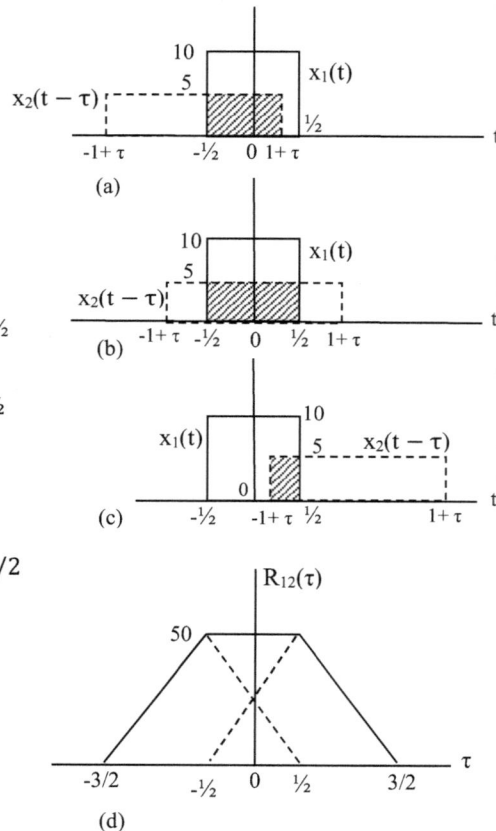
Fig.6.27. a) $-3/2 \leq \tau \leq -\frac{1}{2}$,
b) $-\frac{1}{2} \leq \tau \leq \frac{1}{2}$, c) $\frac{1}{2} \leq \tau \leq 3/2$,
d) The cross-correlation function $R_{12}(\tau)$.

, the cross-correlation function $R_{12}(\tau)$ is obtained by taking the inverse Fourier transform of the cross energy spectral density $\Psi_{12}(f)$, Eq.(6.28), $R_{12}(\tau)$ is given by

$$R_{12}(\tau) = 50 \text{ tri}(\tau - \tfrac{1}{2}) + 50 \text{ tri}(\tau + \tfrac{1}{2}) \qquad (6.29)$$

, Fig.6.27d. shows the cross-correlation function $R_{12}(\tau)$, which co-incidences with the three previous separate equations of the three different time period regions. The two signals $x_1(t)$ and $x_2(t)$ are not orthogonal because $R_{12}(0) \neq 0$.

Problem 6.17

For the two energy signals $x_1(t)$ and $x_2(t)$, where $x_1(t)$ is a doublet pulse and $x_2(t)$ is a rectangular pulse, Fig.6.28a,b:

i. Find the cross-correlation function $R_{12}(\tau)$.

ii. Find the cross energy spectral density $\Psi_{12}(f)$.

iii. Check the orthogonality of the two energy signals $x_1(t)$ and $x_2(t)$.

Solution

The cross-correlation function $R_{12}(\tau)$ can be obtained, using the time domain, Eq.(6.19), Fig.6.29a,b,c, $R_{12}(\tau)$ is given by

$$R_{12}(\tau) = \int_{-2}^{1+\tau} 2\, dt = 6 + 2\tau \quad \text{for} -3 \leq \tau \leq -1$$

$$, R_{12}(\tau) = \int_{-1+\tau}^{0} 2\, dt - \int_{0}^{1+\tau} 2\, dt$$

$$= -4\tau \qquad \text{for} -1 \leq \tau \leq 1$$

$$, R_{12}(\tau) = -\int_{-1+\tau}^{2} 2\, dt$$

$$= -6 + 2\tau \qquad \text{for} \quad 1 \leq \tau \leq 3$$

, the cross-correlation function $R_{12}(\tau)$ can also be obtained, using the frequency domain, where the doublet pulse $x_1(t)$ can be expressed by

$$x_1(t) = \text{rect}\left[\frac{t+1}{2}\right] - \text{rect}\left[\frac{t-1}{2}\right]$$

, also $x_2(t) = 2 \text{ rect}(t/2)$, and their Fourier transforms $X_1(f)$ and $X_2(f)$ are given by

$$X_1(f) = 2 \text{ sinc}(2f)\, e^{j2\pi f} - 2 \text{ sinc}(2f)\, e^{-j2\pi f}$$

, $X_2(f) = 4 \text{ sinc}(2f)$, and the complex conjugate is given by

$$X_2^*(f) = 4 \text{ sinc}(2f)$$

, then the cross energy spectral density is

$$\Psi_{12}(f) = X_1(f)\ X_2^*(f)$$

, or equivalently

$$\Psi_{12}(f) = 8 \text{ sinc}^2(2f)\, e^{j2\pi f} - 8 \text{ sinc}^2(2f)\, e^{-j2\pi f}$$

, the cross-correlation function $R_{12}(\tau)$ is obtained by taking the inverse Fourier transform of the energy cross spectral density, $R_{12}(\tau)$ will be

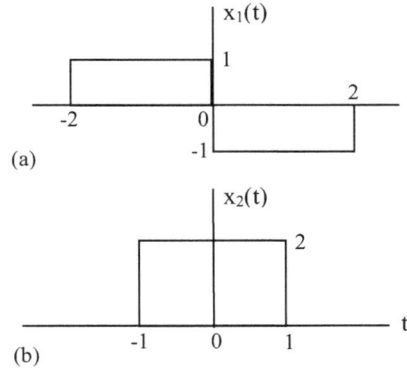

(a)

(b)

Fig.6.28. Two energy signals.

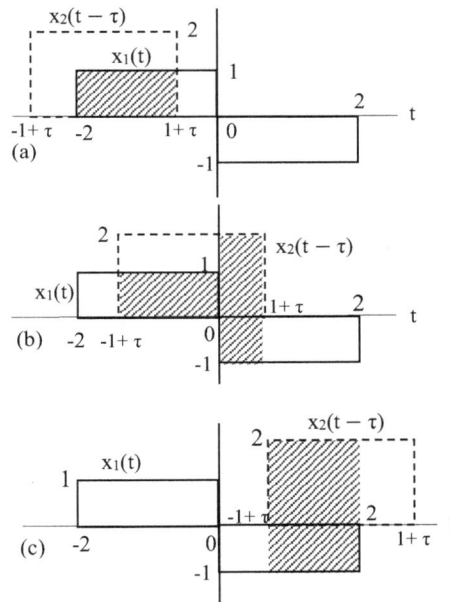

(a)

(b)

(c)

Fig.6.29. a) $-3 \leq \tau \leq -1$, b) $-1 \leq \tau \leq 1$, c) $1 \leq \tau \leq 3$.

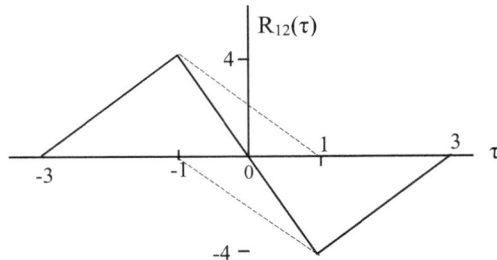

Fig.6.30. The cross-correlation function $R_{12}(\tau)$ of the doublet pulse $x_1(t)$ and $x_2(t)$, Fig.6.28.

$$R_{12}(\tau) = 4 \, \text{tri}\left[\frac{\tau+1}{2}\right] - 4 \, \text{tri}\left[\frac{\tau-1}{2}\right] \qquad\qquad (6.30)$$

, Fig.6.30. shows the cross-correlation function $R_{12}(\tau)$, which co-incidences with the three previous separate equations of the three different time periods. The two signals $x_1(t)$ and $x_2(t)$ are orthogonal because $R_{12}(0) = 0$.

Problem 6.18

For the two energy pulses $x_1(t)$ and $x_2(t)$, Fig.6.31a,b, are given by

$$x_1(t) = \text{rect}(t/2)$$

, and $x_2(t) = t \, \text{rect}(t/2)$.

i. Find the cross-correlation functions $R_{12}(\tau)$ and $R_{21}(\tau)$.

ii. Check the orthogonality of the two energy pulses $x_1(t)$ and $x_2(t)$.

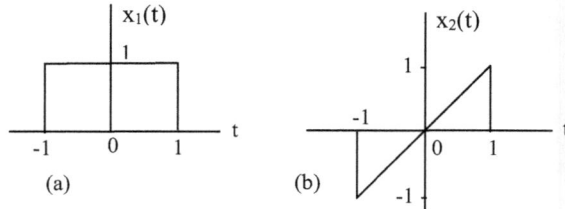

Fig.6.31. Two energy signals.

Solution

The cross-correlation function $R_{12}(\tau)$, can be obtained, using the time domain, Eq.(6.19), Fig.6.32a,b, is given by

$$R_{12}(\tau) = \int_{-1}^{1+\tau} (t - \tau) \, dt = -\tfrac{1}{2}\tau^2 - \tau \qquad \text{for} \;\; -2 \leq \tau \leq 0$$

, and

$$R_{12}(\tau) = \int_{-1+\tau}^{1} (t - \tau) \, dt = \tfrac{1}{2}\tau^2 - \tau \qquad \text{for} \;\; 0 \leq \tau \leq 2$$

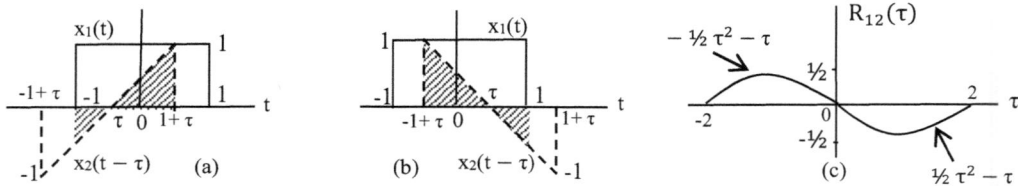

Fig.6.32. a) $-2 \leq \tau \leq 0$, b) $0 \leq \tau \leq 2$, c) The cross-correlation function $R_{12}(\tau)$.

, while the cross-correlation function $R_{21}(\tau)$, using the time domain, Eq.(6.21), Fig.6.33a,b, is given by

$$R_{21}(\tau) = \int_{-1}^{1+\tau} t \, dt = \tfrac{1}{2}\tau^2 + \tau \qquad \text{for} \;\; -2 \leq \tau \leq 0$$

, and

$$R_{12}(\tau) = \int_{-1+\tau}^{1} t \, dt = -\tfrac{1}{2}\tau^2 + \tau \qquad \text{for} \;\; 0 \leq \tau \leq 2$$

, the two signals $x_1(t)$ and $x_2(t)$ are orthogonal because $R_{12}(0) = 0$, also $R_{21}(0) = 0$.

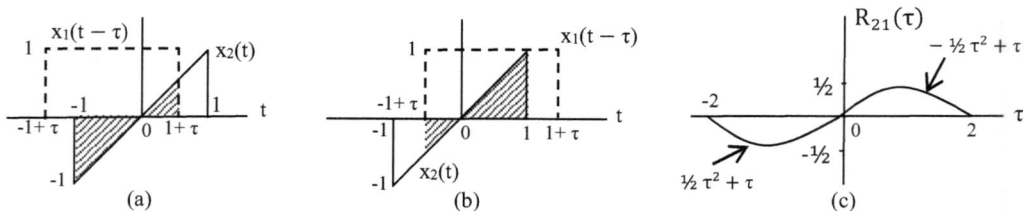

Fig.6.33. a) $-2 \leq \tau \leq 0$, b) $0 \leq \tau \leq 2$, c) The cross-correlation function $R_{21}(\tau)$.

Problem 6.19

For the two energy pulses $x_1(t)$ and $x_2(t)$, Fig.6.34a,b, given by

$$x_1(t) = 2\,\text{tri}(t/4)$$

, and $x_2(t) = \text{rect}(t/4)$

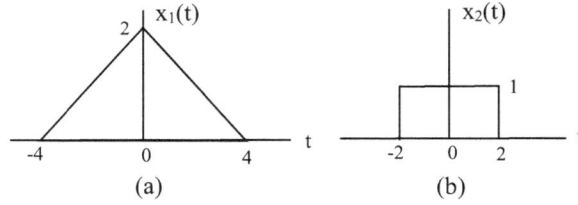

i. Find the cross-correlation function $R_{12}(\tau)$.
ii. Check the orthogonality of the two energy pulses $x_1(t)$ and $x_2(t)$.

(a)   (b)

Fig.6.34. Two energy signals.

Solution

The cross-correlation function $R_{12}(\tau)$, using time domain, Eq.(6.19), Fig.6.35a,b,c, $R_{12}(\tau)$ is given by

$$R_{12}(\tau) = \int_{-4}^{2+\tau} (\tfrac{1}{2}t + 2)\,dt = \tfrac{1}{4}\tau^2 + 3\tau + 9 \qquad \text{for } -6 \le \tau \le -2$$

$$R_{12}(\tau) = \int_{-2+\tau}^{0} (\tfrac{1}{2}t + 2)\,dt + \int_{0}^{2+\tau} (-\tfrac{1}{2}t + 2)\,dt = -\tfrac{1}{2}\tau^2 + 6 \qquad \text{for } -2 \le \tau \le 2$$

, and

$$R_{12}(\tau) = \int_{-2+\tau}^{4} (-\tfrac{1}{2}t + 2)\,dt = \tfrac{1}{4}\tau^2 - 3\tau + 9 \qquad \text{for } 2 \le \tau \le 6$$

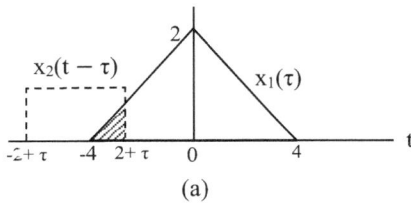

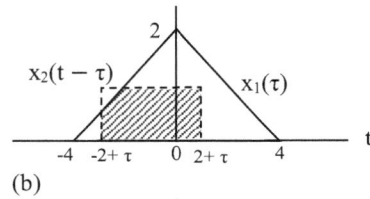

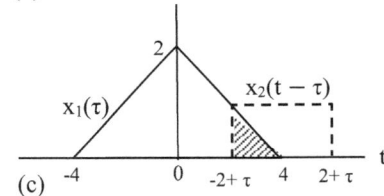

(a)   (b)

Fig.6.35. a) $-6 \le \tau \le -2$,
b) $-2 \le \tau \le 2$,
c) $2 \le \tau \le 6$.

(c)

, Fig.6.36. shows the cross-correlation function $R_{12}(\tau)$, which co-incidences with the three previous separate equations of the three different time period regions. The two signals $x_1(t)$ and $x_2(t)$ are not orthogonal because $R_{12}(0) \ne 0$.

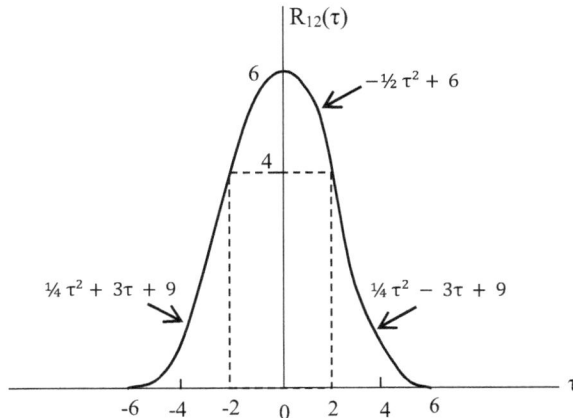

Fig.6.36. The cross-correlation function $R_{12}(\tau)$.

6.4.5. Cross-Correlation Function of Power Signals

Let $x_{1p}(t)$ and $x_{2p}(t)$ are two different complex power signals (unperiodic), both having the same periodic time T_o. The Cross-correlation function $R_{12p}(\tau)$ of this pair of signals, involves the integrating product of $x_{1p}(t)$ and the complex conjugate of the signal $x_{2p}(t)$ delayed by variable amount of time delay τ, $R_{12p}(\tau)$ is given by

$$R_{12p}(\tau) = \frac{1}{T_o} \int_{-T_o/2}^{T_o/2} x_{1p}(t) \, x_{2p}^*(t - \tau) \, dt \tag{6.31}$$

, or

$$R_{12p}(\tau) = \frac{1}{T_o} \int_{-T_o/2}^{T_o/2} x_{1p}(t + \tau) \, x_{2p}^*(t) \, dt \tag{6.32}$$

, and the cross-correlation function $R_{21p}(\tau)$ is given by

$$R_{21p}(\tau) = \frac{1}{T_o} \int_{-T_o/2}^{T_o/2} x_{2p}(t) \, x_{1p}^*(t - \tau) \, dt$$

, or

$$R_{21p}(\tau) = \frac{1}{T_o} \int_{-T_o/2}^{T_o/2} x_{2p}(t + \tau) \, x_{1p}^*(t) \, dt$$

, the cross-correlation functions of the power signals, $R_{12p}(\tau)$ and $R_{21p}(\tau)$ show that $R_{12p}(\tau) = R_{21p}^*(-\tau)$, the cross-correlation process does not obey the commutative law of algebra, where $R_{12p}(\tau) \neq R_{21p}(\tau)$, unlike the convolution function principle, Eq.(3.43a).

6.4.6. Periodicity of Cross-Correlation Function for Power Signals

Since each cycle of the two periodic signals, behaves as an independent energy signals, and have their cross-correlation function $R_{12}(\tau)$, then the cross-correlation function of the two periodic signals $R_{12p}(\tau)$ is also periodic within the same periodic time T_o of the two periodic signals itself. Consider two periodic signals $x_{1p}(t)$ and $x_{2p}(t)$ of periodic time T_o, they can be expressed by

$$x_{1p}(t) = \sum_{m=-\infty}^{\infty} x_1(t - mT_o) \qquad \text{for all } t, \quad m = 0, 1, 2, 3, \dots$$

,

$$x_{2p}(t) = \sum_{m=-\infty}^{\infty} x_2(t - mT_o) \qquad \text{for all } t, \quad m = 0, 1, 2, 3, \dots$$

, or equivalently

$$x_{1p}(t - \tau) = \sum_{m=-\infty}^{\infty} x_1(t - \tau - mT_o) \qquad \text{for all } t, \quad m = 0, 1, 2, 3, \dots$$

,

$$x_{2p}(t - \tau) = \sum_{m=-\infty}^{\infty} x_2(t - \tau - mT_o) \qquad \text{for all } t, \quad m = 0, 1, 2, 3, \dots$$

, where $x_1(t)$ and $x_2(t)$ are the generating functions of the two periodic signals $x_{1p}(t)$ and $x_{2p}(t)$. Then the cross-correlation function of the two periodic signals $R_{12p}(\tau)$, Eq.(6.31) yields

$$R_{12p}(\tau) = \frac{1}{T_o} \int_{-T_o/2}^{T_o/2} x_{1p}(t) \left\{ \sum_{m=-\infty}^{\infty} x_2^*(t - \tau - mT_o) \right\} dt$$

, interchange the order of summation and integration, $R_{12p}(\tau)$ yields

$$R_{12p}(\tau) = \frac{1}{T_o} \sum_{m=-\infty}^{\infty} \int_{-T_o/2}^{T_o/2} x_{1p}(t) \, x_2^*(t - \tau - mT_o) \, dt$$

, or equivalently

$$R_{12p}(\tau) = \frac{1}{T_o} \sum_{m=-\infty}^{\infty} R_{12}(\tau - mT_o) \qquad , m = 0, 1, 2, 3, \dots \qquad (6.33)$$

Then the cross-correlation function $R_{12p}(\tau)$ of the two periodic signals, is also periodic within the same periodic time T_o of the periodic signals, and $R_{12}(\tau)$ is the cross-correlation function of the two generating functions $x_1(t)$ and $x_2(t)$ for the periodic signals $x_{1p}(t)$ and $x_{2p}(t)$.

6.4.7. Fourier Transform of Cross-Correlation Function for Power Signals

Using the Fourier transform technique of the periodic signals, Eq.(5.1), and making use of Eq.(6.24), the Fourier transform pair of the cross-correlation function $R_{12p}(\tau)$ is given by

$$\frac{1}{T_o} \sum_{m=-\infty}^{\infty} R_{12}(\tau - mT_o) \quad \rightleftharpoons \quad \frac{1}{T_o^2} \sum_{n=-\infty}^{\infty} X_1(\frac{n}{T_o}) \, X_2^*(\frac{n}{T_o}) \, \delta(f - \frac{n}{T_o}) \qquad (6.34)$$

, where $X_{1p}(n/T_o)$ and $X_{2p}(n/T_o)$ are the Fourier transforms of the generating functions $x_1(t)$ and $x_2(t)$ for the periodic signals $x_{1p}(t)$ and $x_{1p}(t)$, at the discrete frequencies $f = n/T_o$. The Cross Power Spectral Density CPSD $\Omega_{12}(f)$, is given by

$$\Omega_{12}(f) = \frac{1}{T_o^2} \sum_{n=-\infty}^{\infty} X_1(\frac{n}{T_o}) \, X_2^*(\frac{n}{T_o}) \, \delta(f - \frac{n}{T_o}) \qquad (6.35)$$

Then the cross-correlation function $R_{12p}(\tau)$ of the periodic signals and the Cross Power Spectral Density CPSD $\Omega_{12}(f)$, constitute a Fourier transform pair, or equivalently

$$\Omega_{12}(f) = \int_{-\infty}^{\infty} R_{12}(\tau) \, e^{-j2\pi f\tau} \, d\tau \qquad (6.36)$$

, and $\qquad R_{12}(\tau) = \int_{-\infty}^{\infty} \Omega_{12}(f) \, e^{j2\pi f\tau} \, df \qquad (6.37)$

6.4.8. Cross Spectral Density of Power Signals

Since the cross correlation function of the power signals $R_{12p}(\tau)$ and the Cross Power Spectral Density CPSD $\Omega_{12}(f)$, Eq.(6.34), constitute the following Fourier transform pair

$$R_{12p}(\tau) \quad \rightleftharpoons \quad \frac{1}{T_o^2} \sum_{n=-\infty}^{\infty} X_1(\frac{n}{T_o}) X_2^*(\frac{n}{T_o}) \, \delta(f - \frac{n}{T_o}) \qquad (6.38)$$

, then the cross power spectral density of the two power signals $x_{1p}(t)$ and $x_{2p}(t)$, $\Omega_{12}(f)$ is given by Eq.(6.35).

The evaluation of the cross-correlation function $R_{12p}(\tau)$ of the two power signals $x_{1p}(t)$ and $x_{2p}(t)$ having the same periodic time T_o , may be obtained by three ways, either using the time domain Eq.(6.31) to get $R_{12p}(\tau)$ directly for varying τ, or using the frequency domain where the Fourier transform $X_1(n/T_o)$ and $X_2^*(n/T_o)$ of the two energy signals $x_1(t)$ and $x_2(t)$ at $f = n/T_o$ which generate the two periodic signals $x_{1p}(t)$ and $x_{2p}(t)$, are obtained, then the cross power spectral density $\Omega_{12}(f)$, Eq.(6.35) can be obtained, and then again return to the time domain by taking the inverse Fourier transform, Eq.(6.37) to get the cross-correlation function $R_{12p}(\tau)$. The third way to get the cross-autocorrelation function $R_{12p}(\tau)$ by obtaining the cross-correlation function $R_{12}(\tau)$ of the two generating signals $x_1(t)$ and $x_2(t)$ which generate the two power signals $x_{1p}(t)$ and $x_{2p}(t)$, and then get the cross-correlation function $R_{12p}(\tau)$ of the two periodic signals in terms of the cross-correlation function $R_{12}(\tau)$, Eq.(6.33).

6.4.9. Orthogonal Power Signals in terms of Cross-Correlation Function

The origin value of the cross-correlation $R_{12p}(\tau)$, Eq.(6.31), is given by

$$R_{12p}(0) = \frac{1}{T_o} \int_{-T_o/2}^{T_o/2} x_{1p}(t) \, x_{2p}^*(t) \, dt \qquad (6.39)$$

, if the origin value $R_{12p}(0) = 0$, then the two complex periodic signals $x_{1p}(t)$ and $x_{2p}(t)$ are orthogonal in the periodic time T_o , Eq.(2.7).

Problem 6.20

For the two energy pulses $x_1(t)$ and $x_2(t)$, Fig.6.26a,b, problem 6.16. If $x_1(t)$ and $x_2(t)$ are repeated periodically every T_o ($T_o \geq 3$ seconds), Fig.6.37a,b. Find the cross-correlation function $R_{12p}(\tau)$ of the two periodic signals $x_{1p}(t)$ and $x_{2p}(t)$.

Solution

The cross-correlation function $R_{12p}(\tau)$, Fig.6.37c, can be obtained, where $T_o = 3$ seconds, in terms of the cross-correlation function $R_{12}(\tau)$, Eq.(6.29) in Eq.(6.33), is given by

$$R_{12p}(\tau) = \frac{50}{3} \sum_{m=-\infty}^{\infty} [\, tri(\tau - \tfrac{1}{2} - 3m) + tri(\tau + \tfrac{1}{2} - 3m) \,] \qquad , \qquad m = 0, 1, 2, 3, \ldots$$

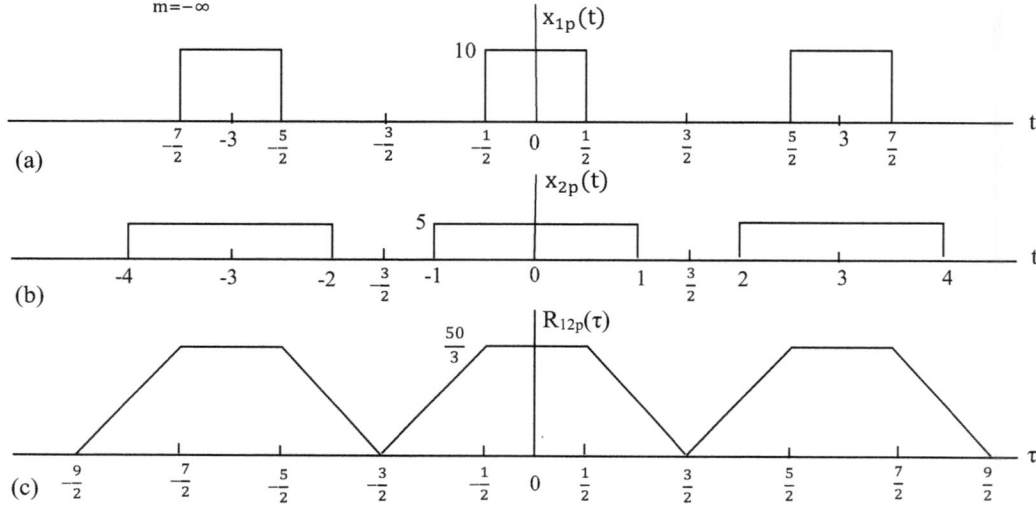

Fig.6.37. a) Periodic signal $x_{1p}(t)$, b) Periodic signal $x_{2p}(t)$, c) The cross-correlation $R_{12p}(\tau)$, $T_o = 3$ seconds.

Problem 6.21

For the two pulses $x_1(t)$ and $x_2(t)$, Fig.6.28a,b, problem 6.17. If $x_1(t)$ and $x_2(t)$ are repeated periodically every T_o ($T_o \geq 6$ seconds), Fig.6.38a,b. Find the cross-correlation function $R_{12p}(\tau)$ of the two periodic signals $x_{1p}(t)$ and $x_{2p}(t)$.

Solution

The cross-correlation function $R_{12p}(\tau)$, Fig.6.38c, can be obtained, where $T_o = 6$ seconds, in terms of the cross-correlation function $R_{12}(\tau)$, Eq.(6.30) in Eq.(6.33), is given by

$$R_{12p}(\tau) = \frac{2}{3} \sum_{m=-\infty}^{\infty} \left\{ tri\left[\frac{\tau + 1 - 6m}{2}\right] - tri\left[\frac{\tau - 1 - 6m}{2}\right] \right\} \qquad , \qquad m = 0, 1, 2, 3, \ldots$$

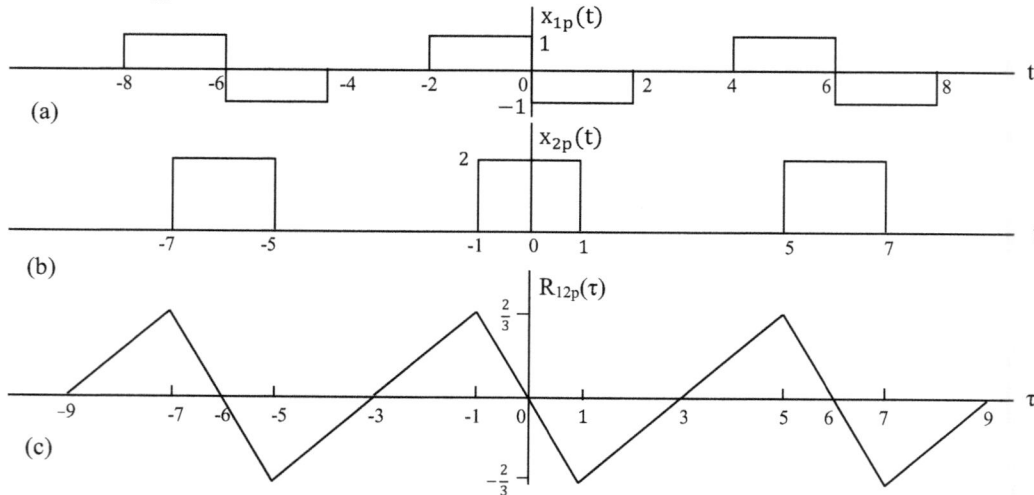

Fig.6.38. a) Periodic signal $x_{1p}(t)$, b) Periodic signal $x_{2p}(t)$, c) The cross-correlation $R_{12p}(\tau)$, $T_o = 6$ seconds.

Problem 6.22

For the two pulses $x_1(t)$ and $x_2(t)$, Fig.6.31a,b, problem 6.18. If $x_1(t)$ and $x_2(t)$ are repeated periodically every T_o ($T_o \geq 4$ seconds), Fig.6.39a,b. Sketch the cross-correlation functions $R_{12p}(\tau)$ and $R_{21p}(\tau)$ of the two periodic signals $x_{1p}(t)$ and $x_{2p}(t)$.

Solution

The cross-correlation functions $R_{12p}(\tau)$ and $R_{21p}(\tau)$ are shown in Fig.6.39c,d. , for $T_o = 4$ seconds.

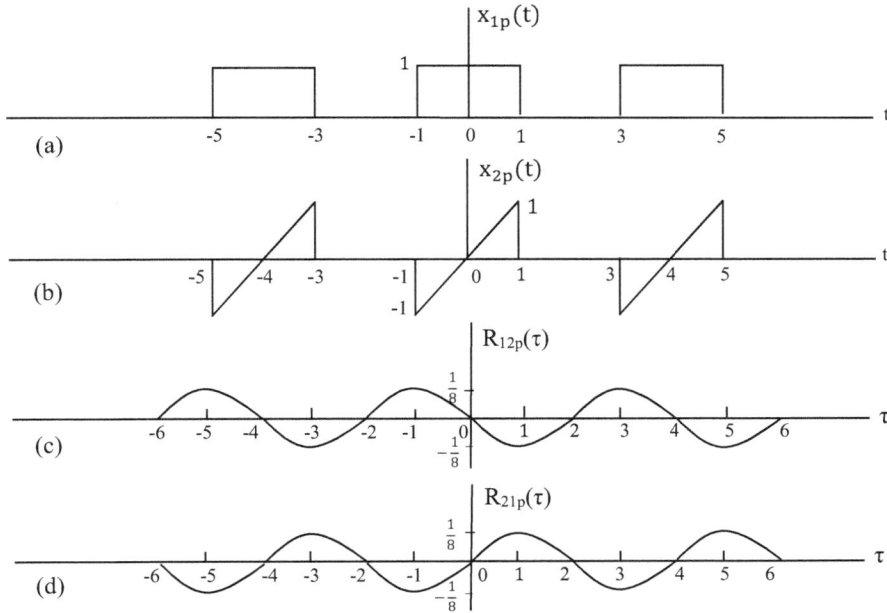

Fig.6.39. a) The periodic signal $x_{1p}(t)$, b) The periodic signal $x_{2p}(t)$,
c) The cross-correlation $R_{12p}(\tau)$, d) The cross-correlation $R_{21p}(\tau)$, $T_o = 4$ seconds.

Problem 6.23

For the two energy pulses $x_1(t)$ and $x_2(t)$, Fig.6.34a,b, problem 6.19. If $x_1(t)$ and $x_2(t)$ are repeated periodically every T_o ($T_o \geq 12$ seconds), Fig.6.40a,b. Sketch the cross-correlation function $R_{12p}(\tau)$ of the two periodic signals $x_{1p}(t)$ and $x_{2p}(t)$.

Solution

The cross-correlation function $R_{12p}(\tau)$ is shown in Fig.6.40c, for $T_o = 12$ seconds.

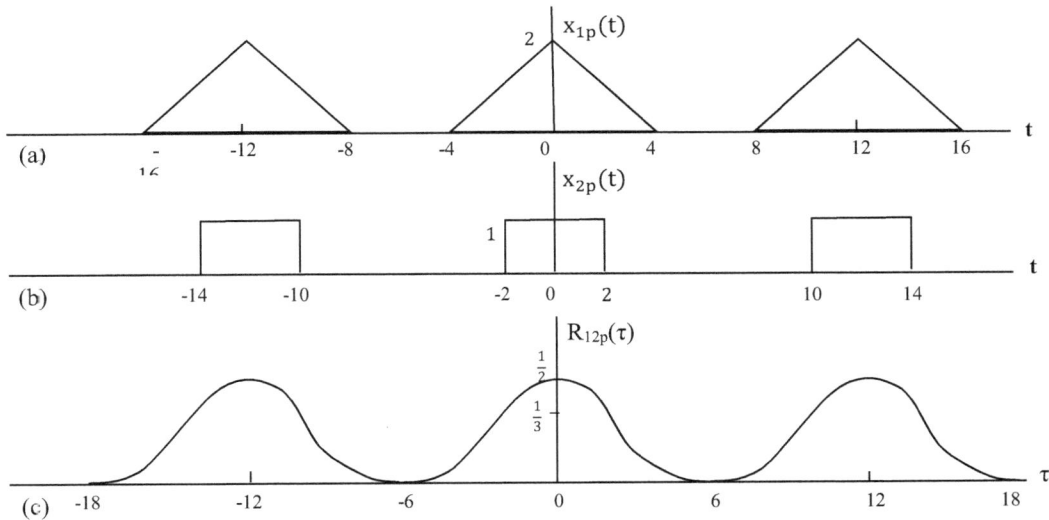

Fig.6.40. a) Periodic signal $x_{1p}(t)$, b) Periodic signal $x_{2p}(t)$, c) The cross-correlation $R_{12p}(\tau)$, $T_o = 12$ seconds.

Problems

1. A voltage across a resistor R ohm, is $v(t) = t\,e^{-t}\,u(t)$. What is the percentage of the energy content dissipated in the resistor associated within the frequency band $0 \le f \le \dfrac{\sqrt{3}}{2\pi}$ Hz.

2. For the energy signal $x(t) = e^{-at}\,u(t)$. Evaluate:
    i. The energy spectral density $\Psi(f)$ of $x(t)$.
    ii. The autocorrelation function $R(\tau)$ of $x(t)$.
    iii. The energy content of $x(t)$ from (ii).

3. For the following sinusoidal power signal

$$x_p(t) = A\,\cos(2\pi f_c t + \theta)$$

    Evaluate:
    i. The power spectral density $\Omega(f)$ of $x(t)$.
    ii. The autocorrelation function $R_p(\tau)$ of $x(t)$.
    iii. The average power of $x(t)$ from (ii).

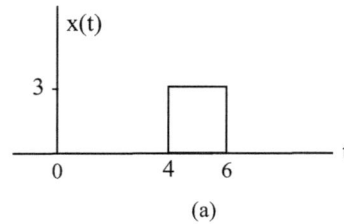
(a)

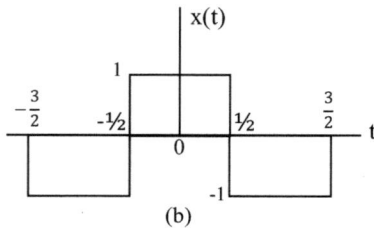
(b)

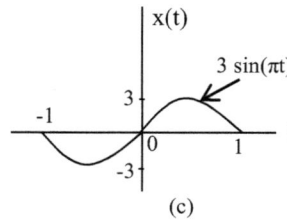
(c)

Fig.6.41. Energy signals.

4. For the energy signals, Fig.6.41a,b,c. Evaluate:
    i. The energy spectral density $\Psi(f)$ of $x(t)$.
    ii. The autocorrelation function $R(\tau)$ of $x(t)$.
    iii. The energy content of $x(t)$ from (ii).

5. For the double sided exponential energy signal, Fig.6.42. Evaluate:
    i. The autocorrelation function $R(\tau)$ of $x(t)$.
    ii. The energy content of $x(t)$ from (i).

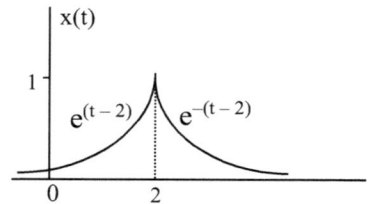
Fig.6.42. Energy signal.

6. For the following sinc spectrum

$$X(f) = 5\,\text{sinc}[3(f - 10^8)]$$

    Evaluate:
    i. The energy spectral density $\Psi(f)$ of $X(f)$.
    ii. The autocorrelation function $R(\tau)$ of $X(f)$.
    iii. The energy content of $x(t)$ from (ii).

7. For the following Gaussian monocycle pulse

$$x(t) = \frac{1}{\sqrt{2\pi}\,\sigma}\,e^{-(t-\mu)^2/2\sigma^2}$$

, where μ is the mean and σ is the variance of the statistical distribution of the pulse, are constant parameters. The first five derivatives of this Gaussian pulse, are the most widely used pulses in the Ultra Wide Band UWB communication systems. Find the autocorrelation functions of the first five derivatives of this Gaussian pulse. Note : consider zero order Gaussian pulse ($\mu = 0$).

8. For the two energy signals $x_1(t)$ and $x_2(t)$, Fig.6.43a,b:
    i. Find the cross-correlation function $R_{12}(\tau)$.
    i. Check the orthogonality of the two energy signals $x_1(t)$ and $x_2(t)$.

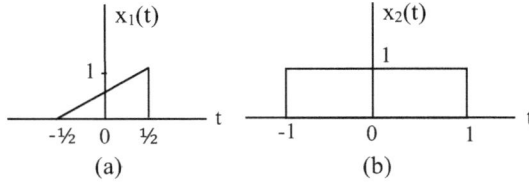

Fig.6.43. Two energy signals.

9. For the two energy signals $x_1(t)$ and $x_2(t)$, , Fig.6.44a,b:
    i. Find the cross-correlation function $R_{12}(\tau)$.
    ii. Find the cross energy spectral density $\Psi_{12}(f)$.
    iii. Check the orthogonality of the two energy signals $x_1(t)$ and $x_2(t)$.

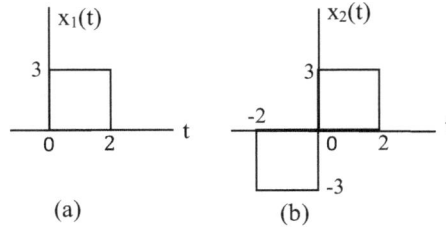

Fig.6.44. Two energy signals.

10. For the two energy signals $x_1(t)$ and $x_2(t)$, Fig.6.45a,b:
    i. Find the cross-correlation function $R_{21}(\tau)$.
    ii. Find the cross energy spectral density $\Psi_{21}(f)$.
    iii. Check the orthogonality of the two energy signals $x_1(t)$ and $x_2(t)$.

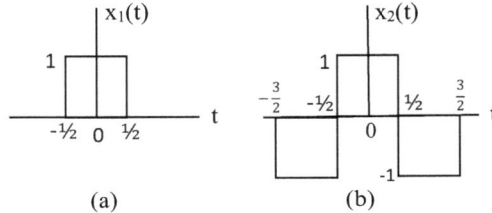

Fig.6.45. Two energy signals.

11. For the two energy signals $x_1(t)$ and $x_2(t)$, Fig.6.46a,b:
    i. Find the cross-correlation function $R_{12}(\tau)$.
    ii. Find the cross energy spectral density $\Psi_{12}(f)$.
    iii. Check the orthogonality of the two energy signals $x_1(t)$ and $x_2(t)$.

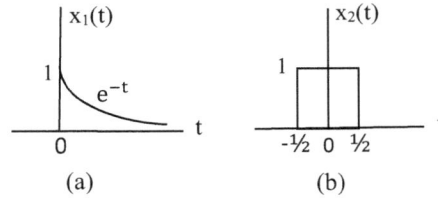

Fig.6.46. Two energy signals.

12 For the single sided decaying and rising exponential pulses $x_1(t) = e^{-t}\,u(t)$, and $x_2(t) = e^{t}\,u(-t)$, Fig.1.9a,b.
    i. Find the cross-correlation function $R_{12}(\tau)$.
    ii. Find the cross energy spectral density $\Psi_{12}(f)$.
    iii. Check the orthogonality of the two energy signals $x_1(t)$ and $x_2(t)$.

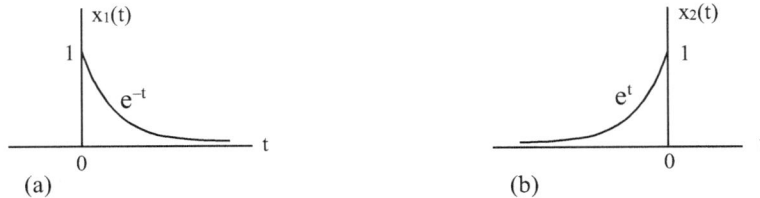

Fig.1.9. a) Single sided decaying signal, b) Single sided rising signal.

13. If the energy signals, Fig.6.41a,b,c, are repeated periodically every $T_o = 6$ seconds. Find the autocorrelation function $R_p(\tau)$ and the average power of the periodic signals $x_p(t)$.

14. If the energy signals $x_1(t)$ and $x_2(t)$, Fig.6.44, are repeated periodically every $T_o = 6$ seconds. Find the cross-correlation function $R_{12p}(\tau)$ of the periodic signals $x_{1p}(t)$ and $x_{2p}(t)$.

15. If the energy signals $x_1(t)$ and $x_2(t)$, Fig.6.45, are repeated periodically every $T_o = 6$ seconds. Find the cross-correlation function $R_{21p}(\tau)$ of the periodic signals $x_{1p}(t)$ and $x_{2p}(t)$.

Send Orders for Reprints to reprints@benthamscience.net

CHAPTER 7

Signal Transmission and Systems

Abstract: Linear system is a physical device that produce an output signal y(t) in response to an input signal x(t), such as filters or communication channels operating in their linear region. The filter is a frequency selective device that is used to limit the spectrum of a signal to some band of frequencies while the communication channel is a transmission medium that connects the transmitter to the receiver through a certain band of frequencies. The linear system refers to superposition principle, where the response of a linear system to a number of different inputs applied simultaneously, equals the sum of the responses of the system when each input is applied individually. While, time invariant system refers to the principle of, if the input signal is delayed by certain delay, the output signal will be delayed by the same amount of delay.

Keywords: Systems; Impulse response; Transfer function; Distortionless system; Distortion System.

It is important to investigate, how the transmitting signals are processed through these linear time invariant filters and communication channels. The investigation may be carried out, either in the time domain where the system is described in terms of its impulse response h(t), or in the frequency domain where the system is described in terms of its transfer function H(f). The impulse response h(t) and the transfer function H(f) of a system constitute the following Fourier transform pair

$$h(t) \; \rightleftharpoons \; H(f)$$

, or equivalently

$$H(f) = \int_{-\infty}^{\infty} h(t)\, e^{-j2\pi ft}\, dt \qquad (7.1a)$$

, and

$$h(t) = \int_{-\infty}^{\infty} H(f)\, e^{j2\pi ft}\, df \qquad (7.1b)$$

7.1. Impulse Response

In the time domain, a linear time invariant system is described by its impulse response. The impulse response of a system h(t) is defined as the response of the system to a Dirac delta function $\delta(t)$ (unit impulse) applied at the input of the system (excitation), Fig.7.1a. Due to time invariant principle, if a delayed Dirac delta function $\delta(t - t_o)$ is applied at the input of the system, then the impulse response of the system will be $h(t - t_o)$ which is the impulse response delayed by the same amount of delay t_o , Fig.7.1b. As a general case, if the system is subjected to an arbitrary driving input function x(t), Fig.7.1c, and it is required to determine its response y(t). Let the signal x(t) is approximated by a staircase function composed of narrow rectangular pulses, each duration $\Delta\tau$. Considering a typical pulse (shaded) occurs at $t = \tau$, then the area of each pulse equals $x(\tau)\, \Delta\tau$. The approximation becomes better for smaller $\Delta\tau$, and as $\Delta\tau$ approaches to zero, each pulse approaches in the limit to a delta function weighted by a factor equals the area of the pulse $x(\tau)\, \Delta\tau$ [1,10].

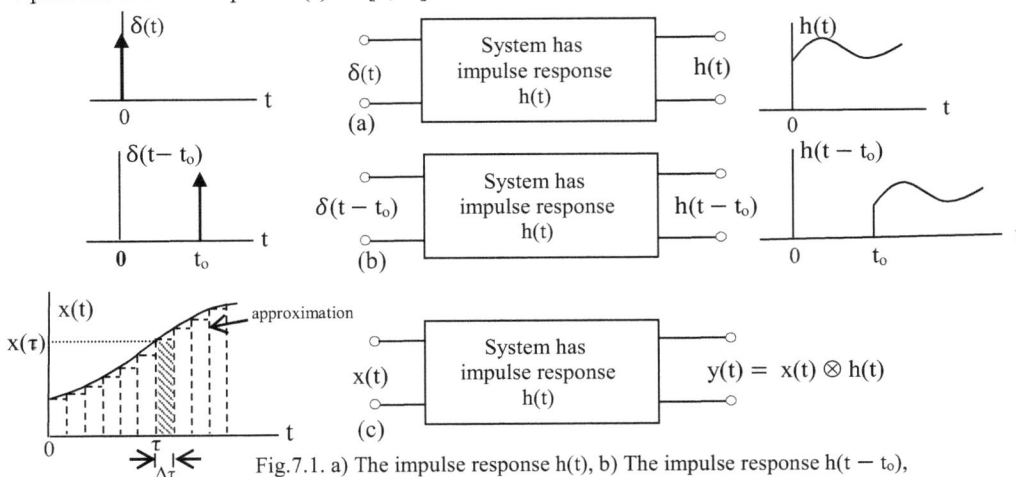

Fig.7.1. a) The impulse response h(t), b) The impulse response h(t − t_o), c) The response of the system y(t) if the input signal is x(t).

Since the response of the system to a unit impulse occurring at t = 0, is the impulse response of the system h(t), and according to the time invariant principle where the response of the system to a delta function weighted by x(τ) Δτ occurring at t = τ, equals x(τ) Δτ h(t − τ). Then due to the linearity principle , the total response y(t) of the system is the summation of the various infinitesimal responses of the various input pulses, and in the limit as Δτ approaches to zero, the input signal x(t) is expressed by

$$x(t) = \lim_{\Delta\tau \to 0} \sum_{n=-\infty}^{\infty} x(\tau)\, \Delta\tau\, \delta(t - n\tau)$$

, and the response y(t) is given by

$$y(t) = \lim_{\Delta\tau \to 0} \sum_{n=-\infty}^{\infty} x(\tau)\, \Delta\tau\, h(t - n\tau)$$

, or equivalently

$$y(t) = \int_{-\infty}^{\infty} x(\tau)\, h(t - \tau)\, d\tau$$

, which is the convolution principle of the driving function x(t) and the impulse response of the system h(t). Then the response y(t) of a system is given by

$$y(t) = x(t) \otimes h(t) \tag{7.2}$$

7.2. Transfer Function
In the frequency domain, the system is described in terms of its transfer function H(f). The evaluation of the transfer function is obtained by taking the Fourier transform of Eq.(7.2), where the multiplication of two signals in any domain is mapped into the convolution of their individual Fourier transforms in the other domain, and Y(f) yields

$$Y(f) = X(f)\, H(f)$$

, or equivalently

$$H(f) = \frac{Y(f)}{X(f)} \tag{7.3}$$

, where X(f) and Y(f) are the Fourier transforms of the driving function x(t) and the system response y(t) respectively. The transfer function of the system H(f) is a complex function and is considered a characteristic property of the linear time invariant system and is given by

$$H(f) = |H(f)|\, e^{j\beta(f)} \tag{7.4}$$

, where $|H(f)|$ is the amplitude (dimensionless) and β(f) is the phase angle (radian) of the transfer function. If the impulse response h(t) is a real valued function, then the transfer function H(f) exhibits conjugate symmetry, where $|H(f)| = |H(-f)|$ is an even function and β(f) = −β(−f) is an odd function.

In most applications of communication systems, it is preferable to work with the logarithm of the transfer function rather than the transfer function itself. Since H(f) is a linear quantity, then the gain or the loss α(f) of the system may be expressed by

$$\alpha(f) = 20 \log_{10} |H(f)| \qquad\qquad \text{dB}$$

, and $\qquad\qquad \alpha'(f) = \alpha(f)/8.686 \qquad\qquad$ Neper

, where α(f) is the gain or the loss of the system in dB and α'(f) is the same gain or the loss in Neper, where

$$1 \text{ Neper} = 8.686 \text{ dB}$$

The transfer function H(f) is the ratio between the output Y(f) to the input X(f) of the system in the frequency domain, and the impulse response h(t) does not equal the ratio between the output y(t) to the input x(t) of the system in the time domain because y(t) and x(t) are related by the convolution formula, Eq.(7.2). On the other hand, there is only one condition satisfy that the transfer function H(f) equals the ratio between the output y(t) to the input x(t) in the time domain when the input signal is given by the complex exponential function x(t) = $e^{j2\pi f t}$, where in this case, the response y(t), Eq.(7.2), yields

$$y(t) = \int_{-\infty}^{\infty} h(\tau)\, e^{j2\pi f(t-\tau)}\, d\tau$$

, or equivalently

$$y(t) = e^{j2\pi f t} \int_{-\infty}^{\infty} h(\tau)\, e^{-j2\pi f\tau}\, d\tau$$

, using Eq.(7.1a), y(t) yields

$$y(t) = e^{j2\pi f t}\, H(f)$$

, then the transfer function H(f) equals the ratio between the output y(t) to the input x(t) in the time domain, only in one condition when the input signal x(t) equals the complex exponential $e^{j2\pi f t}$.

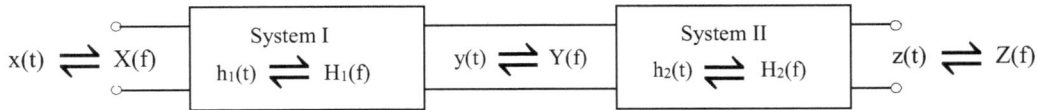

Fig.7.2. Two cascades linear time invariant systems.

7.3. Cascaded Systems

Let two cascaded linear time invariant systems, Fig.7.2, their impulse responses are $h_1(t)$ and $h_2(t)$ and their transfer functions are $H_1(f)$ and $H_2(f)$ respectively. If x(t) and y(t) are the driving function and the response of the first system, then y(t) will be the driving function of the second system, and let z(t) is the response of the second system. The transfer functions of the two systems are $H_1(f)$ and $H_2(f)$ respectively, are given by

$$H_1(f) = \frac{Y(f)}{X(f)} \quad , \quad H_2(f) = \frac{Z(f)}{Y(f)}$$

, and their impulse responses are $h_1(t)$ and $h_2(t)$ will be related to the driving functions x(t) and y(t) by

$$y(t) = x(t) \otimes h_1(t) \quad , \qquad z(t) = y(t) \otimes h_2(t)$$

, if the two systems are equivalent to one system with equivalent impulse response $h_{eq}(t)$ and equivalent transfer function $H_{eq}(f)$, then the equivalent transfer function $H_{eq}(f)$ is given by

$$H_{eq}(f) = \frac{Z(f)}{X(f)} = \frac{Z(f)}{Y(f)}\frac{Y(f)}{X(f)}$$

, or equivalently

$$H_{eq}(f) = H_1(f)\, H_2(f) \qquad\qquad (7.5)$$

, taking the inverse Fourier transform for Eq.(7.5), where the multiplication of two functions in any domain is equivalent to the convolution of their transforms in the other domain, then the equivalent impulse response $h_{eq}(t)$ is given by

$$h_{eq}(t) = h_1(t) \otimes h_2(t) \qquad\qquad (7.6)$$

In general, the equivalent transfer function of any number of cascaded time invariant systems equals the product of their individual transfer functions, and the equivalent impulse response will be the convolution processes of their impulse responses [1].

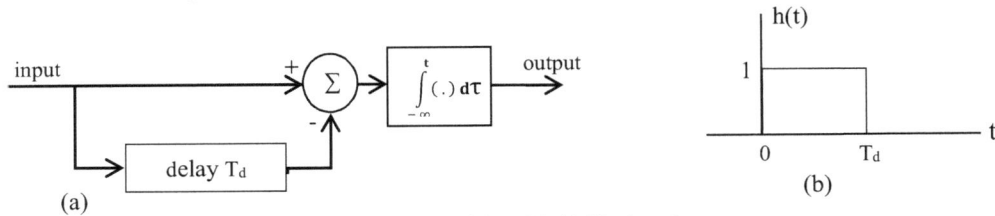

(a)　　　(b)

Fig.7.3. a) The system, problem 7.1, b) The impulse response.

Problem 7.1
Find the impulse response and the transfer function of the linear system, Fig.7.3a.
Solution
Consider x(t) is an input signal, is applied to the system, Fig.7.3a, the signal after the summation quals $[x(t) - x(t - T_d)]$, and the output signal y(t) is given by

$$y(t) = \int_{-\infty}^{t} [x(\tau) - x(\tau - T_d)] \, d\tau \tag{7.7}$$

, making use of Eq.(3.36) to obtain the Fourier transform for both sides of Eq.(7.7), yields

$$Y(f) = \frac{X(f) - X(f) \, e^{-j2\pi f \, T_d}}{j2\pi f} + \frac{[X(f) - X(f) \, e^{-j2\pi f \, T_d}]_{at \, f = 0} \, \delta(f)}{2}$$

, or equivalently

$$Y(f) = \frac{X(f) - X(f) \, e^{-j2\pi f \, T_d}}{j2\pi f}$$

, then the transfer function H(f), Eq.(7.3), will be

$$H(f) = \frac{1 - e^{-j2\pi f \, T_d}}{j2\pi f} \tag{7.8}$$

, the transfer function H(f), Eq.(7.8), can be written in the form

$$H(f) = \frac{e^{-j\pi f \, T_d} [\, e^{j\pi f \, T_d} - e^{-j\pi f \, T_d} \,]}{j2\pi f}$$

, express the exponential terms by sine function, and in terms of sinc function, H(f) yields

$$H(f) = T_d \, sinc(f \, T_d) \, e^{-j\pi f \, T_d} \tag{7.9}$$

, taking the inverse Fourier transform for Eq.(7.9), the impulse response h(t), Fig.7.3b, of the system, is given by

$$h(t) = rect[\frac{t - \frac{1}{2} T_d}{T_d}]$$

Problem 7.2
Find the impulse response and the transfer function of the digital filter, Fig.7.4a.
Solution
Consider x(t) is an input signal, is applied to the system, Fig.7.4a, the output signal y(t) is given by

$$y(t) = x(t) - 2 \, x(t - T_d) + x(t - 2T_d) \tag{7.10}$$

, taking the Fourier transform for both sides of Eq.(7.10), yields

$$Y(f) = X(f) - 2 \, X(f) \, e^{-j2\pi f \, T_d} + X(f) \, e^{-j2\pi f \, 2T_d}$$

, or equivalently

$$Y(f) = X(f) [\, 1 - 2 \, e^{-j2\pi f \, T_d} + e^{-j2\pi f \, 2T_d} \,]$$

, the transfer function H(f), Eq. (7.3), will be

$$H(f) = 1 - 2 \, e^{-j2\pi f \, T_d} + e^{-j2\pi f \, 2T_d} \tag{7.11}$$

, taking the inverse Fourier transform for Eq.(7.11), the impulse response h(t) of the system, Fig.7.4b, is given by

$$h(t) = \delta(t) - 2 \, \delta(t - T_d) + \delta(t - 2T_d)$$

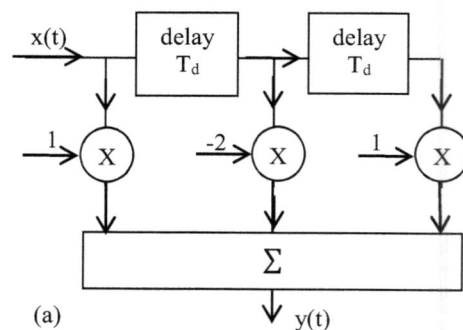

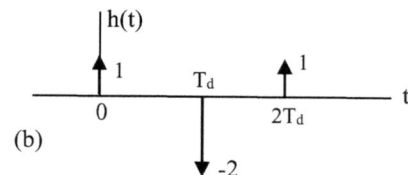

Fig.7.4. a) Digital filter,
b) The impulse response.

Problem 7.3

Find the equivalent impulse response and the equivalent transfer function of the two cascaded systems, Fig.7.5.

Solution

The transfer functions of the two systems are given by

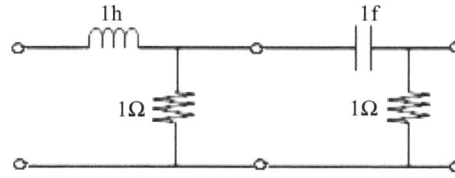

Fig.7.5. Two cascaded low and high-pass filters.

$$H_1(f) = \frac{1}{1 + j2\pi f} \quad , and \quad H_2(f) = \frac{j2\pi f}{1 + j2\pi f}$$

, then the equivalent transfer function $H_{eq}(f)$, Eq.(7.5), is given by

$$H_{eq}(f) = \frac{j2\pi f}{(1 + j2\pi f)^2} \tag{7.12}$$

, and the impulse responses of the two systems are given by

$$h_1(t) = e^{-t} u(t) \quad , and \quad h_2(t) = \frac{d}{dt} [e^{-t} u(t)] = \delta(t) - e^{-t} u(t)$$

, then the equivalent impulse response $h_{eq}(t)$, Eq.(7.6), is given by

$$h_{eq}(t) = e^{-t} u(t) \otimes [\delta(t) - e^{-t} u(t)]$$

, or equivalently

$$h_{eq}(t) = [e^{-t} u(t) \otimes \delta(t)] - [e^{-t} u(t) \otimes e^{-t} u(t)]$$

, solving the second convolution, yields

$$h_{eq}(t) = [e^{-t} u(t) \otimes \delta(t)] - \int_0^t e^{-\tau} e^{-(t-\tau)} d\tau$$

, due to the convolution principle of the Dirac delta function, Eq.(4.4c), the first convolution function equals $e^{-t} u(t)$, and the second convolution function equals $t e^{-t} u(t)$, then the equivalent impulse response $h_{eq}(t)$ is given by

$$h_{eq}(t) = e^{-t} u(t) - t e^{-t} u(t) \tag{7.13}$$

, another way to obtain the equivalent impulse response $h_{eq}(t)$ from the equivalent transfer function $H_{eq}(f)$, Eq.(7.12) directly by taking the inverse Fourier transform, using the differentiation property in the time domain, Eq.(3.25), yields

$$h_{eq}(t) = \frac{d}{dt} [e^{-t} u(t) \otimes e^{-t} u(t)] = \frac{d}{dt} [\int_0^t e^{-\tau} e^{-(t-\tau)} d\tau]$$

, or equivalently

$$h_{eq}(t) = \frac{d}{dt} [t e^{-t} u(t)] = e^{-t} u(t) - t e^{-t} u(t) = (1 - t) e^{-t} u(t)$$

, which is the same equivalent impulse response $h_{eq}(t)$, Eq.(7.13).

Problem 7.4

For the High Pass Filter, Fig.7.6a.
Find:
i. The transfer function and the impulse response.
ii. The unit step response.

Solution

i. Consider $x(t)$ is an input voltage waveform, is applied to the system, Fig.7.6a, then

$$X(f) = I(f) [1 + \frac{j2\pi f}{1 + j2\pi f}]$$

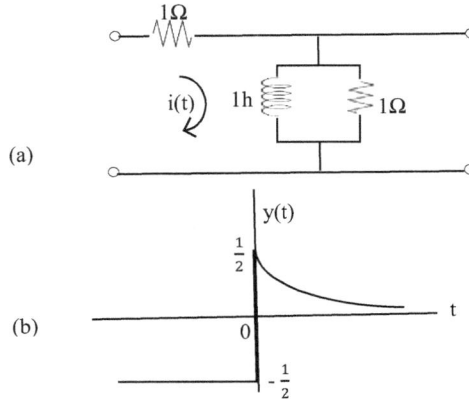

(a)

(b)

Fig.7.6. a) High-pass filter, b) Unit step response.

, and
$$Y(f) = I(f) \frac{j2\pi f}{1 + j2\pi f}$$

, where $X(f)$ and $Y(f)$ are the Fourier transforms of the input and the output voltages waveforms respectively, and $I(f)$ is the Fourier transform of the current $i(t)$, then the transfer function $H(f)$, Eq.(7.3) can be obtained directly, the division of $Y(f)$ by $X(f)$, $H(f)$ yields

$$H(f) = \frac{j2\pi f}{1 + j4\pi f}$$

, taking the inverse Fourier transform using the differentiation property in time domain , Eq.(3.25), the impulse response $h(t)$ is given by

$$h(t) = \frac{d}{dt}\left[\frac{1}{2} e^{-\frac{1}{2}t} u(t)\right] = \frac{1}{2}\delta(t) - \frac{1}{4} e^{-\frac{1}{2}t} u(t)$$

ii. The unit step response $y(t)$ is given by

$$y(t) = u(t) \otimes h(t) = u(t) \otimes \left[\frac{1}{2}\delta(t) - \frac{1}{4} e^{-\frac{1}{2}t} u(t)\right]$$

$$= \left[u(t) \otimes \frac{1}{2}\delta(t)\right] - \left[u(t) \otimes \frac{1}{4} e^{-\frac{1}{2}t} u(t)\right]$$

, solving the second convolution function using the time domain, $y(t)$ yields

$$y(t) = \frac{1}{2} u(t) - \frac{1}{4}\int_0^t e^{-\frac{1}{2}\tau} d\tau$$

, then $y(t)$, Fig.7.6b, is given by

$$y(t) = \frac{1}{2} u(t) + \frac{1}{2}[e^{-\frac{1}{2}t} - 1]$$

7.4. Causal and Non-Causal Systems

In the time domain, the physically realizable system cannot have a response before the driving function is applied, this is known as the causality condition in the time domain. The system is said to be causal, if it does not respond before the driving input function is applied, Fig.7.7a. The causal system has a causal impulse response $h(t)$, where $h(t) = 0$ for $t < 0$. But if $h(t) \neq 0$ for $t < 0$, the system is said to be non-causal system and has a non-causal impulse response, Fig.7.7b. The causality principle is very important in the design of communication systems because if the system is causal, it should be physically realizable and if the system is non-causal, it should be not physically realizable, and in this case, is there any way to make the system realizable ?.

Fig.7.7. a) Causal impulse response, b) Non-causal impulse response.

In the frequency domain, the Paley-Wiener criterion [1,3] provides us with a means of knowing whether or not the magnitude of the transfer function $|H(f)|$, corresponds to some causal impulse response, where

$$\int_{-\infty}^{\infty} |H(f)|^2 \, df < \infty$$

, Paley-Wiener criterion gives the frequency domain equivalence for the causality condition, which is $|H(f)|$ must satisfy the following condition

$$\int_{-\infty}^{\infty} \frac{\left|\ln|H(f)|\right|}{1 + f^2}\, df < \infty \qquad (7.14)$$

, where $\ln|H(f)|$ equals the Gain $\alpha(f)$ and the Paley-Wiener criterion states that provide the gain $\alpha(f)$ with a suitable phase $\beta(f)$ satisfies the condition, Eq.(7.14) such that the resulting system has a causal impulse response h(t). Eq.(7.14) shows that, at discrete points of frequencies, the magnitude of the transfer function $|H(f)|$ may be equals zero (infinite attenuation), and cannot be zero over a finite band of frequencies, this means that, no filter having a stop band with infinite attenuation can ever be constructed, as will be seen in all the ideal filters (the transfer function abrupt and fall off faster), falls in this category and consequently are not realizable. Although the ideal filters cannot be realized, the magnitude of their transfer function $|H(f)|$ may be approached and then may not fall off faster than a function of exponential orders such as the high order maximally flat Butterworth filter design, where as the order becomes very large and the approximation becomes quite excellent, Fig.7.8. [1,4].

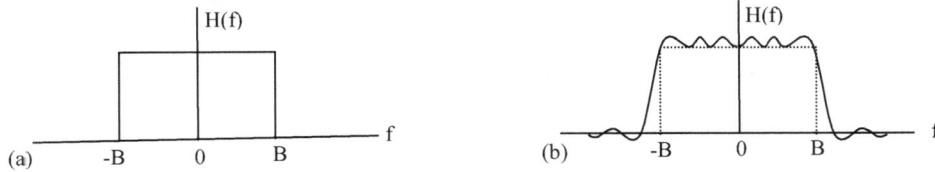

Fig.7.8. a) Ideal low-pass filter, b) Approximated ideal low-pass filter.

7.5. Stable and Non-Stable Systems
The system is said to be stable if the output signal is bounded for all bounded input signals. If x(t) is the driving function of the system, and is bounded, then $|x(t)| \le M$, where M is positive real and finite number. Then using the convolution principle, Eq.(7.2), the output y(t) of the system yields

$$|y(t)| \le M \int_{-\infty}^{\infty} |h(\tau)|\, d\tau \qquad (7.15)$$

If the impulse response h(t) is real valued function, then $H^*(f) = H(-f)$, where H(f) is the Fourier transform of h(t), Eq.(7.1a). The linear filters equations can be generalized to filters having multiple inputs and outputs, and it is assumed that all inputs are bounded. Then for a linear time invariant system to be stable, the magnitude response $|y(t)|$ will be bounded, where $|y(t)| < \infty$, using Eq.(7.15), then the necessary and sufficient condition for stability in the time domain, is the impulse response h(t) of the stable system, must be absolutely integrable, where

$$\int_{-\infty}^{\infty} |h(\tau)|\, d\tau < \infty$$

7.6. Bandwidth of Low-Pass and Band-Pass Systems
The bandwidth of a system is defined as the interval of frequencies over which the magnitude of the transfer function $|H(f)|$ remains within $1/\sqrt{2}$ times its value at the mid-band H(0), where the constancy of the magnitude $|H(f)|$ in a system is usually specified by its bandwidth. The constancy of the magnitude of the transfer function $|H(f)|$ or gain of a system is one of the conditions necessary for the transmission of a signal with no distortion. Fig.7.9. shows the bandwidth of the low-pass filter is B Hz while the bandwidth of the band-pass filter is 2B Hz. The points at which the bandwidth is determined, are called the half power points where the output power has dropped to half of its mid-band level, or called the -3 dB points or the 0.707 points where this occurs when the output voltage has dropped by 3 dB, and the band-pass system will have two half power points, whilst the low pass system will have only one point [1,10].

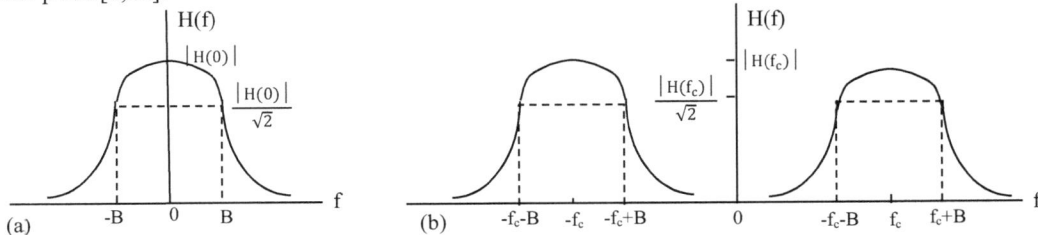

Fig.7.9. a) Bandwidth of low-pass filter, b) Bandwidth of band-pass filter.

7.7. Relation between Input and Output Energy Spectral Densities

Consider an energy signal x(t), is a driving input function, is applied to a linear time invariant system of impulse response h(t) and transfer function H(f), the output waveform is y(t). Then the input energy spectral density $\Psi_i(f)$ is given by

$$\Psi_i(f) = |X(f)|^2 \qquad\qquad \text{joule/Hz}$$

, where X(f) is the Fourier transform of x(t), and the output energy spectral density $\Psi_o(f)$ is given by

$$\Psi_o(f) = |Y(f)|^2 \qquad\qquad \text{joule/Hz}$$

, where Y(f) is the Fourier transform of y(t). Squaring the magnitude of Eq.(7.3), the output energy spectral density $\Psi_o(f)$ yields

$$\Psi_o(f) = |H(f)|^2 \; \Psi_i(f) \qquad\qquad (7.16)$$

Then the output energy spectral density $\Psi_o(f)$ which contains certain interval of frequency band due to output filter selection, is obtained by multiplying the squared magnitude of the transfer function by the input energy spectral density $\Psi_i(f)$ which contains the whole frequency band $-\infty \leq f \leq \infty$, in the condition of the following Paley-Weiner criteria

$$\int_{-\infty}^{\infty} |H(f)|^2 \; df < \infty$$

The autocorrelation functions of the input and output signals can be obtained by taking the inverse Fourier transform of the input and output spectral densities $\Psi_i(f)$ and $\Psi_o(f)$ respectively [8,10].

7.8. Distortionless System

The response (output) of the distortionless transmission system is a replica of the driving input signal. Consider an energy signal x(t) is applied to a linear time invariant distortionless system. Of course, the signal will be subjected to some change in amplitude and to sometime delay only associated with its replica, and the response is given by

$$y(t) = k \, x(t - t_o) \qquad\qquad (7.17)$$

, where k accounts the change in amplitude and t_o account the delay in transmission. Taking the Fourier transform for both sides of Eq.(7.17), yields

$$Y(f) = k \, X(f) \; e^{-j2\pi f t_o}$$

, where X(f) and Y(f) are the Fourier transforms of the driving input function x(t) and the response y(t) of the system respectively. Then the transfer function H(f) of the distortionless system is given by

$$H(f) = k \, e^{-j2\pi f t_o}$$

, or equivalently

$$H(f) = k \, e^{-j2\pi f t_o \pm n\pi} \qquad , n = 0, 1, 2, 3, \dots \qquad (7.18)$$

, if the transfer function H(f) is expressed by $H(f) = |H(f)| \, e^{j\beta(f)}$, Eq.(7.18) illustrates that there are two important conditions for the transfer function of the distortionless system, these are: i) the magnitude $|H(f)|$ is constant and equals k, where the frequency components of the input signal are transmitted with the same weight k, and ii) the phase angle $\beta(f)$ is a linear function of frequency f and is given by

$$\beta(f) = -2\pi f t_o \pm n\pi$$

, the output spectrum of the distortionless filter, will be limited to a band of frequencies, and the two conditions are satisfied by the distortionless system for that band of frequencies. Well-known examples of practical systems that approach of a distortionless transmission systems are waveguides, transmission lines, quality delay lines, and some amplifiers. Two well examples of the distortionless system are the ideal low-pass filter and the ideal band-pass filter [1].

7.9. Ideal Low-Pass Filter

The Ideal Low-pass Filter ILPF is a frequency selective device which used to limit the spectrum of a signal to some specified band of low frequencies. The transfer function of the low-pass filter $H(f)_{ILPF}$, magnitude and phase angle, Fig.7.10a, is characterized by a pass-band of low frequencies, inside this pass-band, the low frequencies are transmitted without distortion (k = 1, n = 0, and with linear phase angle $\beta(f) = -2\pi t_o f$, Eq.(7.18), and $H(f)_{ILPF}$ yields

$$H(f)_{ILPF} = rect(\frac{f}{2B})\ e^{-j2\pi ft_o} \tag{7.19}$$

, where B is the bandwidth of the low-pass filter. The impulse response of the ideal low-pass filter $h(t)_{ILPF}$, is the inverse Fourier transform of the transfer function $H(f)_{ILPF}$, is given by

$$h(t)_{ILPF} = 2B\ sinc[2B(t - t_o)] \tag{7.20}$$

, Fig.7.10b. shows the impulse response $h(t)_{ILPF}$ is a sinc function extends from $-\infty$ to ∞, and centered at the time shift t_o, then the ideal low-pass filter is non-causal system because $h(t) \neq 0$ for $t < 0$, this means that the ideal low pass filter is not physically realizable.

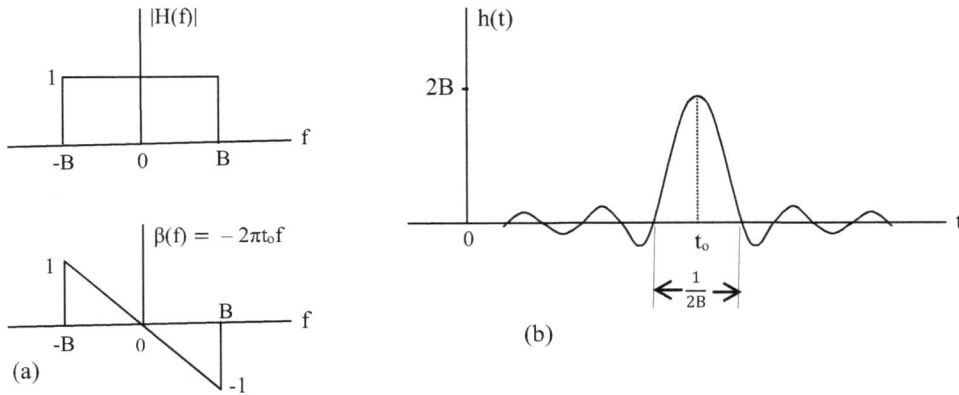

Fig.7.10. Ideal low-pass filter, a) Transfer function (magnitude and phase angle), b) Impulse response.

Problem 7.5

Find the response of the ideal low-pass filter when the input driving function is an rectangular pulse given by $x(t) = A\ rect(t/T)$.

Solution

The response of the ideal low-pass filter is given by the following convolution

$$y(t) = A\ rect(\frac{t}{T}) \otimes h(t)_{ILPF}$$

, using Eq.(7.20), y(t) yields

$$y(t) = A\ rect(\frac{t}{T}) \otimes 2B\ sinc[2B(t - t_o)] \tag{7.21}$$

, making use of Eq.(3.42), y(t) yields

$$y(t) = 2AB \int_{-\infty}^{\infty} rect(\frac{\tau}{T})\ sinc[2B(t - \tau - t_o)]\ d\tau \tag{7.22}$$

, since the function $rect(\tau/T)$ extends from $-T/2$ to $T/2$, Eq.(7.22) yields

$$y(t) = 2AB \int_{-T/2}^{T/2} sinc[2B(t - \tau - t_o)]\ d\tau \tag{7.23}$$

, or equivalently

$$y(t) = 2AB \int_{-T/2}^{T/2} \frac{sin[2\pi B(t - \tau - t_o)]}{[2\pi B(t - \tau - t_o)]}\ d\tau \tag{7.24}$$

, let $x = [2\pi B(t - \tau - t_o)]$, then $dx = -2\pi B\ d\tau$, as $\tau \rightarrow -T/2$ then $x_1 \rightarrow [2\pi B(t + \frac{1}{2} T - t_o)]$, and as $\tau \rightarrow T/2$, then $x_2 \rightarrow [2\pi B(t - \frac{1}{2}T - t_o)]$, Eq.(7.24) yields

$$y(t) = -\frac{A}{\pi} \int_{x_1}^{x_2} \frac{sin(x)}{x}\ dx \tag{7.25}$$

, where $x_1 = [2\pi B(t + \frac{1}{2} T - t_o)]$ and $x_2 = [2\pi B(t - \frac{1}{2} T - t_o)]$, and y(t) can be expressed by

$$y(t) = \frac{A}{\pi} \left[\int_0^{x_1} \frac{\sin(x)}{x} dx - \int_0^{x_2} \frac{\sin(x)}{x} dx \right] \qquad (7.26)$$

, in terms of sine integral function Si(u), Eq.(3.45), Fig.3.43, Eq.(7.26) yields

$$y(t) = \frac{A}{\pi} \left\{ Si \left[2\pi B(t + \frac{T}{2} - t_o) \right] - Si \left[2\pi B(t - \frac{T}{2} - t_o) \right] \right\}$$

, Fig.7.11. shows the response y(t) of the ideal low-pass filter if the input is a rectangular pulse x(t) = A rect(t/T), where Fresnel ripples in the pass-band and Gibbs ripples in the stop-band, and Si(u) is the sine integral function, Fig.3.43, Eq.3.45, is extensively tabulated in standard tables.

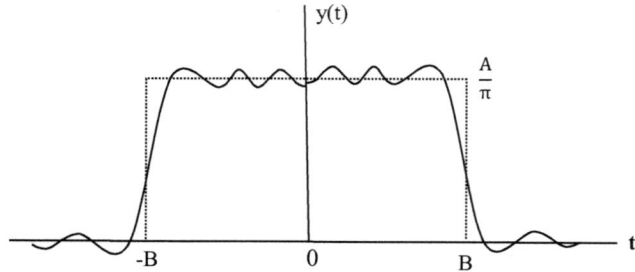

Fig.7.11. The response of the ideal low-pass filter y(t) when the input is x(t) = A rect(t/T).

Problem 7.6

An energy signal $x(t) = e^{-t} u(t)$, is applied to an ideal low-pass filter with cut-off frequencies $-1/2\pi \leq f \leq 1/2\pi$ Hz. Find:

    i. The autocorrelation function of the input signal.
    ii. The energy spectral density of the output signal.
    iii. The energy content of the output signal.

Solution
The autocorrelation function $R_i(\tau)$, Eq.(6.1), of the input signal $x(t) = e^{-t} u(t)$, is given by

$$R_i(\tau) = \int_\tau^\infty e^{-t} e^{-(t-\tau)} dt = \frac{1}{2} e^{-\tau} u(\tau) \qquad \text{for positive } \tau$$

$$R_i(\tau) = \int_0^\infty e^{-t} e^{-(t-\tau)} dt = \frac{1}{2} e^{\tau} u(-\tau) \qquad \text{for positive } \tau$$

, since x(t) is real valued function, the function $R_i(\tau)$, will be even function , Fig.7.12a, is given by

$$R_i(\tau) = \frac{1}{2} e^{-|\tau|}$$

, the energy spectral density $\Psi_i(f)$, Fig.7.12b, is the Fourier transform of the autocorrelation function $R_i(\tau)$, is given by

$$\Psi_i(f) = \frac{1}{1 + (2\pi f)^2} \qquad \text{for } -\infty \leq f \leq \infty$$

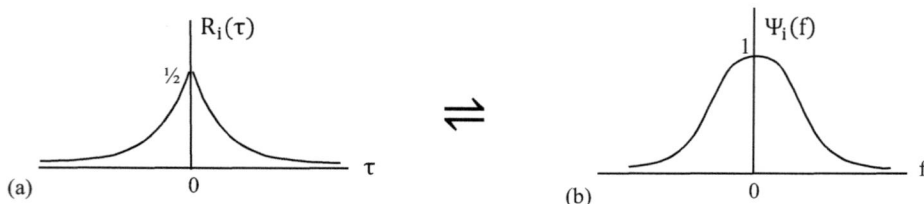

Fig.7.12. a) The input autocorrelation function, b) The input energy spectral density.

, since the output energy spectral density $\Psi_o(f)$, Eq.(7.16), contains certain interval of the frequency band $-1/2\pi \leq f \leq 1/2\pi$ Hz, due to output filter selection, and is obtained by multiplying the squared magnitude of the transfer function for the ideal low pass filter $|H(f)| = 1$, by the input energy spectral density $\Psi_i(f)$ which contains the whole frequency band $-\infty \leq f \leq \infty$, then $\Psi_o(f)$ is given by

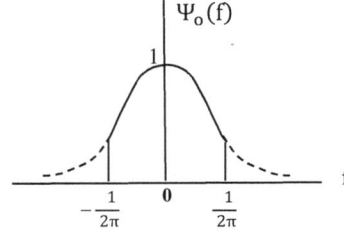

Fig.7.13. The output energy spectral density.

$$\Psi_o(f) = \frac{1}{1 + (2\pi f)^2} \quad \text{for} \quad -1/2\pi \leq f \leq 1/2\pi$$

, the energy content of the output signal will be

$$E_o = \int_{-1/2\pi}^{1/2\pi} \frac{1}{1 + (2\pi f)^2} \, df$$

, let $\lambda = 2\pi f$, then $d\lambda = 2\pi \, df$, as $f \rightarrow -1/2\pi$, then $\lambda \rightarrow -1$, and as $f \rightarrow 1/2\pi$, then $\lambda \rightarrow 1$, E_o yields

$$E_o = \frac{1}{2\pi} \int_{-1}^{1} \frac{1}{1 + \lambda^2} \, d\lambda$$

, since $\Psi_o(f)$ is an even function, E_o yields

$$E_o = \frac{1}{\pi} \int_{o}^{1} \frac{1}{1 + \lambda^2} \, d\lambda$$

$$= \left[\frac{1}{\pi} \tan^{-1}(\lambda) \right]_{\lambda=0}^{\lambda=1} = \frac{1}{\pi} \left[\frac{\pi}{4} - 0 \right] = 0.25 \quad \text{joule}$$

Problem 7.7

An energy signal has autocorrelation function $R_i(\tau) = k \, \delta(\tau)$, is applied to an ideal low-pass filter of bandwidth B Hz. Evaluate the output autocorrelation function and the output spectral density of the filter.

Solution

Since the energy spectral density and the autocorrelation function constitute a Fourier transform pair, then the energy spectral density is given by

$$\Psi_i(f) = k \qquad\qquad \text{for} \quad -\infty \leq f \leq \infty$$

, and the output energy spectral density, Equations (7.16) and (7.19), is given by

$$\Psi_o(f) = k \, \text{rect}(\frac{f}{2B})$$

, and the output autocorrelation function $R_o(\tau)$ is the inverse Fourier transform of the output spectral density $\Psi_o(f)$, $R_o(\tau)$ yields

$$R_o(\tau) = 2B \, k \, \text{sinc}(2B\tau)$$

, where k is constant account the change of the amplitude and B is the bandwidth of the low pass filter.

7.10. Ideal Band-Pass Filter

The Ideal Band-Pass Filter IBPF is a frequency selective device which used to limit the spectrum of a signal to some specified band of certain high frequencies components. The transfer function of the band-pass filter $H(f)_{IBPF}$ (magnitude and phase angle), Fig.7.14a, is characterized by a pass-band of certain high frequency components. Inside this band-pass, the high frequencies are transmitted without distortion ($k = 1$, $n = 0$, and with linear phase angle $\beta(f)$), Eq.(7.18), and $H(f)_{IBPF}$ yields

$$H(f)_{IBPF} = \text{rect}\left[\frac{f + f_c}{2B}\right] e^{-j2\pi f t_o} + \text{rect}\left[\frac{f - f_c}{2B}\right] e^{-j2\pi t t_o} \qquad (7.27)$$

, where $2B$ is the bandwidth of the band-pass filter and f_c is the center frequency of the pass-band (carrier frequency). The impulse response of the ideal band-pass filter $h(t)_{IBPF}$, is the inverse Fourier transform of the transfer function $H(f)_{IBPF}$, is given by

$$h(t)_{IBPF} = 2B \, \text{sinc}[2B(t - t_o)] \, e^{-j2\pi f_c t} + 2B \, \text{sinc}[2B(t - t_o)] \, e^{j2\pi f_c t}$$

, express the exponential terms by cosine function, $h(t)_{IBPF}$ yields

$$h(t)_{IBPF} = 4B \, \text{sinc}[2B(t - t_o)] \, \cos(2\pi f_c t) \qquad (7.28)$$

, Eq.(7.28) is an Amplitude Modulated Double Side Band Suppressed Carrier AM-DSBSC wave, consists of a carrier wave $\cos(2\pi f_c t)$ multiplied by a low-pass signal equals $4B \, \text{sinc}[2B(t - t_o)]$. Fig.7.14b. shows the impulse response $h(t)_{IBPF}$, is a cosine function $\cos(2\pi f_c t)$, amplitude modulated by a low-pass signal sinc function extends from $-\infty$ to ∞, and centered at the time shift t_o, then the ideal band-pass filter is non-causal system because $h(t) \neq 0$ for $t < 0$, this means that the ideal band-pass filter is not physically realizable [1,6].

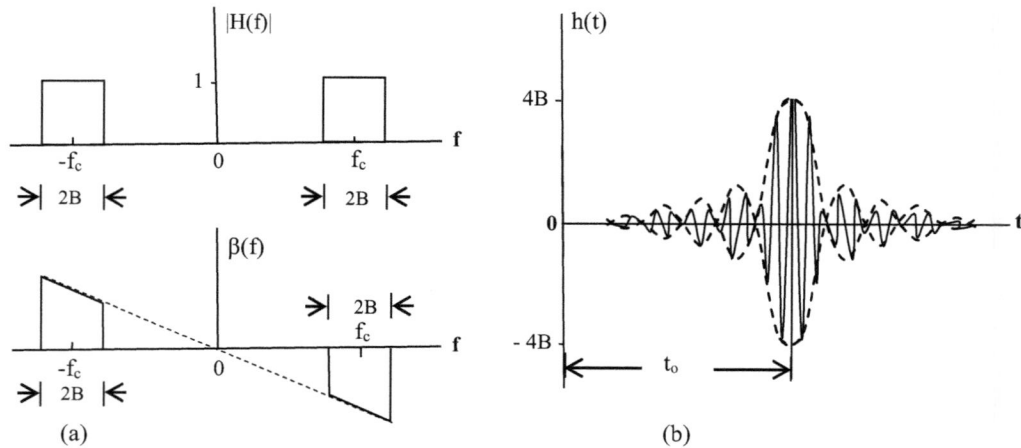

Fig.7.14. Ideal band-pass filter, a) Transfer function, b) Impulse response.

Problem 7.8

Find the transfer function of an ideal band-pass filter has a truncated impulse response with length T, Fig.7.15, using convolution principle.

Solution

The impulse response of the ideal band-pass filter, Eq.(7.28), extends from $-\infty$ to ∞. Then the impulse response of the ideal band-pass filter has a truncated impulse response with length T, is given by

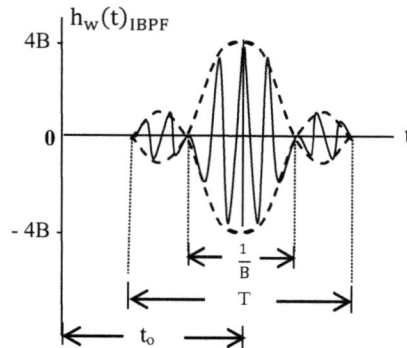

Fig.7.15. Truncated Impulse response.

$$h_w(t)_{IBPF} = 4B \, sinc[2B(t - t_o)] \, cos(2\pi f_c t) \, rect[\frac{t - t_o}{T}]$$

, express the cosine function by exponential terms, the transfer function $H_w(f)_{IBPF}$ of the truncated impulse response $h_w(t)_{IBPF}$ of the ideal band-pass filter, is given by

$$H_w(f)_{IBPF} = \left\{ rect\left[\frac{f + f_c}{2B}\right] \otimes T \, sinc(fT) + rect\left[\frac{f - f_c}{2B}\right] \otimes T \, sinc(fT) \right\} e^{-j2\pi f t_o}$$

, in terms of sine integral function Si(u), Eq.(3.45), Fig.3.43, and using the same steps of problem 7.5, $H_w(f)_{IBPF}$ yields

$$H_w(f)_{IBPF} = \frac{1}{\pi} \{ Si[\pi(f - f_c + B)T] - Si[\pi(f - f_c - B)T]$$
$$+ Si[\pi(f + f_c + B)T] - Si[\pi(f + f_c - B)T] \} e^{-j2\pi f t_o} \quad (7.29)$$

Typical examples of the ideal band-pass filters (Surface Acoustic Wave SAW filters) are simulated by the author [11,12], using the convolution principle technique, with carrier frequencies 100 MHz and 240 MHz, bandwidths of 46.7 MHz and 140 MHz, and the truncation transducer length T of 1.05 micro second and 350 nano second respectively. Fig.7.16 shows the magnitude of the transfer functions $H_w(f)_{IBPF}$ of the truncated impulse responses $h_w(t)_{IBPF}$ of these typical ideal band pass filters. Also a model of an ideal SAW duplexer is simulated by the author [13] for the American Interim Standard IS'95, Global System for Mobile GSM communication, and Universal Mobile Telecommunication System UMTS, using the convolution principle technique.

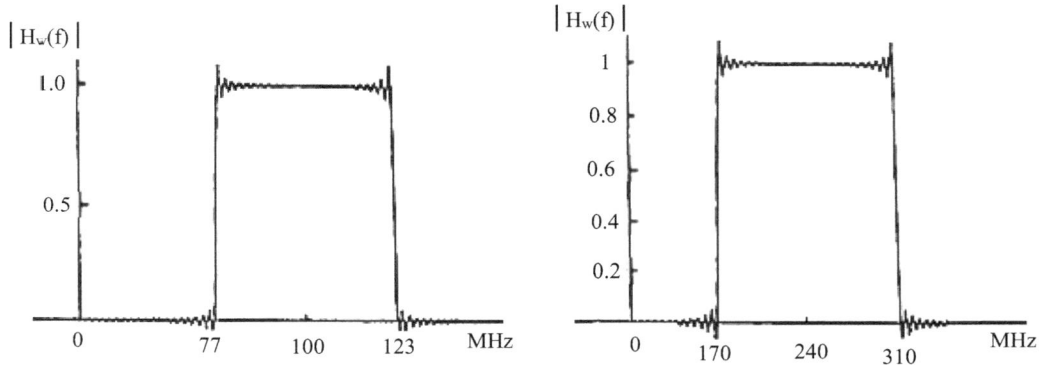

Fig.7.16. Transfer functions of two typical band-pass filters, using the convolution technique.

The transfer function $H_w(f)$ is characterized by the following five parameters:
 i. The center frequency (carrier) f_c.
 ii. The pass-bandwidth B_2.
 iii. The transition bandwidth B_1 (skirt steepness).
 iv. The pass-band ripples level r_1 (Fresnel ripples).
 v. The sidelobe ripples level r_2 (Gibbs ripples).

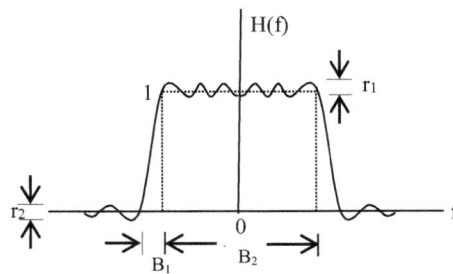

Fig.7.17. The skirt selectivity is specified by the Shape Factor SF = (2B₁+B₂)/B₁.

The skirt selectivity is specified by the Shape Factor SF, Fig.7.17, is given by

$$SF = (2B_1 + B_2)/B_2$$

7.11. Distortion System

In practice, the conditions of distortionless transmission systems can only satisfied approximately, that is means, there is always a certain amount of distortion, is present in the output signal and the system is said to be dispersive system, where the non-constant of the amplitude causes amplitude distortion, and the nonlinear phase angle also causes phase distortion.

7.11.1. Amplitude Distortion
The amplitude of the transfer function $|H(f)|$ is not constant within the frequency band, where the frequency components of the input signal are transmitted with different amounts of nonlinearity, Fig.7.18a.

7.11.2. Phase Distortion
The phase angle of the transfer function $\beta(f)$ is nonlinear with frequency, where a set of frequency components (narrow band) is subjected to different delay, and the output signal has a different waveform from the input signal, Fig.7.18b.

Fig.7.18. Transfer function of dispersive low-pass filter, a) Magnitude, b) Phase angle.

Problem 7.9
For the RC low-pass filter, Fig.7.19a. Find:
  i. The transfer function of the filter and, is a distortion caused by the filter ?
  ii. The impulse response of the filter and, is the filter a causal system?
  iii. The response of the filter if the input signal is $x(t) = (1/T)\,rect(t/T)$, and by taking the lim T tends to zero, show that the response equals the impulse response of the filter part (ii), (let the time constant value RC = 5 second).

Solution
i. Consider $x(t)$ is an input voltage waveform, is applied to the RC low-pass filter, Fig.7.19a. If the output voltage waveform is $y(t)$ and using Kirchhoff's law for the sum of the voltages around the loop, then

$$x(t) = R\,i(t) + y(t) \qquad (7.30)$$

, where

$$i(t) = C\,\frac{d\,y(t)}{dt}$$

, Eq.(7.30), yields

$$x(t) = RC\,\frac{d\,y(t)}{dt} + y(t) \qquad (7.31)$$

, taking the Fourier transform for both sides of Eq.(7.31), yields

$$X(f) = RC\,j2\pi f\,Y(f) + Y(f)$$

, or equivalently

$$X(f) = [1 + j2\pi f\,RC]\,Y(f)$$

, where $X(f)$ and $Y(f)$ are the Fourier transforms of the input and the output voltages waveforms respectively, then the transfer function of the RC low-pass filter, Eq.(7.3), $H(f)$ is given by

$$H(f) = \frac{1}{1 + j2\pi f\,RC} \qquad (7.32)$$

, or equivalently

$$H(f) = \frac{1}{\sqrt{1 + (2\pi f\,RC)^2}}\;e^{-j\,\tan^{-1}(2\pi f\,RC)}$$

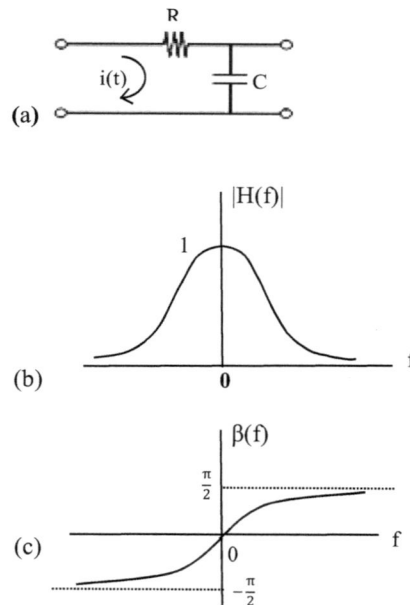

Fig.7.19. Transfer function of RC low-pass filter, a) Magnitude, b) Phase angle.

, Fig.7.19b,c. show the amplitude and phase angle of the transfer function $H(f)$.

, another way to get the transfer function, Eq.(7.32), using the frequency domain, where X(f) and Y(f) are given by

$$X(f) = \left[R + \frac{1}{j2\pi f\,C} \right] I(f) \qquad , \text{and} \qquad Y(f) = \left[\frac{1}{j2\pi f\,C} \right] I(f)$$

, where I(f) is the Fourier transform of the current i(t), then the transfer function H(f), Eq.(7.3) can be obtained directly, the division of Y(f) by X(f), Eq.(7.32)

The RC low-pass filter, Fig.7.19a, will cause some distortion because the amplitude of the transfer function $|H(f)|$, Fig.7.19b, is not constant within the frequency band of the -3 dB points (half power points), where the frequency components of the input signal are transmitted with different change in amplitude causing some amounts of nonlinearity, and the phase angle $\beta(f)$, Fig.7.19c, is nonlinear function with frequency, where a set of frequency components (narrow band) is subjected to different delay and the output signal has a different waveform from the input signal [1,4].

ii. The impulse response of the RC low-pass filter h(t), Fig.7.19a, can be obtained by taking the inverse Fourier transform of Eq.(7.32), h(t) is given by

$$h(t) = \frac{1}{RC}\,e^{-t/RC}\,u(t) \qquad (7.33)$$

, where u(t) is the unit step function. Fig.7.20a. shows the impulse response h(t). This RC low-pass filter is a causal system because it has a causal impulse response h(t), where h(t) = 0 for t < 0.

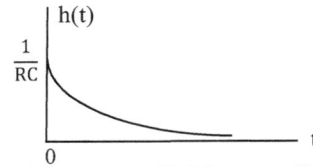

a) Impulse response of RC low-pass filter.

iii. If the input signal is x(t) = (1/T) rect(t/T) , and let the time constant value RC = 5 second, the response y(t) is given by

$$y(t) = x(t) \otimes h(t)$$

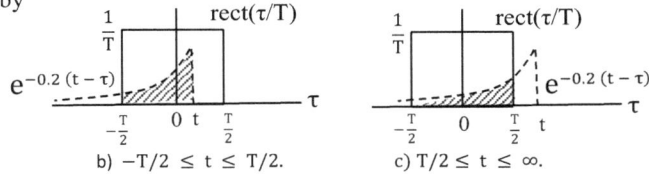

b) $-T/2 \le t \le T/2$.

c) $T/2 \le t \le \infty$.

$$= \frac{1}{T} \operatorname{rect}\!\left(\frac{t}{T}\right) \otimes 0.2\,e^{-0.2t}\,u(t)$$

Fig.7.20. a) Impulse response of RC low-pass filter, b) $-T/2 \le t \le T/2$, c) $T/2 \le t \le \infty$.

, using the time domain, y(t) yields

$$y(t) = \frac{1}{T} \int_{-T/2}^{t} 0.2\,e^{-0.2(t-\tau)}\,d\tau$$

$$= \frac{1}{T} \left[1 - e^{-0.2(t + \frac{1}{2}T)} \right] \qquad \text{for} \quad -T/2 \le t \le T/2$$

, and $$y(t) = \frac{1}{T} \int_{-T/2}^{T/2} 0.2\,e^{-0.2(t-\tau)}\,d\tau$$

$$= \frac{2}{T}\,e^{-0.2t}\,\sinh(0.1T) \qquad \text{for} \quad T/2 \le t \le \infty \qquad (7.34)$$

, taking the lim T tends to zero, Eq.(7.34) yields

$$\lim_{T \to 0} y(t) = \lim_{T \to 0} \left[\frac{2}{T}\,e^{-0.2t}\,\sinh(0.1T) \right] \qquad \text{for} \quad 0 \le t \le \infty \qquad (7.35)$$

, Eq.(7.35) has a undetermined value (zero/zero), then differentiating Eq.(7.35) yields

$$\lim_{T \to 0} y(t) = \lim_{T \to 0} 0.2\,e^{-0.2t}\,\cosh(0.1T) = 0.2\,e^{-0.2t}\,u(t)$$

, which is the same impulse response h(t) of the RC low-pass filter of part (i).

Problem 7.10

For the RC low-pass filter, Fig.7.19a. Find the unit step response of the filter (let the time constant value RC = 5 second).

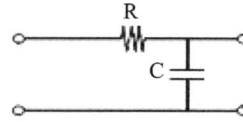

Fig.7.19a. RC low-pass filter.

Solution

Considering the time constant value RC = 5 sec, and using the time domain, where the impulse response h(t) is given by Eq.(7.33), the unit step response y(t) is given by

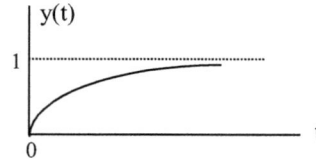

$$y(t) = h(t) \otimes u(t) = 0.2\, e^{-0.2t}\, u(t) \otimes u(t)$$

$$= \int_0^t 0.2\, e^{-0.2\tau}\, d\tau \qquad (7.36)$$

Fig.7.21. Unit step response of RC low-pass filter.

, or equivalently

$$y(t) = [\, 1 - e^{-0.2t}\,]\, u(t) \qquad (7.37)$$

, Fig.7.21. shows the response y(t) of the RC low pass filter for the input unit step function u(t).

Another way to obtain the unit step response y(t), using the frequency domain, where the Fourier transform pair of Eq.(7.36) is given by

$$0.2\, e^{-0.2t}\, u(t) \otimes u(t) \quad \rightleftharpoons \quad \frac{1}{1 + j10\pi f}\, [\, \frac{1}{j2\pi f} + \frac{\delta(f)}{2}\,]$$

, or equivalently

$$0.2\, e^{-0.2t}\, u(t) \otimes u(t) \quad \rightleftharpoons \quad \frac{1}{j2\pi f\, (1 + j10\pi f)} + \frac{\delta(f)}{2}$$

, using the partial fraction technique, yields

$$0.2\, e^{-0.2t}\, u(t) \otimes u(t) \quad \rightleftharpoons \quad \frac{1}{j2\pi f} - \frac{5}{(1 + j10\pi f)} + \frac{\delta(f)}{2}$$

, taking the inverse Fourier transform, yields

$$u(t) \otimes e^{-0.2t}\, u(t) \quad \rightleftharpoons \quad \frac{1}{j2\pi f} + \frac{\delta(f)}{2} - \frac{5}{(1 + j10\pi f)}$$

, where the time domain is the same output waveform y(t), Eq.(7.37).

Problem 7.11

For the RC low-pass filter, Fig.7.19a, if the input voltage waveform is x(t) = 15 e^{-5t} u(t), let the time constant value RC = 0.1 sec. Find the 1 Ω energy available at the input and at the output of the filter ?

Solution

The 1 Ω energy available at the input, using time domain, Eq.(1.10), E_i is given by

$$E_i = \int_0^\infty |15\, e^{-5t}|^2\, dt = 225 \int_0^\infty e^{-10t}\, dt = 22.5 \qquad \text{joule}$$

, since the time constant value RC = 0.1 sec, the cut-of frequency of the RC low pass filter, Fig.7.19a, is given by

$$f_c = \frac{1}{2\pi RC} = \frac{5}{\pi} \quad \text{Hz}$$

, then the transfer function H(f) of the RC low-pass filter is given by

$$H(f) = \frac{1}{1 + j(f\,/\,f_c)} = \frac{10}{10\, +\, j2\pi f}$$

, and the Fourier transform of the input voltage waveform $x(t) = 15\,e^{-5t}\,u(t)$, is given by

$$X(f) = \frac{15}{5 + j2\pi f}$$

, then the Fourier transform $Y(f)$ of the output voltage waveform $y(t)$ is given by

$$Y(f) = H(f)\,X(f) = \frac{150}{(5 + j2\pi f)\,(10 + j2\pi f)}$$

, and the output energy spectral density $\Psi_o(f)$, Eq.(7.16), is given by

$$\Psi_o(f) = |Y(f)|^2 = |H(f)|^2\ |X(f)|^2$$

, or equivalently

$$\Psi_o(f) = \frac{22500}{[25 + (2\pi f)^2]\,[100 + (2\pi f)^2]}$$

, since the cut-off frequency is $f_c = \pm\,5/\pi$ Hz, the one Ω energy available at the output of the filter E_o, will be within the frequency range $-5/\pi \le f \le 5/\pi$ Hz, and using the partial fraction technique, Eq.(3.62), E_o yields

$$E_o = \int_{-5/\pi}^{5/\pi} \left[\frac{300}{[25 + (2\pi f)^2]} - \frac{300}{[100 + (2\pi f)^2]} \right] df$$

, since $\Psi_o(f)$ is even function, E_o will be

$$E_o = 2\int_{0}^{5/\pi} \left[\frac{300}{[25 + (2\pi f)^2]} - \frac{300}{[100 + (2\pi f)^2]} \right] df$$

, let $w = 2\pi f$, then $dw = 2\pi\,df$, as $f\to 0$, then $w\to 0$, and as $f\to 5/\pi$, then $w\to 10$, E_o yields

$$E_o = \frac{300}{\pi} \int_{0}^{10} \left[\frac{1}{[25 + w^2]} - \frac{1}{[100 + w^2]} \right] dw$$

$$= \frac{300}{\pi} \left[\frac{1}{5}\tan^{-1}(\frac{w}{5}) - \frac{1}{10}\tan^{-1}(\frac{w}{10}) \right]_{w=0}^{w=10}$$

$$= \frac{300}{\pi} \left\{ \frac{1}{5}\left[\tan^{-1}(2) - \tan^{-1}(0)\right] - \frac{1}{10}\left[\tan^{-1}(1) - \tan^{-1}(0)\right] \right\}$$

$$= \frac{30}{\pi}\left[\frac{2\pi}{2,84} - \frac{\pi}{4} \right] = 14.64 \qquad \text{joule}$$

Problem 7.12

For the RC high-pass filter, Fig.7.22a. Find:

i. The transfer function of the filter, and is a distortion caused by the filter?.

ii. The impulse response of the filter and, is the filter a causal system?.

Solution

i. Consider $x(t)$ is an input voltage waveform is applied to the RC high-pass filter, Fig.7.22a. If the output voltage waveform is $y(t)$, then the voltage difference across the capacitor C equals $[x(t) - y(t)]$, and the current in the capacitor C and in the resistor R is $i(t)$, is given by

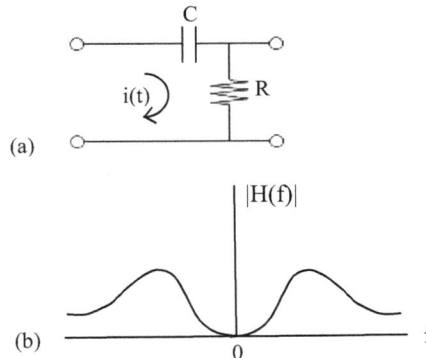

(a)

(b)

Fig.7.22. a) RC high-pass filter,
   b) The transfer function
   of RC high-pass filter.

$$i(t) = C \frac{d}{dt} [x(t) - y(t)] = \frac{y(t)}{R} \qquad (7.38)$$

, or equivalently

$$y(t) = RC \frac{d}{dt} [x(t) - y(t)] \qquad (7.39)$$

, taking the Fourier transform for both sides of Eq.(7.39), yields

$$Y(f) = j2\pi f \, RC \, [\, X(f) - Y(f) \,]$$

, or equivalently

$$[1 + j2\pi f \, RC] \, Y(f) = j2\pi f \, RC \, X(f)$$

, where X(f) and Y(f) are the Fourier transforms of the input and the output voltages waveforms respectively, then the transfer function of the RC high-pass filter, Eq.(7.3), H(f) is given by

$$H(f) = \frac{j2\pi f \, RC}{1 + j2\pi f \, RC} \qquad (7.40)$$

, another way to get the transfer function, using the frequency domain, where the Fourier transforms X(f) and Y(f) of the input and the output voltages waveforms are given by

$$X(f) = \left[R + \frac{1}{j2\pi f \, C} \right] I(f) \qquad , \qquad \text{and} \qquad Y(f) = R \, I(f)$$

, where I(f) is the Fourier transform of the current i(t), then the transfer function H(f), Eq.(7.3) can be obtained directly, the division of Y(f) by X(f), Eq.(7.40).

The RC high-pass filter, Fig.7.22a, will cause some distortion because the amplitude of the transfer function $|H(f)|$, Fig.7.22b, is not constant within the frequency band of the -3 dB points (half power points), where the frequency components of the input signal are transmitted with different change in amplitude, causing some amounts of nonlinearity, and the phase angle $\beta(f)$ is nonlinear function with frequency, where a set of frequency components (narrow band) is subjected to different delay, and the output signal has a different waveform from the input signal [1,4].

ii. The impulse response h(t) can be obtained by taking the inverse Fourier transform of Eq.(7.40), h(t) is given by

$$h(t) = RC \frac{d}{dt} [\frac{1}{RC} \, e^{-\frac{t}{RC}} \, u(t) \,] \qquad (7.41)$$

, making use of Eq.(4.21b), Eq.(7.41) yields

$$h(t) = \delta(t) - \frac{1}{RC} \, e^{-\frac{t}{RC}} \, u(t) \qquad (7.42)$$

, where u(t) is the unit step function. Fig.7.23. shows the impulse response h(t) of the RC high-pass filter, Fig.7.22a. The RC high-pass filter is a causal system because it has a causal impulse response h(t), where h(t) = 0 for t < 0.

, another way to get the impulse response h(t), is the transfer function H(f), Eq.(7.40) can be written in the form

$$H(f) = 1 - \frac{1}{1 + j2\pi f \, RC} \qquad (7.43)$$

, taking the inverse Fourier transform for Eq.(7.43), the same impulse response h(t), Eq.(7.42) is obtained.

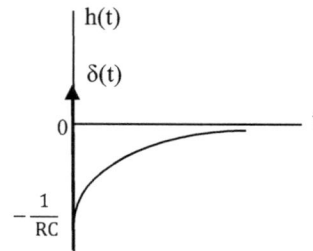

Fig.7.23. Impulse response of RC high-pass filter, Fig.7.22a.

Problem 7.13

For the RC high-pass filter, Fig.7.22a. Find the unit step response of the filter (let the time constant value RC = 5 second).

Solution

Considering the time constant value RC = 5 sec, where the impulse response h(t), Eq.(7.42), the unit step response y(t) is given by

$$y(t) = h(t) \otimes u(t)$$

$$= [\, \delta(t) - 0.2\ e^{-0.2t}\ u(t)\,] \otimes u(t)$$

, using frequency domain and Eq.(7.40), a Fourier transform pair is given by

$$[\, \delta(t) - 0.2\ e^{-0.2t}\ u(t)\,] \otimes u(t) \ \rightleftharpoons\ \frac{j10\pi f}{1 + j10\pi f}\ [\frac{1}{j2\pi f} + \frac{\delta(f)}{2}]$$

, or equivalently

$$[\, \delta(t) - 0.2\ e^{-0.2t}\ u(t)\,] \otimes u(t) \ \rightleftharpoons\ \frac{5}{1 + j10\pi f}$$

, taking the inverse Fourier transform, the response y(t) is given by

$$y(t) = [\, \delta(t) - 0.2\ e^{-0.2t}\ u(t)\,] \otimes u(t)$$

$$= e^{-0.2t}\ u(t)$$

, Fig.7.24. shows the unit step response y(t) for the RC high-pass filter, Fig.7.22a.

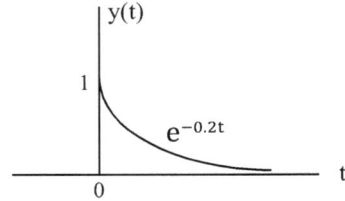

Fig.7.24. Unit step response of the RC high-pass filter, Fig.7.22a.

Problem 7.14

For the RC high-pass filter, Fig.7.22a. Find the response of the filter if the input signal is x(t) = rect(t) , let the time constant value RC = 5 second.

Solution

Considering the time constant value RC = 5 sec, where the impulse response h(t), Eq.(7.42), the response of the RC high-pass filter y(t) when the input signal is x(t) = rect(t), is given by

$$y(t) = h(t) \otimes rect(t)$$

$$= [\, \delta(t) - 0.2\ e^{-0.2t}\ u(t)\,] \otimes rect(t)$$

, taking the Fourier transform, Y(f) is given by

$$Y(f) = \left[1 - \frac{1}{1 + j10\pi f} \right]\ sinc(f)$$

, express the sinc function in terms of exponential terms, yields

$$Y(f) = sinc(f) - \left[\frac{e^{j\pi f} - e^{-j\pi f}}{j2\pi f\ (1 + j10\pi f)} \right]$$

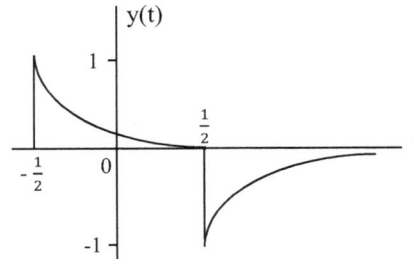

Fig.7.25. Response of RC high-pass filter, Fig.7.22a, for input pulse x(t) = rect(t).

, using the partial fraction technique, Y(f) yields

$$Y(f) = sinc(f) - \left\{ \left[\frac{1}{j2\pi f} - \frac{5}{1 + j10\pi f} \right] [\, e^{j\pi f} - e^{-j\pi f}\,] \right\}$$

, or equivalently

$$Y(f) = \frac{5}{1 + j10\pi f}\ [\, e^{j\pi f} - e^{-j\pi f}\,]$$

, taking the inverse Fourier transform, y(t) yields

$$y(t) = e^{-0.2(t + \frac{1}{2})}\ u(t + \frac{1}{2}) - e^{-0.2(t - \frac{1}{2})}\ u(t - \frac{1}{2})$$

, Fig.7.25 shows the response y(t) of the RC high-pass filter, Fig.7.22a, when the input signal is rec(t).

Problem 7.15

An input voltage $x(t) = 120\ e^{-24t}\ u(t)$, is applied to an ideal band-pass filter, passes frequencies between $12/\pi$ and $24/\pi$ Hz, and rejects all frequencies outside this pass-band. Find the $1\ \Omega$ energy available at the input and the output of the filter.

Solution

The Fourier transform of the input voltage is

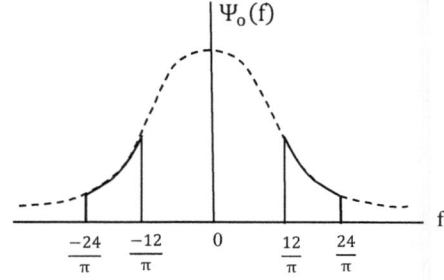

Fig.7.26. Output energy spectral density.

$$X(f) = \frac{120}{24 + j2\pi f}$$

, the input spectral density is given by

$$\Psi_i(f) = \frac{14400}{(24)^2 + (2\pi f)^2} \qquad \text{for} \quad -\infty \le f \le \infty$$

, the $1\ \Omega$ energy available at the input of the filter, is given by

$$E_i = \int_{-\infty}^{\infty} \frac{14400}{(24)^2 + (2\pi f)^2}\ df \qquad\qquad (7.44)$$

, since $\Psi_i(f)$ is even function, E_i yields

$$E_i = 2 \int_{0}^{\infty} \frac{14400}{(24)^2 + (2\pi f)^2}\ df$$

, let $w = 2\pi f$, then $dw = 2\pi\ df$, as $f \to 0$, then $w \to 0$, and as $f \to \infty$, then $w \to \infty$, E_i yields

$$E_i = \frac{1}{\pi} \int_{0}^{\infty} \frac{14400}{(24)^2 + w^2}\ dw = \frac{14400}{\pi} \left[\frac{1}{24} \tan^{-1}\left(\frac{W}{24}\right) \right]_{w=0}^{w=\infty}$$

$$= \frac{14400}{\pi} \left\{ \frac{1}{24} \left[\frac{\pi}{2} - 0 \right] \right\} = 300 \qquad \text{joule}$$

, for the ideal band-pass filter, the magnitude of the transfer function $|H(f)| = 1$, then the output energy spectral density, and using Eq.7.16, $\Psi_o(f)$, Fig.7.26, is given by

$$\Psi_o(f) = \frac{14400}{(24)^2 + (2\pi f)^2} \qquad \text{for} \quad 12/\pi \le f \le 24/\pi$$

, and the $1\ \Omega$ energy available at the output of the filter, E_o is given by

$$E_o = \int_{-24/\pi}^{-12/\pi} \frac{14400}{(24)^2 + (2\pi f)^2}\ df + \int_{12/\pi}^{24/\pi} \frac{14400}{(24)^2 + (2\pi f)^2}\ df$$

, since $\Psi_o(f)$ is even function, E_0 yields

$$E_o = 2 \int_{12/\pi}^{24/\pi} \frac{14400}{(24)^2 + (2\pi f)^2}\ df$$

, let $w = 2\pi f$, then $dw = 2\pi\ df$, as $f \to 12/\pi$, then $w \to 24$, and as $f \to 24/\pi$, then $w \to 48$, E_o yields

$$E_o = \frac{1}{\pi} \int_{24}^{48} \frac{14400}{(24)^2 + w^2} \, dw = \frac{14400}{\pi} \left[\frac{1}{24} \tan^{-1}(\frac{w}{24}) \right]_{w=24}^{w=48}$$

$$= \frac{600}{\pi} \left[\frac{\pi}{2.84} - \frac{\pi}{4} \right] = 61.45 \quad \text{joule}$$

Problem 7.16
A Gaussian filter has the following transfer function

$$H(f) = e^{-(4\pi f^2 + j6\pi f)}$$

Find the impulse response of this Gaussian filter and state with reasons whether the filter is physically realizable or not, and is there a way to make realizable if it is not ?.

Solution
The transfer function of this Gaussian filter H(f) can be written in the form

$$H(f) = e^{-4\pi f^2} e^{-j6\pi f}$$

, taking the inverse Fourier transform, the impulse response h(t) is given by

$$h(t) = \frac{1}{2} e^{-\pi(t-3)^2/2}$$

, it is not physically realizable, because it is non-causal filter, since $h(t) \neq 0$, for $t < 0$, and no way.

7.12. Uniformly Distributed Resistance-Capacitance Interconnect Systems

The uniformly distributed Resistance-Capacitance RC interconnects along the wire network, is one of the models for the analysis of the waveform interconnects of the Bipolar Complementary Metal Oxide Semiconductor BiCMOS and Emitter Coupled Logic ECL gates. Models of open, shorted and resistively and inductively loaded RC distributed interconnects thin film structure, for input step and pulse voltages functions with low and high transient response, are simulated by the author [14−18]. In this section, the transfer function H(f) and the impulse response h(t) for different sections of the uniformly distributed RC low-pass filter are obtained, and also the general formulas of the equivalent transfer function $H_{eq}(f)$ and the equivalent impulse response $h_{eq}(t)$ of the multi-sections are obtained .

In the case of the distributed one RC section, Fig.7.19a. The transfer function $H_1(f)$ and the impulse response $h_1(t)$ are

$$H_1(f) = \frac{1}{1 + j2\pi f RC} \qquad (7.32)$$

$$h_1(t) = \frac{1}{RC} e^{-t/RC} u(t) \qquad (7.33)$$

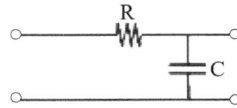

Fig.7.19a. One RC section.

In the case of the two distributed RC sections, Fig.7.27. The transfer function $H_2(f)$ and the impulse response $h_2(t)$, are given by

$$H_2(f) = \frac{1}{(1 + j2\pi f RC)^2}$$

$$h_2(t) = \frac{1}{RC} e^{-t/RC} u(t) \otimes \frac{1}{RC} e^{-t/RC} u(t)$$

, or equivalently

Fig.7.27. Two distributed RC sections.

$$h_2(t) = \frac{1}{(RC)^2} \int_0^t e^{-\tau/RC} e^{-(t-\tau)/RC} \, d\tau = \frac{1}{(RC)^2} t \, e^{-t/RC} u(t) \qquad (7.45)$$

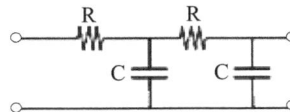

In the case of the three distributed RC sections, Fig.7.28. The transfer function $H_3(f)$ and the impulse response $h_3(t)$, are given by

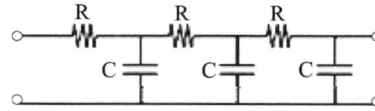

Fig.7.28. Three distributed RC sections.

$$H_3(f) = \frac{1}{(1 + j2\pi f\,RC)^3}$$

$$, h_3(t) = h_2(t) \otimes \frac{1}{RC}\,e^{-t/RC}\,u(t)$$

, or equivalently

$$h_3(t) = \frac{1}{(RC)^2}\,t\,e^{-t/RC}\,u(t) \otimes \frac{1}{RC}\,e^{-t/RC}\,u(t)$$

$$= \frac{1}{(RC)^3}\int_0^t \tau\,e^{-\tau/RC}\,e^{-(t-\tau)/RC}\,d\tau = \frac{1}{2(RC)^3}\,t^2\,e^{-t/RC}\,u(t) \qquad (7.46)$$

In the case of the four distributed RC sections, Fig.7.29. The transfer function $H_4(f)$ and the impulse response $h_4(t)$ are given by

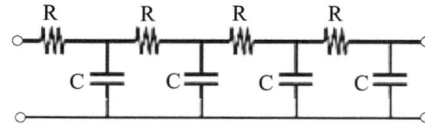

Fig.7.29. Four distributed RC sections.

$$H_4(f) = \frac{1}{(1 + j2\pi f\,RC)^4}$$

$$, h_4(t) = h_3(t) \otimes \frac{1}{RC}\,e^{-t/RC}\,u(t)$$

, or equivalently

$$h_4(t) = \frac{1}{2(RC)^3}\,t^2\,e^{-t/RC}\,u(t) \otimes \frac{1}{RC}\,e^{-t/RC}\,u(t)$$

$$= \frac{1}{2(RC)^4}\int_0^t \tau^2\,e^{-\tau/RC}\,e^{-(t-\tau)/RC}\,d\tau = \frac{1}{2\times3}\frac{1}{(RC)^4}\,t^3\,e^{-t/RC}\,u(t) \qquad (7.47)$$

In the case of the five distributed RC sections, Fig.7.30. The transfer function $H_5(f)$ and the impulse response $h_5(t)$, are given by

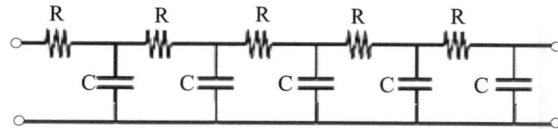

Fig.7.30. Five distributed RC sections.

$$H_5(f) = \frac{1}{(1 + j2\pi f\,RC)^5}$$

$$, h_5(t) = h_4(t) \otimes \frac{1}{RC}\,e^{-t/RC}\,u(t)$$

$$= \frac{1}{2\times3\,(RC)^4}\,t^3\,e^{-t/RC}\,u(t) \otimes \frac{1}{RC}\,e^{-t/RC}\,u(t)$$

, or equivalently

$$h_5(t) = \frac{1}{2 \times 3 \ (RC)^5} \int_0^t \tau^3 \ e^{-\tau/RC} \ e^{-(t-\tau)/RC} \ d\tau$$

$$= \frac{1}{2 \times 3 \times 4} \ \frac{1}{(RC)^5} \ t^4 \ e^{-t/RC} \ u(t) \qquad\qquad (7.48)$$

In the general case of N distributed RC sections, the equivalent impulse response $h_N(t)$, Fig.7.31, using Equations (7.45), (7.46), (7.47), and (7.48), is given by

$$h_N(t) = \frac{1}{(N-1)!} \ \frac{1}{(RC)^{N-1}} \ t^{N-1} \ h_1(t) \qquad , \qquad \text{and} \quad N > 1$$

, where $h_1(t)$ is the impulse response of the distributed one RC section, Eq.(7.33), and $(N-1)!$ is the vectorial of $(N-1)$, is given by

$$(N-1)! = (N-1)\ (N-2)\ (N-3) \dots \dots \dots$$

, and the equivalent transfer function $H_N(f)$, Fig.7.32, is given by

$$H_N(f) = \frac{1}{(1 + j2\pi f \ RC)^N}$$

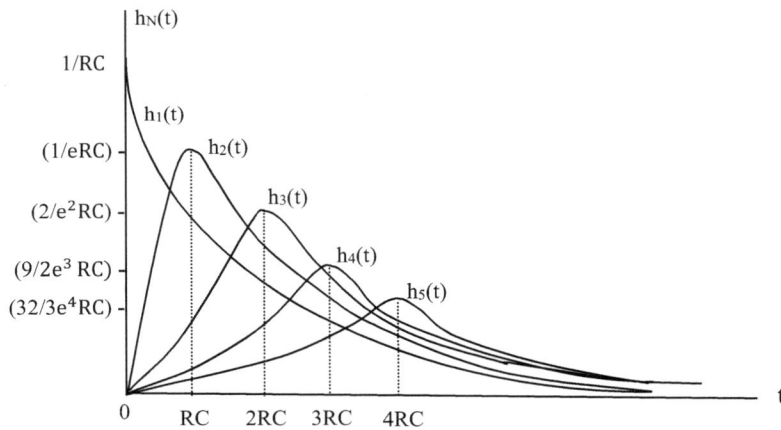

Fig.7.31. The impulse response of N sections distributed RC low-pass filter.

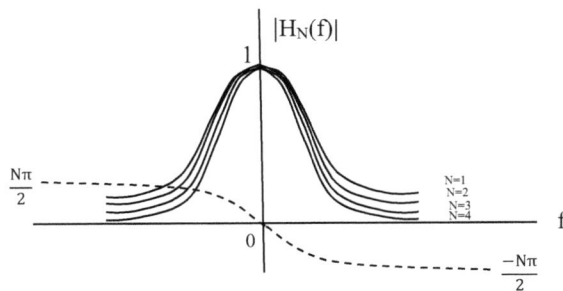

Fig.7.32. The equivalent transfer function $H_N(f)$.

Problems

1. Find the impulse response and the transfer function of the following filters:
   - i. The ideal high-pass filter.
   - ii. The ideal band-elimination filter.

   , are these filters causal systems ?

2. For the system, Fig.7.33. Find:
   - i. The transfer function and the impulse response.
   - ii. The unit step response.

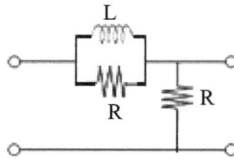

Fig.7.33. L= 1 henry , R= 1 ohm.

3. Find the transfer functions and the impulse responses of the low-pass and high-pass filters, Fig.7.34a,b, respectively.

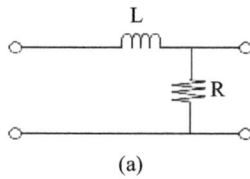

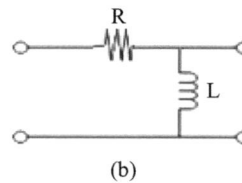

(a)   (b)

Fig.7.34. RL filters.

4. Find the response of the RC low-pass Filter, Fig.7.19a, if the input signal is given by

   $$x(t) = 40 \ \text{sgn}(t)$$

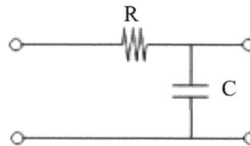

Fig.7.19a. RC low-pass filter.

5. Find the equivalent impulse response and the equivalent transfer function of the two cascaded systems, Fig.7.35.

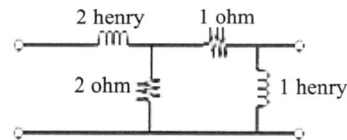

Fig.7.35. Two cascaded filters.

6. A train of pulses, Fig.7.36, is applied to the low-pass Filter, Fig.7.19a. Find the response of the system at the time t = 3.5 sec, (let the time constant value RC = 1 second).

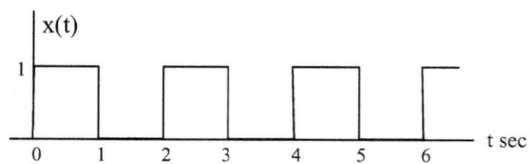

Fig.7.36. Train of pulses.

7. For the RC low-pass filter, Fig.7.19a. Find the output auto-correlation function and the output spectral density in the two following cases:
   - i. A signal x(t) has autocorrelation function $R_i(\tau) = A\ \delta(\tau)$, is applied at the filter.
   - ii. A signal x(t) has autocorrelation function $R_i(\tau) = A\ \delta(\tau - 3)$, is applied at the filter.

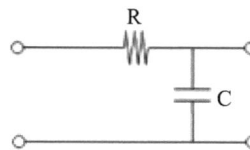

Fig.7.19a. RC low-pass filter.

8. For the RC high-pass filter, Fig.7.22a.
   Find the output auto-correlation function and the output spectral density in the two following cases:
   i. A signal x(t) has autocorrelation function $R_i(\tau) = k\,\delta(\tau)$, is applied at the filter.
   ii. A signal x(t) has autocorrelation function $R_i(\tau) = k\,\delta(\tau - 5)$, is applied at the filter.

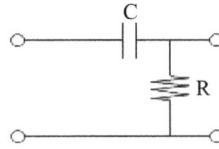

Fig.7.22a. RC high-pass filter.

9. Find the output auto-correlation function and the output spectral density of the ideal low-pass filter, in the two following cases:
   i. A signal x(t) has autocorrelation function $R_i(\tau) = k\,\delta(\tau)$, is applied at the filter.
   ii. A signal x(t) has autocorrelation function $R_i(\tau) = k\,\delta(\tau + 2)$, is applied at the filter.

10. Find the output auto-correlation function and the output spectral density of the ideal band-pass filter, in the two following cases:
    i. A signal x(t) has autocorrelation function $R_i(\tau) = k\,\delta(\tau)$, is applied at the filter.
    ii. A signal x(t) has autocorrelation function $R_i(\tau) = k\,\delta(\tau + 4)$, is applied at the filter.

11 A low pass filter has a transfer function, magnitude and phase angle, Fig.7.37. Find the impulse response of the filter.

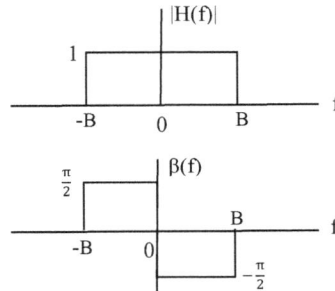

Fig.7.37. Transfer function, magnitude and phase angle

12 Find the transfer functions and the impulse responses of the RL and RLC filters, Fig.7.38a,b,c,d.

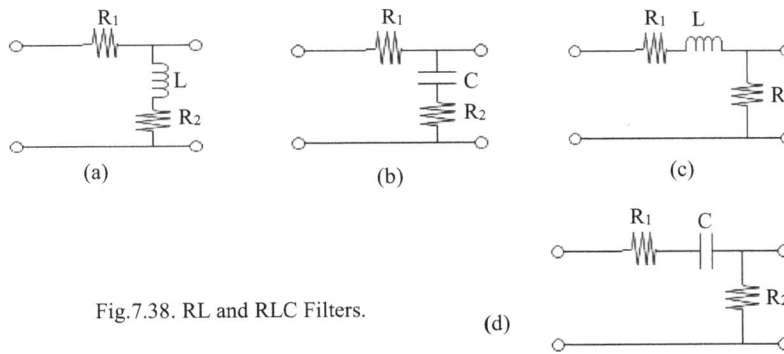

Fig.7.38. RL and RLC Filters.

13. For the RLC low-pass filter, Fig.7.39.
    Find:
    i. The transfer function of the filter.
    ii. The impulse response of the filter.

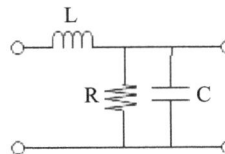

Fig.7.39. RLC filter.

14. Find the transfer function and the impulse response of the RLC band-pass filter, Fig.7.40.

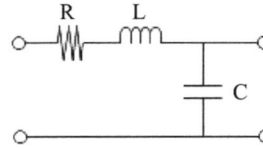

Fig.7.40. RLC filter.

15. Find the transfer function and the impulse response of the RLC band-pass filter, Fig.7.41.

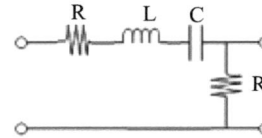

Fig.7.41. RLC filter.

Send Orders for Reprints to reprints@benthamscience.net

CHAPTER 8

Hilbert Transform

Abstract: The Fourier transform is used for evaluating the frequency content of the energy signal (chapter III) and in a limiting sense the power signal (chapter V). The Fourier transform analysis provides a means of analyzing and designing the frequency selective filters for the separation of signals on the basis of their frequency content, where the frequency selectivity plays the main role in the design of the communication systems. The Hilbert transform plays the same role depending on the phase selectivity. The Hilbert transform uses the phase shifts between the signals to achieve the desired separation, where the phase angles of all components of a given signal are shifted by ±90°, the resulting function of time is known as the Hilbert transform of the signal.

Keywords: Hilbert transform; Orthogonality; Convolution principle; Narrow band-pass signals.

The Hilbert transform of a real-valued function $x(t)$ is another real-valued function in the same domain $\hat{x}(t)$. The Hilbert transform is named after the renowned mathematician David Hilbert. The signal $x(t)$ and its Hilbert transform $\hat{x}(t)$ form a very useful analytic representation, is defined by the pre-envelope $x_+(t)$, (chapter IX). In the signal processing, Hilbert transform plays an important part in the analysis of the band-pass system which can be establishing an analogy between low-pass system equivalent to band-pass system, this analogy is based on the use of the Hilbert transform for facilitating the representation of the band-pass signals [1,4].

The Hilbert transform uses the phase shifts between the signals to achieve the desired separation, where the phase angles of all components of a given signal are shifted by ±90°. The signum function in the frequency domain $sgn(f)$ plays the role of the separation of the signals in the Hilbert transform, where the amplitude of the signal at all the frequency components is not affected by the transmission. The Fourier transform pair of the Hilbert transform is given by

$$\frac{1}{\pi t} \rightleftharpoons -j\ sgn(f) \tag{8.1}$$

, where the positive frequency components are shifted by −90°, the negative frequency components are shifted by +90°, and the transfer function $H(f)$ of the Hilbert transform is given by

$$H(f) = -j\ sgn(f) \tag{8.2}$$

, or equivalently

$$
\begin{aligned}
H(f) &= +j & \text{for}\ \ f < 0 \tag{8.3}\\
&= 0 & \text{for}\ \ f = 0 \\
&= -j & \text{for}\ \ f > 0 \tag{8.4}
\end{aligned}
$$

, and the impulse response $h(t)$ of the Hilbert transform is given by

$$h(t) = \frac{1}{\pi t} \tag{8.5}$$

, then for a given energy signal $x(t)$, its Hilbert transform $\hat{x}(t)$ is obtained by passing $x(t)$ through a linear two port device having transfer function $H(f) = -j\ sgn(f)$, Fig.8.1. This linear device produces a phase shift of −90° for all the positive frequencies of the input signal $x(t)$, and +90° for all the negative frequencies of the input signal $x(t)$, Equations (8.3) and (8.4). The amplitude of the signal at all the frequency components is not affected by the transmission through the Hilbert transform device. The Hilbert transform of the signal $x(t)$ is given by

$$\hat{x}(t) = x(t) \otimes \frac{1}{\pi t}$$

, or equivalently

$$\hat{x}(t) = \frac{1}{\pi} \int_{-\infty}^{\infty} \frac{x(\tau)}{(t-\tau)}\ d\tau \tag{8.6}$$

, or

$$\hat{x}(t) = \frac{1}{\pi} \int_{-\infty}^{\infty} \frac{x(t-\tau)}{\tau}\ d\tau \tag{8.7}$$

, since $\hat{x}(t)$ is the Hilbert transform of $x(t)$, then the Hilbert transform of $\hat{x}(t)$ is $x(t)$ with negative sign "$-x(t)$ ", and the inverse Hilbert transform is given by

$$x(t) = -\frac{1}{\pi} \int_{-\infty}^{\infty} \frac{\hat{x}(t)}{(t-\tau)} d\tau$$

$$, \text{or} \quad x(t) = -\frac{1}{\pi} \int_{-\infty}^{\infty} \frac{\hat{x}(t-\tau)}{\tau} d\tau$$

, where $\hat{x}(t)$ is the Hilbert transform of x(t) and may be viewed as the quadrature function of x(t). Also the magnitude of the transfer function H(f) is unity while the phase of H(f) is given by Fig.8.1b.

On the other hand, in the frequency domain, if the Fourier transform of the input signal x(t) is X(f), then the Fourier transform of $\hat{x}(t)$ is $\hat{X}(f)$, is given by

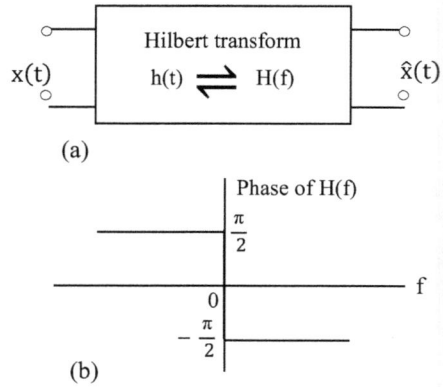

(a)

(b)

Fig.8.1. a) Hilbert transform is a linear two port device of impulse response h(t) = 1/πt, b) The phase of the transfer function H(f).

$$\hat{X}(f) = -j \operatorname{sgn}(f) X(f) \tag{8.8}$$

The evaluation of the Hilbert transform $\hat{x}(t)$ of the signal x(t) may be obtained by two ways, either in the time domain using Eq.(8.6) to get $\hat{x}(t)$ directly for varying t, or in the frequency domain where get the Fourier transform X(f) of the signal x(t) and then use Eq.(8.8) to get the Hilbert transform $\hat{X}(f)$ of the Fourier transform X(f) and then get the inverse Fourier transform to obtain $\hat{x}(t)$ [1].

8.1. Hilbert transform of Sinusoidal Functions

According to the definition of the Hilbert transform, the Hilbert transform of the steady cosine function "$\cos(2\pi f_c t)$" is the steady sine function "$\sin(2\pi f_c t)$", and the Hilbert transform of the steady sine function "$\sin(2\pi f_c t)$, is the steady cosine function with negative sign "$-\cos(2\pi f_c t)$".

proof

The Hilbert transform of the cosine function $x(t) = \cos(2\pi f_c t)$, using the time domain, Eq.(8.6), is given by

$$\hat{x}(t) = \frac{1}{\pi} \int_{-\infty}^{\infty} \frac{\cos[(2\pi f_c(t-\tau)]}{\tau} d\tau \tag{8.9}$$

, use the formula "$\cos(x - y) = \cos(x)\cos(y) + \sin(x)\sin(y)$", $\hat{x}(t)$ yields

$$\hat{x}(t) = \frac{1}{\pi} \int_{-\infty}^{\infty} \frac{\cos(2\pi f_c t)\cos(2\pi f_c \tau)}{\tau} d\tau + \frac{1}{\pi} \int_{-\infty}^{\infty} \frac{\sin(2\pi f_c t)\sin(2\pi f_c \tau)}{\tau} d\tau$$

, the principal value of the first integral is zero since the integrand is odd because of the function "$1/\tau$" is odd function, $\hat{x}(t)$ yields

$$\hat{x}(t) = \frac{1}{\pi}\sin(2\pi f_c t) \int_{-\infty}^{\infty} \frac{\sin(2\pi f_c \tau)}{\tau} d\tau$$

, let $\lambda = 2\pi f_c \tau$, then $d\lambda = 2\pi f_c \, d\tau$, as $\tau \to -\infty$ then $\lambda \to -\infty$, and as $\tau \to \infty$ then $\lambda \to \infty$, then $\hat{x}(t)$ yields

$$\hat{x}(t) = \frac{1}{\pi}\sin(2\pi f_c t) \int_{-\infty}^{\infty} \frac{\sin(\lambda)}{\lambda} d\lambda$$

, since the principal value of this integral equals π, Eq.(3.53), then the Hilbert transform $\hat{x}(t)$ is given by

$$\hat{x}(t) = \sin(2\pi f_c t) \tag{8.10}$$

, another way to get the Hilbert transform of the steady cosine function $x(t) = \cos(2\pi f_c t)$, using the frequency domain, where its Fourier transform $X(f)$ is given by

$$X(f) = \tfrac{1}{2} [\delta(f - f_c) + \delta(f + f_c)]$$

, then the Hilbert transform $\widehat{X}(f)$, Eq.(8.8), is given by

$$\widehat{X}(f) = -j \ \text{sgn}(f) \ \tfrac{1}{2} [\delta(f - f_c) + \delta(f + f_c)]$$

, or equivalently

$$\widehat{X}(f) = -j \ \tfrac{1}{2} [\delta(f - f_c) - \delta(f + f_c)]$$

, taking the inverse Fourier transform $\hat{x}(t)$, then the Hilbert transform $\hat{x}(t) = \sin(2\pi f_c t)$, Eq.(8.10), is obtained.

Using the same steps in the time and the frequency domains, the Hilbert transform of the steady sine function $x(t) = \sin(2\pi f_c t)$, is $\hat{x}(t) = -\cos(2\pi f_c t)$.

Numerical examples

i. The Hilbert transform of the signal $x(t) = 10 \cos(2\pi 10^6 t)$ is $\hat{x}(t) = 10 \sin(2\pi 10^6 t)$

ii. The Hilbert transform of the signal $x(t) = 4 \sin(2\pi 10^3 t)$ is $\hat{x}(t) = -4 \cos(2\pi 10^3 t)$

8.2. Hilbert transform and Orthogonality

Since the Hilbert transform uses the phase shifts between the signals to achieve the desired frequency separation, where the phase angles of all the frequency components of a given signal are shifted by $\pm$ 90°. Then the energy signal $x(t)$ and its Hilbert transform $\hat{x}(t)$ are orthogonal functions. Checking the orthogonality, is obtained by the aid of the cross-correlation function of the signal $x(t)$, Eq.(6.27), or by the orthogonality condition, Eq.(2.7), where

$$\int_{-\infty}^{\infty} x^*(t) \ \hat{x}(t) \ dt = 0$$

Proof:

The cross-correlation function $R_{12}(\tau)$ of the energy signal $x(t)$ and its Hilbert transform $\hat{x}(t)$, has the following Fourier transform pair Eq.(6.24),

$$R_{12}(\tau) \quad \rightleftharpoons \quad \widehat{X}(f) \ X^*(f)$$

, where $\widehat{X}(f)$ is the Hilbert transform of $X(f)$, $X^*(f)$ is the complex conjugate of $X(f)$, and $X(f)$ is the Fourier transform of $x(t)$. According to the Hilbert transform in frequency domain, Eq.(8.8), the Fourier transform pair yields

$$R_{12}(\tau) \quad \rightleftharpoons \quad -j \ \text{sgn}(f) \ X(f) \ X^*(f)$$

, or equivalently

$$R_{12}(\tau) \quad \rightleftharpoons \quad -j \ \text{sgn}(f) \ |X(f)|^2$$

, then the cross-correlation function $R_{12}(\tau)$ can be written in the form

$$R_{12}(\tau) = \int_{-\infty}^{\infty} -j \ \text{sgn}(f) \ |X(f)|^2 \ e^{j2\pi f \tau} \ df$$

, and at $\tau = 0$, $R_{12}(0)$ yields

$$R_{12}(0) = \int_{-\infty}^{\infty} -j \ \text{sgn}(f) \ |X(f)|^2 \ df$$

, since the function $|X(f)|^2$ is an even function and $\text{sgn}(f)$ is an odd function, and the integration along odd function for complete period equals zero, then $R_{12}(0) = 0$, and then the energy signal $x(t)$ and its Hilbert transform $\hat{x}(t)$ are orthogonal functions [1].

8.3. Hilbert transform and Convolution Principle

Consider the following convolution function

$$z(t) = x(t) \otimes y(t)$$

, the Hilbert transform $\hat{z}(t)$ is given by

$$\hat{z}(t) = x(t) \otimes \hat{y}(t) \qquad \text{, or} \qquad \hat{z}(t) = \hat{x}(t) \otimes y(t)$$

Proof:

The Hilbert transform $\hat{z}(t)$, Eq.(8.6), is given by

$$\hat{z}(t) = \frac{1}{\pi} \int_{-\infty}^{\infty} \frac{z(\tau)}{(t - \tau)} \, d\tau$$

, since z(t) is the convolution function of x(t) and y(t), z(t) is given by

$$z(t) = \int_{-\infty}^{\infty} x(\eta) \, y(t - \eta) \, d\eta \qquad , \text{or} \qquad z(\tau) = \int_{-\infty}^{\infty} x(\eta) \, y(\tau - \eta) \, d\eta$$

, then $\hat{z}(t)$ yields

$$\hat{z}(t) = \frac{1}{\pi} \int_{-\infty}^{\infty} \int_{-\infty}^{\infty} \frac{x(\eta) \, y(\tau - \eta)}{(t - \tau)} \, d\eta \, d\tau$$

, let $\xi = \tau - \eta$, then $d\xi = d\tau$, as $\tau \to -\infty$ then $\xi \to -\infty$, and as $\tau \to \infty$ then $\xi \to \infty$, and $\hat{z}(t)$ yields

$$\hat{z}(t) = \frac{1}{\pi} \int_{-\infty}^{\infty} \int_{-\infty}^{\infty} \frac{x(\eta) \, y(\xi)}{(t - \eta - \xi)} \, d\eta \, d\xi$$

, or equivalently

$$\hat{z}(t) = \int_{-\infty}^{\infty} x(\eta) \left[\frac{1}{\pi} \int_{-\infty}^{\infty} \frac{y(\xi)}{(t - \eta - \xi)} \, d\xi \right] d\eta$$

, the inner integral between brackets equals $\hat{y}(t - \eta)$, Eq.(8.6), then $\hat{z}(t)$ yields

$$\hat{z}(t) = \int_{-\infty}^{\infty} x(\eta) \, \hat{y}(t - \eta) \, d\eta$$

, or equivalently

$$\hat{z}(t) = x(t) \otimes \hat{y}(t) \qquad\qquad (8.11a)$$

, also due to the convolution commutative process, Eq.(3.43a), it may be similarly shown that

$$\hat{z}(t) = \hat{x}(t) \otimes y(t) \qquad\qquad (8.11b)$$

8.4. Hilbert transform of Narrow Band-Pass Signals

The band-pass signals arise from a modulated signal. Consider the following modulated signal

$$s(t) = m(t) \, c(t) \qquad\qquad (8.12)$$

, where m(t) is the baseband signal (low-pass signal) designates the band of the low frequency components, and is defined by its maximum frequency components W Hz, Fig.1.13a, which cannot propagate very long distances such as the audio frequency from 16 Hz to 20 KHz, the low frequency components of the video signal range, and the baseband signal of the voice channel from 300 to 3400 Hz. While c(t) is the carrier wave has single high frequency f_c, and is generated by an oscillator in the transmitter. This carrier frequency f_c is much higher than the maximum frequency of the baseband signal W Hz. The modulated signal s(t) is the carrier wave c(t) modulated by the baseband signal m(t) inside the modulator of the transmitter, forming the band-pass signal s(t). This band-pass signal s(t) is a modulated signal contain certain band of high frequency components around the carrier frequency f_c and can be transmitted in the transmission medium (channel) over a much longer distance convoying the information m(t). Fig.1.13b. shows the Fourier transform S(f) of the modulated wave s(t), where the carrier wave c(t) is represented by a cosine function $\cos(2\pi f_c t)$. Since the modulated wave s(t) is the product of the baseband signal m(t) by the carrier wave $\cos(2\pi f_c t)$, then the Fourier transform S(f) of the modulated signal s(t) is given by

$$S(f) = M(f) \otimes \tfrac{1}{2} \, [\delta(f - f_c) + \delta(f + f_c)]$$

, or equivalently

$$S(f) = \tfrac{1}{2} \, [M(f - f_c) + M(f + f_c)] \qquad\qquad (8.13)$$

This kind of modulation is defined by Amplitude Modulation Double Side-Band Suppressed Carrier AM-DSB-SC [1,4,6].

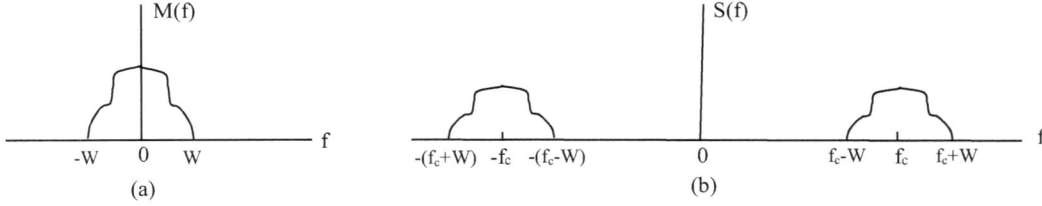

Fig.1.13. a) Low-pass signal, b) Band-pass signal.

The Hilbert transform $\hat{s}(t)$ of the product of non-overlapping low-pass signal m(t) and the high frequency signal c(t), Eq.(8.12), is to multiply the low-pass signal m(t) by the Hilbert transform $\hat{c}(t)$ of the high frequency signal c(t). In other words, for the modulated signal s(t), the low-pass signal m(t) is considered the amplitude of the high-frequency signal c(t), then the Hilbert transform $\hat{s}(t)$ is given by

$$\hat{s}(t) = m(t) \ \hat{c}(t) \tag{8.14}$$

Proof:

The Fourier transform of the modulated signal s(t), Eq.(8.12) is given by

$$S(f) = M(f) \otimes C(f)$$

, where M(f) and C(f) are the Fourier transforms of m(t) and c(t) respectively, S(f) yields

$$S(f) = \int_{-\infty}^{\infty} M(\lambda) \ C(f - \lambda) \, d\lambda$$

, and the Hilbert transform $\hat{S}(f)$ of S(f), Eq.(8.8), is given by

$$\hat{S}(f) = -j \ \text{sgn}(f) \ S(f)$$

, or equivalently

$$\hat{S}(f) = \int_{-\infty}^{\infty} [\,-j \ \text{sgn}(f)\,] \ M(\lambda) \ C(f - \lambda) \, d\lambda$$

, the inverse Fourier transform $\hat{s}(t)$, is given by

$$\hat{s}(t) = \int_{-\infty}^{\infty} \hat{S}(f) \ e^{j2\pi ft} \, df$$

, or equivalently

$$\hat{s}(t) = \int_{-\infty}^{\infty} \left\{ \int_{-\infty}^{\infty} -j \ \text{sgn}(f) \ M(\lambda) \ C(f - \lambda) \, d\lambda \right\} e^{j2\pi ft} \, df$$

, let $\xi = f - \lambda$, then $d\xi = df$, as $f \to -\infty$ then $\xi \to -\infty$, and as $f \to \infty$ then $\xi \to \infty$, and $\hat{s}(t)$ yields

$$\hat{s}(t) = \int_{-\infty}^{\infty} \int_{-\infty}^{\infty} [\,-j \ \text{sgn}(\xi + \lambda)] \ M(\lambda) \ C(\xi) \ e^{j2\pi(\xi + \lambda)t} \, d\xi \, d\lambda$$

, since the product $M(\lambda) \ C(\xi)$ is nonzero for $|\lambda| < W$ and $|\xi| > W$, then $\text{sgn}(\xi + \lambda)$ equals $\text{sgn}(\xi)$, and $\hat{s}(t)$ may be written in the form

$$\hat{s}(t) = \int_{-\infty}^{\infty} M(\lambda) \ e^{j2\pi\lambda t} \, d\lambda \int_{-\infty}^{\infty} [\,-j \ \text{sgn}(\xi)] \ C(\xi) \ e^{j2\pi\xi t} \, d\xi$$

, or equivalently

$$\hat{s}(t) = m(t) \ \hat{c}(t) \tag{8.14}$$

, which show that the Hilbert transform of the product of the non-overlapping low-pass signal m(t) and the high frequency signal c(t) is to multiply the low-pass signal m(t) by the Hilbert transform $\hat{c}(t)$ of the high frequency signal c(t).

Another way to get the Hilbert transform of the narrow band-pass signal s(t), Eq.(8.12), consider c(t) is a steady cosine function cos(2πf$_c$t), using Equations (8.8) and (8.13), the Hilbert transform Ŝ(f) yields

$$\hat{S}(f) = -j \text{ sgn}(f) \; \tfrac{1}{2} [M(f - f_c) + M(f + f_c)]$$

, or equivalently

$$\hat{S}(f) = -j \; \tfrac{1}{2} [M(f - f_c) - M(f + f_c)] \tag{8.15}$$

, Eq.(8.15) can be expressed by

$$\hat{S}(f) = M(f) \otimes -j \; \tfrac{1}{2} [\delta(f - f_c) - \delta(f + f_c)]$$

, then taking the inverse Fourier transform ŝ(t), will be the Hilbert transform of s(t), and is given by

$$\hat{s}(t) = m(t) \sin(2\pi f_c t)$$

Numerical examples

i. The Hilbert transform of x(t) = 10 rect(t/4) cos(2π10^{6}t)   is x̂(t) = 10 rect(t/4) sin(2π10^{6}t)

ii. The Hilbert transform of x(t) = 10 tri(t/5) sin(2π10^{3}t)   is x̂(t) = − 10 tri(t/5) cos(2π10^{3}t)

iii. The Hilbert transform of x(t) = 10 sin(200πt) cos(2π10^{7}t) is x̂(t) = 10 sin(200πt) sin(2π10^{7}t)

iv. The Hilbert transform of x(t) = 10 sin(500t) sin(2π10^{6}t)   is x̂(t) = − 10 sin(500 t) cos(2π10^{6}t)

v. The Hilbert transform of x(t) = sin(100t) sin((2π10^{5}t) + cos(50t) cos(2π10^5 t), is given by

$$\hat{x}(t) = - \sin(100t) \cos((2\pi10^5 t) + \cos(50t) \sin(2\pi10^5 t)$$

8.5. Some Important Hilbert Transforms

Since the relationship between the signal x(t) and its Hilbert transform x̂(t) is achieved, in the time domain using Eq.(8.6) and in the frequency domain using Eq.(8.8). Then the following results of the Hilbert transform are deduced

i. The Hilbert transform of the constant is zero.

ii. The double Hilbert transform of the signal x(t) is "− x(t) ".

iii. The Hilbert transform of a real function is a real function.

iv. The signal x(t) and its Hilbert transform x̂(t) have the same spectral density functions and the same auto-correlation functions.

v. The Hilbert transform of the derivative of the function x(t) is the derivative of the Hilbert transform x̂(t) where

if $y(t) = \dfrac{d}{dt} x(t)$ $\hat{y}(t) = \dfrac{d}{dt} \hat{x}(t)$

vii. The cross-correlation function of the signal x(t) and its Hilbert transform x̂(t) is zero.

viii. The Hilbert transform of the convolution of two functions is the convolution of one function with the Hilbert transform of the other function.

ix. The Hilbert transform of the Dirac delta function δ(t) is " 1/πt ".

x. The Hilbert transform of the function x(t) = rect(t/T), using the time domain, Eq.(8.6), x̂(t) is given by

$$\hat{x}(t) = \frac{1}{\pi} \int_{-T/2}^{T/2} \frac{1}{(t-\tau)} d\tau = -\left[\frac{1}{\pi} \ln(t-\tau) \right]_{\tau=-T/2}^{\tau=T/2}$$

$$= -\frac{1}{\pi} \ln \left[\frac{t - \tfrac{1}{2}T}{t + \tfrac{1}{2}T} \right] \tag{8.16}$$

Problem 8.1

Prove that the Hilbert transform of the function x(t) = sinc(t) is given by

$$\hat{x}(t) = \frac{1 - \cos(\pi t)}{\pi t}$$

Solution

Using the frequency domain, the Fourier transform of the function x(t) = sinc(t), is X(f) = rect(f), and the Hilbert transform of X(f), Eq.(8.8), is given by X̂(f) = −j sgn(f) rect(f), from the graphical multiplication of the function sgn(f) by the function rect(f), Fig.8.2, X̂(f) yields

$$\widehat{X}(f) = -j \left\{ \text{rect}\left[\frac{f - \frac{1}{4}}{\frac{1}{2}}\right] - \text{rect}\left[\frac{f + \frac{1}{4}}{\frac{1}{2}}\right] \right\}$$

, then the Hilbert transform $\hat{x}(t)$ is the inverse Fourier transform of $\widehat{X}(f)$, is given by

$$\hat{x}(t) = -j\left[\tfrac{1}{2} \text{sinc}(\tfrac{1}{2} t)\, e^{j\pi\, t/2} - \tfrac{1}{2} \text{sinc}(\tfrac{1}{2} t)\, e^{-j\pi\, t/2} \right]$$

, express the exponential terms by sine function, $\hat{x}(t)$ yields

$$\hat{x}(t) = 2\, \frac{\sin^2(\tfrac{1}{2}\,\pi t)}{\pi t}$$

, using the formula "$2 \sin^2(x) = 1 - \cos(2x)$", $\hat{x}(t)$ yields

$$\hat{x}(t) = \frac{1 - \cos(\pi t)}{\pi t}$$

Problem 8.2

Prove that the Hilbert transform of the function

$$x(t) = \frac{1}{1 + t^2}$$

, is given by

$$\hat{x}(t) = \frac{t}{1 + t^2}$$

Solution

Using the frequency domain, the Fourier transform of The function $x(t) = 1/(1 + t^2)$, is given by the double sided exponential function $X(f) = \pi\, e^{-|2\pi f|}$, and the Hilbert transform of $X(f)$, Eq.(8.8), is given by

$$\widehat{X}(f) = -j\pi\, \text{sgn}(f)\, e^{-|2\pi f|}$$

, or

$$\widehat{X}(f) = -j\,\pi\, \text{sgn}(f)\, [\, e^{-2\pi f} U(f) + e^{2\pi f} U(-f)\,]$$

, where $U(f)$ is the unit step function in the frequency domain. The graphical multiplication of the function $\text{sgn}(f)$ by the double sided exponential function, Fig.8.3, $\widehat{X}(f)$ yields

$$\widehat{X}(f) = -j\pi\, [\, e^{-2\pi f} U(f) - e^{2\pi f} U(-f)\,]$$

, then the Hilbert transform $\hat{x}(t)$ is the inverse Fourier transform of $\widehat{X}(f)$, is given by

$$\hat{x}(t) = -j\tfrac{1}{2}\left[\frac{1}{1 + jt} - \frac{1}{1 - jt} \right]$$

, or equivalently

$$\hat{x}(t) = \frac{t}{1 + t^2}$$

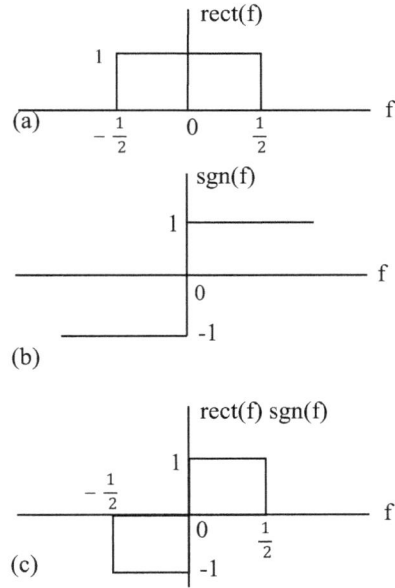

Fig.8.2. Graphical multiplication of the function $\text{sgn}(f)$ by the function $\text{rect}(f)$.

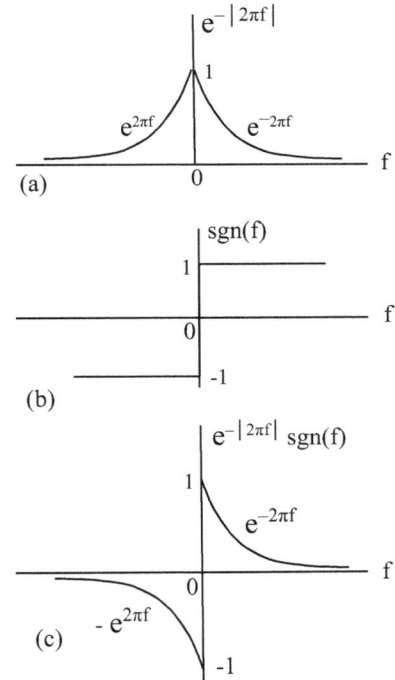

Fig.8.3. Graphical multiplication of the function $\text{sgn}(f)$ by the pulse $e^{-|2\pi f|}$.

Problem 8.3

For the linear system, Fig.8.4. Find

i. The impulse response and the transfer function of the system.

ii. The output y(t) of the system if the input signal is $x(t) = rect(t)$

iii. The energy content of y(t) from part (ii).

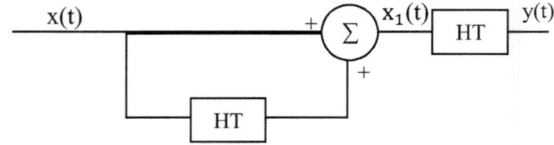

Fig.8.4. Linear system, problem 8.3.

Solution

Consider an energy signal x(t) is applied to the system, its Fourier transform is X(f) , the signal after the summation will be

$$x_1(t) = x(t) + \hat{x}(t)$$

, then the Fourier transform X(f) is given by

$$X_1(f) = [\, 1 - j\, sgn(f)\,]\, X(f)$$

, and the Fourier transform Y(f) of the output signal y(t) is given by

$$Y(f) = -j\, sgn(f)\, [\, 1 - j\, sgn(f)\,]\, X(f)$$

, or equivalently

$$Y(f) = -[\, 1 + j\, sgn(f)\,]\, X(f)$$

, then the transfer function of the system, Eq.(7.3), is given by

$$H(f) = -[\, 1 + j\, sgn(f)\,]$$

, and the impulse response h(t) of the system is the inverse Fourier transform of H(f), h(t) is given by

$$h(t) = \frac{1}{\pi t} - \delta(t)$$

, if the input signal is x(t), the output signal y(t) is given by

$$y(t) = x(t) \otimes [\frac{1}{\pi t} - \delta(t)\,]$$

, or equivalently

$$y(t) = \hat{x}(t) - x(t) \tag{8.17}$$

, considering the Hilbert transform of the input signal $x(t) = rect(t)$, Eq.(8.16), y(t) yields

$$y(t) = \left\{ -\frac{1}{\pi}\, \ln\left[\frac{t - \frac{1}{2}}{t + \frac{1}{2}} \right] \right\} - rect(t) \tag{8.18}$$

, the energy content of the output signal y(t) equals the energy content of the signal x(t) plus the energy content of its Hilbert transform $\hat{x}(t)$, Eq.(8.17). Since the energy content of the function rect(t) is one joule, then the energy content of its Hilbert transform is also one joule, and the total energy content of the output signal y(t) will be two joule.

Problem 8.4

For the linear system, Fig.8.5. Find the output waveform y(t) when the input signal is given by

$$x(t) = m(t)\, \cos(2\pi f_c t) + \hat{m}(t)\, \sin(2\pi f_c t)$$

, where m(t) is the low-pass signal with band-limited W Hz, Fig.1.13a, and f_c is the high frequency, is much higher than the maximum frequency of the baseband signal W Hz.

Solution

The Hilbert transform of the signal x(t) is given by

$$\hat{x}(t) = m(t) \sin(2\pi f_c t) - \hat{m}(t) \cos(2\pi f_c t)$$

, the signal after the multiplier is expressed by

$$x_1(t) = [\, m(t) \sin(2\pi f_c t) - \hat{m}(t) \cos(2\pi f_c t) \,] \sin(2\pi f_c t)$$

, using the formulas " $2 \sin^2(2\pi f_c t) = 1 - \cos(2\pi 2 f_c t)$ "
, and $\sin(2\pi 2 f_c t) = 2 \sin(2\pi f_c t) \cos(2\pi f_c t)$, $x_1(t)$ yields

$$x_1(t) = \tfrac{1}{2} m(t) - \tfrac{1}{2} m(t) \cos(2\pi 2 f_c t) - \tfrac{1}{2} \hat{m}(t) \sin(2\pi 2 f_c t)$$

, after the low pass filter, the output signal y(t) is given by

$$y(t) = \tfrac{1}{2} m(t)$$

, then the system, Fig.8.5, is simply demodulator.

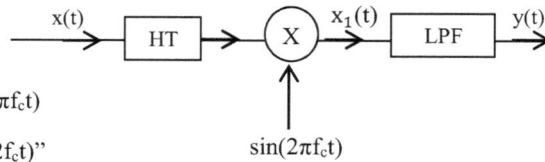

Fig.8.5. Linear system, problem 8.4.

Problems

1. Find the Hilbert transform of the following band-pass signals
    i. $x(t) = \sin(2\pi 10^3 t) \cos^3(2\pi 10^6 t)$
    ii. $x(t) = \sin(100\,t) \cos(2\pi 10^7 t) - \cos(50t) \sin(2\pi 10^7 t)$
    iii. $x(t) = \text{rect}(t) \cos^2(2\pi 10^5 t)$
    iv. $x(t) = \delta(t - 5)$

2. Find the Hilbert transform of the following band-pass signal

$$x(t) = m(t)\ e^{\,j 2\pi 10^6 t}$$

, where m(t) is the low-pass signal with band-limited W Hz, Fig.1.13a, and f_c is much higher than the maximum frequency of the baseband signal W Hz.

3. For the linear system, Fig.8.6.
Find the output waveform y(t)
of the system if the input signal
x(t) is given by
    i. $x(t) = m(t) \cos(2\pi f_c t)$
    ii. $x(t) = m(t) \cos(2\pi f_c t) - \hat{m}(t) \sin(2\pi f_c t)$
    iii. $x(t) = A[1 + m(t)] \cos(2\pi f_c t) - A\,\hat{m}(t) \sin(2\pi f_c t)$

, where m(t) is the low-pass signal with band-limited W Hz, Fig.1.13a, and f_c is much higher than W Hz , and A is the carrier amplitude.

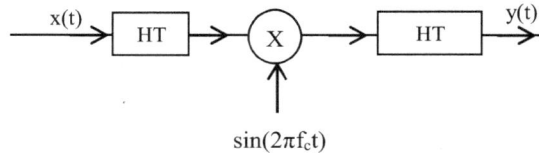

Fig.8.6. Linear system.

4. For the linear system, Fig.8.7.
Find the output waveform y(t)
of the system if the input signal
x(t) is given by
    $x(t) = \cos^3(2\pi 10^6 t)$

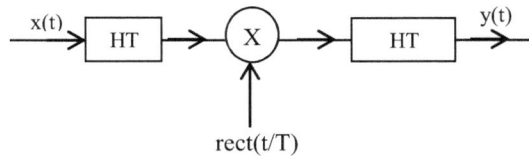

Fig.8.7. Linear system.

5. For the linear system, Fig.8.8.
Find the output waveform y(t)
If the input signal x(t) is given
by
    $x(t) = 10 \text{ tri}(t/5)$

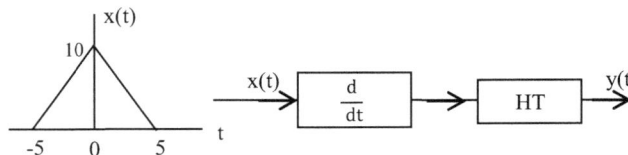

Fig.8.8. Linear system.

6. Prove that the Hilbert transform of the following energy signal

$$x(t) = \frac{t}{1 + t^2}$$

, is given by

$$\hat{x}(t) = -\frac{1}{1 + t^2}$$

7. Prove that the Hilbert transform of the following energy signal

$$x(t) = \frac{1 - \cos(\pi t)}{\pi t}$$

, is given by

$$\hat{x}(t) = -\operatorname{sinc}(t)$$

8. Prove that the Hilbert transform of the triangle function $x(t) = \operatorname{tri}(t)$, is given by

$$\hat{x}(t) = \frac{1}{\pi} \ln\left[\frac{t + 1}{t - 1}\right] - \frac{1}{\pi} \ln[t^2 - 1] - \frac{2}{\pi(t^2 - 1)}$$

9. Find the Hilbert transform of the following energy signals

i. $x(t) = \dfrac{1}{1 + j2\pi t}$ ii. $x(t) = \dfrac{1}{1 - j2\pi t}$

Send Orders for Reprints to reprints@benthamscience.net

CHAPTER 9

Narrow Band-Pass Signals and Systems

Abstract: In the communication systems, many information bearing signals are transmitted by service type of carrier modulation. The channel over which the signal is transmitted is limited in bandwidth to an interval of frequencies centered at the carrier as in the double side-band modulation techniques, Fig.1.13b, or adjacent to the carrier as in the single side-band modulation techniques, Fig.9.1a,b.
Keywords: Pre-envelope; Complex envelope; Natural envelope; Kham-Shen criteria; Group delay.

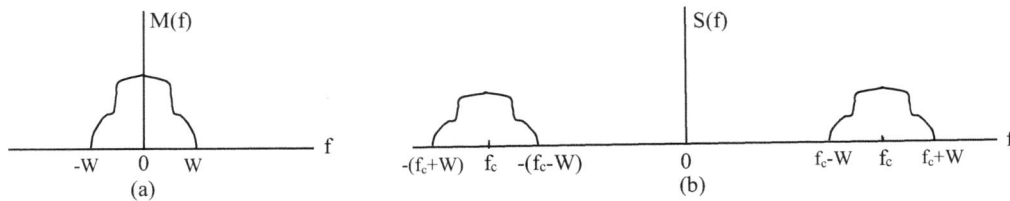

Fig.1.13. a) Low-pass signal, b) Band-pass signal

Signals and channels which satisfy the condition that their bandwidth is much smaller than the carrier frequency, are termed by the narrow band signals. All the band-pass signals arise from a modulated signal, interfering signals, or noise.

The mathematical complex representation of the narrow band-pass signals is handling by the pre-envelope of the signals while the complex-envelope is useful in handling the low-pass signals [1,8].

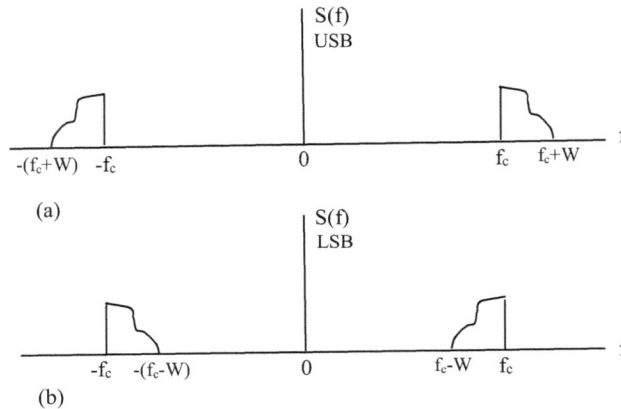

Fig.9.1. a) Spectrum of Upper Side-Band USB,
b) Spectrum of Lower Side-Band LSB.

9.1. Pre-Envelope

Since the pre-envelope is useful in handling the band-pass signals. Consider a narrow band-pass signal $x(t)$. the pre-envelope is the complex representation of $x(t)$ and is denoted by $x_+(t)$. The pre-envelope $x_+(t)$ is defined by the following complex valued function

$$x_+(t) = x(t) + j\,\hat{x}(t) \qquad (9.1)$$

, where $\hat{x}(t)$ is the Hilbert transform of $x(t)$ and may be viewed as the quadrature function of $x(t)$. Taking the Fourier transform of both sides of Eq.(9.1), yields

$$X_+(f) = X(f) + j\,\hat{X}(f) \qquad (9.2)$$

, where $X_+(f)$ is the Fourier transform of the pre-envelope $x_+(t)$. Using Eq.(8.8), Eq.(9.2) yields

$$X_+(f) = X(f) + j\,[-j\,\mathrm{sgn}(f)\,]\,X(f)$$

, or equivalently

$$X_+(f) = X(f) + \mathrm{sgn}(f)\,X(f)$$

, according to the definition of the signum function, $X_+(f)$ yields

$$
\begin{aligned}
X_+(f) &= 2\,X(f) && \text{for } f > 0 \\
&= X(0) && \text{for } f = 0 \\
&= 0 && \text{for } f < 0
\end{aligned}
$$

, and

, which means that the pre-envelope $x_+(t)$ has spectrum only in the positive frequencies and no frequency components in the negative frequencies, or equivalently

$$X_+(f) = 2 X(f) U(f) \tag{9.3}$$

, where U(f) is the unit step function in the frequency domain. Taking the inverse Fourier transform for Eq.(9.3), the pre-envelope $x_+(t)$ is given by

$$x_+(t) = 2 \int_0^\infty X(f) \, e^{j2\pi f t} \, df \tag{9.4}$$

, then the evaluation of the pre-envelope $x_+(t)$ for the narrow band-pass signal x(t) may be obtained by two ways, either using the time domain by getting the Hilbert transform $\hat{x}(t)$ and then use Eq.(9.1) to get $x_+(t)$ directly, or using the frequency domain by getting the Fourier transform X(f) of the signal x(t) and use Eq.(9.3) to get $X_+(f)$, and then Eq.(9.4) to obtain the inverse Fourier transform $x_+(t)$.

9.2. Complex Envelope

The complex-envelope $\tilde{x}(t)$ is useful in handling the mathematical complex representation of the low-pass signals. The information is a low-pass signal has frequency components from or near zero up to some finite value (baseband signal) and is defined by its maximum frequency components W Hz, Fig.1.13a. The complex envelope $\tilde{x}(t)$ is given by

$$\tilde{x}(t) = I_x(t) + j \, Q_x(t) \tag{9.5}$$

, where $I_x(t)$ and $Q_x(t)$ are the inphase and quadrature components of the complex envelope and the spectrum of the complex envelope is limited to the band W Hz adjacent to the origin, Fig.1.13a, while the spectrum of the pre-envelope is limited to the frequency band $(f_c - W) \leq f \leq (f_c + W)$ and centered at the carrier frequency f_c, Fig.1.13b. as in the double side band modulation techniques or is limited to the frequency band $f_c \leq f \leq (f_c + W)$ as in the Upper Side Band USB modulation technique, Fig.9.1a, and the frequency band $(f_c - W) \leq f \leq f_c$, as in the Lower Side Band LSB modulation technique, Fig.9.1b. Then the pre-envelope $x_+(t)$ in terms of the complex envelope, is given by

$$x_+(t) = \tilde{x}(t) \, e^{j2\pi f_c t} \tag{9.6}$$

, where the pre-envelope $x_+(t)$ of the band-pass signal x(t) equals the complex envelope $\tilde{x}(t)$ of x(t) multiplied by the exponential envelope $e^{j2\pi f_c t}$, where f_c is the carrier frequency of the band-pass signal. Eq.(9.6) shows that the pre-envelope $x_+(t)$ equals a frequency shifted version of the complex envelope $\tilde{x}(t)$ at the carrier frequency f_c. Since the band-pass signal x(t) is the real part of the pre-envelope $x_+(t)$, Eq.(9.1), then x(t) is given by

$$x(t) = \text{Re}[\, \tilde{x}(t) \, e^{j2\pi f_c t} \,] \tag{9.7}$$

, or equivalently

$$x(t) = \text{Re}\{\, [\, I_x(t) + j \, Q_x(t) \,] \, [\, \cos(2\pi f_c t) + j \sin(2\pi f_c t)] \,\}$$

, then the band-pass signal x(t) can be expressed by the following canonical form

$$x(t) = I_x(t) \cos(2\pi f_c t) - Q_x(t) \sin(2\pi f_c t) \tag{9.8}$$

, where $I_x(t)$ and $Q_x(t)$ are the inphase and quadrature components of the complex envelope $\tilde{x}(t)$. Also, Eq.(9.6) can be written in the form

$$\tilde{x}(t) = x_+(t) \, e^{-j2\pi f_c t} \tag{9.9}$$

Eq.(9.6) shows the relation between the low-pass envelope $\tilde{x}(t)$ and the band-pass envelope $x_+(t)$, that means if the low-pass envelope x(t) is known, its band-pass envelope can be obtained by multiplying $\tilde{x}(t)$ by $e^{j2\pi f_c t}$, and then taking the real part of the pre-envelope $x_+(t)$, to obtain the band-pass signal x(t), Eq.(9.7), this is known by low-pass to band-pass transformation. Also, Eq.(9.9) shows that if the band-pass envelope $x_+(t)$ is known, its low pass signal component can be obtained by putting f_c equals zero, this is known by band-pass to low-pass transformation. Then the analysis of the band-pass systems can be establishing an analogy between low-pass equivalent system and band-pass system. The pre-envelope $x_+(t)$ is a complex band-pass signal whose value depends on the carrier frequency f_c. While the complex envelope $\tilde{x}(t)$ is a complex low-pass signal whose value is independent of the choice of the carrier frequency f_c [1].

Numerical example

Consider s(t) is a single tone amplitude modulated wave, given by

$$s(t) = [1 + 0.3 \cos(2\pi10^3t)] \cos(2\pi10^7t)$$

, the term $[1 + 0.3 \cos(2\pi10^3t)]$ is a low-pass term and is considered the amplitude of the high frequency term $\cos(2\pi10^7t)$, then due to Eq.(8.14), the Hilbert transform ŝ(t) of s(t), is given by

$$ŝ(t) = [1 + 0.3 \cos(2\pi10^3t)] \sin(2\pi10^7t)$$

, using the time domain, Eq.(9.1), the pre-envelope $s_+(t)$ of s(t), is given by

$$s_+(t)) = [1 + 0.3 \cos(2\pi10^3t)] \cos(2\pi10^7t) + j [1 + 0.3 \cos(2\pi10^3t)] \sin(2\pi10^7t)$$

, or equivalently

$$s_+(t)) = [1 + 0.3 \cos(2\pi10^3t)] \, e^{j2\pi10^6t}$$

, and the complex envelope, Eq.(9.6) is given by

$$\tilde{s}(t) = [1 + 0.3 \cos(2\pi10^3t)]$$

9.3. Natural envelope

Since the complex-envelope $\tilde{x}(t)$ is represented by its inphase and quadrature components, Eq.(9.5). , then using the polar coordinates, the complex envelope $\tilde{x}(t)$ can also be expressed by

$$\tilde{x}(t) = a(t) \, e^{j\varphi(t)}$$

, where a(t) and $\varphi(t)$ are both real valued low pass functions, and the pre-envelope $x_+(t)$ is given by

$$x_+(t) = a(t) \, e^{j\varphi(t)} \, e^{j2\pi f_ct}$$

So the band-pass signal x(t), Eq.(9.7), is defined by

$$x(t) = a(t) \cos[2\pi f_ct + \varphi(t)]$$

, where a(t) is called the natural envelope of the band-pass signal x(t), and $\varphi(t)$ is the phase of the signal. The natural envelope is a real low-pass signal, and equals the magnitude of the complex envelope $\tilde{x}(t)$ and also equals the magnitude of the pre-envelope $x_+(t)$.

Problem 9.1

Find the different envelopes of the Radio Frequency RF pulse $x(t) = A \, rect(t/T) \cos(2\pi f_ct)$, where the carrier frequency f_c is much higher than $1/T$, and T is the pulse width of the RF pulse.

Solution

Using the time domain, the Hilbert transform of the RF pulse is $\hat{x}(t) = A \, rect(t/T) \sin(2\pi f_ct)$. The pre-envelope $x_+(t)$ of x(t), Eq.(9.1), is given by

$$x_+(t) = A \, rect \left(\frac{t}{T}\right) \cos(2\pi f_ct) + j \, A \, rect \left(\frac{t}{T}\right) \sin(2\pi f_ct)$$

, or equivalently

$$x_+(t) = A \, rect \left(\frac{t}{T}\right) \, e^{j2\pi f_ct} \qquad (9.10)$$

, this pre-envelope can also be obtained, using the frequency domain, where the Fourier transform X(f) of the RF pulse x(t), Eq.(3.8), is given by

$$X(f) = \tfrac{1}{2} AT \, sinc[T(f - f_c)] + \tfrac{1}{2} AT \, sinc[T(f + f_c)] \qquad (3.8)$$

, since the pre-envelope $x_+(t)$ has spectrum only in the positive frequencies and has no frequency components in the negative frequencies, then the Fourier transform $X_+(f)$, Fig.9.2, of the pre-envelope $x_+(t)$, Eq.(9.3), is given by

$$X_+(f) = AT \, sinc[T(f - f_c)] \qquad (9.11)$$

, taking the inverse Fourier transform for Eq.(9.11), the same pre-envelope $x_+(t)$, Eq.(9.10) is obtained.

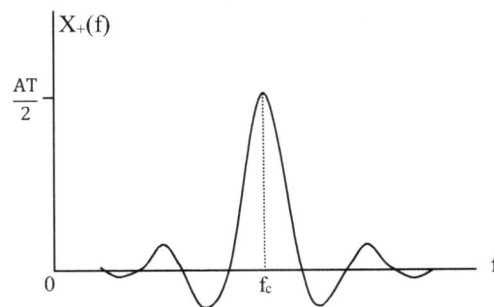

Fig.9.2. Spectrum of pre-envelope.

, then the complex envelope $\tilde{x}(t)$, Equations (9.6) and (9.10), is given by

$$\tilde{x}(t) = A \, rect \left(\frac{t}{T} \right)$$

, and the natural envelope $a(t)$ is the magnitude of the complex envelope, is given by

$$a(t) = \left| \tilde{x}(t) \right| = \left| x_+(t) \right| = A \, rect \left(\frac{t}{T} \right)$$

Problem 9.2

Find the different envelopes of the following Quadrature Amplitude Modulated QAM wave, given by

$$s(t) = m_1(t) \cos(2\pi f_c t) - m_2(t) \sin(2\pi f_c t) \qquad (9.12)$$

, where $m_1(t)$ and $m_2(t)$ are the low-pass signals with maximum frequencies W_1 and W_2 Hz respectively, and f_c is the carrier frequency is much higher than W_1 and W_2 Hz.

Solution

Since the spectrum of the low-pass signals and the carrier frequency f_c are non-overlapped, then the Hilbert transform $\hat{s}(t)$ of $s(t)$, Eq.(8.14), is given by

$$\hat{s}(t) = m_1(t) \sin(2\pi f_c t) + m_2(t) \cos(2\pi f_c t)$$

, using the time domain, Eq.(9.1), the pre-envelope of $s(t)$, $s_+(t)$ is given by

$$s_+(t) = [m_1(t) \cos(2\pi f_c t) - m_2(t) \sin(2\pi f_c t)] + j \, [m_1(t) \sin(2\pi f_c t) + m_2(t) \cos(2\pi f_c t)]$$

, or equivalently

$$s_+(t) = m_1(t) \, [\cos(2\pi f_c t) + j \sin(2\pi f_c t)] + j \, m_2(t) \, [\cos(2\pi f_c t) + j \sin(2\pi f_c t)]$$

, the pre-envelope can be expressed by

$$s_+(t) = [\, m_1(t) + j \, m_2(t) \,] \, e^{j2\pi f_c t} \qquad (9.13)$$

, this pre-envelope can also be obtained, using the frequency domain, where the Fourier transform $S(f)$ of the QAM wave $s(t)$, Eq.(9.12), $S(f)$ is given by

$$S(f) = \frac{1}{2} \, [M_1(f - f_c) + M_1(f + f_c)] - \frac{1}{2j} \, [M_2(f - f_c) - M_2(f + f_c)]$$

, and the pre-envelope $s_+(t)$ has spectrum only in the positive frequencies and has no frequency components in the negative frequencies, then the Fourier transform of the pre-envelope $s_+(t)$, Eq.(9.3), $S_+(f)$ is given by

$$S_+(f) = M_1(f - f_c) + j \, M_2(f - f_c) \qquad (9.14)$$

, taking the inverse Fourier transform for Eq.(9.14), the same pre-envelope $s_+(t)$, Eq.(9.13) is obtained.

The complex envelope $\tilde{s}(t)$, Equations (9.6) and (9.13), is given by

$$\tilde{s}(t) = m_1(t) + j \, m_2(t)$$

, the natural envelope $a(t)$ equals the magnitude of the complex envelope $\tilde{s}(t)$ equals also the magnitude of the pre-envelope $s_+(t)$, is given by

$$a(t) = \sqrt{[m_1(t)]^2 + [m_2(t)]^2}$$

9.4. Band-Pass Systems

Two different main techniques are used to represent and analyze the narrow band-pass systems, namely: the equivalent low-pass technique and the input/output pre-envelope technique.

9.4.1. Equivalent Low-Pass Technique

The band-pass systems are analyzed using the equivalent low pass technique by establishing an analogy between the low-pass equivalent filter to the band-pass filter, which are related by the pre-envelope and complex envelope, Equations (9.7) and (9.9). This analogy is based on the use of the Hilbert transform for the representation of the band-pass signals [1].

Consider x(t) is a narrow band-pass signal, its Fourier transform is X(f), is limited to the frequencies within 2W Hz around the carrier frequency f_c, and for which the ratio of the center frequency f_c to the bandwidth 2W is quite large, as in the double side band modulation techniques, Fig.1.13b, x(t) can be represented by Eq.(9.7), where $\tilde{x}(t)$ is the complex envelope of x(t), Eq.(9.5), $I_x(t)$ and $Q_x(t)$ are the inphase and quadrature components of the complex envelope, and are low-pass functions to the frequency band W Hz adjacent to the origin.

If x(t) is applied to a linear time invariant band-pass system with impulse response h(t) and transfer function H(f) limited to frequency band 2B Hz around the carrier frequency f_c. The bandwidth of the system 2B Hz is slightly larger than or at least equals the bandwidth 2W Hz of the input signal x(t).

Then the impulse response h(t) of the system can be represented by

$$h(t) = Re[\ \tilde{h}(t)\ e^{j2\pi f_c t}\ ] \tag{9.15}$$

, where $\tilde{h}(t)$ is the complex envelope of h(t), can be represented by

$$\tilde{h}(t) = I_h(t) + j\ Q_h(t) \tag{9.16}$$

, where $I_h(t)$ and $Q_h(t)$ are the inphase and quadrature components of the complex envelope $\tilde{h}(t)$, and are low-pass functions to the frequency band B Hz adjacent to the origin, and $\tilde{h}(t)$ is given by

$$\tilde{h}(t) = h_+(t)\ e^{-j2\pi f_c t} \tag{9.17}$$

, where $h_+(t)$ is the pre-envelope of h(t). The complex envelope $\tilde{h}(t)$, Eq.(9.17), can be obtained in terms of the complex envelope $\tilde{H}(f)$ of the transfer function H(f), yields

$$\tilde{h}(t) = \int_{-\infty}^{\infty} \tilde{H}(f)\ e^{j2\pi ft}\ df \tag{9.18}$$

, where

$$\tilde{H}(f) = H_+(f)\ e^{-j2\pi f_c t}$$

, and $H_+(f)$ is the Fourier transform of the pre-envelope of the impulse response $h_+(t)$ of the system.

The output signal y(t) of the system is derived from

$$y(t) = Re[\ \tilde{y}(t)\ e^{j2\pi f_c t}\ ] \tag{9.19}$$

, where $\tilde{y}(t)$ is the complex envelope of the output signal y(t), is obtained in terms of h(t) and x(t), Equations (9.16) and (9.5) from

$$\tilde{y}(t) = \tilde{h}(t) \otimes \tilde{x}(t) \tag{9.20}$$

, in terms of the inphase and quadrature components of the output y(t), the complex envelope $\tilde{y}(t)$ can also be expressed by

$$\tilde{y}(t) = I_y(t) + j\ Q_y(t)$$

, then Eq.(9.20) can be written in terms of Equations (9.5) and (9.16), yields

$$[\ I_y(t) + j\ Q_y(t)] = [\ I_h(t) + j\ Q_h(t)\ ] \otimes [\ I_x(t) + j\ Q_x(t)\ ]$$

, or equivalently

$$[\ I_y(t) + j\ Q_y(t)] = [\ I_h(t) \otimes I_x(t) - Q_h(t) \otimes Q_x(t)] + j\ [Q_h(t) \otimes I_x(t) + I_h(t) \otimes Q_x(t)]$$

, then the inphase and quadrature components of the output y(t), are given by

$$I_y(t) = [\ I_h(t) \otimes I_x(t) - Q_h(t) \otimes Q_x(t)\ ]$$
$$Q_y(t) = [\ Q_h(t) \otimes I_x(t) + I_h(t) \otimes Q_x(t)\ ]$$

, then the output waveform y(t) of the band-pass system is given by Eq.(9.19).

The equivalent low-pass technique model provides a practical basis for the simulation of the band-pass filters or communication channels on a digital computer. This simulation may be carried out using the Fast Fourier transform algorithm (chapter X), performing the convolution operations by the computer saving computation time [1].

Problem 9.3

Find the transfer function of an Ideal Band-Bass Filter IBPF has a truncated impulse response with length T, using the equivalent low-pass technique.

Solution

The transfer function and the impulse response of the ideal band-pass filter, Fig.7.14. are given by Equations (7.27) and (7.28), where 2B is the bandwidth of the band-pass filter, and f_c is the center frequency of the band-pass (carrier frequency). Then the impulse response of the truncated impulse response with length T, is given by

$$h_w(t)_{IBPF} = 4B \; sinc[2B(t - t_o)] \; cos(2\pi f_c t) \; rect[\frac{t - t_o}{T}]$$

, where t_o accounts the delay in transmission caused by the ideal band-pass filter. Using the same procedure of problems 9.1. and 9.2, then the complex envelope $\tilde{h}_w(t)_{IBPF}$ can be obtained, and is given by

$$\tilde{h}_w(t)_{IBPF} = 4B \; sinc[2B(t - t_o)] \; rect[\frac{t - t_o}{T}]$$

, and using the same procedure of problem 3.21. and problem 7.5, the Fourier transform $\tilde{H}_w(f)_{IBPF}$ of the complex envelope $\tilde{h}_w(t)_{IBPF}$ is given by

$$\tilde{H}_w(f)_{IBPF} = \frac{1}{\pi} \left\{ Si[\pi(f + B)T] - Si[\pi(f - B)T] \right\} e^{-j2\pi f t_o}$$

, where Si(u) is the sine integral function, Fig.3.43, Eq.3.45, extensively tabulated in standard tables.

Due to the Fourier transform of the real part of a function, Equations (3.66a) and (3.67a), and using conjugate property, Eq.(3.57). Then the Fourier transform $H_w(f)_{IBPF}$ is given by

$$H_w(f)_{IBPF} = \tilde{H}_w(f - f_c) + \tilde{H}_w[-(f - f_c)]$$

, or equivalently

$$H_w(f)_{IBPF} = \frac{1}{\pi} \left\{ \left[Si[\pi(f - f_c + B)T] - Si[\pi(f - f_c - B)T] \right] e^{-j2\pi(f-f_c)t_o} \right.$$
$$\left. + \left[Si[\pi(f + f_c + B)T] - Si[\pi(f + f_c - B)T] \right] e^{-j2\pi(f+f_c)t_o} \right\} \qquad (9.21)$$

Typical examples Models of the ideal band-pass filters (Surface Acoustic Wave SAW filters) are simulated by the author [11,12], using the pre-envelope and the equivalent low-pass filter technique, with carrier frequencies 100 MHz and 240 MHz, bandwidths of 46.7 MHz and 140 MHz, and the truncation transducer length T of 1.05 micro second and 350 nano second respectively. Fig.9.3. shows the magnitude of the transfer functions $H_w(f)_{IBPF}$ of these typical ideal band pass filters. These simulations are identical with the simulations of the same typical band-pass filters, using convolution principle technique, Fig.7.16, problem 7.8.

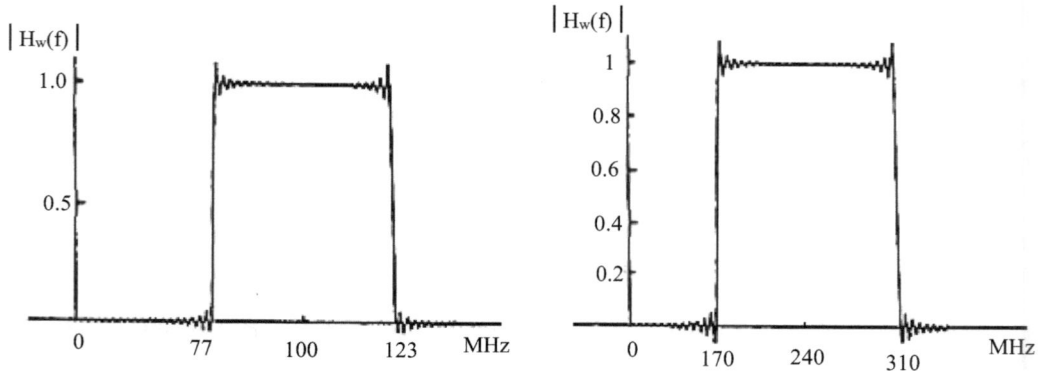

Fig.9.3. Transfer functions of typical band-pass filters, using the equivalent low-pass filter technique.

9.4.1.1. Kham-Shen Criteria for Ideal Band-Pass Systems

A new design criterion for the ideal band-pass filters have truncated impulse response using the Surface Acoustic Wave filters, is obtained by the author [11,12]. The transfer function $H_w(f)_{IBPF}$ of the truncated impulse response $h_w(t)_{IBPF}$, with length T sec, carrier frequency f_c Hz, and bandwidth 2B Hz, using the convolution principle technique, is given by Eq.(7.29), problem 7.8, and using the equivalent low-pass filter technique, is given by Eq.(9.21), problem 9.3. Comparing Equations (7.29) and (9.21), a new design criterion can be deduced and should be taken into consideration during the design procedures, that is $e^{\pm j2\pi f_c t_o} = 1$, that is $f_c t_o = N$, where N = 1, 2, 3, ... , in another meaning, the transmission time delay t_o of the filter, is a multiple of the periodic time ($1/f_c$) of the carrier wave. This criterion improves the performance of the ideal band-pass digital signal processing in communication systems.

Problem 9.4

An ideal slope circuit, Fig.9.4a, has the following transfer function H(f)

$$H(f) = j2\pi a\,[f - (f_c - B)] \qquad (f_c - B) \le f \le (f_c + B)$$
$$= j2\pi a\,[f + (f_c - B)] \qquad -(f_c + B) \le f \le -(f_c - B)$$
$$= 0 \qquad\qquad\qquad \text{otherwise}$$

, and , where the value $2\pi a$ is the slope of the circuit, and the "a" is constant. If an angle modulated wave $s(t) = A_c\,\cos[2\pi f_c t + k\,m(t)]$, is applied to the slope circuit, where A_c and f_c are the amplitude and the frequency of the carrier wave respectively, while m(t) is the low-pass signal and k is a constant. Find the output of the slope circuit using the equivalent low pass technique.

Solution

The transfer function H(f) of the slope circuit, Fig.9.4a, can be written in the form

$$H(f) = \text{Re}[\,\tilde{H}(f)\,e^{j2\pi f_c t}\,]$$

, then the complex envelope $\tilde{H}(f)$, Fig.9.4c, is given by

$$\tilde{H}(f) = j2\pi a(f + B) \quad \text{for } -B \le f \le B$$
$$= 0 \qquad\qquad\qquad \text{otherwise}$$

, also the input signal s(t), can be written in the form

$$s(t) = \text{Re}[\,A_c\,e^{jk\,m(t)}\,e^{j2\pi f_c t}\,]$$

, then the complex envelope $\tilde{s}(t)$, yields

$$\tilde{s}(t) = A_c\,e^{jk\,m(t)}$$

, using the equivalent low-pass filter technique, the complex envelope of the output signal in the frequency domain, is given by

$$\tilde{Y}(f) = \tilde{H}(f)\,\tilde{S}(f)$$

, or equivalently

$$\tilde{Y}(f) = j2\pi a\,(f + B)\,\tilde{S}(f) \qquad \text{for } -B \le f \le B \qquad (9.22)$$
$$= 0 \qquad\qquad\qquad\qquad \text{otherwise}$$

, taking the inverse Fourier transform for Eq.(9.22), the complex envelope $\tilde{y}(t)$ yields

$$\tilde{y}(t) = a\,\frac{d\,\tilde{s}(t)}{dt} + j2\pi aB\,\tilde{s}(t)$$

, or equivalently

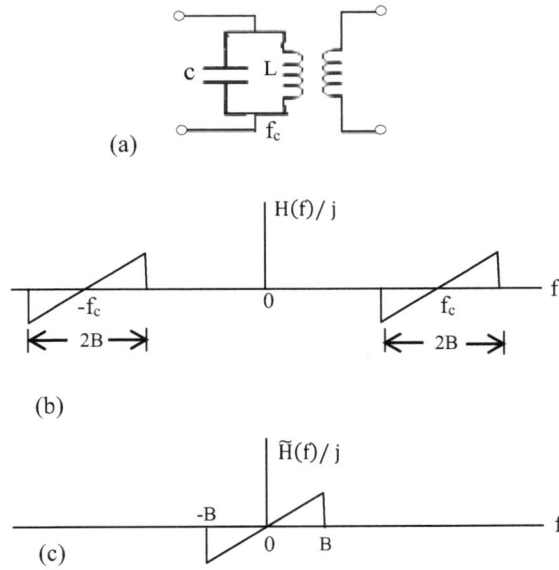

(a)

(b)

(c)

Fig.9.4. a) Ideal slope circuit, b) Transfer function, c) Equivalent low-pass transfer function.

$$\tilde{y}(t) = j \, A_c \left[ak \, \frac{dm(t)}{dt} + 2\pi aB \right] e^{jk \, m(t)}$$

, then the output waveform y(t), Eq.(9.19), is given by

$$y(t) = j \, A_c \, [ak \, m'(t) + 2\pi aB] \, \cos[2\pi f_c t + k \, m(t)]$$

, or equivalently

$$y(t) = A_c \left[ak \, m'(t) + 2\pi aB \right] \cos[\, 2\pi f_c t + k \, m(t) + \frac{\pi}{2} \,]$$

, where m'(t) is the first derivative of m(t), and m(t) is the baseband low-pass signal.

Problem 9.5

Find the response of an ideal band-pass filter, has a transfer function H(f) of narrow bandwidth 2B Hz, centered at the carrier frequency f_c Hz, and f_c is much higher than the bandwidth 2B Hz. If the input signal is given by the following RF pulse

$$x(t) = A \, \text{rect}(\frac{t}{T}) \, \cos(2\pi f_c t)$$

, where the carrier frequency f_c is much higher than 1/T, using the equivalent low-pass technique.

Solution

The transfer function of the ideal band-pass filter, Fig.7.14. is given by Eq.(7.27), where 2B is the bandwidth of the band-pass filter, and f_c is the center frequency of the band-pass (carrier frequency). Then the complex envelope transfer function $\tilde{H}(f)$ of the low-pass equivalent filter from Eq.(7.27), is given by

$$\tilde{H}(f) = 2 \, \text{rect}(\frac{f}{2B}) \, e^{-j2\pi f t_o} \tag{9.23}$$

, where t_o accounts the delay in transmission, caused by the ideal band-pass filter. Then the complex envelope impulse response $\tilde{h}(t)$ of the low-pass equivalent filter, is the inverse Fourier transform of $\tilde{H}(f)$, Eq.(9.23), $\tilde{h}(t)$ yields

$$\tilde{h}(t) = 4B \, \text{sinc}[2B(t - t_o)] \tag{9.24}$$

Also, the complex envelope $\tilde{x}(t)$ of the input RF pulse is given by

$$\tilde{x}(t) = A \, \text{rect}(\frac{t}{T})$$

, then the complex envelope $\tilde{y}(t)$ of the output signal y(t) , is obtained in terms of $\tilde{h}(t)$ and $\tilde{x}(t)$ by

$$\tilde{y}(t) = \tilde{x}(t) \otimes \tilde{h}(t)$$

, or equivalently

$$\tilde{y}(t) = A \, \text{rect}(\frac{t}{T}) \otimes 4B \, \text{sinc}[2B(t - t_o)] \tag{9.25}$$

, using the same procedure of problem 7.5, the output complex envelope $\tilde{y}(t)$ is given by

$$\tilde{y}(t) = \frac{A}{\pi} \left\{ \text{Si} \left[2\pi B(t + \frac{T}{2} - t_o) \right] - \text{Si} \left[2\pi B(t - \frac{T}{2} - t_o) \right] \right\}$$

, where $S_i(u)$ is the sine integral function, Fig.3.43, Eq.3.45, extensively tabulated in standard tables.
, then the output signal y(t) of the band-pass filter, Eq.(9.19) is given by

$$y(t) = \frac{A}{\pi} \left\{ \text{Si} \left[2\pi B(t + \frac{T}{2} - t_o) \right] - \text{Si} \left[2\pi B(t - \frac{T}{2} - t_o) \right] \right\} \cos(2\pi f_c t) \tag{9.26}$$

, or equivalently

$$y(t) = m(t) \, \cos(2\pi f_c t) \tag{9.27}$$

, where m(t) is the low-pass signal given by

$$m(t) = \frac{A}{\pi} \left\{ \text{Si} \left[2\pi B(t + \frac{T}{2} - t_o) \right] - \text{Si} \left[2\pi B(t - \frac{T}{2} - t_o) \right] \right\} \tag{9.28}$$

, Eq.(9.26) is an Amplitude Modulated Double Side Band Suppressed Carrier AM-DSBSC wave, consists of a carrier wave $\cos(2\pi f_c t)$ multiplied by a low-pass signal m(t). Fig.9.5. shows the cosine function $\cos(2\pi f_c t)$ is amplitude modulated by a low-pass signal m(t) extends from $-\infty$ to ∞ [1,4].

Another way to obtain the output y(t) of this ideal band-pass filter is given by the following convolution process

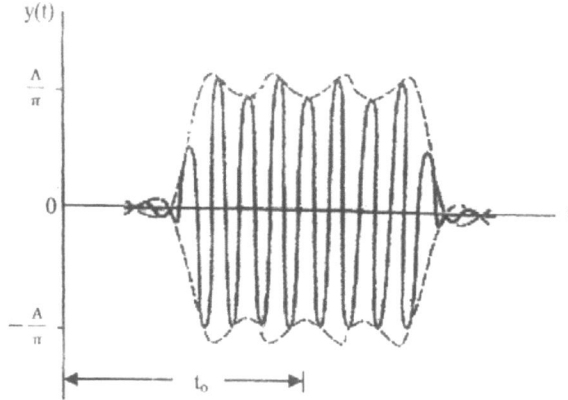

Fig.9.5. Waveform of the modulated wave, Eq.(9.26).

$$y(t) = \left[A \operatorname{rect}(\frac{t}{T}) \cos(2\pi f_c t) \right] \otimes h(t)_{IBPF}$$

, in terms of the impulse response of the ideal band-pass filter, Eq.(7.28), y(t) yields

$$y(t) = \left[A \operatorname{rect}(\frac{t}{T}) \cos(2\pi f_c t) \right] \otimes \{4B \operatorname{sinc}[2B(t - t_o)] \cos(2\pi f_c t)\}$$

, or equivalently

$$y(t) = \left[A \operatorname{rect}(\frac{t}{T}) \otimes 4B \operatorname{sinc}[2B(t - t_o)] \right] \cos(2\pi f_c t)$$

, the convolution term between brackets is identical to the convolution term in the case of ideal low-pass filter of Eq.7.21, Problem 7.5. The same response y(t) of the ideal band-pass filter, Eq.(9.26) will be obtained.

9.2.2. Input/Output Pre-Envelope Technique

Another way of representing and analyzing the band-pass systems, is the input/output pre-envelope technique. Consider x(t) is a real narrow band-pass signal, its Fourier transform is X(f) is limited to the frequencies within 2W Hz around the carrier frequency f_c, and for which the ratio of the center frequency f_c to the bandwidth 2W is quite large, and x(t) can be represented by Eq.(9.7), where $\tilde{x}(t)$ is the complex envelope of x(t), Eq.(9.5), $I_x(t)$ and $Q_x(t)$ are the inphase and quadrature low-pass functions limited to the frequency band W Hz adjacent to the origin. If x(t) is applied to a linear time invariant band-pass system with impulse response h(t) and transfer function H(f) limited to frequency band 2B Hz around the carrier frequency f_c, and the bandwidth of the system 2B Hz is slightly larger than or at least equals the bandwidth of the input signal 2W Hz. Then in terms of the input pre-envelope $x_+(t)$, the output pre-envelope $y_+(t)$ is given by

$$y_+(t) = h(t) \otimes x_+(t)$$

Proof:
The output of the band-pass filter y(t) is given by

$$y(t) = h(t) \otimes x(t)$$

, according to the Hilbert transform and convolution principle, Equations.(8.11a) and (8.11b), and using Eq.(9.1), the pre-envelope of the output signal y(t) is given by

$$y_+(t) = [h(t) \otimes x(t)] + j [h(t) \otimes \hat{x}(t)]$$

, since the convolution process obeys the distributive law of algebra, Eq.(3.43b), $y_+(t)$ yields

$$y_+(t) = h(t) \otimes [x(t) + j \hat{x}(t)]$$

, or equivalently

$$y_+(t) = h(t) \otimes x_+(t) \qquad (9.29)$$

Problem 9.6

A narrow band-pass RF pulse x(t) is applied at a band-pass filter having an impulse response h(t), are given by

$$x(t) = A\cos(2\pi f_c t) \qquad \text{for} \quad 0 \le t \le T$$

, and

$$h(t) = \cos(2\pi f_c t) \qquad \text{for} \quad 0 \le t \le T_f$$

, where T and T_f are the width of the input RF pulse x(t) and the impulse response h(t) respectively. Determine the pre-envelope of the output signal of the band-pass filter, in the following three cases:

     i. $T_f < T$ ii. $T_f = T$ iii. $T_f > T$

Solution

The input RF pulse x(t), and the impulse response h(t), Fig.9.6a,b, can be written in the form

$$x(t) = A\,\text{rect}\left[\frac{t - \tfrac{1}{2}\,T}{T}\right]\cos(2\pi f_c t)$$

, and

$$h(t) = \text{rect}\left[\frac{t - \tfrac{1}{2}\,T_f}{T_f}\right]\cos(2\pi f_c t)$$

The pre-envelope of the input RF pulse x(t) is given by

$$x_+(t) = A\,\text{rect}\left[\frac{t - \tfrac{1}{2}\,T}{T}\right]e^{j2\pi f_c t}$$

, according to the input/output pre-envelope technique, the pre-envelope of the output signal, Eq.(9.29), is given by

(a) x(t) = A cos(2πf_c t).

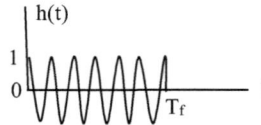

(b) h(t) = cos(2πf_c t).

Fig.9.6. a) An input energy signal x(t), b) An impulse response h(t).

$$y_+(t) = \left\{A\,\text{rect}\left[\frac{t - \tfrac{1}{2}\,T}{T}\right]e^{j2\pi f_c t}\right\} \otimes \left\{\text{rect}\left[\frac{t - \tfrac{1}{2}\,T_f}{T_f}\right]\cos(2\pi f_c t)\right\} \qquad (9.30)$$

, or equivalently

$$y_+(t) = A\int_a^b e^{j2\pi f_c \tau}\cos[2\pi f_c(t - \tau)]\,d\tau \qquad (9.31)$$

, the limits of integration a and b, are those regions of τ which satisfies the constrains $0 \le \tau \le T$, and $0 \le t - \tau \le T_f$, express the cosine function in terms of the exponential terms, Eq.(9.31) yields

$$y_+(t) = \tfrac{1}{2}A\left\{e^{j2\pi f_c t}\int_a^b d\tau + e^{-j2\pi f_c t}\int_a^b e^{j2\pi f_c 2\tau}\,d\tau\right\}$$

, for t is much more than $1/f_c$, the second integral may be neglected since it is small compared to the first integral. The pre-envelope $y_+(t)$ can be expressed by

$$y_+(t) \approx \tfrac{1}{2}A\,e^{j2\pi f_c t}\int_a^b d\tau$$

, and the real value of the output signal y(t) will be

$$y(t) \approx \tfrac{1}{2}A\cos(2\pi f_c t)\int_a^b d\tau \qquad (9.32)$$

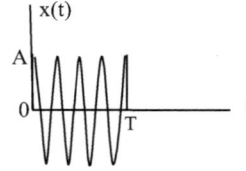

i. For the case $T_f < T$, Fig.9.7a, Eq.(9.32) yields

$$y(t) \approx \tfrac{1}{2} A \cos(2\pi f_c t) \int_0^t d\tau = \tfrac{1}{2} A\, t \cos(2\pi f_c t) \qquad\qquad \text{for } 0 < t < T_f$$

$$, \quad y(t) \approx \tfrac{1}{2} A \cos(2\pi f_c t) \int_{t-T_f}^t d\tau = \tfrac{1}{2} A\, T_f \cos(2\pi f_c t) \qquad\qquad \text{for } T_f < t < T$$

, and

$$y(t) \approx \tfrac{1}{2} A \cos(2\pi f_c t) \int_{t-T_f}^T d\tau = \tfrac{1}{2} A\, [T + T_f - t] \cos(2\pi f_c t) \qquad \text{for } T < t < (T + T_f)$$

ii. For the case $T_f = T$, Fig.9.7b, Eq.(9.32) yields

$$y(t) \approx \tfrac{1}{2} A \cos(2\pi f_c t) \int_0^t d\tau = \tfrac{1}{2} A\, t \cos(2\pi f_c t) \qquad\qquad \text{for } 0 < t < T \qquad (9.33a)$$

, and

$$y(t) \approx \tfrac{1}{2} A \cos(2\pi f_c t) \int_{t-T}^T d\tau = \tfrac{1}{2} A\, [2T - t] \cos(2\pi f_c t) \qquad\qquad \text{for } T < t < 2T \qquad (9.33b)$$

iii. For the case $T_f > T$, Fig.9.7c, Eq.(9.32) yields

$$y(t) \approx \tfrac{1}{2} A \cos(2\pi f_c t) \int_0^t d\tau = \tfrac{1}{2} A\, t \cos(2\pi f_c t) \qquad\qquad \text{for } 0 < t < T$$

$$, \quad y(t) \approx \tfrac{1}{2} A \cos(2\pi f_c t) \int_0^T d\tau = \tfrac{1}{2} A\, T \cos(2\pi f_c t) \qquad\qquad \text{for } T < t < T_f$$

, and

$$y(t) \approx \tfrac{1}{2} A \cos(2\pi f_c t) \int_{t-T_f}^T d\tau = \tfrac{1}{2} A\, [T + T_f - t] \cos(2\pi f_c t) \qquad \text{for } T_f < t < (T + T_f)$$

, and the values of the integrations are zero outside these regions, Fig.(9.7).

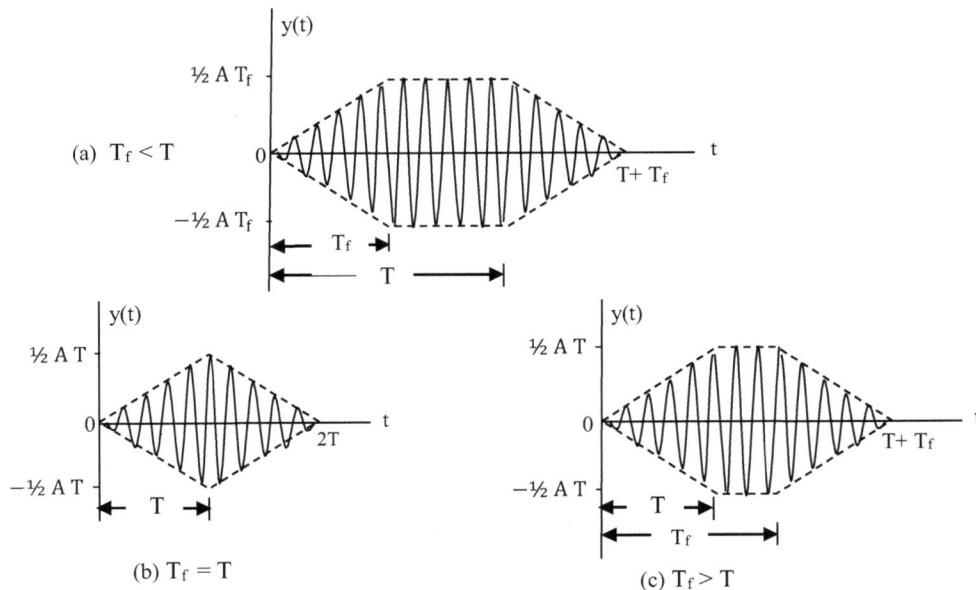

Fig.9.7. The response y(t) for: a) $T_f < T$, b) $T_f = T$, c) $T_f > T$.

, another way to verify the solution, using the equivalent low-pass filter technique, in the case of $T = T_f$, where the complex envelope $\tilde{h}(t)$ of the impulse response h(t) is given by

$$\tilde{h}(t) = \text{rect}\left[\frac{t - \frac{1}{2}T}{T}\right] \qquad \text{for } T = T_f$$

, and the complex envelope $\tilde{x}(t)$ of the input RF pulse x(t) is given by

$$\tilde{x}(t) = \text{rect}\left[\frac{t - \frac{1}{2}T}{T}\right] \qquad \text{for } T = T_f$$

, then the complex envelope $\tilde{y}(t)$ of the output signal y(t), is obtained by

$$\tilde{y}(t) = A\,\text{rect}\left[\frac{t - \frac{1}{2}T}{T}\right] \otimes \text{rect}\left[\frac{t - \frac{1}{2}T}{T}\right] \qquad \text{for } T = T_f \qquad (9.34)$$

, using the frequency domain, the Fourier transform pair is given by

$$A\,\text{rect}[\frac{t - \frac{1}{2}T}{T}] \otimes \text{rect}\,[\frac{t - \frac{1}{2}T}{T}] \rightleftharpoons AT\,\text{sinc}(fT)\,e^{-j\pi fT} \times T\,\text{sinc}(fT)\,e^{-j\pi fT}$$

, or equivalently

$$A\,\text{rect}[\frac{t - \frac{1}{2}T}{T}] \otimes \text{rect}\,[\frac{t - \frac{1}{2}T}{T}] \rightleftharpoons AT^2\,\text{sinc}^2(fT)\,e^{-j2\pi fT}$$

, taking the inverse Fourier transform, yields

$$AT\,\text{tri}[\frac{t - T}{T}] \rightleftharpoons AT^2\,\text{sinc}^2(fT)\,e^{-j2\pi fT}$$

, then the convolution function $\tilde{y}(t)$, Eq.(9.34), is given by

$$\tilde{y}(t) = A\,\text{rect}\left[\frac{t - \frac{1}{2}T}{T}\right] \otimes \text{rect}\left[\frac{t - \frac{1}{2}T}{T}\right] = AT\,\text{tri}\left[\frac{t - T}{T}\right] \quad \text{for } T = T_f$$

, then the pre-envelope $y_+(t)$, Eq.(9.1), is given by

$$y_+(t) = AT\,\text{tri}\left[\frac{t - T}{T}\right] e^{j2\pi f_c t} \qquad \text{for } T = T_f$$

, and the real value y(t), Eq.(9.19), is given by

$$y(t) = AT\,\text{tri}\left[\frac{t - T}{T}\right] \cos(2\pi f_c t) \qquad \text{for } T = T_f \qquad (9.35)$$

, which is an accurate response y(t) when $T = T_f$, and has the same shape response of the approximated y(t) obtained using the input/output pre-envelope technique, Equations (9.33a,b).

Problem 9.7
A signal x(t) has the autocorrelation function $R(\tau) = \delta(\tau)$, is applied to an ideal band-pass filter, is limited to the frequencies within 2B Hz around the carrier frequency f_c, and for which the ratio of the center frequency f_c to the bandwidth 2B is quite large. Evaluate the pre-envelope and complex envelope of the output autocorrelation function.

Solution

Since the energy spectral density $\Psi_i(f)$ and the autocorrelation function $R(\tau)$ constitute a Fourier transform pair, then the input energy spectral density, Fig.9.8a, is given by

$$\Psi_i(f) = 1 \qquad \text{for} \quad -\infty \leq f \leq \infty$$

, the output energy spectral density $\Psi_o(f)$ equals the product of the input spectral density by the magnitude square of the transfer function H(f), Eq.(7.16). Then $\Psi_o(f)$, Fig.9.8c, is given by

$$\Psi_o(f) = \text{rect}\left[\frac{f + f_c}{2B}\right] + \text{rect}\left[\frac{f - f_c}{2B}\right]$$

, and the output autocorrelation function $R_o(\tau)$ is the inverse Fourier transform of the output spectral density $\Psi_o(f)$, then $R_o(\tau)$ yields

$$R_o(\tau) = 2B \text{ sinc}(2B\tau) \left[e^{-j2\pi f_c t} + e^{j2\pi f_c t} \right]$$

, express the exponential terms by cosine function, $R_o(\tau)$, Fig.9.9a, yields

$$R_o(\tau) = 4B \text{ sinc}(2B\tau) \cos(2\pi f_c \tau)$$

, the Hilbert transform of $R_o(\tau)$ is given by

$$\widehat{R}_o(\tau) = 4B \text{ sinc}(2B\tau) \sin(2\pi f_c \tau)$$

, and the pre-envelope of the output correlation function $R_o(\tau)$, Eq.(9.1), is given by

$$[R_o]_+(\tau) = R_o(\tau) + j\,\widehat{R}_o(\tau)$$

, or equivalently

$$[R_o]_+(\tau) = 4B \text{ sinc}(2B\tau) \, e^{j2\pi f_c t}$$

, then the complex envelope of the output correlation function $R_o(\tau)$, Fig.9.9b, is given by

$$\widetilde{R}_o(\tau) = 4B \text{ sinc}(2B\tau)$$

, where B Hz is the bandwidth of the ideal band-pass filter.

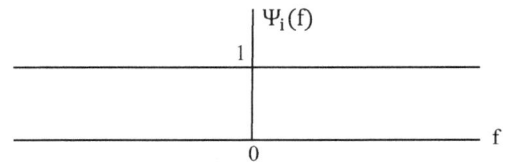

(a) Input energy spectral density.

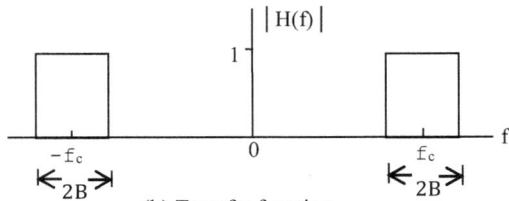

(b) Transfer function.

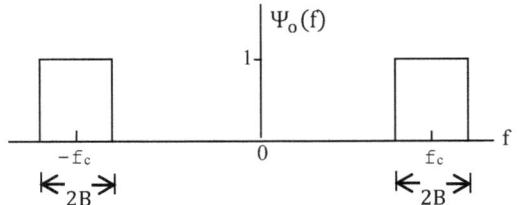

(c) Output energy spectral density.

Fig.9.8. a) Input energy spectral density,
b) Transfer function,
c) Output energy spectral density.

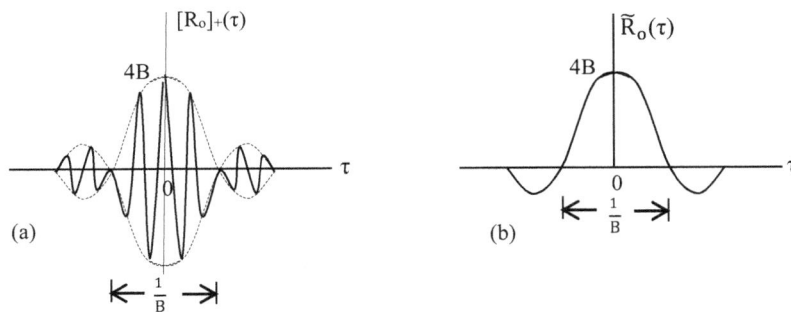

Fig.9.9. a) Pre-envelope of the output correlation function $R_o(\tau)$,
b) Complex envelope of the output correlation function $R_o(\tau)$.

9.5. Envelope Delay (Group Delay) and Dispersive System

Consider a steady (does not carry information) sinusoidal signal of frequency f_c , $x(t) = \cos(2\pi f_c t)$, is processed through a dispersive transmission channel which has a total phase shift of $\beta(f_c)$. The output signal of the channel is a phasor lags the phasor of the input signal by $\beta(f_c)$ radians, Fig.9.10. The phase delay (carrier delay) of the channel is the time taken by the signal inside the channel and is given by

$$\tau_p = -\frac{\beta(f_c)}{2\pi f_c} \qquad \text{radians} \qquad (9.36)$$

But in the case of the sinusoidal signal carry information m(t), the information causes some appropriate change to the sinusoidal wave. Consider a slowly varying baseband signal m(t) is multiplied by the sinusoidal wave, and is given by

$$s(t) = m(t)\cos(2\pi f_c t)$$

, where m(t) is a low-pass function with spectrum limited to the frequency interval $|f| \le W$ Hz, Fig.1.13a, and the carrier frequency f_c is much higher than W Hz. s(t) is a modulated wave consists of narrow group of high frequencies around the carrier frequency f_c with limited frequency bandwidth 2W Hz. During the processing of s(t) inside the channel, a true signal delay τ_g is called the group delay (envelope delay) between the envelope of the input signal and the envelope of the output signal of the channel which represents the true signal delay. Since, in the dispersive channel, the phase $\beta(f)$ is a nonlinear function of frequency f, the group delay is given by

$$\tau_g = -\frac{1}{2\pi}\frac{\partial \beta(f)}{\partial f}\bigg|_{\text{at } f = f_c} \qquad \text{radians} \qquad (9.37)$$

, where $\beta(f)$ is the output phase shift of the signal s(t). The phase shift $\beta(f)$ can be obtained in terms of the phase delay τ_p , by expanding the phase $\beta(f)$ using Taylor series at the frequency $f = f_c$, and approximate $\beta(f)$ by retaining only the first two terms, $\beta(f)$ yields

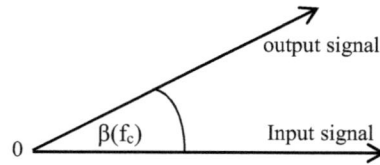

Fig.9.10. Phase delay (carrier delay).

$$\beta(f) \approx \beta(f_c) + (f - f_c)\frac{\partial \beta(f)}{\partial f}\bigg|_{\text{at } f = f_c}$$

, using Equations (9.36) and (9.37), $\beta(f)$ yields

$$\beta(f) \approx -2\pi f_c \tau_p - 2\pi(f - f_c)\tau_g$$

, since the transfer function of the channel is given by

$$H(f) = K\ e^{j\beta(f)}$$

, or equivalently

$$H(f) \approx k\ e^{\,j[-2\pi f_c \tau_p - 2\pi(f - f_c)\tau_g]}$$

, using the equivalent low-pass filter technique to obtain the output signal of the channel, the complex envelope $\widetilde{H}(f)$ is given by

$$\widetilde{H}(f) \approx 2k\ e^{-j2\pi(f_c\tau_p + f\tau_g)}$$

, and the complex envelope of the input signal is given by

$$\tilde{s}(t) = m(t)$$

, and the Fourier transform of the complex envelope $\tilde{s}(t)$ is given by

$$\tilde{S}(f) = M(f)$$

, then the Fourier transform of the complex envelope of the output signal from the channel is given by

$$\tilde{Y}(f) = \tfrac{1}{2}\, \tilde{H}(f)\, \tilde{S}(f) \approx k\, e^{-j2\pi(f_c\tau_p + f\tau_g)}\, M(f)$$

, or equivalently

$$\tilde{Y}(f) \approx k\, e^{-j2\pi f_c\tau_p}\, e^{-j2\pi f\tau_g}\, M(f) \qquad\qquad (9.38)$$

, taking the inverse Fourier transform for Eq.(9.38), the complex envelope of the output signal of the dispersive channel y(t) is given by

$$\tilde{y}(t) \approx k\, e^{-j2\pi f_c\tau_p}\, m(t - \tau_g)$$

, and the real value of the output signal from the channel, Eq.9.19, y(t) is given by

$$y(t) \approx k\, m(t - \tau_g)\, \cos[2\pi f_c(t - \tau_p)]$$

, where τ_p is the carrier delay (phase delay), τ_g is the group delay (envelope delay), and m(t) is the low-pass signal of maximum frequency W Hz is much lower than the carrier frequency f_c .

Problems

1. Find the different envelopes of the following band-pass signals
    i. $x(t) = \text{rect}(t)\, \sin(2\pi 10^7 t)$
    ii. $x(t) = \text{sinc}(t)\, \sin(2\pi 10^6 t)$
    iii. $x(t) = [1 + \sin(2\pi 10^3 t)]\, \sin(2\pi 10^6 t)$
    iv. $x(t) = \sin(100t)\, \cos(2\pi 10^7 t) - \cos(50t)\, \sin(2\pi 10^7 t)$
    v. $x(t) = [1 + \sin(200t)]\, \cos(2\pi 10^7 t) + \sin(100t)\, \sin(2\pi 10^7 t)$

2. Find the pre-envelope of the following band-pass signals:
    i. $x(t) = \text{rect}(t)\, \cos^2(2\pi 10^5 t)$
    ii. $x(t) = \sin(2\pi 10^3 t)\, \cos^3(2\pi 10^6 t)$
    vi. $x(t) = \text{rect}(t)\, \cos^3(2\pi 10^5 t)$

3. A narrow band-pass RF pulse x(t) is applied to band-pass filter having an impulse response h(t), are given by

$$x(t) = A\, \cos(2\pi f_c t) \qquad,\qquad -T/2 \le t \le T/2$$

, and

$$h(t) = \cos(2\pi f_c t) \qquad,\qquad -T_f/2 \le t \le T_f/2$$

, where T and T_f are the width of the input RF pulse x(t) and the impulse response h(t) respectively. Determine the pre-envelope of the output signal of the band-pass filter, in the following three cases:

    i. $T_f < T$ ii. $T_f = T$ iii. $T_f > T$

4. A signal x(t) has an autocorrelation function $R(\tau) = \delta(\tau - t_0)$, is applied to an ideal band-pass filter, is limited to the frequencies within 2B Hz around the carrier frequency f_c , and for which the ratio of the center frequency f_c to the bandwidth 2B is quite large, Evaluate the pre-envelope and complex envelope of the output autocorrelation function.

5. A narrow band-pass signal is given by

$$s(t) = [1 + m(t)] \cos(2\pi f_c t)$$

, is processed through a dispersive transmission channel, where m(t) is a low-pass function with spectrum limited to the frequency interval $|f| \leq W$ Hz, Fig.1.13a, and the carrier frequency f_c is much higher than W. If the transfer function of the channel is given by

$$H(f) = K \; e^{j\beta(f)}$$

Find an expression of the output signal y(t) in terms of the phase delay τ_p and the group delay τ_g.

Communication Theory and Signal Processing for Transform Coding, 2014, 193-250   **193**

CHAPTER 10

Numerical Computation of Transform Coding
(Compression Techniques)

Abstract: Numerical computation of Fourier transform is particularly well suited for use on a Digital computer, and is defined by Discrete Fourier Transform DFT and Fast Fourier Transform FFT. Also, more efficient transforms such as Discrete Cosine Transform DCT, Discrete Wavelet Transform DWT, and Contourlet Transform CT are developed.

Keywords: Ideal sampling; Discrete Fourier transform; Fast Fourier transform; Discrete Wavelet transform; Contourlet transform; Huffman encoding; Run length encoding; Lempel-Ziv-Wekh encoding; Predictive Coding; Delta Encoding; Audio Compression; Video Compression; MPEG-2; MPEG-4; JPEG-2000; Transform-domain weighted interleave Vector Quantization TwinVQ.

Discrete Fourier Transform DFT requires a discrete input function whose non-zero values have limited (finite) duration. Therefore it is often said that the DFT is a transform for the Fourier analysis of finite domain, such inputs are often created by sampling the continuous function (sampling theorem). Since the input function is a finite sequence of real or complex numbers, the DFT is ideal for processing information stored in computers. The DFT is widely employed in signal processing and related fields to analyze the frequencies contained in a sampled signal, and to solve the partial differential equations and also to perform other operations such as convolution processes.

Fast Fourier Transform FFT algorithms are commonly employed to compute the Discrete Fourier Transform DFT. FFT and DFT are often used interchangeably in colloquial settings, although there is a clear distinction that DFT refers to a mathematical transformation, regardless of how it is computed, while FFT refers to any one of several efficient algorithms for the DFT. In other words, FFT reduces the number of computations dramatically and made the digital computation of the frequency spectra a practical reality.

Discrete Cosine Transform DCT is the real part (cosine function) of the discrete Fourier transform, and is important to numerous applications in science and engineering from lossy compression of audio and images, to spectral methods for the numerical solution of partial differential equations. The use of cosine function rather than sine function (the imaginary part of the Fourier transform), is critical in these applications: for compression, it turns out that the cosine functions are much more efficient, where as for differential equations the cosines express a particular choice of boundary conditions.

Discrete Wavelet Transform DWT is a windowed Fourier transform, where the signal is analyzed locally. This is achieved by multiplying the signal with a moving window function before computing its Fourier transform. However, the Short Time Fourier Transform STFT uses a fixed size window in its original domain (time domain) for the large and the small components of the signal. But what is really needed, is a wider window to analyze the large scale components of the signal and a narrow window to detect the small scale features of the signal. This exactly what is provided by the wavelet transform. The Short Time Fourier Transform STFT analyses the signal into frequency bands through frequency modulation of the same window size while the Wavelet transform projects the signal onto scaled versions of a limited-size oscillatory function. Wavelet Transform is useful for a number of applications such as data compression, feature extraction and denoising, human vision, and Radar.

Contourlet Transform CT avoids the drawbacks of the wavelet transform for image processing (compression algorithms), by capturing the directional information such as the smooth contours and the directional edges of the image, where the contourlet transform uses the Laplacian Pyramid LP technique to capture the point discontinuities and followed by a Directional Filter Bank DFB to link these point discontinuities into linear structures.

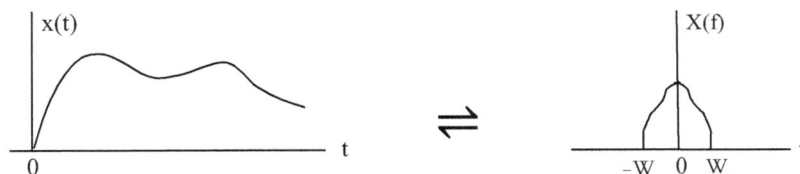

Fig.10.1. Band limited signal x(t) with maximum frequency W Hz.

Khamies El-Shennawy

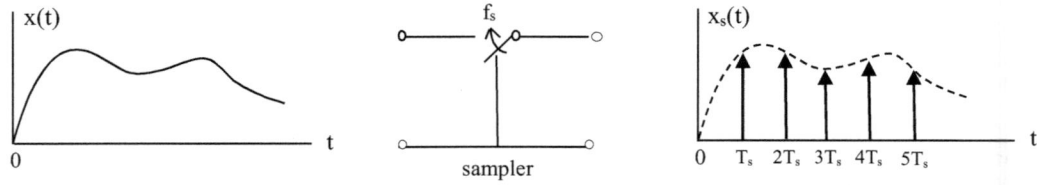

Fig.10.2. Sampling process.

10.1. Sampling Theorem (Ideal Sampling)

Sampling theorem is the very foundation of all pulse and digital modulation systems, where the pulse analog modulation of Pulse Amplitude Modulation PAM, Pulse Width Modulation PWM, Pulse Position Modulation PPM, and where the Pulse Code Modulation of Amplitude Shift Keying ASK, Frequency Shift Keying FSK, Phase Shift Keying PSK, and also Delta Modulation. The validity of all these modulation techniques are demonstrated in the digital Communication systems [2,5,7].

The AND gate is a sampler when the continuous band-limited energy signal x(t) is fed to one input of the AND gate and the other input is fed by pulses having sampling frequency f_s. The output of the AND gate is then the discrete time sampling signal $x_s(t)$ which are passed through a pulse-shaping network which gives them flat tops.

Consider an arbitrary band limited energy signal x(t) and its Fourier transform X(f) having maximum frequency W Hz, Fig.10.1, x(t) is uniformly sampled instantaneously (equally spaced) with sampling period T_s, and at uniform sampling rate $1/T_s$. As a result of the sampling process, Fig.10.2, an infinite sequence of numbers spaced T_s seconds apart and denoted by $x(nT_s)$ where the index n takes on all possible integer values, then the ideal sampled signal $x_s(t)$ is given by

$$x_s(t) = x(t) \sum_{n=-\infty}^{\infty} \delta(t - nT_s) \qquad (10.1)$$

, where $\delta(t)$ is the Dirac delta function. Due to the spectral analysis of periodic signals using the Fourier transform, Eq.(5.6), the Fourier transform of the sampled signal $x_s(t)$, $X_s(f)$ is given by

$$X_s(f) = X(f) \otimes \frac{1}{T_s} \sum_{n=-\infty}^{\infty} \delta(f - \frac{n}{T_s})$$

, according to the convolution property, Eq.(4.5a), $X_s(f)$ yields

$$X_s(f) = \frac{1}{T_s} \sum_{n=-\infty}^{\infty} X(f - \frac{n}{T_s}) \qquad (10.2)$$

, then $X_s(f)$ represents a continuous spectrum which is periodic with a period equals $1/T_s$, Fig.10.3a, In other words, the Fourier transform of a uniform sampling period T_s seconds, is a periodic spectrum in the frequency domain with a period equals the sampling rate $1/T_s$, and to recover the original signal without distortion, the sampling rate $1/T_s$ must equal to or greater than twice the highest frequency component W Hz of the signal x(t), (Shannon`s theorem)

$$\frac{1}{T_s} \geq 2W \quad Hz$$

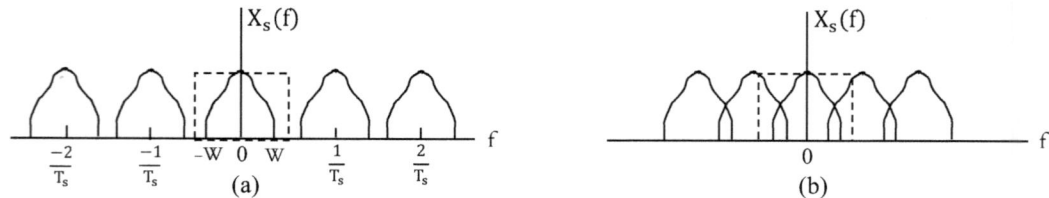

Fig.10.3. a) The sampling rate $(1/T_s) > 2W$ Hz to recover the signal without distortion,
b) The sampling rate $(1/T_s) < 2W$ Hz, where the spectrum $X_s(f)$ will be overlapped.

, when the sampling rate $(1/T_s) = 2W$ Hz, this is called the Nyquist rate, where the repetition of the signal spectrum $X(f)$ being without overlap and without any free interval between the successive $X(f)$ where the practical low pass filter used in the demodulator must be with very sharp cut-off characteristics. In practice, it is customary to use a sampling rate slightly higher than the Nyquist rate so as to assure the physical realizability of the reconstruction filter in the demodulator of the receiver which need not have a sharp cut-off characteristics. On the other hand, if the sampling rate is less than the Nyquist rate $2W$ Hz, the frequency shifted versions of $X(f)$ of the spectrum $X_s(f)$ will be overlapped, Fig.10.3b, this sampling version causes aliasing distortion and the original signal $x(t)$ cannot be recovered from its sampled version because information is thereby lost in the sampling process due to the aliasing distortion [1,<,8].

Equation(10.1), can be written in the form

$$x_s(t) = \sum_{n=-\infty}^{\infty} x(nT_s)\ \delta(t - nT_s) \tag{10.3}$$

, the Fourier transform of Eq.(10.3) is given by

$$X_s(f) = \sum_{n=-\infty}^{\infty} x(nT_s)\ e^{-j2\pi f\,nT_s} \tag{10.4}$$

, assume that the sampling rate is the Nyquist rate, where $(1/T_s) = 2W$ Hz, $X_s(f)$ yields

$$X_s(f) = \sum_{n=-\infty}^{\infty} x(\frac{n}{2W})\ e^{-j2\pi f\frac{n}{2W}} \tag{10.5}$$

, and also Eq.(10.2) yields

$$X_s(f) = 2W \sum_{n=-\infty}^{\infty} X(f - 2nW) \tag{10.6}$$

, Eq.(10.6) in the frequency band $-W \le f \le W$, Fig.10.3a, where $n = 0$, $X_s(f)$ yields

$$X_s(f) = 2W\ X(f) \qquad\qquad \text{for}\quad -W \le f \le W$$

, and hence, $X(f)$ is given by

$$X(f) = \frac{1}{2W}\ X_s(f) \qquad\qquad \text{for}\quad -W \le f \le W$$

, using Eq.(10.5), Fourier transform of the original signal is $X(f)$ yields

$$X(f) = \frac{1}{2W} \sum_{n=-\infty}^{\infty} x(\frac{n}{2W})\ e^{-j2\pi f\frac{n}{2W}} \qquad \text{for}\quad -W \le f \le W \tag{10.7}$$

, taking the inverse Fourier transform, Eq.(10.7), the original signal $x(t)$ yields

$$x(t) = \int_{-W}^{W} \left[\frac{1}{2W} \sum_{n=-\infty}^{\infty} x(\frac{n}{2W})\ e^{-j2\pi f\frac{n}{2W}} \right] e^{j2\pi f t}\ df$$

, interchanging the order of summation and integration

$$x(t) = \sum_{n=-\infty}^{\infty} x(\frac{n}{2W}) \left[\frac{1}{2W} \int_{-W}^{W} e^{j2\pi f(t - \frac{n}{2W})}\ df \right]$$

$$= \sum_{n=-\infty}^{\infty} x(\frac{n}{2W})\ \frac{\sin[\pi(2Wt - n)]}{\pi(2Wt - n)} \qquad \text{for}\quad -W \le f \le W$$

, in terms of sinc function the original signal $x(t)$, Fig.10.4, will be

$$x(t) = \sum_{n=-\infty}^{\infty} x(\frac{n}{2W})\ \text{sinc}(2Wt - n) \qquad \text{for}\quad -W \le f \le W$$

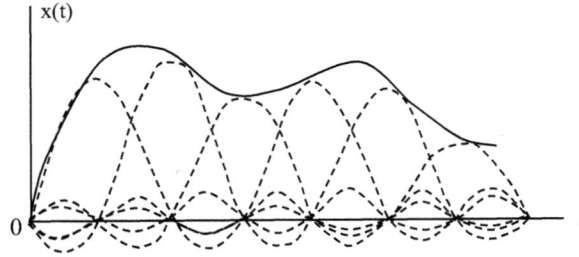

Fig.10.4. The original signal x(t) may be recovered exactly from the sequence of the samples x(n/2W).

, by inspection the spectrum X(f), Fig.10.4, the original signal x(t) may be recovered exactly from the sequence of the samples x(n/2W) by passing x(t) through an ideal low pass filter of bandwidth W Hz, which is the dotted amplitude response of this ideal reconstruction filter, Fig.10.3a.

10.2. Discrete Fourier Transform

Since the Discrete Fourier Transform DFT requires a discrete input function. Let x(t) is a continuous signal with finite duration, and the uniform sampling of x(t) is done to obtain a finite sequence of samples $x(nT_s)$, with sampling period T_s, and $1/T_s$ is the sampling rate must equal to or greater than twice the highest frequency component of the signal x(t). If N is the number of samples, then $x(nT_s)$ is the data array sequence, denoted by: $x(0)$, $x(T_s)$, $x(2T_s)$, ... , $x[(N-1)T_s]$, where the index n takes the values from 0 to $(N-1)$.

The discrete Fourier transform of the data array sequence $x(nT_s)$ may be defined by $X(kf_s)$, and can be obtained by using the trapezoidal rule, Fig.10.5, for approximating the integral which defines the Fourier transform of the given function x(t), where the dashed area equals " $T_s x(nT_s)$ ". The Fourier transform $X(kf_s)$ consists of another data array sequence of samples N separated in the frequency by f_s Hz, is given by

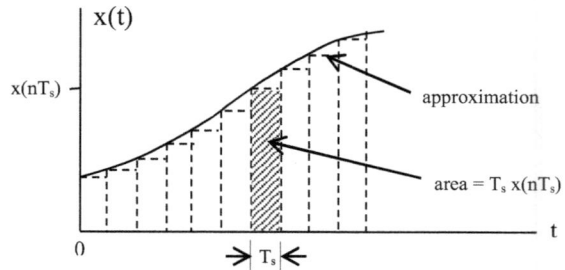

Fig.10.5. An approximated $x(nT_s)$.

$$X(kf_s) = T_s \sum_{n=0}^{N-1} x(nT_s)\, e^{-j2\pi\, nT_s kf_s} \qquad , \quad k = 0, 1, 2, 3, ... , (N-1) \qquad (10.8)$$

, since N denote the number of frequency samples contained in an interval f_s. Hence, the frequency resolution involved in the numerical computation of the Fourier transform is defined by

$$\Delta f = \frac{f_s}{N} = \frac{1}{NT_s} = \frac{1}{T_N}$$

, where T_N is the total duration of the signal x(t), then the product of the parameters T_s and f_s, are related by $T_s f_s = 1/N$, and $X(kf_s)$ yields

$$X(kf_s) = T_s \sum_{n=0}^{N-1} x(nT_s)\, e^{-j2\pi \frac{nk}{N}} \qquad , \quad k = 0, 1, 2, 3, ... , (N-1) \qquad (10.9)$$

, the inverse discrete Fourier transform $x(nT_s)$ can be obtained by multiplying both sides of Eq.(10.9) by $e^{j2\pi\, mk/N}$, and sum over the index k from 0 to $(N-1)$, yields

$$\sum_{k=0}^{N-1} X(kf_s)\, e^{j2\pi\, mk/N} = T_s \sum_{n=0}^{N-1} \sum_{k=0}^{N-1} x(nT_s)\, e^{j2\pi \frac{k(m-n)}{N}} \qquad (10.10)$$

, interchange the order of the summation, and using the fact that

$$\sum_{k=0}^{N-1} e^{j2\pi \frac{k(m-n)}{N}} = N \qquad \text{for} \quad m = n$$

$$= 0 \qquad \text{otherwise}$$

, Eq.(10.10) yields

$$x(nT_s) = \frac{1}{NT_s} \sum_{k=0}^{N-1} X(kf_s) \, e^{j2\pi \frac{nk}{N}} \qquad , n = 0, 1, 2, 3, \dots , (N-1) \qquad (10.11)$$

, $x(nT_s)$ and $X(kf_s)$ constitute a discrete Fourier transform pair and each of them is called the mate of the other, and each of them can be obtained in terms of the other, Equations (10.9) and (10.11). The symbol $X(kf_s) = DFT[x(nT_s)]$ indicates the discrete Fourier transform operation, Eq.(10.9), and the symbol $x(nT_s) = DFT^{-1}[X(kf_s)]$ indicates the inverse discrete Fourier transform operation, Eq.(10.11).

Another most convenient symbol more frequently used, indicates the discrete Fourier transform pair, is given by

$$x(nT_s) \;\rightleftharpoons\; X(kf_s)$$

An important feature of the DFT is that, the signal $x(nT_s)$ and its spectrum $X(kf_s)$ are both in discrete form, and since the DFT is a transform for Fourier analysis of finite domain, and requires a discrete input function whose non-zero values have limited (finite) duration. Then $x(nT_s)$ and $X(kf_s)$ are periodic with the period of either one, consisting of a finite number of samples N, that is

$$x(nT_s) = x(nT_s + NT_s)$$

, and also its spectrum

$$X(kf_s) = X(kf_s + Nf_s)$$

, this discrete Fourier transform pair is an exact and faithful relationship. There is no error in going from the data sequence $x(nT_s)$ to the data sequence $X(kf_s)$ and vice versa. The sequence $X(kf_s)$ is periodic of period N, as are previously explained and as one can easily verify. The process of sampling the period N whether or not, the original sequence $x(nT_s)$ is periodic. This easily verified by calculating $x(nT_s + NT_s)$, as follow from Eq.(10.11)

$$x(nT_s + NT_s) = \frac{1}{NT_s} \sum_{k=0}^{N-1} X(kf_s) \, e^{j2\pi \frac{(n+N)k}{N}}$$

$$= \frac{1}{NT_s} \sum_{k=0}^{N-1} X(kf_s) \, e^{j2\pi \frac{nk}{N}} \, e^{j2\pi k} \qquad (10.12)$$

, since the value of the exponent $e^{j2\pi k}$ equals unity, where $k = 0, 1, 2, 3, \dots , (N-1)$, Eq.(10.12) yields

$$x(nT_s + NT_s) = \frac{1}{NT_s} \sum_{k=0}^{N-1} X(kf_s) \, e^{j2\pi \frac{nk}{N}}$$

, or equivalently

$$x(nT_s + NT_s) = x(nT_s) \qquad , n = 0, 1, 2, 3, \dots , (N-1) \qquad (10.13)$$

, N is the number of the samples of the data sequence, in other words, the sampling in one domain induces periodicity in the other domain. When the samples in a time series, its spectrum becomes periodic. In another meaning, the sampling in one domain induces periodicity in the other domain. On the other hand, when sampling a spectrum of a time series, the time series is periodically extended. The discrete Fourier transform $X(kf_s)$, Eq.(10.9), and the inverse discrete Fourier transform $x(nT_s)$, Eq.(10.11), are complex quantities, they contain the real cosine terms and the imaginary sine terms, so they may be have magnitude discrete spectrum and phase discrete spectrum [1,6].

Considering the sampling period T_s equals unity, Equations (10.9) and (10.11), yield

$$X(k) = \sum_{n=0}^{N-1} x(n) \, e^{-j2\pi \frac{nk}{N}} \qquad , k = 0, 1, 2, 3, \dots , (N-1) \qquad (10.14)$$

, and

$$x(n) = \frac{1}{N} \sum_{k=0}^{N-1} X(k) \, e^{j2\pi \frac{nk}{N}} \qquad , n = 0, 1, 2, 3, \dots , (N-1) \qquad (10.15)$$

, Equations (10.14) and (10.15) illustrate that any algorithm for computing X(k) from the sequence x(n), can also be used to compute x(n) from the sequence X(k), where it is necessary only to multiply by (1/N) and change the sign of the exponent in the summation.

Problem 10.1

Evaluate the discrete Fourier transform of the following data array sequence "1 , 0 , 1 , 0", consider the sampling period T_s equals unity.

Solution

The number of data samples N equals 4, and Eq.(10.14) yields

$$X(k) = \sum_{n=0}^{3} x(n) \ e^{-j\pi nk/2} \qquad , \quad k = 0, 1, 2, \text{and } 3.$$

, then the discrete frequency domain X(0), X(1), X(2), and X(3), are given by

$$X(0) = \sum_{n=0}^{3} x(n) \qquad\qquad = 1 + 0 + 1 + 0 = 2$$

$$X(1) = \sum_{n=0}^{3} x(n) \ e^{-j\pi n/2} \ = 1 + 0 - 1 + 0 = 0$$

$$X(2) = \sum_{n=0}^{3} x(n) \ e^{-j\pi n} \quad = 1 + 0 + 1 + 0 = 2$$

, and

$$X(3) = \sum_{n=0}^{3} x(n) \ e^{-j3\pi n/2} = 1 + 0 - 1 + 0 = 0$$

, then the discrete Fourier transform of the data array sequence "1 , 0 , 1 , 0", is another data array sequence "2 , 0 , 2 , 0".

Problem 10.2

Evaluate the inverse discrete Fourier transform of the following data array sequence "2 , 0 , 2 , 0", consider the sampling period T_s equals unity.

Solution

The number of samples N = 4, and Eq.(10.15) yields

$$x(n) = \frac{1}{4} \sum_{k=0}^{3} X(k) \ e^{j\pi nk/2} \qquad , \quad n = 0, 1, 2, \text{and } 3.$$

, then the discrete time domain x(0), x(1), x(2), and x(3), are given by

$$x(0) = \frac{1}{4} \sum_{k=0}^{3} X(k) \qquad\qquad = \frac{1}{4} [2 + 0 + 2 + 0] = 1$$

$$x(1) = \frac{1}{4} \sum_{k=0}^{3} X(k) \ e^{j\pi k/2} \ = \frac{1}{4} [2 + 0 - 2 + 0] = 0$$

$$x(2) = \frac{1}{4} \sum_{k=0}^{3} X(k) \ e^{j\pi k} \ = \frac{1}{4} [2 + 0 + 2 + 0] = 1$$

, and

$$x(3) = \frac{1}{4} \sum_{k=0}^{3} X(k) \ e^{j3\pi k/2} \ = \frac{1}{4} [2 + 0 - 2 + 0] = 0$$

, then the inverse discrete Fourier transform of the data array sequence "2 , 0 , 2 , 0", is another data array sequence "1 , 0 , 1 , 0" , which is the inverse discrete Fourier transform of problem 10.1.

Problem 10.3
Evaluate the discrete Fourier transform of the following data array sequence "1 , 1 , 0 , 0", consider the sampling period T_s equals unity.

Solution
The number of data samples N equals 4, and Eq.(10.14) yields

$$X(k) = \sum_{n=0}^{3} x(n)\, e^{-j\pi nk/2} \qquad , \quad k = 0, 1, 2, \text{and } 3.$$

, then the discrete frequency domain X(0), X(1), X(2), and X(3), are given by

$$X(0) = \sum_{n=0}^{3} x(n) \qquad = 1 + 1 + 0 + 0 = 2$$

,
$$X(1) = \sum_{n=0}^{3} x(n)\, e^{-j\pi n/2} = 1 - j + 0 + 0 = 1 - j$$

,
$$X(2) = \sum_{n=0}^{3} x(n)\, e^{-j\pi n} = 1 - 1 + 0 + 0 = 0$$

, and
$$X(3) = \sum_{n=0}^{3} x(n)\, e^{-j3\pi n/2} = 1 + j + 0 + 0 = 1 + j$$

, then the discrete Fourier transform of the data array sequence "1 , 1 , 0 , 0", is another data array sequence "2 , (1 − j) , 0 , (1 + j)".

Problem 10.4
Evaluate the discrete Fourier transform of the following data array sequence "1 , 1 , 1 , 1", consider the sampling period T_s equals unity.

Solution
The number of data samples N equals 4, and Eq.(10.14) yields

$$X(k) = \sum_{n=0}^{3} x(n)\, e^{-j2\pi nk/4} \qquad , \quad k = 0, 1, 2, \text{and } 3.$$

, then the discrete frequency domain X(0), X(1), X(2), and X(3), are given by

$$X(0) = \sum_{n=0}^{3} x(n) \qquad = 1 + 1 + 1 + 1 = 4$$

,
$$X(1) = \sum_{n=0}^{3} x(n)\, e^{-j\pi n/2} = 1 - j - 1 + j = 0$$

,
$$X(2) = \sum_{n=0}^{3} x(n)\, e^{-j\pi n} = 1 - 1 + 1 - 1 = 0$$

, and
$$X(3) = \sum_{n=0}^{3} x(n)\, e^{-j3\pi n/2} = 1 + j - 1 - j = 0$$

, then the discrete Fourier transform of the data array sequence "1 , 1 , 1 , 1", is another data array sequence "4 , 0 , 0 , 0". The physical meaning of the sequence "1 , 1 , 1 , 1", may be considered a rectangular function x(t) = rect(t/4), and its Fourier transform will be X(f) = 4 sinc(4f), where the first discrete value 4 of the sequence represents the origin value of the sinc function, while the other three digits "0 , 0 , 0", represent the crossing points of the sinc function.

10.3. Discrete Fourier Transform DFT Properties
The properties of the DFT are similar to those of the continuous Fourier transform (chapter III), except the periodicity of the data array sequences, however, introduces certain basic differences [1,10].

10.3.1. DFT Linearity Property
Let the following discrete Fourier transform pairs

$$x_1(nT_s) \quad \rightleftharpoons \quad X_1(kf_s)$$

, and

$$x_2(nT_s) \quad \rightleftharpoons \quad X_2(kf_s)$$

, where N is the number of data sequence, $n = 0, 1, 2, \ldots , (N-1)$, and $k = 0, 1, 2, \ldots , (N-1)$. The linearity property states that, the DFT of the sum of two discrete data sequences, is the sum of the DFT of the two data sequences, then the linearity property DFT pair is

$$a_1 \, x_1(nT_s) + a_2 \, x_2(nT_s) \quad \rightleftharpoons \quad a_1 \, X_1(kf_s) + a_2 \, X_2(kf_s)$$

, where the "a_1 and a_2" are constants.

Problem 10.5
Based on the following discrete Fourier transform pairs of problems 10.1, and 10.4

$$1, 0, 1, 0 \quad \rightleftharpoons \quad 2, 0, 2, 0$$

, and

$$1, 1, 1, 1 \quad \rightleftharpoons \quad 4, 0, 0, 0$$

Prove that their addition can be given by the following discrete Fourier transform pair

$$2, 1, 2, 1 \quad \rightleftharpoons \quad 6, 0, 2, 0$$

Solution
The discrete Fourier transform of the data sequence "2 , 1 , 2 , 1", Eq.(10.14), is given by

$$X(k) = \sum_{n=0}^{3} x(n) \, e^{-j\pi nk/2} \qquad , \quad k = 0, 1, 2, \text{and } 3.$$

, then the discrete frequency domain X(0), X(1), X(2), and X(3), are given by

$$X(0) = \sum_{n=0}^{3} x(n) \qquad = 2 + 1 + 2 + 1 = 6$$

$$X(1) = \sum_{n=0}^{3} x(n) \, e^{-j\pi n/2} \quad = 2 - j - 2 + j = 0$$

$$X(2) = \sum_{n=0}^{3} x(n) \, e^{-j\pi n} \quad = 2 - 1 + 2 - 1 = 2$$

, and

$$X(3) = \sum_{n=0}^{3} x(n) \, e^{-j3\pi n/2} \ = 2 + j - 2 - j = 0$$

, then the discrete Fourier transform of the data array sequence "2 , 1 , 2 , 1", is another data array sequence "6 , 0 , 2 , 0".

Problem 10.6
Based on the following discrete Fourier transform pairs of problems 10.1, and 10.4

$$1,0,1,0 \rightleftharpoons 2,0,2,0$$

, and

$$1,1,1,1 \rightleftharpoons 4,0,0,0$$

Prove that their subtraction can be given by the following discrete Fourier transform pair

$$0,-1,0,-1 \rightleftharpoons -2,0,2,0$$

Solution
The discrete Fourier transform of the data sequence "0, −1, 0, −1", Eq.(10.14), is given by

$$X(k) = \sum_{n=0}^{3} x(n) \, e^{-j\pi nk/2} \qquad , \ k = 0, 1, 2, \text{and } 3.$$

, then the discrete frequency domain X(0), X(1), X(2), and X(3), are given by

$$X(0) = \sum_{n=0}^{3} x(n) \qquad = 0 - 1 + 0 - 1 = -2$$

,

$$X(1) = \sum_{n=0}^{3} x(n) \, e^{-j\pi n/2} = 0 + j + 0 - j = 0$$

,

$$X(2) = \sum_{n=0}^{3} x(n) \, e^{-j\pi n} = 0 + 1 + 0 + 1 = 2$$

, and

$$X(3) = \sum_{n=0}^{3} x(n) \, e^{-j3\pi n/2} = 0 - j + 0 + j = 0$$

, then the discrete Fourier transform of the data array sequence "0 , −1 , 0 , −1", is another data array sequence " −2 , 0 , 2 , 0".

10.3.2. DFT Shifting Property
Let the following discrete Fourier transform pair

$$x(nT_s) \rightleftharpoons X(kf_s)$$

, where N is the number of data array sequence, n = 0, 1, 2, ... , (N − 1), and k = 0, 1, 2, ... , (N − 1), the shifting property discrete Fourier transform pair is

$$x[(n - i)T_s] \rightleftharpoons X(kf_s) \, e^{-j2\pi ki/N}$$

, and

$$x[(n + i)T_s] \rightleftharpoons X(kf_s) \, e^{j2\pi ki/N}$$

, where iT_s is the time shifting of the data array sequence $x(nT_s)$. Since $x(nT_s)$ is a discrete periodic function of sequence N samples, and the shift iT_s may be greater than NT_s (multiple of NT_s and remainder), then the discrete Fourier transform operation can be written in the form

$$\text{DFT}\{x[n - (i \bmod N)T_s]\} = X(kf_s) \, e^{-j2\pi ki/N} \qquad (10.16)$$

, where the notation (i Mod N) means that, to divide i by N and retain the remainder only. This is necessary because a shift larger than N of integer number and remainder, leaves the sequence unchanged because $x(nT_s)$ is periodic of N.

Problem 10.7
Based on the following discrete Fourier transform pair (problems 10.1. and 10.2.)

$$1,0,1,0 \rightleftharpoons 2,0,2,0$$

Find the discrete Fourier transform of the data array sequence " 1 , 0 , 1 , 0 ", shifted by 27 samples.

Solution

According to the shifting property of the discrete Fourier transform, Eq.(10.16) yields

$$x[n - (27 \text{ Mod } 4)] \quad \rightleftharpoons \quad X(k) \ e^{-j2\pi \ ki/4}$$

, due to the sampling in a domain, causing periodicity in the other domain, Eq.(10.13), yields

$$x(n - 3) \quad \rightleftharpoons \quad X(k) \ e^{-j\pi \ k3/2}$$

, or equivalently

$$0, 1, 0, 1 \quad \rightleftharpoons \quad 2, 0, 2\, e^{-j3\pi}, 0$$

, then

$$0, 1, 0, 1 \quad \rightleftharpoons \quad 2, 0, -2, 0$$

10.3.3. DFT Convolution Property (Circular Convolution)

Let the following discrete Fourier transform pairs

$$x_1(nT_s) \quad \rightleftharpoons \quad X_1(kf_s)$$

, and

$$x_2(nT_s) \quad \rightleftharpoons \quad X_2(kf_s)$$

, and if N is the number of data array sequence, $n = 0, 1, 2, \dots , (N - 1)$, and $k = 0, 1, 2, \dots , (N - 1)$, the convolution property discrete Fourier transform pair is

$$x_1(nT_s) \otimes x_2(nT_s) \quad \rightleftharpoons \quad X_1(kf_s)\, X_2(kf_s)$$

, in the discrete time domain, the convolution signal $x_{12}(nT_s)$ is given by

$$x_{12}(nT_s) = x_1(nT_s) \otimes x_2(nT_s)$$

$$= T_s \sum_{i=0}^{2N-1} x_1(iT_s) \ x_2[(n - i)T_s] \quad , \ n = 0, 1, 2, \dots , (2N - 1) \qquad (10.17)$$

, the indices i and n of the convolution function $x_{12}(nT_s)$, Eq.(10.17), ly between zero and $(2N - 1)$ because since the bandwidth sequence of $x_1(nT_s)$ and $x_2(nT_s)$ is N samples, $n = 0, 1, 2, \dots, (N - 1)$, then the bandwidth sequence of $x_{12}(nT_s)$ is $(2N - 1)$, where $n = 0, 1, 2, \dots , (2N - 1)$.

But if the bandwidth sequences of $x_1(nT_s)$ and $x_2(nT_s)$ are N_1 and N_2 respectively. Then the bandwidth sequence of the convoluted signal $x_{12}(nT_s)$ is $N_c = N_1 + N_2 - 1$, where $n = 0, 1, 2, \dots , (N_1 + N_2 - 1)$.

For the evaluation of the circular convolution function $x_{12}(nT_s) = x_1(nT_s) \otimes x_2(nT_s)$, since $x_1(nT_s)$ and $x_2(nT_s)$ are periodic in their N_1 and N_2 data sequences respectively, in this case, the circular convolution function $x_{12}(nT_s)$ of the successive periods can overlap in the convolution sum, so some additional thought must be taken into consideration by adding a guard band of zeroes at the ends of their sequences so that successive periods can not overlap in the convolution sum. The length of the guard band depends on the data sequence N of $x(nT_s)$ and equals $(N - 1)$, in this case the convolution function $x_{12}(nT_s)$ can be obtained and repeated periodically, Fig.10.6, problem 10.8.

Problem 10.8

Evaluate the circular convolution signal $x_{12}(nT_s) = x_1(nT_s) \otimes x_2(nT_s)$, where $x_1(nT_s)$ and $x_2(nT_s)$ are given by the following data array sequences: $x_1(nT_s) = 1, 3, 2, 1$, and $x_2(nT_s) = 1, 2, 3, 4$, and consider the sampling period T_s equals unity.

Solution

The bandwidth sequences of $x_1(n)$ and $x_2(n)$ are N_1 equals 4 and N_2 also equals 4, where $n = 0, 1, 2,$ and 3. To avoid the overlap of the successive periods in the convolution sum, a guard band of three zeroes $(N - 1)$ at the ends of the sequences $x_1(n)$ and $x_2(n)$ are added, Fig.10.6a,b. The length of the guard band is three zeroes for both $x_1(n)$ and $x_2(n)$. The bandwidth sequence of the circular convoluted signal $x_{12}(n)$ will be $N_c = N_1 + N_2 - 1 = 7$, where $n = 0, 1, 2, 3, 4, 5, 6$. The convolution function $x_{12}(n)$, Fig.10.6c. can be obtained and repeated periodically, then using Fig.10.6a,b, Eq.(10.17) yields

$$x_{12}(n) = \sum_{i=0}^{6} x_1(i) \ x_2(n-i) \qquad , \quad n = 0, 1, 2, 3, 4, 5 \text{ and } 6$$

, then the discrete time domain $x_{12}(0)$, $x_{12}(1)$, $x_{12}(2)$, $x_{12}(3)$, $x_{12}(4)$, $x_{12}(5)$, and $x_{12}(6)$, are given by

$$x_{12}(0) = \sum_{i=0}^{6} x_1(i) \ x_2(-i) \quad = 1 \times 1 + 3 \times 0 + 2 \times 0 + 1 \times 0 + 0 \times 4 + 0 \times 3 + 0 \times 2 = 1$$

$$x_{12}(1) = \sum_{i=0}^{6} x_1(i) \ x_2(1-i) = 1 \times 2 + 3 \times 1 + 2 \times 0 + 1 \times 0 + 0 \times 0 + 0 \times 4 + 0 \times 3 = 5$$

$$x_{12}(2) = \sum_{i=0}^{6} x_1(i) \ x_2(2-i) = 1 \times 3 + 3 \times 2 + 2 \times 1 + 1 \times 0 + 0 \times 0 + 0 \times 0 + 0 \times 4 = 11$$

$$x_{12}(3) = \sum_{i=0}^{6} x_1(i) \ x_2(3-i) = 1 \times 4 + 3 \times 3 + 2 \times 2 + 1 \times 1 + 0 \times 1 + 0 \times 2 + 0 \times 0 = 18$$

$$x_{12}(4) = \sum_{i=0}^{6} x_1(i) \ x_2(4-i) = 1 \times 0 + 3 \times 4 + 2 \times 3 + 1 \times 2 + 0 \times 1 + 0 \times 0 + 0 \times 0 = 20$$

$$x_{12}(5) = \sum_{i=0}^{6} x_1(i) \ x_2(5-i) = 1 \times 0 + 3 \times 0 + 2 \times 4 + 1 \times 3 + 0 \times 2 + 0 \times 1 + 0 \times 0 = 11$$

$$x_{12}(6) = \sum_{i=0}^{6} x_1(i) \ x_2(6-i) = 1 \times 0 + 3 \times 0 + 2 \times 0 + 1 \times 4 + 0 \times 3 + 0 \times 2 + 0 \times 1 = 4$$

, then the circular discrete convolution signal $x_{12}(n)$, Fig.10.6c, is given by

$$x_{12}(n) = 1, 3, 2, 1 \otimes 1, 2, 3, 4 = 1, 5, 11, 18, 20, 11, 4$$

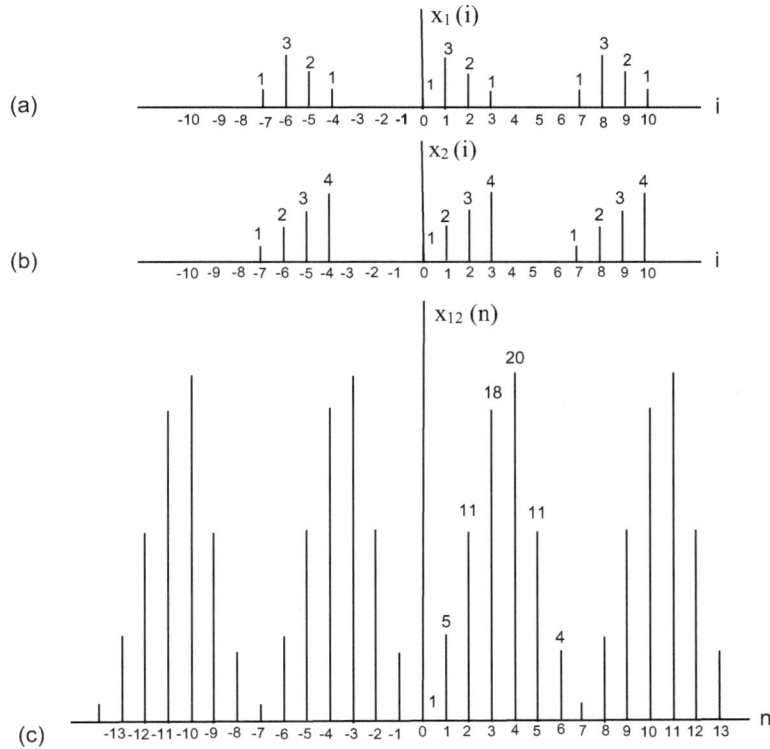

Fig.10.6. a) $x_1(i)$ with guard band, b) $x_2(i)$ with guard band, c) The circular convolution $x_{12}(n)$.

10.4. Fast Fourier Transform

Fast Fourier Transform FFT algorithms are commonly employed to compute the Discrete Fourier Transform DFT. FFT refers to any one of several efficient algorithms for the DFT. FFT reduces the number of computations dramatically and made the digital computation of the frequency spectra a practical reality. Fast Fourier transform algorithm was independently discovered by Cooley and Tukey in 1965. FFT chooses the length of the data sequence N to be a power of 2, where it is possible to reduce the number of multiplications in the arithmetic operations from N^2 in the brute computation of the discrete Fourier transform to approximately $Nlog_2N$ in the FFT [19].

An example, consider a periodic data sequence of length N equals 4096 samples (2^{12}), the Discrete Fourier Transform DFT will require N^2 arithmetic multiplications of 16777216 directly, while the Fast Fourier Transform algorithm will require $Nlog_2N$ arithmetic multiplications of 49152, which reduces the required DFT arithmetic multiplications by a factor of approximately 341.

Since the FFT algorithms are efficient because they use a greatly reduced number of arithmetic operations as compared to the brute force computation of the DFT. Basically, an FFT algorithm attains its computational efficiency by following a "divide and conquer" strategy, where the original DFT computation is decomposed successively into smaller DFT computations. The following algorithm is a one version of a popular FFT algorithms. Consider a data sequence array x(n) where the sampling period T_s equals unity, its DFT, Eq.(10.14), can be written in the form

$$X(k) = \sum_{n=0}^{N-1} x(n) \, e^{-j\frac{2\pi}{N} nk} \qquad , \; k = 0, 1, 2, 3, \dots , (N-1)$$

, or equivalently

$$X(k) = \sum_{n=0}^{N-1} x(n) \, W^{nk} \tag{10.18}$$

, where $\quad W = e^{-j2\pi/N}$

Let $N = 2^L$, then $L = Log_2N$, where N is even integer and N/2 is integer. Eq.(10.18) can be written in the form

$$X(k) = \sum_{n=0}^{(N/2)-1} x(n) \, W^{nk} + \sum_{n=N/2}^{N-1} x(n) \, W^{nk}$$

, since x(n) is periodic of period N/2 in both terms, X(k) can be written in the form

$$X(k) = \sum_{n=0}^{(N/2)-1} x(n) \, W^{nk} + \sum_{n=0}^{(N/2)-1} x(n + \frac{N}{2}) \, W^{(n + \frac{N}{2}) k}$$

, or equivalently

$$X(k) = \sum_{n=0}^{(N/2)-1} \left[x(n) + x(n + \frac{N}{2}) W^{Nk/2} \right] W^{nk}, \qquad k = 0, 1, 2, 3, \dots , (N-1) \tag{10.19}$$

, since $\quad W^N = e^{-j2\pi} = +1 \quad$, and $\quad W^{N/2} = e^{-j\pi} = -1$

, then $\quad W^{Nk/2} = (-1)^k = +1 \quad$ if k is even

$\qquad\qquad\qquad = -1 \quad$ if k is odd

In the case of k is even, let $k = 2\ell$, and $W^{kN/2} = + 1$. While in the case of k is odd, let $k = 2\ell + 1$, and $W^{kN/2} = - 1$, where $\ell = 0, 1, 2, 3, \dots , [(N/2) - 1]$. Eq.(6.19) yields

$$X(2\ell) = \sum_{n=0}^{(N/2)-1} \left[x(n) + x(n + \frac{N}{2}) \right] (W^2)^{\ell n} \qquad \text{for k is even} \tag{10.20a}$$

, and

$$X(2\ell + 1) = \sum_{n=0}^{(N/2)-1} \left\{ \left[x(n) - x(n + \frac{N}{2}) \right] W^n \right\} (W^2)^{\ell n} \qquad \text{for k is odd} \tag{10.20b}$$

, where $\ell = 0, 1, 2, 3, \dots , [(N/2) - 1]$, $k = 0, 1, 2, 3, \dots , (N-1)$, and $W^2 = e^{-j4\pi/N} = e^{-j2\pi/\frac{1}{2} N}$.

Then the right hand side, Equations(10.20a) and (10.20b), is the sum of N/2 point DFT of the data sequence $[x(n) + x(n + N/2)]$ and $\{[x(n) - x(n + N/2)] W^n\}$ respectively. The parameter W^n associated with $[x(n) - x(n + N/2)]$ is called a twiddle factor. Equations (10.20a) and (10.20b), show that the even and odd valued samples of the transform sequence X(k) can be obtained from the N/2 point DFT of the sequence $[x(n) + x(n + N/2)]$ and $\{[x(n) - x(n + N/2)] W^n\}$ respectively [19].

The problem of computing an N point DFT is reduced to that of computing two N/2 point DFT`s. This procedure is repeated a second time, whereby, an N/2 point is decomposed into two N/4 point DFT`s. The decomposition procedure is continued in this fashion until reaching to the trivial case of N single point DFT`s (after $L = Log_2N$ stages).

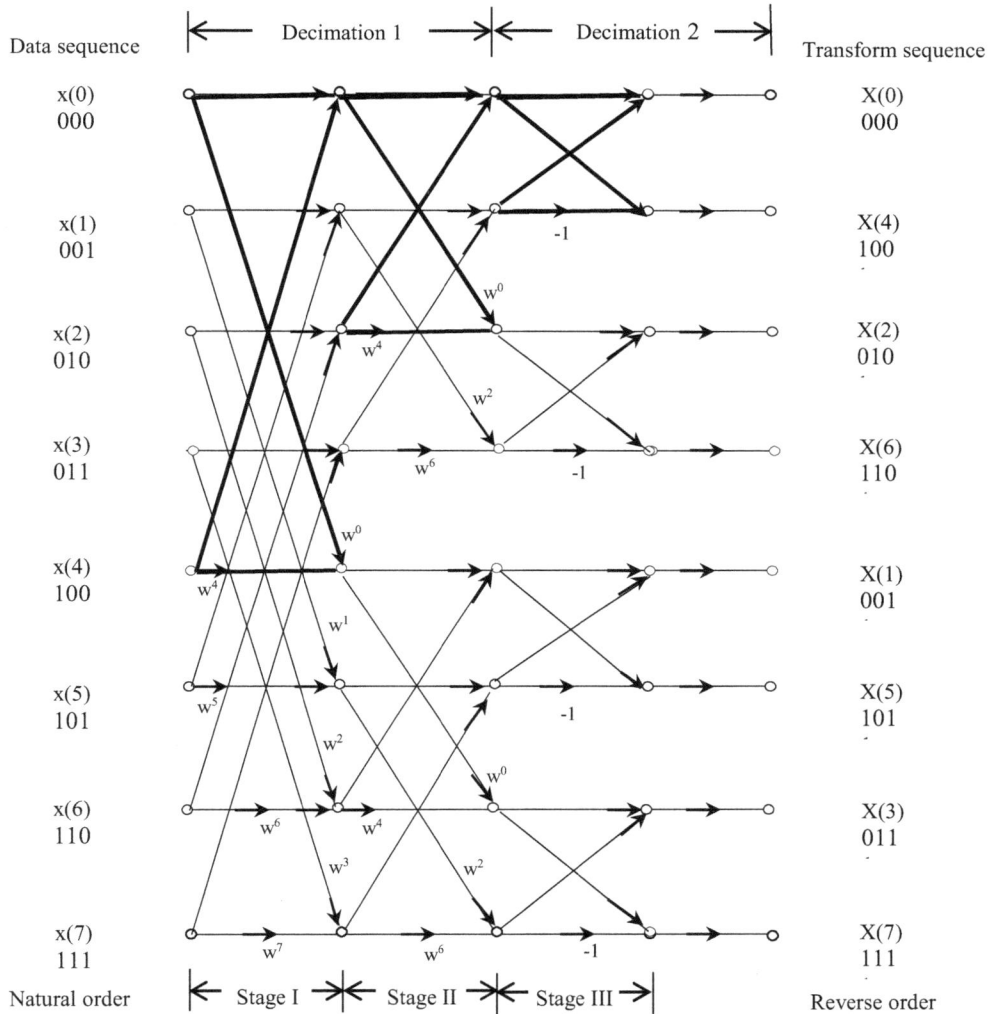

Fig.10.7. Decimation-in-frequency Fast Fourier transform algorithm.

A complete signal flow graph for the computation of a discrete data array sequence $N = 8 = 2^3$, where $L = 3$ stages of decomposition, Fig.10.7. In the first stage, the computation of the 8 point DFT is reduced to that of two 4 point DFT`s. In the second stage, the computation of the 4 point DFT is reduced to that of two 2 point DFT`s. Finally, the third stage is the computation of the 2 point DFT. A repetitive structure called a butterfly, can be discerned in the FFT algorithm, Fig. 10.7. The butterfly has two inputs and two outputs. Butterflies for the three stages of the algorithm are shown by bold-faced lines. In this example, the FFT algorithm requires $NLog_2N = 8Log_28 = 24$ arithmetic multiplications, while the brute force computation of the DFT requires $N^2 = 8^2 = 64$.

The divide-and-conquer approach where the original DFT computation is decomposed successively into smaller DFT computations, has led to the development of the FFT algorithms. The FFT algorithm, Fig.10.7, is referred to as a decimation-in-frequency algorithm because the frequency transform X(k) is divided successively into smaller sub-frequencies, where the input data sequence x(n) is in natural order of n, while the output data sequence X(k) is in bit-reversed order of k. The reverse is true for decimation-in-time because the time data sequence x(n) is divided successively into smaller subsequences, except that the butterfly for decimation-in-time is slightly different from that for decimation-in-frequency [19].

For the general case of $N = 2^L$, the algorithm requires $L = Log_2N$ stages of computation. Each stage requires N/2 butterflies. Each butterfly involves one complex multiplication and two complex additions (one addition and one subtraction). Then the FFT structure requires $(N/2)Log_2N$ complex multiplications and $NLog_2N$ complex additions. Since all twiddle factors $W^0 = 1$ and $W^{N/2} = -1$, $W^{N/4} = j$, $W^{3N/4} = -j$, where it is significantly smaller than that of N^2 complex multiplications and $N(N - 1)$ complex additions required for the direct computation of the DFT. The computational savings made possible by the FFT algorithm become more substantial as the data array length N increases.

The FFT algorithm can be used to handle the computation of both the IDFT and the DFT. This technique is illustrated in Fig.10.8, where the IDFT x(n) of the discrete data sequence X(k) is given by

$$x(n) = \frac{1}{N} \sum_{k=0}^{N-1} X(k) W^{-nk} \quad , \quad n = 0, 1, 2, 3, \dots , (N - 1) \quad (10.21)$$

, where $W = e^{-j2\pi/N}$, then taking the complex conjugate and multiply by N, Eq.(10.21) yields

$$N x^*(n) = \sum_{k=0}^{N-1} X^*(k) W^{nk} \quad , \quad n = 0, 1, 2, 3, \dots , (N - 1) \quad (10.22)$$

, the right hand side of Eq.(10.22), is recognized as the N point DFT of the complex conjugated sequence $X^*(k)$.

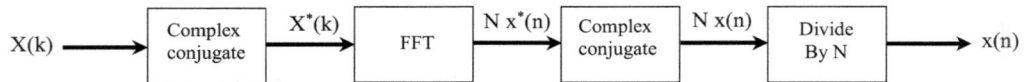

Fig.10.8. Use of Fast Fourier transform for computing the inverse Fourier transform.

10.5. Sine and Cosine Transforms
The energy signal x(t) and its continuous Fourier transform X(f), Equations (3.2) and (3.1), in terms of Euler`s formula, X(f) and x(t) can be expressed by

$$X(f) = \int_{-\infty}^{\infty} x(t) [\cos(2\pi ft) - j \sin(2\pi ft)] dt$$

,

$$x(t) = \int_{-\infty}^{\infty} X(f) [\cos(2\pi ft) + j \sin(2\pi ft)] df$$

, or equivalently

$$X(f) = \int_{-\infty}^{\infty} x(t) \cos(2\pi ft) dt - j \int_{-\infty}^{\infty} x(t) \sin(2\pi ft) dt \quad (10.23a)$$

,

$$x(t) = \int_{-\infty}^{\infty} X(f) \cos(2\pi ft) df + j \int_{-\infty}^{\infty} X(f) \sin(2\pi ft) df \quad (10.23b)$$

, if x(t) is an even function, the product "x(t) cos(2πft)" is also even value, whilst the product "x(t) sin(2πft)" is odd value, and the integration along the odd function from −∞ to ∞ equals zero , Equations (10.23a) and (10.23b) yield

$$X(f) = \int_{-\infty}^{\infty} x(t)\ \cos(2\pi ft)\ dt \qquad \text{for x(t) even function} \qquad (10.24a)$$

$$x(t) = \int_{-\infty}^{\infty} X(f)\ \cos(2\pi ft)\ df \qquad \text{for x(t) even function} \qquad (10.24b)$$

, Equations (10.24a) and (10.24b) express the Fourier cosine transform and the inverse Fourier cosine transform for an even function x(t).

Also, if x(t) is an odd function, the product "x(t) cos(2πft)" is also odd value, whilst the product "x(t) sin(2πf t)" is even value, and the integration along the odd function from −∞ to ∞ equal zero, Equations (10.23a) and (10.23b) yield

$$X(f) = -j \int_{-\infty}^{\infty} x(t)\ \sin(2\pi ft)\ dt \qquad \text{for x(t) odd function} \qquad (10.25a)$$

$$x(t) = j \int_{-\infty}^{\infty} X(f)\ \sin(2\pi ft)\ df \qquad \text{for x(t) odd function} \qquad (10.25b)$$

, Equations (10.25a) and (10.25b) express the Fourier sine transform and the inverse Fourier sine transform for an odd function x(t).

Note that the numerical factors in the transforms are defined uniquely only by their product, as discussed for general continuous Fourier transforms. For this reason, the imaginary units j and −j, of Equations (10.25a) and (10.25b), can be omitted.

10.6. Discrete Cosine Transform

Discrete cosine transform DCT is important to numerous applications in science and engineering, from lossy compression of audio and images (where small high-frequency components can be discarded), to spectral methods for numerical solution of partial differential equations. The use of cosine functions rather than sine functions is critical in these applications: for compression, it turns out that cosine functions are much more efficient, whereas for differential equations the cosines express a particular choice of boundary conditions [8,10].

Digital speech and image compression is a field that studies methods for reducing the total number of bits required to represent this speech or image. Data compression is defined as the process of encoding the data using a representation that reduces the overall size of the data. This reduction is possible when the original dataset contains some type of redundancy. This can be achieved by eliminating various types of redundancy that exist in the pixel values. With the advance development in internet and multimedia technologies, the amount of information that is handled by computers has grown exponentially over the past decades. This information requires large amount of storage space and transmission bandwidth that the current technology is unable to be handled technically and economically. One of the possible solutions to this problem is to compress the information so that the storage space and transmission time can be reduced.

The discrete cosine transform converts a time domain function into a frequency domain function. The functions are sampled at equally spaced discrete points. In the one dimensional Discrete Cosine Transform DCT, if N is the number of samples, and assuming the sampling period T_s is unity, then the sampling rate $1/T_s$ is unity. The finite discrete data array sequence x(n) are denoted by: x(0), x(1), x(2), x(3), ... , x(N − 1), where the index n takes the values from 0 to (N − 1). Since the discrete cosine transform is the real part of the discrete Fourier transform, then the one dimensional discrete cosine transform X(k) is given by

$$X(k) = \sum_{n=0}^{N-1} x(n) \cos(2\pi \frac{nk}{N}) \qquad , \qquad k = 0, 1, 2, 3, \dots, (N-1) \qquad (10.26)$$

, and the inverse DCT is given by

$$x(n) = \frac{1}{N} \sum_{k=0}^{N-1} X(k) \cos(2\pi \frac{nk}{N}) \qquad , \qquad n = 0, 1, 2, 3, \dots, (N-1) \qquad (10.27)$$

, due to Ahmed and Rao [20], and Chen et al [21], similar to the idea of the fast Fourier transform, by adding an angle $\pi k/N$ and the whole argument is divided by 2, a Fast Discrete Cosine Transform FDCT can be obtained, and if the Discrete Cosine Transform DCT, Eq.(10.26), is typically multiplied by $\alpha(k)$, a modified version of Equations (10.26) and (10.27) yield

$$X(k) = \alpha(k) \sum_{n=0}^{N-1} x(n) \cos\left(\pi \frac{(2n+1)k}{2N}\right) \qquad , \qquad k = 0, 1, 2, 3, \dots, (N-1) \qquad (10.28)$$

, and the inverse DCT is given by

$$x(n) = \frac{1}{N} \sum_{k=0}^{N-1} X(k) \cos\left(\pi \frac{(2n+1)k}{2N}\right) \qquad , \qquad n = 0, 1, 2, 3, \dots, (N-1) \qquad (10.29)$$

, where $\alpha(k)$ equals $\sqrt{1/N}$ for $k = 0$ and equals $\sqrt{2/N}$ for $k \neq 0$. The one dimensional discrete cosine transform is useful in processing one dimensional signals such as speech waveforms. Discrete cosine transform is a Fourier-related transform similar to discrete Fourier transform, but using only real numbers. The algorithms for discrete Fourier transform and discrete cosine transform are similar transforms and are all so closely related.

Also, the two dimensional discrete cosine transform is useful in processing two-dimensional signals such as image processing, where a N×N matrix. The two dimensional discrete cosine transform is computed in a simple way, where the one dimensional discrete cosine transform is applied to each row of the matrix and then to each column of the result. The discrete cosine transform $X(k_1,k_2)$ of the matrix, is given by

$$X(k_1, k_2) = \sum_{n_2=0}^{N-1} \sum_{n_1=0}^{N-1} x(n_1, n_2) \cos(2\pi \frac{n_1 k_1}{N}) \cos(2\pi \frac{n_2 k_2}{N}) \qquad (10.30)$$

, where $k_1 = 0$ to $(N-1)$, $k_2 = 0$ to $(N-1)$, and in this case the inverse DCT is given by

$$x(n_1, n_2) = \frac{1}{N^2} \sum_{k_2=0}^{N-1} \sum_{k_1=0}^{N-1} X(k_1, k_2) \cos(2\pi \frac{n_1 k_1}{N}) \cos(2\pi \frac{n_2 k_2}{N}) \qquad (10.31)$$

, where $n_1 = 0$ to $(N-1)$, $n_2 = 0$ to $(N-1)$. Due to Ahmed and Rao [20], and Chen et al [21], a Fast Discrete Cosine Transform FDCT can be obtained, and if the Discrete Cosine Transform DCT, Eq.(10.30), is typically multiplied by $\alpha(k_1)$ and $\alpha(k_2)$, a modified version of Equations (10.30) and (10.31) yield

$$X(k_1, k_2) = \alpha(k_1) \alpha(k_2) \sum_{n_2=0}^{N-1} \sum_{n_1=0}^{N-1} x(n_1, n_2) \cos\left(\pi \frac{(2n_1+1)k_1}{2N}\right) \cos\left(\pi \frac{(2n_2+1)k_2}{2N}\right) \qquad (10.32)$$

, where $k_1 = 0$ to $(N-1)$, $k_2 = 0$ to $(N-1)$, and in this case the inverse DCT is given by

$$x(n_1, n_2) = \frac{1}{N^2} \sum_{k_2=0}^{N-1} \sum_{k_1=0}^{N-1} X(k_1, k_2) \cos\left(\pi \frac{(2n_1+1)k_1}{2N}\right) \cos\left(\pi \frac{(2n_2+1)k_2}{2N}\right) \qquad (10.33)$$

, where $n_1 = 0$ to $(N-1)$, $n_2 = 0$ to $(N-1)$, and $\alpha(k_i)$ equals $\sqrt{1/N}$ for $k_i = 0$ and equals $\sqrt{2/N}$ for $k_i \neq 0$, and $i = 1$ or 2.

The two dimensional DCT transforms an image from the spatial domain to the frequency domain with an input image $x(n_1, n_2)$, and the coefficients $X(k_1,k_2)$ are the output discrete cosine transform. If the input image is N_1 pixels wide by N_2 pixels high, then $x(n_1,n_2)$ is the intensity of the pixel in row n_1 and column n_2, while $X(k_1,k_2)$ is the DCT coefficient in row k_1 and column k_2 of the discrete cosine transform matrix. All the discrete cosine transform multiplication are real, this lowers the number of the required multiplications, as compared to the discrete Fourier transform [10].

The rapid growth of the digital imaging applications, including laptop and desktop publishing, multimedia, teleconferencing, and high definition television has increased the need for effective and standardized image compression techniques. Among the emerging standards, the following three standards employ the basic technique of the discrete cosine transform:

i. The Joint Photographic Experts Group JPEG for compression of still images, where individual pictures are compressed without reference to any other pictures. In JPEG, the intra-coding video compression treats each picture independently and absolute picture data are transmitted.

ii. The Moving Picture Experts Group MPEG for compression of motion video, where in the inter-coding video compression, an individual picture may exist only in terms of the difference from a previous picture. This difference between the previous picture and the current picture is transmitted, in a form of differential coding.

iii. The Consultative Committee of International Telegraph and Telephone CCITT where a high resolution still image transmission based on recommendation H.261 (video codec for audio visual services at p×64 Kbits/sec, and p $\leq \pm 15$) for compression of video telephony and teleconferencing.

An example, it is required to get the DCT of an image has spatial domain 8×8 array of integer, the array contains each pixel's gray scale level: 8 bit pixels gave levels from 0 to 255. Then the output array of the DCT coefficients contains integers: these can range from −1024 to 1023. For most images, much of the signal energy lies at low frequencies, these appear in the upper left corner of the DCT. The lower right values represent higher frequencies and are often small-small enough to be neglected with little visible distortion.

Mathematically the DCT is practically reversible and no image definition is lost until coefficient quantization is started. Two types of DCT can be performed: the full image DCT domain and the block-wise DCT domain. Authors distinguish strictly between the two types, they are in fact conceptually very close [8].

The drawback of the Discrete Cosine Transform DCT is the severe blocking effects after the image reconstruction at high compression ratios, where in the application of the personal Identification ID fingerprint compression, the DCT did not perform well because it produced these blocking effects which made it impossible to follow the ridge lines in the fingerprints after reconstruction at the high compression ratios [22]. This drawback disappear in the Discrete wavelet transform due to its property of retaining the details present in the data, where the wavelet transforms are used to compress the fingerprint pictures for storage in their data bank and hence the prominent applications of the personal Identification ID fingerprint compression standard [23].

Problem 10.9
Evaluate the Discrete Cosine Transform DCT of the one dimensional data array sequence "1, 0, 1, 0", consider the sampling period T_s equals unity.

Solution
The number of data samples N equals 4, and Eq.(10.26) yields

$$X(k) = \sum_{n=0}^{3} x(n) \cos(\frac{\pi n k}{2}) \qquad , \ k = 0, 1, 2, \text{and } 3.$$

, then the discrete frequency domain X(0), X(1), X(2), and X(3), are given by

$$X(0) = \sum_{n=0}^{3} x(n) \cos(0) \quad = 1 + 0 + 1 + 0 = 2$$

$$, \quad X(1) = \sum_{n=0}^{3} x(n) \cos(\frac{\pi n}{2}) \ = 1 + 0 - 1 + 0 = 0$$

$$, \quad X(2) = \sum_{n=0}^{3} x(n) \cos(\pi n) \ = 1 + 0 + 1 + 0 = 2$$

, and

$$X(3) = \sum_{n=0}^{3} x(n) \cos(\frac{3\pi n}{2}) = 1 + 0 - 1 + 0 = 0$$

Then the DCT of the data array sequence "1, 0, 1, 0", is another data array sequence " 2, 0, 2, 0", and identical to the discrete Fourier transform for the same sequence, problem 10.1.

Problem 10.10

Evaluate the inverse discrete cosine transform of the one dimensional data array sequence "2, 0, 2, 0",
, consider the sampling period T_s equals unity.

Solution

The number of samples N = 4, and Eq.(10.27) yields

$$x(n) = \frac{1}{4} \sum_{n=0}^{3} X(k) \cos(\frac{\pi n k}{2}) \qquad , \quad n = 0, 1, 2, \text{ and } 3.$$

, then the discrete time domain x(0), x(1), x(2), and x(3), are given by

$$x(0) = \frac{1}{4} \sum_{k=0}^{3} X(k) \cos(0) \quad = \frac{1}{4} [2 + 0 + 2 + 0] = 1$$

$$x(1) = \frac{1}{4} \sum_{k=0}^{3} X(k) \cos(\frac{\pi n}{2}) = \frac{1}{4} [2 + 0 - 2 + 0] = 0$$

$$x(2) = \frac{1}{4} \sum_{k=0}^{3} X(k) \cos(\pi n) \quad = \frac{1}{4} [2 + 0 + 2 + 0] = 1$$

, and

$$x(3) = \frac{1}{4} \sum_{k=0}^{3} X(k) \cos(\frac{3\pi n}{2}) = \frac{1}{4} [2 + 0 - 2 + 0] = 0$$

Then the inverse discrete cosine transform of the data array sequence 2, 0, 2, 0, is another data array sequence "1, 0, 1, 0", which is the inverse discrete cosine transform of problem 10.9, and identical to the discrete Fourier transform for the same sequence, problem 10.2.

Problem 10.11

Evaluate the Discrete Cosine Transform DCT of the two dimensional data array sequence 2×2, given by: $\begin{vmatrix} 0 & 1 \\ 1 & 0 \end{vmatrix}$, consider the sampling period T_s equals unity.

Solution

The number of row data samples N_1 equals 2 and $n_1 = 0$ and 1, also the number of column data samples N_2 equals 2 and $n_2 = 0$ and 1, Eq.(10.30) yields

$$X(k_1, k_2) = \sum_{n_2=0}^{1} \sum_{n_1=0}^{1} x(n_1, n_2) \cos(\pi n_1 k_1) \cos(\pi n_2 k_2)$$

, where $k_1 = 0$ and 1, $k_2 = 0$ and 1. Then the discrete frequency domain X(0, 0), X(0, 1), X(1, 0), and X(1, 1), are given by

$$X(0, 0) = \sum_{n_2=0}^{1} \sum_{n_1=0}^{1} x(n_1, n_2) \cos(0) \cos(0) \quad = 0 + 1 + 1 + 0 = 2$$

$$X(0, 1) = \sum_{n_2=0}^{1} \sum_{n_1=0}^{1} x(n_1, n_2) \cos(0) \cos(\pi n_2) \quad = 0 - 1 + 1 + 0 = 0$$

$$X(1, 0) = \sum_{n_2=0}^{1} \sum_{n_1=0}^{1} x(n_1, n_2) \cos(\pi n_1) \cos(0) \quad = 0 + 1 - 1 + 0 = 0$$

, and

$$X(1, 1) = \sum_{n_2=0}^{1} \sum_{n_1=0}^{1} x(n_1, n_2) \cos(\pi n_1) \cos(\pi n_2) = 0 - 1 - 1 + 0 = -2$$

, then the discrete cosine transform of the two dimensional data array sequence $\begin{vmatrix} 0 & 1 \\ 1 & 0 \end{vmatrix}$

, is another two dimensional data array sequence $\begin{vmatrix} 2 & 0 \\ 0 & -2 \end{vmatrix}$.

Problem 10.12
Evaluate the Inverse Discrete Cosine Transform IDCT of the two dimensional data array sequence
2×2, given by $\begin{vmatrix} 2 & 0 \\ 0 & -2 \end{vmatrix}$, consider the sampling period T_s equals unity.

Solution
The number of row data samples N_1 equals 2 and $k_1 = 0$ and 1, also the number of column data samples N_2 equals 2 and $k_2 = 0$ and 1, Eq.(10.31) yields

$$x(n_1, n_2) = \frac{1}{4} \sum_{k_2=0}^{1} \sum_{k_1=0}^{1} X(k_1, k_2) \cos(\pi n_1 k_1) \cos(\pi n_2 k_2)$$

, where $n_1 = 0$ and 1, $n_2 = 0$ and 1. Then the discrete time domain $x(0, 0)$, $x(0, 1)$, $x(1, 0)$, and $x(1, 1)$, are given by

$$x(0, 0) = \frac{1}{4} \sum_{k_2=0}^{1} \sum_{k_1=0}^{1} X(k_1, k_2) \cos(0) \cos(0) = \frac{1}{4}[2 + 0 + 0 - 2] = 0$$

$$x(0, 1) = \frac{1}{4} \sum_{k_2=0}^{1} \sum_{k_1=0}^{1} X(k_1, k_2) \cos(0) \cos(\pi k_2) = \frac{1}{4}[2 + 0 + 0 + 2] = 1$$

$$x(1, 0) = \frac{1}{4} \sum_{k_2=0}^{1} \sum_{k_1=0}^{1} X(k_1, k_2) \cos(\pi k_1) \cos(0) = \frac{1}{4}[2 + 0 + 0 + 2] = 1$$

, and

$$x(1, 1) = \frac{1}{4} \sum_{k_2=0}^{1} \sum_{k_1=0}^{1} X(k_1, k_2) \cos(\pi k_1) \cos(\pi k_2) = \frac{1}{4}[-2 + 0 + 0 + 2] = 0$$

, then the inverse discrete cosine transform of the two dimensional data array sequence $\begin{vmatrix} 2 & 0 \\ 0 & -2 \end{vmatrix}$, is
another two dimensional data array sequence $\begin{vmatrix} 0 & 1 \\ 1 & 0 \end{vmatrix}$, which is the inverse discrete cosine transform of problem 10.11.

10.7. Drawbacks of Fourier Transform
Fourier transform technique works only when the statistical properties of a signal such as the mean value and the autocorrelation function (chapter VI) do not vary with time [24]. Because the basis functions (series of sine and cosine functions) used for the Fourier transform extend over an infinite time interval, they are not well suited for non-stationary signals (the frequency content of which varies with time), which are often the signals encountered in real-world applications. The time domain representation of a signal does not provide quantitative information on frequency content of the signal, while the Fourier transformed frequency representation provides frequency content without any indication of time localization of the frequency components. Consequently, analyzing a non-stationary signal requires a transformation technique that can provide a two dimensional time-frequency representation.

So, Fourier transform is a good method for analyzing stationary data, with small scale (high frequency) features representing the detail or noise originates in the signal, and also with large scale (low frequency) features representing the basic shapes. Since Fourier transform maps a signal from its time domain $x(t)$ to its frequency domain $X(f)$, However, Fourier transform has the disadvantages that the frequency information is global because its basis functions are infinite duration of sine and cosine functions, where the signal is expanded in terms of the orthonormal basis functions of sine and cosine waves of infinite duration. This is not satisfactory when searching for localized features in the signal. The difficult in the Fourier transform is, in signals with components localized in time and the transform does not generally convey any information pertaining to translation of the signal in time. The drawbacks of the Fourier transform can be summarized in: a) It uses infinite sinusoidal basis to analyze finite signals, b) Its localization is only in one domain, and the time information is lost, c) Its poor recovery of non-stationary signals, d) It is unable to analyze trends, drift, abrupt changes, and the beginnings and the ends of events, and e) The abrupt change in signals is spread out. These problems are partially avoided by using the windowed Fourier transform (short time Fourier transform), and the wavelet transform, where the signal is analyzed locally.

10.8. Short Time Fourier Transform

Short Time Fourier Transform STFT is used to analyze the non-stationary signals. Short time Fourier transform is achieved by multiplying the signal with a moving window function before computing its Fourier transform, Fig.10.9a. Short time Fourier transform uses a fixed size window in its original domain (time domain) for large and small components of the signal. Short Time Fourier Transform STFT analyses the signal into frequency bands through frequency modulation of the same window size for all frequencies and consequently the resolution of the analysis is the same at all locations in the time-frequency resolutions. Short Time Fourier Transform STFT has been used extensively for pre-processing data with localized features. Windowing is used as a means of converging the sample data to zero at the end points, the conventional STFT is used to combat this convergence. The STFT positions a shifted window function w(t) at τ on the time axis, Fig.10.9b, and thus the Fourier transform of this window function for the signal x(t) is given by

$$X(f, \tau) = \int_{-\infty}^{\infty} x(t)\, w^*(t - \tau)\, e^{-j2\pi ft}\, dt \qquad (10.34)$$

, where τ is the translation parameter, the positions of the window function $w(t - \tau)$.

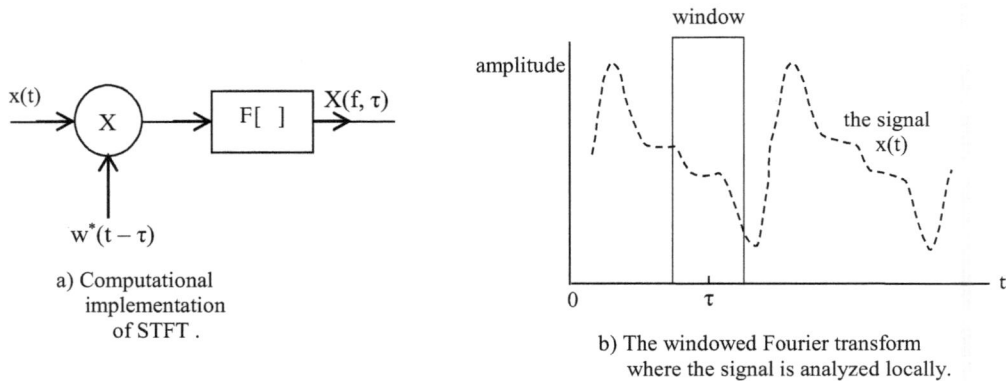

a) Computational
implementation
of STFT.

b) The windowed Fourier transform
where the signal is analyzed locally.

Fig.10.9. a) Implementation of STFT, b) The windowed Fourier transform where the signal is analyzed locally.

The STFT $X(f, \tau)$ is dependent on the choice of the window function w(t). Many different window shapes are used in practice. Typically, they are symmetric, unimodal, and smooth such as the Gaussian function and the raised cosine (single period). Eq.(10.34) shows that the inner (scalar) product of the signal x(t) with a two-parameter family of basis functions, $w(t - \tau)\, e^{-j2\pi ft}$, for varying τ and f. It is important to note that these basis functions do not constitute an orthonormal set [19].

Many of the properties of the Fourier transform are carried over to the Short Time Fourier Transform STFT. In particular, the two signal preserving shifting properties in the time and frequency domains, and the spectrogram. For the STFT preserves shifting in the time domain, except for a linear modulation, let x(t) and $X(f, \tau)$ constitute a STFT pair. Then the STFT of the time shifting function $x(t - t_0)$ is $X(f, \tau - t_0)\, e^{-j2\pi ft_0}$. Also, for the STFT preserves shifting in the frequency domain, the Inverse STFT of the shifting function $X(f - f_c, \tau)$ is $x(t)\, e^{j2\pi f_c t}$, while the spectrogram $\Psi(f, \tau)$ is obtained by getting the following squared modulus of the STFT $X(f, \tau)$

$$\Psi(f, \tau) = |X(f, \tau)|^2$$

, the spectrogram $\Psi(f, \tau)$ represents a simple and yet powerful extension of the classical Fourier theory. In physical terms, it provides a measure of the signal energy in the time-frequency plane. The spectrogram $\Psi(f, \tau)$ is used extensively in the analysis of speech signals and other nonstationary signals for the physical insight it provides.

In the STFT, the fixed time width of the window function w(t) is the limitation of the STFT because of the fixed time-frequency resolution. This technique of STFT, Fig.10.9b, maps the signal x(t) into two dimensional of time and frequency X(f, τ), Fig.10.10a. Gabor [25] tried to analyze small sections of signal at a time (windowing), this STFT is a Gabor function when the window is a Gaussian function.

The drawbacks of the Short Time Fourier Transform STFT can be summarized in: a) Its frequency time resolution constraint, leads to a limited precision, b) The precision is determined by the size of the window which is fixed, c) The window is the same for all the frequencies, d) The analysis windows distort the signal and smears its power. These drawbacks are partially avoided by using the wavelet transform which uses a variable window size function depend on the scale components of the signal, and leads to multi-resolution analysis, Fig.10.10b. This approach enables the time resolution becomes arbitrarily good at high frequencies while the frequency resolution becomes arbitrarily good at low frequencies. This is the foundation of the wavelet transform. Wavelet transform technique has been applied in such diverse fields such as digital communication, remote sensing, biomedical signal processing, medical imaging, astronomy, and numerical analysis. Engineers and scientists are now using wavelet transform to compress the digital signals and images, to speed up their fundamental scientific algorithms, and to get ride the digital signal from the noise.

10.9. Wavelet Transform

Wavelet Transform WT has been an area of active research for the non-stationary signal processing. As a result, on the various time-frequency techniques, wavelet transform provides information about a signal in the time-scale domain simultaneously through a series of correlation operations between the signal being analyzed and the base wavelet function [24]. In the Wavelet Transform WT, the size of the window function vary, this gives them an infinite set of possible basis functions, and what is really needed, is a long window to analyze the large scale components of the signal and a narrow window to detect the small scale features of the signal. This exactly what is provided by the wavelet transform. Wavelet transform projects the signal onto scaled versions of a limited-size oscillatory window function. Gross and Morlet [26] used finite functions called wavelets, instead of sine and cosine basis. They discovered that the wavelet can be made short at high frequency (pick abrupt change) and long at low frequency (pick overall trends). The wavelet analysis is an irregular and asymmetric waveform of effectively limited duration. The wavelet transform at high frequencies gives good time resolution and poor frequency resolution, while at low frequencies gives good frequency resolution and poor time resolution.

Generally, If the input energy signal is x(t), the wavelet transform of x(t) is given by

$$WT(s, \tau) = \int_{-\infty}^{\infty} x(t) \, W^*_{s,\tau}(t) \, dt \tag{10.35}$$

, where $W^*_{s,\tau}(t)$ is the complex conjugate of the wavelet analysis function $W_{s,\tau}(t)$. The s and τ are the scale and shift parameters of the wavelet analysis function $W_{s,\tau}(t)$ respectively. Then the wavelet transform uses multi-resolution analysis for which different frequencies are analyzed with different resolutions by breaking the signal into shifted and scaled version of the original (mother) wavelet analyses function.

Just as the STFT has signal-preserving properties of its own, so does the wavelet transform. In particular the signal preserving shifting property in the time domain, the time scaling property, and the scalogram. For the Wavelet transform preserves the shifting property in the time domain, let x(t) and WT(s, τ) constitute a Wavelet transform pair. Then the Wavelet transform of the time shifting function x(t – t₀) is WT(s, τ – t₀). However, unlike the STFT, the Wavelet transform does not preserve frequency shifts. And for the Wavelet transform preserves time scaling property, the Wavelet transform of the time-scaled signal x(at) is [(1/√s) WT(s/a , aτ)]. While the scalogram Ψ(s, τ) is obtained by getting the following squared modulus of the Wavelet transform WT(s,τ) of the energy signal x(t)

$$\Psi(s, \tau) = |WT(s, \tau)|^2$$

, the scalogram Ψ(s, τ) represents a distribution of the energy of the signal in the time-scale plane.

In both the scalogram and spectrogram, phase information is lost: neither one of them can therefore be inverted in general. Also, both are bilinear functions of the signal under analysis, with the result that " cross-terms" appear as interference patterns, which are undesirable [19].

The wavelet analysis function $W_{s,\tau}(t)$ must be localized in time and frequency domains and has an average power equals zero (the average power of the energy signal, chapter I). Regardless of the scale and magnitude of the wavelet analysis function $W_{s,\tau}(t)$. $W^*_{s,\tau}(t)$ is admissible as a wavelet if and only if the time-scale distribution is given by the following admissibility condition

$$\int_{-\infty}^{\infty} W_{s,\tau}(t) \ dt \ = \ 0 \qquad\qquad (10.36)$$

, for which it is sufficient that the mean value of $W_{s,\tau}(t)$ is vanish. If the wavelet analysis function $W_{s,\tau}(t)$ satisfies the admissibility condition, Eq.(10.36), Eq.(10.35) can be inversely transformed to the original signal x(t). The admissibility condition, Eq.(10.36) means that the wavelet analysis function $W_{s,\tau}(t)$ has some oscillations due to its translation by amounts proportional to their width, Fig.10.10b. This is the main difference between the Short Time Fourier Transform STFT and the Wavelet Transform WT, where the wavelet analysis function of the STFT is fixed size window, Fig.10.10a. The validation of the admissibility condition, Eq.(10.36), that the mean value of $W_{s,\tau}(t)$ is vanish (zero or say 0.0001), this depend on the chosen parameters of $W_{s,\tau}(t)$ [25].

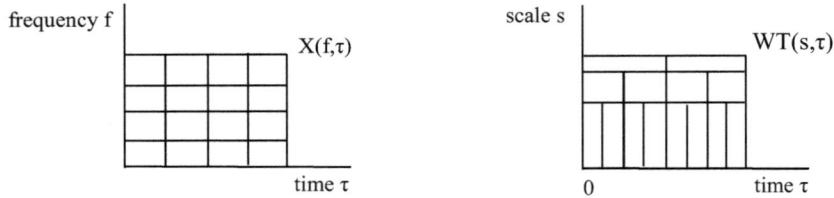

Fig.10.10. a) The time-frequency of STFT, b) The time-scale of the wavelet transform.

Additional optional properties of the wavelets are the following higher moments

$$\int_{-\infty}^{\infty} t^n \ W_{s,\tau}(t) \ dt$$

may also vanish. This makes the wavelet blind to polynomial behavior of degree $(n - 1)$, which can be helpful to study singularities.

On the other hand, the Continuous Wavelet Transform CWT, is the sum over all time of the signal x(t) multiplied by scaled and shifted versions of the mother wavelet, or it tries to find a resemblance index (wavelet coefficients) between the signal x(t) and wavelet, Eq.(10.35) yields

$$CWT(s, \tau) \ = \ \frac{1}{\sqrt{s}} \ \int_{-\infty}^{\infty} x(t) \ W^* \left[\frac{t - \tau}{s} \right] dt$$

, where the wavelet analysis function $W_{s,\tau}(t)$ in Eq.(10.35), is given by

$$W_{s,\tau}(t) \ = \ \frac{1}{\sqrt{s}} \ W \left[\frac{t - \tau}{s} \right] \qquad\qquad (10.37)$$

, and $1/\sqrt{s}$ is the normalization factor for conversation of energy, where the transform kernel $W_{s,\tau}(t)$ is a zero mean band-pass function, and the s and τ are the scaling and the dilation of the wavelet analysis function $W_{s,\tau}(t)$ respectively. The CWT is continuous in its usage of every possible scale and shift. The scale is related to the frequency where in the low scale, a compressed wavelet of rapidly changing details with high frequency, while in the high scale, a stretched wavelet of slowly changing and coarse features with low frequency.

The computation of the Continuous Wavelet Transform CWT which is particularly well-suited for use on a digital computer, is obtained by discretizing the CWT, then the sampling of the time-scale plane is occurred. The sampling rate can be changed accordingly with scale change without violating the Nyquist criterion. The Nyquist criterion states that the minimum sampling rate that allows the reconstruction of the original signal is 2W Hz, where W is the maximum frequency of the signal x(t). Therefore, as the time-scale goes higher (low frequencies), the sampling rate can be decreased thus reducing the number of computations [24].

10.10. Discrete Wavelet Transform

Discrete Wavelet Transform DWT is a subset of the Continuous Wavelet Transform CWT by choosing certain scales and positions to reduce the digital computations and the processing data. The Discrete wavelet transform algorithm was independently discovered by Croiser, Esteban, and Galand in 1976. The Discrete wavelet transform is based on the sub-band coding and is found to yield a fast computation of the wavelet transform. It is easy to be implemented and reduces the computation time and the resources required. Many improvements were made to these coding schemes which resulted in efficient multi-resolutions analysis schemes. In the CWT, the signals are analyzed using a set of basis functions which relate to each other by simple scaling and translation while in the DWT, a time-scale representation of the digital signal is obtained using digital filtering techniques, where the signal to be analyzed is passed through filters with different cut-off frequencies at different scales.

Mallat [27,28] showed that if the scales s and the positions τ are chosen to be based on powers of 2 (dyadic), the analysis will be efficient and just as accurate. Mallat suggested the use of filters banks algorithm to give multi-resolution analysis. In this algorithm, the signal is passed through a low-pass filter to yield the approximation function and high-pass filter to yield the details function, where the two band orthonormal filter bank and the wavelet theory are strongly linked. The dyadic sub-band tree structure servers as the Fast Wavelet Transform FWT algorithm if the proper initialization at the top resolution level is performed. Discrete wavelet transform maps the original signal vector into a new vector, which is filled sequentially with the wavelet coefficients of the different levels (scales).

Then when the wavelet analysis function $W_{s,\tau}(t)$ is sampled in dyadic grid, that is $s = 2^{-i}$, and $\tau = ks = k\,2^{-i}$, where i and k are belong to a set of positive integers, Eq.(10.37) yields

$$W_{i,k}(t) = \frac{1}{2^{-i/2}}\ W\left[\frac{t - k\,2^{-i}}{2^{-i}}\right]$$

, or equivalently

$$W_{i,k}(t) = 2^{i/2}\ W[\,2^i\,t - k\,]$$

The wavelet family is a dilated mother wavelets of selected s and τ, constitute an orthonormal basis. The sampling of $W_{i,k}(t)$ in dyadic grid, makes this wavelet transform is also called dyadic orthonormal wavelet transform. Due to the orthonormal properties, there is no information redundancy in discrete wavelet transform. With this choice of s and τ, there exists the multi resolution analysis algorithm, which decompose a signal into scales with different time and frequency resolution.

The most interesting dissimilarity of the wavelet transform and Fourier transform, is that an individual wavelet functions are localized in space. This localization feature together with wavelets localization of frequency makes many functions and operators using wavelets, sparse when transformed into the wavelet domain. This sparseness makes wavelets useful for a number of applications such as data compression, feature extraction and denoising. In general, the wavelet transforms can be used in: a) detecting discontinuities, breakdown points, long term evolution, and self similarity, b) identifying pure frequencies, c) suppressing, de-noising, and compressing signals.

The superiority of Discrete Wavelet Transform DWT over discrete Fourier transform, is that DWT `s have the simultaneous localization of frequency and time. The drawback however is in the fact that the frequency divisions in DWT are not in integral steps but instead there are in octave bands.

Also, Fast Fourier Transform FFT and Discrete Wavelet Transform DWT are linear operations that generate a data structure containing $\log_2 N$ segments of various lengths, usually filling and transforming it into a different data vector of length 2^N. The properties of the matrices of the FFT and DWT are similar. The functions are both rotation in function space to a different domain. For the FFT, the new domain contains basis functions that are sines and cosines, while for the DWT the new domain contains more complicated basis functions called wavelets, mother wavelets or analyzing wavelets. Both the basis functions of the transforms are localized in frequency.

The Discrete Wavelet Transform DWT becomes a popular tool for image compression research, although they have yet to make a big impact on image compression standards, most of which still use the Discrete Cosine Transform DCT as their basic energy compaction (or decorrelation) process. DWT may be regarded as equivalent to filtering the input signal with a bank of band-pass filters, whose impulse responses are all approximately given by scaled versions of a mother wavelet. DWT is normally implemented by a binary tree of filters, for the one dimensional case, Fig.10.11.

10.10.1. Filters Bank of Discrete Wavelet Transform

As a tool for spectral analysis, it is useful to break the digital signal into sub-bands using digital filters of different cut-off frequencies in order to analyze the signal at different time scales. These sub-bands are ranges of frequencies over which are interested in the spectral content of the signal. Wavelet transforms can be realized by iteration of the filters with time rescaling. The time scale representation of the signal is done using digital filtering techniques and can be obtained by passing the signal through a series of high-pass filters H to analyze the high frequencies, and low-pass filters G to analyze the low frequencies. The resolution of the signal which is a measure of the amount of detail information in the signal, is determined by filtering operations and the time scale is determined by up-sampling and down-sampling (sub-sampling) operations.

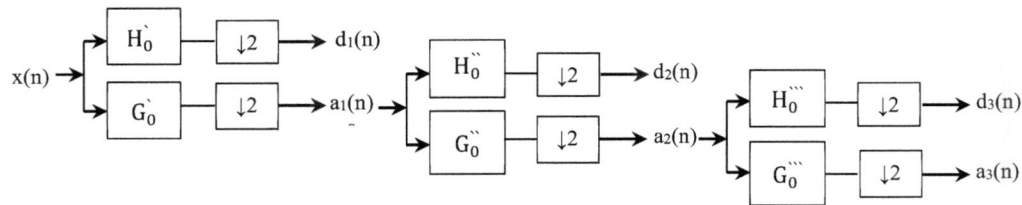

Fig.10.11. Three level wavelet decomposition tree, G and H are low-pass & high-pass analysis filters respectively.

Fig.10.11. is called the Mallat algorithm or Mallat-tree decomposition [27,28], where the discrete wavelet transform is computed by a successive low-pass and high-pass filtering of the discrete time domain signal, The significance of Mallat algorithm is in the manner it connects the continuous time multi-resolution to discrete time filters. Fig.10.11. shows three levels wavelet decomposition tree for the digital signal $x(n)$, and n is an integer. The low-pass and high-pass analyses filters are denoted by G_o and H_o respectively in the three levels of decomposition. At each level, the high-pass filter produces detail information $d_i(n)$ while the low-pass filter associated with scaling function produces coarse approximations $a_i(n)$, and i is the decomposition level number.

At each decomposition level, the half band filters produce signals spanning only half the frequency band where each half frequency band of them is down sampling by two. This technique doubles the frequency resolution as the uncertainity in frequency is reduced by half, then the time scaling is doubles. The decimation by two is the process of reducing the sampling rate by an integer factor of two. In accordance with the Nyquist`s rule, if the original signal has maximum frequency W Hz, which requires a sampling frequency 2W Hz, then the down sampling by two, it now has a maximum frequency of W/2 Hz, and it can now be sampled at a frequency of W Hz, and thus discarding half the samples with no loss of information. This decimation by two, halves the time resolution as the entire signal is now represented by only half the number of samples. Thus, while the half band low-pass filtering removes half of the frequencies and thus halves the resolution, the decimation by two doubles the time scale.

With this approach, the time resolution becomes arbitrarily good at high frequencies, while the frequency resolution becomes arbitrarily good at low frequencies. The time frequency plane is thus resolved as shown in Fig.10.10b. The filtering and decimation process is continued until the desired level is reached. The maximum number of levels depends on the length of the signal. The discrete wavelet transform of the original signal is then obtained by concatenating all the coefficients $a_i(n)$ and $d_i(n)$, starting from the last level of decomposition.

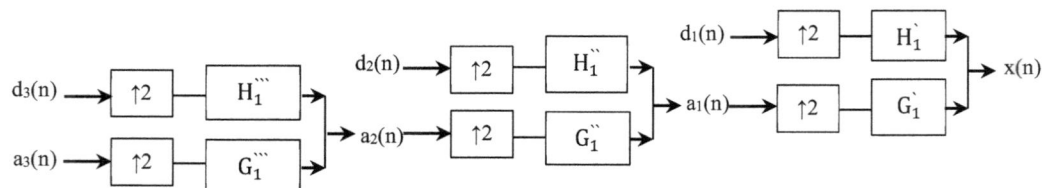

Fig.10.12. Three level wavelet reconstruction tree, G and H are low-pass & high-pass synthesis filters respectively.

Fig 10.12. shows the reconstruction composition of the original signal from the wavelet coefficients $a_i(n)$ and $d_i(n)$. Basically, the reconstruction is the reverse process of the decomposition process. The approximation coefficient $a_i(n)$, and detail coefficient $d_i(n)$ at every level are up-sampled by 2, passed through the low-pass and high-pass synthesis filters G_1 and H_1 respectively and then added. This process is continued through the same number of levels as in the decomposition process to obtain the original digital signal $x(n)$. Mallat algorithm [27,28] works equally well if the analysis filters, G_o and H_o, are exchanged with the synthesis filters G_1 and H_1.

In most wavelet transform applications, it is required that the original signal be synthesized from the wavelet coefficients. To achieve the perfect reconstruction, the analysis and synthesis filters have to satisfy certain conditions, if $x(t)$ is a signal, with the z-transform it yields

$$X(z) = \sum_{k=0}^{\infty} x(k)\, z^{-k}$$

Let $G_o(z)$ and $G_1(z)$ be the transfer functions of the low–pass analysis and low-pass synthesis filters respectively, and $H_o(z)$ and $H_1(z)$ be the transfer functions of the high-pass analysis and high-pass synthesis filters respectively. Then, for perfect reconstruction, the filters have to satisfy the following two conditions

$$G_o(-z)\, G_1(z) + H_o(-z)\, H_1(z) = 0 \qquad (10.38)$$
$$, \qquad G_o(z)\, G_1(z) + H_o(z)\, H_1(z) = 2\, z^{-d} \qquad (10.39)$$

, where d is usually equals ±1. The first condition, Eq.(10.38), implies that the reconstruction is aliasing-free while the second condition, Eq.(10.39), implies that the amplitude distortion has amplitude of unity. It can be observed that the perfect reconstruction condition does not change if the analysis and synthesis filters are switched. There are a number of filters which satisfy these conditions. But not all of them give accurate Wavelet transforms, especially when the filter coefficients are quantized. The accuracy of the wavelet transform can be determined after reconstruction by calculating the Signal to Noise ratio of the signal. Some applications like pattern recognition do not need reconstruction, and in such applications, the above conditions need not apply.

The wavelet filter banks are classified into two classes: orthogonal wavelet filter banks and biorthogonal wavelet filter banks, based on the application, either of them can be used. Each of them has special and independent features.

In the features of the orthogonal wavelet filter banks, the coefficients are real numbers, the filters are of the same length and are not symmetric, the low-pass analysis filter G_o and high-pass analysis filter H_o are related to each other by

$$H_o(z) = z^{-N}\, G_o(-z^{-1}) \qquad (10.40)$$

, and the two filters are alternated flip of each other. The alternating flip automatically gives double-shift orthogonality between the low-pass and high-pass analysis filters, then the scalar product of the analysis filters for a shift by two, is zero, where

$$\sum G(k)\, H(k - 2\ell) = 0 \qquad \text{, and } k, \ell \in Z \qquad (10.41)$$

, filters that satisfy Eq.(10.40) are known as Conjugate Mirror Filters CMF. For perfect reconstruction, the synthesis filters are identical to the analysis filters except for a time reversal. Orthogonal filters offer a high number of vanishing moments, this property is useful in many signals and image processing applications. They have regular structure which leads to easy implementation and scalable architecture.

While, in the features of the biorthogonal wavelet filter banks, the low-pass and high-pass filters do not have the same length. The low-pass filter is always symmetric, while the high-pass filter could be either symmetric or anti-symmetric. The coefficients of the filters are either real numbers or integers. For perfect reconstruction, the biorthogonal filter banks have all odd length filters or all even length filters. The two analysis filters can be symmetric with odd length or one symmetric and the other anti-symmetric with even length. Also the two sets of the analysis and synthesis filters must be dual. The distortionless (linear phase) biorthogonal filters are the most popular filters for data compression applications.

10.10.2. Wavelets Function Families

The wavelet analyses functions $W_{s,\tau}(t)$, Eq.(10.37), are band-pass functions (energy signals) of zero dc components (zero average power) and generated from the transform kernel or mother wavelet. The mother wavelet produces all wavelet functions used in the transformation through the translation τ and the scaling s, it determines the characteristics of the resulting wavelet transform. Therefore the details of the particular application should be taken into account and the appropriate mother wavelet should be chosen in order to use the wavelet transform effectively [23,26]. The mother wavelet function $W_{s,\tau}(t)$ must satisfy the two following properties

$$\int_{-\infty}^{\infty} W_{s,\tau}(t)\ dt = 0$$

, and

$$\int_{-\infty}^{\infty} |W_{s,\tau}(t)|^2\ dt < \infty$$

, which are the mother wavelet function integrates to zero and also its square integrable, or equivalently, has finite energy. There are infinite number of functions that satisfy these properties and thus qualify to be mother wavelet.

Fig.10.13. illustrates some of the commonly used wavelet functions where several mother wavelets were introduced: a) Haar, Daubechies, Biorthogonal, Coiflet, and Symlets (wavelet and scaling functions), b) Morlet and Mexican Hat (no scaling function), c) Meyer (wavelet and scaling functions are determined in the frequency domain. The elegance of the orthonormal wavelet theory developed by Daubechies, its illustration the linkages of the discrete–time sub-band filter banks and the continuous-time wavelet bases via the design of their discrete time counterparts, adds significant flexibility to the design of wavelet transform bases.

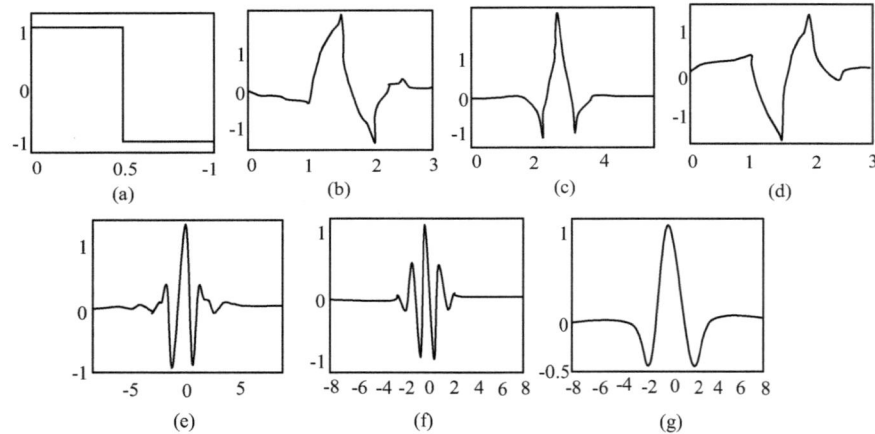

Fig.10.13. Wavelet families: a) Haar, b) Daubechies 4 , c) Coiflet 1,
d) Symlet 2, e) Meyer, f) Morlet, g) Mexican Hat.

The Haar wavelet is one of the oldest and simplest wavelet. Therefore, any discussion of wavelets starts with the Haar wavelet. Daubechies wavelets are the most popular wavelets. They represent the foundations of wavelet signal processing and are used in numerous applications. These are also called Maxflat wavelets as their frequency responses have maximum flatness at frequencies o and π. This is a very desirable property in some applications. The Haar, Daubechies, Symlets and Coiflets are compactly supported orthogonal wavelets. These wavelets along with Meyer wavelets are capable of perfect reconstruction. Meyer, Morlet and Mexican Hat wavelets are symmetric in shape. The wavelets are chosen based on their shape and their ability to analyze the signal in a particular application.

The prime importance in designing the high quality speech coders, is the choice of the mother wavelet function used. Several different criteria can be used in selecting an optimal wavelet function. The objective is to minimize the reconstructed error variance and maximize the Signal to Noise Ratio SNR.

The SNR gives the quality of the reconstructed signal. The higher the SNR, the better the quality of the reconstructed signal, where the SNR is given by

$$SNR \ = \ \log_{10}(\frac{\sigma_x^2}{\sigma_e^2})$$

, where σ_x^2 is the mean square of the speech signal, and σ_e^2 is the mean square difference between the original and the reconstructed signals.

10. 0.3. Applications of Wavelet Transform

There is a wide range of applications for wavelet transform which are applied in different fields ranging from signal processing to biometrics, and the application is still growing. One of the prominent applications is in the Personal Identification ID fingerprint compression standard. WT`s are used to compress the fingerprint pictures for storage in their data bank. The major application domain of the medical imaging technology is the radiology where some of the imaging modalities include computed tomography, magnetic resonance imaging, ultrasound, and positron emission tomography. All the image data compression schemes can be categorized into lossless and lossy groups. Although the lossless one is especially preferred in medical images, it makes necessary the use of the lossy schemes due to the having relatively low achieved compression ratios. This must not cause to have the less diagnostic features of the image. Therefore, the new algorithms can be developed to minimize the effect of data loss on the diagnostic features of the image.

In the Discrete Wavelet Transform DWT, the most prominent information of the signal appears in the high amplitudes and the less prominent information appears in the very low amplitudes. The data compression can be achieved by discarding these low amplitudes. The wavelet transforms enables high compression ratios with good quality of reconstruction. At present, the application of wavelets for the image compression is the hottest areas of research. Recently, the Wavelet transforms have been chosen for the Joint Photographic Experts Group JPEG-2000 compression standard of still images that the individual pictures are compressed without reference to any other pictures, where the intra-coding video compression treats each picture independently and absolute picture data are transmitted.

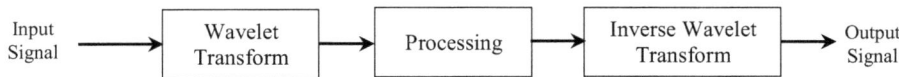

Fig.10.14. Signal processing application.

Compression technique can be achieved by transforming the data, quantizing the coefficients obtained during transformation and then encoding the quantized coefficients. Fig.10.14. shows the general steps followed in a signal processing application. After performing the wavelet transform to the signal, the processing may involve compression, encoding, denoising etc, where the processed data is either stored or transmitted. To avoid redundancy which hinders the compression technique, the transform must be at least biorthogonal. In order to save the Central Processing Unit CPU time, the corresponding algorithm must be fast [20]. The two dimensional wavelet transform satisfies these conditions. Wavelet compression allows the integration of various compression techniques into one algorithm.

The digital signal processing of compression applications, involves quantization and entropy coding to yield a compressed image, the quantization may be scalar quantization or vector quantization, and the entropy encoding may be Run Length Encoding RLE or Huffman encoding or Lempel-Ziv-Wekh LZW encoding or delta encoding. During this processing, all the wavelet coefficients that are below a chosen threshold are discarded. These discarded coefficients are replaced with zeros during the reconstruction of the signal, where the entropy coding is decoded, then quantized and then finally inverse wavelet transformed.

Wavelet transforms are very promising for real time audio and video compression applications, and also have numerous applications in digital communications such as Orthogonal Frequency Division Multiplexing OFDM. Wavelet transforms are used in biomedical imaging. For example, the Electro-Cardio-Gram ECG signals, measured from the heart, are analyzed using wavelets or compressed for storage. The popularity of the Wavelet transform is growing because of its ability to reduce distortion in the reconstructed signal while retaining all the significant features present in the signal. One dimensional Wavelet transforms find application in speech compression, which reduces transmission time in mobile applications, and are used in denoising, edge detection, feature extraction, speech recognition, echo cancellation and others.

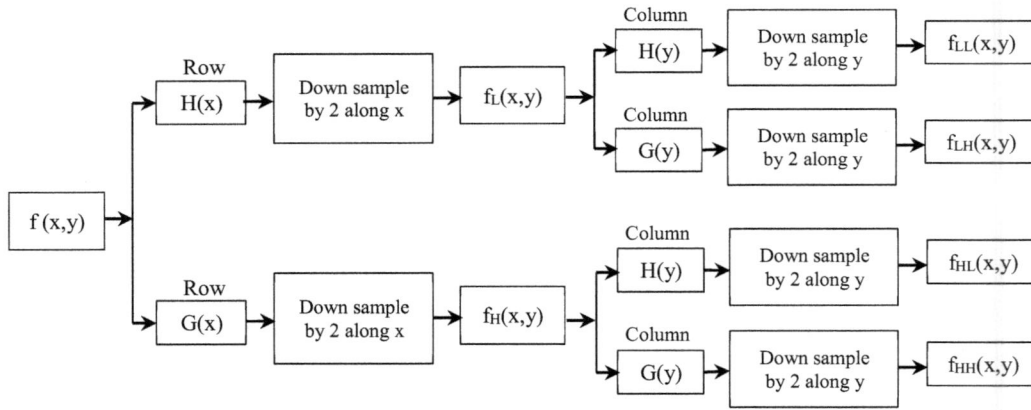

Fig.10.15. The resultant image split into multi-bands.

The two dimensional wavelet transform find application in video compression, Fig.10.15, where it can be accomplished by performing two separate one dimensional wavelet transforms. First, the image is filtered along the x dimension and decimated by two. Then, it is followed by filtering the sub-image along the y direction and decimated by two. Finally, splitting the image into four bands denoted by LL, HL, LH, HH, after one level decomposition, in this case, the decomposition can be achieved by acting upon the LL sub-band successively and the resultant image is split into multiple bands, Fig.10.16. An application of this motivation is the JPEG-2000 project was motivated by Ricoh's submission of the Compression with Reversible Embedded Wavelets CREW algorithm [29], and the Set Partitioning In Hierarchical Trees SPIHT algorithm [30], to an earlier standardization effort for lossless and near-lossless compression, known as JPEG-LS.

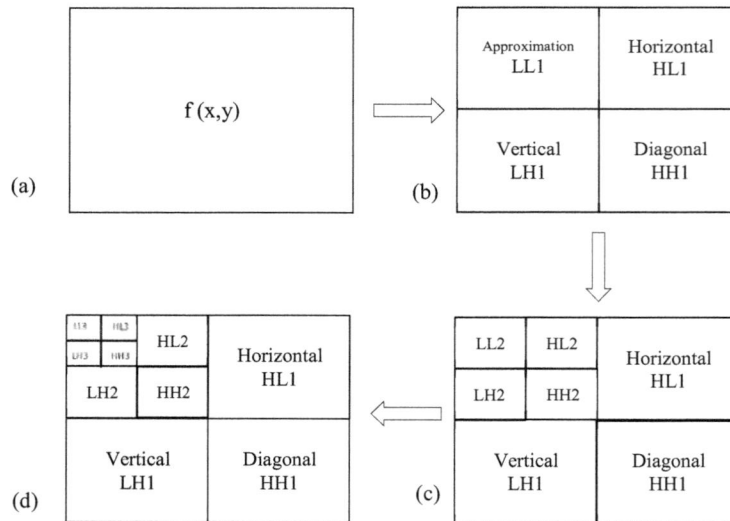

Fig.10.16. Two dimensional discrete wavelet transform sub-band:
a) Original image,
b) One level decomposition,
c) Two levels decomposition,
d) Three levels decomposition.

In accordance with the decomposition at each level in the filter bank of the discrete wavelet transform, the high-pass filter produces the details information $d_i(n)$, and the low-pass filter is associated with the scaling function produces coarse approximation $a_i(n)$, where the i is the number of decomposition level. The low-pass filtering mathematical terms averaging operation, is the inner product between the signal $x(t)$ and the scaling function $S_{j,k}(t)$, Eq.(10.42), whereas the high-pass filtering mathematical terms differencing operation, is the inner product between the signal $x(t)$ and the wavelet function $W_{i,k}(t)$, Eq.10.43). Then the average coefficients $a_i(k)$ and the detail coefficients $d_i(k)$ are given by

$$a_i(k) = <\ x(t), S_{j,k}(t)\ > = \int x(t)\ S_{j,k}(t)\ dt \qquad (10.42)$$

, and

$$d_i(k) = <\ x(t), W_{j,k}(t)\ > = \int x(t)\ W_{j,k}(t)\ dt \qquad (10.43)$$

, where the scaling function $S_{i,k}(t)$ or low pass filter and the wavelet function $W_{i,k}(t)$ or high pass filter are defined by

$$S_{i,k}(t) = 2^{i/2}\ S(2^i\ t - k)$$

, and

$$W_{i,k}(t) = 2^{i/2}\ W(2^i\ t - k)$$

, where the i denotes the discrete scaling index and k denotes the discrete translation index.

The reconstruction of the image can be carried out, Fig.10.17. First up-sampling by a factor two on all the four sub-bands at the coarsest scale, and filter the sub-bands in each dimension. Then sum the four filtered sub-bands to reach the low-low sub-band at the next finer scale. The process is repeated again until the image is fully reconstructed [28,31].

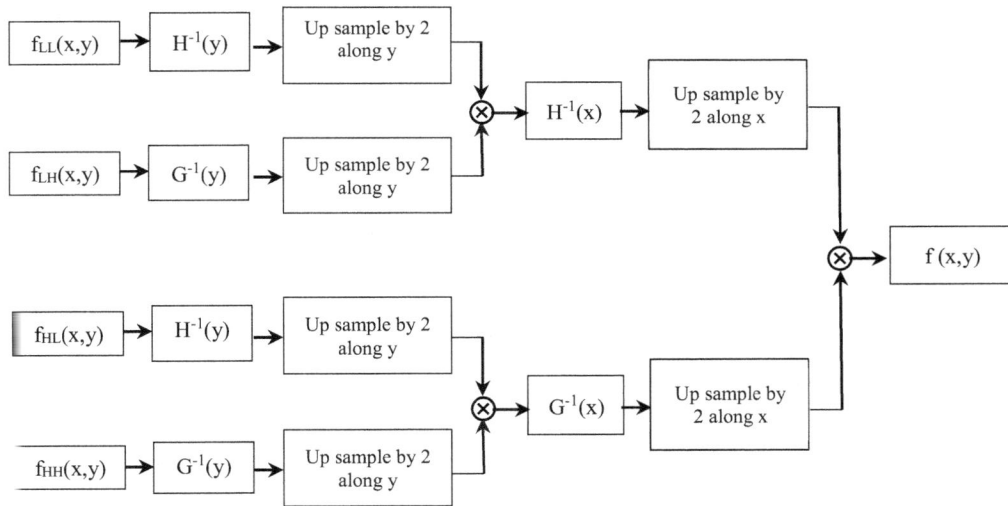

Fig.10.17. Two dimensional inverse discrete wavelet transform (image reconstruction).

Wavelet transforms are popular, powerful and familiar to image compression algorithms. Wavelet transforms have proved the ability of representing the natural images that contain smooth areas separated with edges. However, Wavelet transforms have its own limitations in capturing the directional information such as smooth contours, and directional edges of the image because it cannot efficiently take the advantage of the fact that the edges which usually found in the Personal Identification ID fingerprints and medical images, are smooth curves. This drawback is addressed by directional transforms known as contourlet transforms which have the property of preserving the edges due to its inherent characteristics: directionality and anisotropy scaling [32].

10. 11. Contourlet Transform

Contourlet transform has been a wide interest in the image representations where it is efficiently handle the image geometric structure. The contourlet transform is considered an extension to the wavelet transform in two dimensions using nonseparable and directional filter banks. The contourlet transform is a directional transform, and capable of capturing the contours and the fine details in the images. The contourlet expansion is composed of basis function oriented at various directions in multiple scales, with flexible aspect ratios. These rich sets of basis functions enable the contourlet transform to be an effectively capture the smooth contours that are the dominant feature in the natural images. Two different decompositions are used in the contourlet transform namely: Laplacian Pyramid LP decomposition and Directional Filter Bank DFB decomposition. So the coefficients of the contourlet transform are quantized by multistage vector quantization and then encoded. The combination of the double filter bank is named the Pyramidal Directional Filter Bank PDFB, proposed by MinhDo and Vetterli [32], which overcomes the block-based approach of curvelet transform by a directional filter bank, applied on the whole scale and known as contourlet transform.

The decomposition of the Laplacian Pyramid LP decomposes the images into sub-bands while the decomposition of the Directional Filter Bank DFB analyzes each detail image, Fig.10.18. The grouping of the wavelet coefficients suggests that one can obtain a sparse image expansion by first applying a multi-scale transform and then applying a local directional transform to gather the nearby basis functions at the same scale into linear structure.

In essence, first a wavelet-like transform is used for edge (points) detection, and then a local directional transform for contour segments detection. In this insight, the double filter bank structure can be constructed, Fig.10.18, in which first the Laplacian pyramid is used to capture the point discontinuities and followed by a directional filter bank to link the point discontinuities into linear structures [33]. The overall result is an image expansion with basis images as contour segments, and thus it is named contour transform.

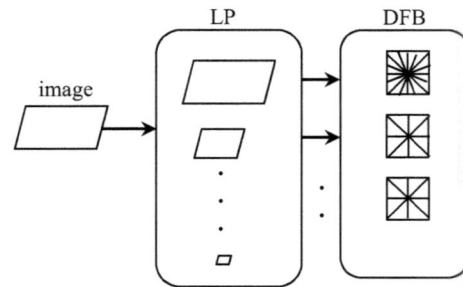

Fig.10.18. Flow graph of contourlet transform.

Fig.10.19 shows a block diagram for the Pyramidal Directional Filter Bank PDFB. The Laplacian Pyramid LP computes a standard multi-scale decomposition into octave bands (low-pass channel and high-pass channel). The low-pass channel is down sampled by two in the two dimensions, and the high-pass channel is not sampled, while the directional decomposition by the Directional Filter Bank DFB is applied to each high-pass channel.

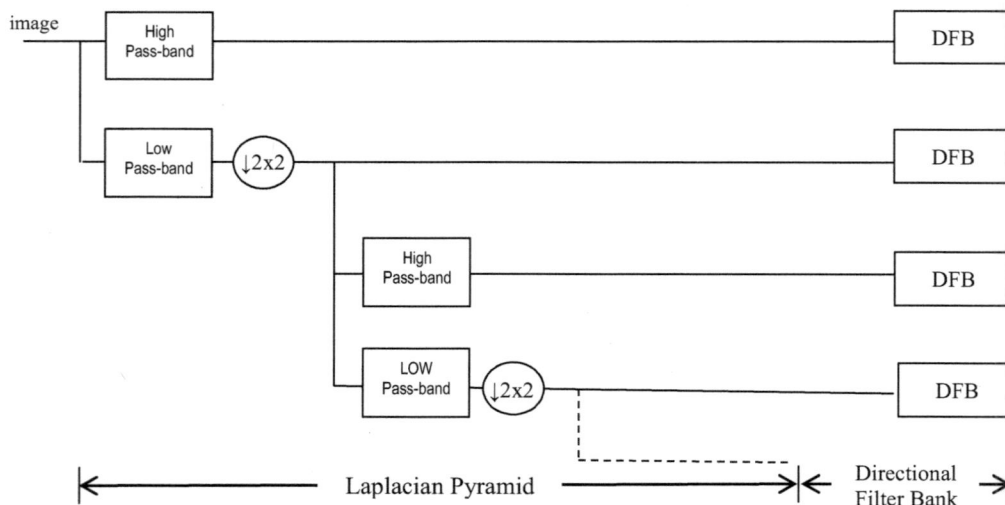

Fig.10.19. Block diagram of the pyramidal directional filter banks, supports for contourlets.

Fig.10.20. shows the Pyramidal Directional Filter Bank PDFB to implement the support sizes for contourlets that satisfies the anisotropy scaling relation. Starting from the upper basis element to the lower basis element, the scale is reduced four times while the number of directions is doubled. The pyramidal directional filter bank allows for different number of directions at each scale/resolution to nearly achieve critical sampling. Since the Directional Filter Bank DFB is designed to capture the high frequency components (representing directionality), the Laplacian pyramid permits the sub-band decomposition to avoid the "leaking" of the low frequencies into several directional sub-bands, so the directional information can be captured efficiently. Generally, the contourlet construction allows for any number of directional filter bank decomposition levels " ℓ_J" to be applied at each Laplacian pyramid level "J". Also, to satisfy the anisotropy scaling relation in the contourlet transform, it is simply needs to impose that the number of directions is doubled at every other finer scale of the pyramid. Fig.10.20. illustrates the graphical representation to depict the supports of the basis functions generated by the pyramidal directional filter bank [32,33].

The two shown Pyramidal levels of the basis elements in Fig.10.20. illustrate that the support size of the Laplacian pyramid is reduced four times while the number of directions in the directional filter bank is doubled. The combination of these two steps shows that the support size of the basis functions in the pyramidal directional filters are changed from one level to the next level in accordance with the curve scaling relation. Each contourlet scheme generation doubles the spatial resolution as well as the angular resolution. Then the PDFB bank provides a frame expansion for the images with frame elements like contour segments, and thus is called the contour transform.

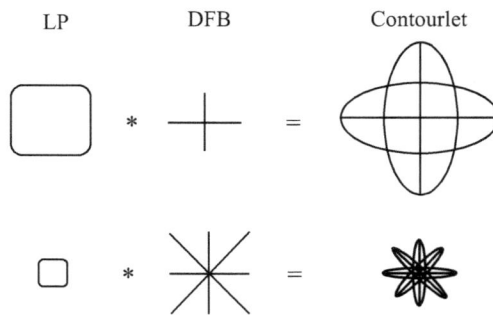

Fig.10.20. The support sizes of contourlet functions.

10.11.1. Laplacian Pyramid

Laplacian pyramid is a one way of achieving the multi-stage decomposition. Laplacian pyramid is introduced by Burt and Adelson [34]. The decomposition of Laplacian pyramid at each level generates a down sampled low-pass version of the original and the difference between the original and the prediction, resulting in a band-pass image, Fig.10.21a, where H and G are the analysis and synthesis low-pass filters respectively while M is the sampling matrix. The process can be iterated on the coarse version. The outputs are the coarse approximation a(n) and the difference b(n) between the original and the prediction. The process can be iterated by decompositing the coarse version repeatedly. The original image is convolved with Gaussian kernel [35]. The resulting image is a low-pass filtered version of the original image. The Laplacian is then computed as the difference between the original image and the low-pass filtered image. This process is continued to obtain a set of band-pass filtered images, each one is the difference between two levels in the Laplacian pyramid.

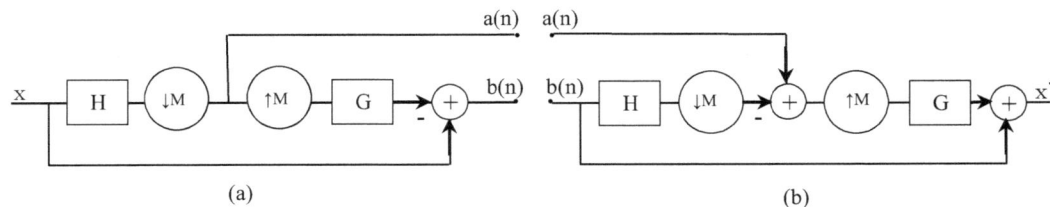

Fig.10.21. Laplacian pyramid scheme, a) Analysis where the outputs are a coarse approximation a(n) and a difference b(n) between the original signal and the prediction, b) the reconstruction scheme for the Laplacian pyramid.

Thus the Laplacian pyramid is a set of band-pass filters, and by repeating these steps several times, a sequence of images, are obtained. If these images are stacked one above another, the result is a tapering pyramid data structure, Fig.10.22, and hence the name. The Laplacian pyramid can thus be used to represent the images as a series of band-pass filtered images, each sampled at successively sparser densities. It is frequently used in image processing and pattern recognition tasks because of its ease of computation.

The drawback of the Laplacian pyramid is the implicit oversampling. However, in contrast to the the critically sampled wavelet scheme, where the Laplacian pyramid has a distinguishing feature that each pyramid level generates only one band-pass image (even for multi-dimensional cases), which does not have scrambled frequency.

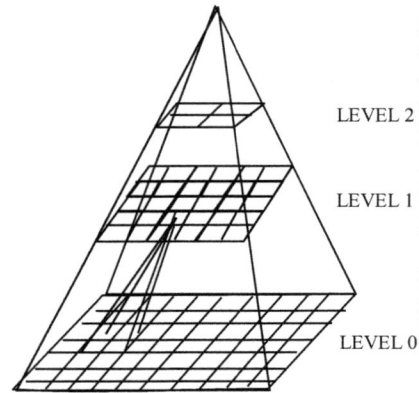

Fig.10.22. Laplacian pyramid structure.

This scrambled frequency happens in the wavelet filter bank when a high-pass channel after down sampling is folded back into the low frequency band, and thus its spectrum is reflected. This effect is avoided by down sampling the low-pass channel only in the Laplacian pyramid [34,35].

10.11.2. Directional Filter Bank

Directional filter bank is designed to capture the high frequency content like smooth contours and directional edges. The directional filter bank is a critically sampled filter bank that can decompose images into any power of 2 number of directions. The directional filter bank is efficiently implemented via a ℓ-level binary tree structured decomposition that leads to number of sub-bands $k = 2^\ell$ with wedge-shaped frequency partitioning, Fig.10.23. The original construction of the directional filter bank involves modulating the input signal and using diamond-shaped filter banks. Furthermore, in order to obtain the desired frequency partition, an involved tree expanding rule has to be followed for finer directional sub-bands. As a result, the frequency regions for the resulting sub-bands do not follow a simple ordering, Fig.10.22, based on the channel indices. Bamberger and Smith [26] introduced a two dimensional filter bank that can be maximally decimated while achieving perfect reconstruction.

Since the directional filter bank is designed to capture the high frequency components (representing directionality) of images, therefore the low frequency components are handled poorly by the directional filter bank. In Fig.10.23, a frequency partitioning where $\ell = 3$, then the number of sub-bands $k = 2^3 = 8$ real wedge-shaped frequency bands. The frequency sub-bands from 0 to 3 correspond to the mostly horizontal directions, while the frequency sub-bands from 4 to 7 correspond to the mostly vertical directions, and then the low frequencies would leak into several directional sub-bands, hence directional filter bank does not provide a sparse representation for images. In order to improve the situation, the low frequencies should be removed before the directional filter bank, this provides another reason to combine the directional filter bank with a multi-resolution scheme. Therefore , the Laplacian pyramid permits further sub-band decomposition to be applied on its band-pass images. Those band-pass images can be fed into a directional filter bank so that the directional information can be captured efficiently. The scheme can be iterated repeatly on the coarse image, Fig.10.19. The end result is a double iterated filter bank structure, named pyramidal directional filter bank which decomposes images into directional sub-bands at multiple scales. The scheme is flexible since it allows for a different number of directions at each scale [36].

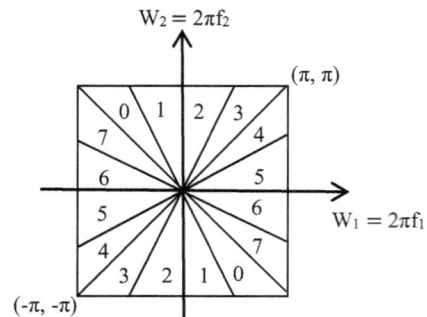

Fig.10.23. Directional filter bank frequency partitioning $(\ell = 3, k = 2^3 = 8)$.

Fig.10.24. shows a contourlet transform decomposition for an image of 256×256 and decomposed into 4 levels where the Laplacian pyramid decomposition $\ell = 0, 2, 3,$ and $4,$ this lead to number of sub-bands k = 1, 4, 8, and 16 directions respectively. The image is decomposed into a low-pass sub-band and several band-pass directional sub-bands, where the contourlets match with both location and direction and produce significant coefficients.

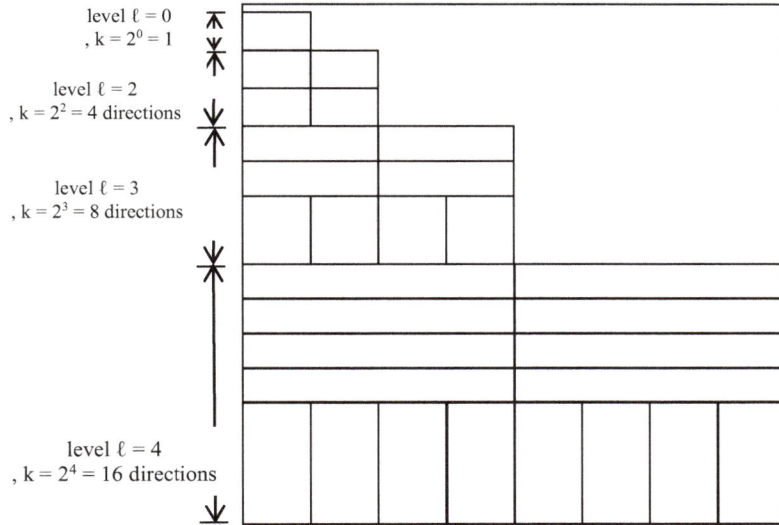

Fig.10.24. Contourlet transform decomposition of an image of 256×256 and decomposed into 4 levels for Laplacian pyramid decomposition $\ell = 0, 2, 3, 4.$

Since the decomposition of each scale into any arbitrary power of 2 number of directions where different scales can be decomposed into the different numbers of directions. This feature makes the contourlets a unique transform that can achieve a high level of flexibility in the decomposition while being close to critically sampled. Thus, the contourlet transform effectively explores the fact, that the edges in images are localized in both location and direction. Other multi-scale directional transforms either have a fixed number of directions or are significantly over complete.

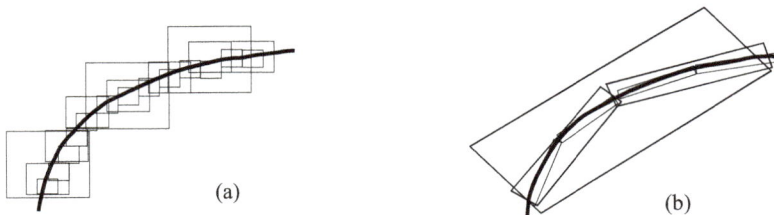

Fig.10.25. a) Wavelet transform presents point of discontinuity, b) Contourlet transform presents smooth contour.

The contourlet transform not only has the multi-scale and time-frequency localization properties of wavelets but also offers a high degree of directionality and anisotropy. Contourlet transform involves basis functions that are oriented at any power of two number of directions with flexible ratios. The Contourlet transforms can represent a smooth contour with fewer coefficients compared with wavelets, Fig.10.25, which showing how wavelets having square supports that can only capture point discontinuities, whereas contourlets having elongated supports that can capture linear segments of contours. The contourlet transform is considered the unique transform that can achieve a high level of flexibility in the decomposition while being close to critically sampled up to 33% over complete which comes from the Laplacian Pyramid and a small redundancy occurs [32,33].

By altering the depth of the Directional Filter Bank DFB decomposition tree, Fig.10.18. at different scales (and even at different orientations in a contourlet packets transform), a rich set of contourlets with variety of support sizes and aspect ratios is obtained. This flexibility allows the contourlet transform and the contourlet packets to fit smooth contours of various curvatures well. Candes and Donoho [37] point out that a key to achieving the correct nonlinear approximation behavior is to select support sizes obeying the parabolic scaling relation for curves: *width×length²*, Fig.10.26. The motivation behind the parabolic scaling is to efficiently approximate a smooth discontinuity curve by "laying" basis elements with elongated supports along the curve.

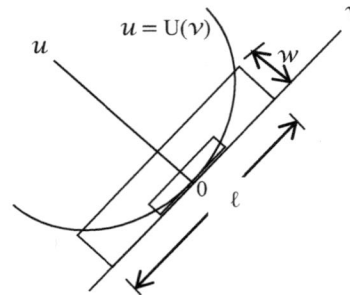

Fig.10.26. Parabolic scaling relation.

In the wavelet transform, the wavelet approximation theory brought a novel condition into filter bank design, which earlier only focused on designing filters with good frequency selection properties. This new condition requires wavelet functions to have a sufficient number of vanishing moments, or equivalently, the high-pass filter in the wavelet filter bank must have enough "zeros at w = 0". The vanishing moments property is the key for the sparse expansion of piecewise smooth signals by wavelets [23]. Intuitively, wavelets with vanishing moments are orthogonal to polynomial signals, and thus only a few wavelet basis functions around the discontinuities and lead to significant coefficients [38]. So Directional Vanishing Moment DVM property also holds in other two dimensional expansions. In particular, two dimensional separable wavelets have directional vanishing moments in the horizontal and vertical directions, which make wavelets especially good in capturing horizontal and vertical edges [39].

While in the contourlet transform, our target for approximation is piecewise smooth images with smooth contours. The key feature of these images is that image edges are localized in both location and direction. More specifically, a local region around a smooth contour can be approximated by two polynomial surfaces separated by a straight line. Thus it is desirable that only few contourlet functions whose supports intersect with a contour and align with the contour local direction would "feel" this discontinuity. One way to achieve this desideratum is to require all the one dimensional slices in a certain direction of contourlet functions to have vanishing moments. This requirement refers to have Directional Vanishing Moment DVM condition [40].

Contourlet transform is proposed as a mean to fix the failure of the wavelet transform in handling the image geometry by the presence of the directional vanishing moments in the contourlet frame element where the vector quantization technique [21] is applied to an ordered set of symbols. The superiority of the vector quantization lies in the block coding gain, the flexibility in partitioning the vector space, and the ability to exploit intra-vector correlations. Multi-space vector quantization divides the encoding task into several stages. The first stage performs a relatively crude encoding of the input vector using a small codebook while the second stage is a quantizer operates on the error vector between the original vector and the quantized first stage output. This quantized vector error provides a refinement to the first approximation, and then the indices obtained by the multistage vector quantizer are encoded [41,42].

In the contourlet transform, the major difficulty in implementing a vector quantizer are the computation and the storage requirements [43]. The complexity of these requirements is the exponential rising function of the number of bits used in quantizing each frame of the spectral information. The storage requirement in the multistage vector quantization is less when compared to full search vector quantization. An example [41], the people Personal Identification ID with massive collection of fingerprints cards, contains more than 200 million cards and is growing at a rate of 30,000 – 50,000 new cards per day. In the digitizing of these ID cards to allow for electronic storage retrieval and transmission, the fingerprints are digitized at a resolution of 500 pixels/inch with 256 gray levels. The single finger print is about 700,000 pixels and needs about 0.6 Mbytes to store. Then the pair of hands requires about 6 Mbytes of storage. So the digitizing of any current archive would result in more than 200 Terabytes of data. Because of the storage requirements of these data and the time needed to send a fingerprint card over a modem, the files must be compressed. Although there are many image compression techniques currently available, there still exists a need to develop faster and more robust algorithms adapted to fingerprints. Fingerprint standards is based on the discrete wavelet transform [44] using the biorthogonal wavelet. Although the performance of this standard is better than that of the cosine transform based the Joint Photographic Experts Group JPEG standard, it still needs enhancements. The contourlet transform [40] and the multistage vector quantization [45] are employed to achieve better quality (high peak signal to noise ratio).

10.12. Some Application of Compression Techniques

Compression, bite rate reduction, and data reduction, all these terms mean basically the same thing, in essence, the same information is carried using a smaller quantity of data or smaller rate of data. Compression is summarized as in Fig.10.27. that the date rate of the source R_s is reduced to the rate R_r by the compressor. In the signal processing, the compressed data either stored or transmitted, and the original source data rate R_s can be obtained by the expander. The compressor and the expander in series are referred to as a compander. Also the compressor may equally well be referred to as a coder and the expander a decoder in which case the tandem pair may be called a codec. The Compression Ratio CR or the coding gain is given by

$$CR = \frac{\text{source data rate } R_s}{\text{compressed (reduced) data rate } R_r}$$

In the computer data storage application where the lossless coding, the compression ratio CR may be around 2:1, such as the Lempel-Ziv-Wekh LZW Encoding. While in digital application where the lossy coding, a greater compression ratio than the lossless codec. Audio compression means a process where the dynamic range of the sound is reduced. In digital audio broadcasting and digital TV transmitter, the compressor is used to reduce the bandwidth needed [46].

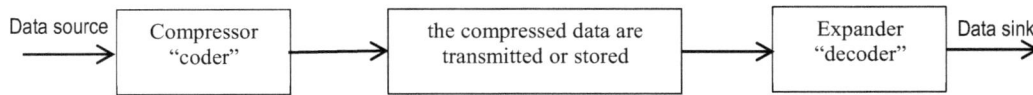

Fig.10.27. Compression system (coder and decoder).

10.12.1. Lossless and Lossy (Perceptive) Coding

Although there are many different coding techniques, all of them fall into one or other of these categories. In the lossless coding, the data from the expander are identical bit-for-bit with the original source data. Lossless coding is popular for data storing in computer disk drive such as Lempel-Ziv- Wekh LZW coding and Huffman encoding. Lossless coding is not popular for audio and video coding where the lossy coding is permissible.

In the lossy coding (perceptive), the data from the expander are not identical with the source data bit-for-bit and as a result compressing the input with the output is bound to reveal differences. Lossy coding is popular for audio and video applications such as predictive coding, JPEG and MPEG coding, where greater compression ratio than lossless coding is achieved. Lossy coding is not popular for data storing in computer desk drive. In digital audio processing, successful lossy coding are those in which the differences are arranged so that a human listener finds them subjectively difficult to detect. Lossy coding is based on understanding of pysochoacoustic perception, so called perceptive coding. In the digital video processing, the generation of color difference signals from Red-Green-Blue RGB represents an application of perceptive coding where the human viewer sees no change in quality although the bandwidth of the color difference is reduced.

Perceptive codes often obtain a coding gain by shortening the word length of the data representing the signal waveform. This must increase the noise level and the strick is to ensure that the resultant noise is placed at frequency where human senses at least able to perceive it. As a result, although the received signal is measurable different from the source data, it can appear the same to the human listener or viewer at moderate compression factors. As these codes rely on the characteristics of human sight and hearing, they can be fully tested subjectively.

10.12.2. Compression Principle

In the pulse code modulation systems, the bit rate is generally constant and equal the product of the sampling rate and the number of bits in each sample, Table 10.3. Since the information rate of a real signal varies, the difference between the information rate and the bit rate is known as the redundancy. Compression systems are designed to eliminate this redundancy. One way in which the redundancy can be done is to exploit the statistical predictability in the signals. The entropy (information content) of a sample of function, of how different this it is from the predicted value. Most signals have some degree of predictability. As an example the sine wave is highly predictable because all the cycles look the same single frequency and carries no information and have no bandwidth (Shannon`s theory). While the noise signal is completely unpredictable. The entropy is the actual area occupied by the signal, which the area must be transmitted if there are to be no subjective differences or artifacts in the received signal, Fig.10.28. The remaining area is called the redundancy because it adds nothing to the information conveyed. An ideal decoder would recreate the original information quite perfectly. As the ideal is approached, the coder complexity rises. Also if the channel capacity is not sufficient for that, then the coder would have to discard some of the entropy. The digital systems need moderate coding gain (compression ratio) and remove redundancy (subjectively lossless) [46].

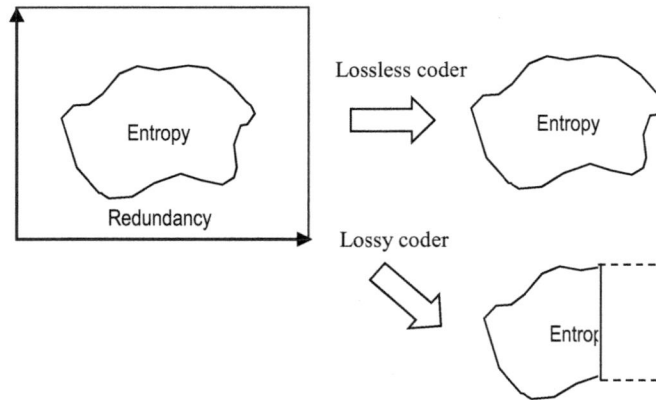

Fig.10.28. The perfect coder removes redundancy from the input signal and results in subjectively lossless coding.

10.12.3. Huffman Encoding (Variable Length Coding)

Huffman Encoding (lossless coding) is a statistical compression technique developed by David Huffman [47]. It has been used in various compression applications such as image compression. In this coding, the probability of the different code values to be transmitted is studied. Variable length coding is used where the frequently used letters are allocated short codes, and the infrequently used letters are allocated long codes. The input source data are assembled in order of descending probability, the two lowest probabilities are distinguished and combined by a single probability. The process of combining probabilities, is continued until the probability of unity is reached and at each stage a bit is used to distinguish the path. The bit will be a 1 for the most probable path and 0 for the least probable path. The compressed output is obtained by reading the bits which describe which path to take going from right to left which is a table of the code word lengths text. Decoding the encoded data of the Huffman code, is the reverse of the encoding process, this is done by sending the table of the code word lengths text to the decoder. The average length L of the Huffman code is given by

$$L = \sum_{i=-\infty}^{\infty} p_i L_i \qquad\qquad \text{binary digits} \qquad (10.44)$$

, where p_i and L_i are the probability and the code word length of the message m_i respectively. Since the information content of each message is given by $Log_2[1/p_i]$. Then the entropy H(m) of the source data is given by

$$H(m) = \sum_{i=-\infty}^{\infty} p_i \, Log_2 \frac{1}{p_i} = - \sum_{i=-\infty}^{\infty} p_i \, Log_2 \, p_i \qquad \text{bits} \qquad (10.45)$$

, the code efficiency η is given by

$$\eta = \frac{\text{the entropy H(m)}}{\text{the average length L of the Huffman code}} \qquad (10.46)$$

, and the redundancy γ is given by

$$\gamma = 1 - \eta \qquad\qquad (10.47)$$

Problem 10.13

Sketch the Huffman coding tree building of a quantized data contain six messages m_1, m_2, m_3, m_4, m_5, and m_6, have the probabilities of 0.08, 0.1, 0.12, 0.15, 0.25, and 0.3 respectively, and then

   i. Deduce the code word of each message.

   ii. Find the average length of this Huffman coding.

   iii. Find the entropy of the source data.

   iv. Find the code efficiency.

   v. Find the redundancy

   vi. Find the compression ratio if each message is represented by 3 fixed bits before compression.

Solution

The Huffman coding tree building is shown in Fig.6.29.

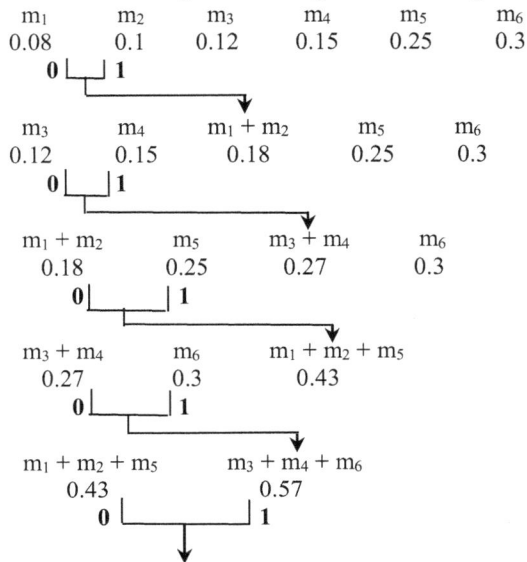

Table 10.1. i. The code word

Message	Probability	Code word
m_1	0.08	000
m_2	0.1	001
m_3	0.1 2	100
m_4	0.15	101
m_5	0.25	01
m_6	0.3	11

Fig.10.29. A Huffman coding tree.

Table 10.1 shows the code word lengths of the six messages m_1, m_2, m_3, m_4, m_5, and m_6, which are assembled in order of descending probability, the two lowest probabilities are distinguished and combined by a single probability. The process of combining probabilities, is continued until the probability of unity is reached and at each stage a bit is used to distinguish the path. The bit will be a 1 for the most probable path and 0 for the least probable path.

ii. The average length of this Huffman code, Eq.(10.44), is given by

$$L = \sum_{i=1}^{6} p_i\, L_i \ = \ 0.08 \times 3 + 0.1 \times 3 + 0.12 \times 3 + 0.15 \times 3 + 0.25 \times 2 + 0.3 \times 2$$

$$= \ 2.45 \quad \text{binary digits}$$

iii. The entropy H(m) of the source data, Eq.(10.45), is given by

$$H(m) = -\sum_{i=1}^{6} p_i\, Log_2\, p_i = -\,[\,0.08\, Log_2\, 0.08\ + 0.1\ Log_2\, 0.1 + 0.12\, Log_2\, 0.12\ +$$

$$0.15\, Log_2\, 0.15\ + 0.25\, Log_2\, 0.25 + 0.3\, Log_2\, 0.3\,]\ = 2.418\ \text{bits}$$

iv. The code efficiency η, Eq.(10.46), is given by

$$\eta\ =\ \frac{H(m)}{L}\ =\ \frac{2.418}{2.48} = 0.976$$

v. The redundancy γ, Eq.(10. 47), is given by

$$\gamma\ =\ 1 -\ \eta\ =\ 1 - 0.976\ =\ 0.024$$

vi. The fixed code length if each message is represented by 3 fixed bits is given by

$$L_f\ =\ 0.08 \times 3 + 0.1 \times 3 + 0.12 \times 3 + 0.15 \times 3 + 0.25 \times 3 + 0.3 \times 3 = 3$$

$$\text{Compression ratio of this code}\ \ =\ \frac{L_f}{L}\ =\ \frac{3}{2.45} = 1.22$$

10.12.4. Run Length Encoding

Run Length Encoding RLE is a lossless data compression algorithm, sometimes called recurrence coding, is one of the simplest data compression algorithms. It is effective for data sets that are comprised of long sequences of a single repeated character and based on the idea of encoding a consecutive occurrence of the same symbol. This is achieved by replacing a series of repeated symbols with a count and the symbol. Suppose the following string of data (16 bytes) has to be compressed:

ABBBBBBBBBCDEEEEF

, using RLE compression, the compressed file takes up 10 bytes and could look like this:

A*8BCD*4EF

, the repetitive strings of data are replaced by a control character * followed by the number of repeated characters and the repetitive character itself. The control character is not fixed, it can differ from implementation to implementation. If the control character itself appears in the file, then one extra character is coded. Also RLE encoding is only effective if there are sequences of 4 or more repeating characters because three characters are used to conduct RLE. So coding two repeating characters would even lead to an increase in the file size [8,46].

There are many different run-length encoding schemes. Sometimes the implementation of the RLE is adapted to the type of data that are being compressed. Therefore, so another RLE technique for encoding long binary bit strings containing mostly zeros, where each k-bit symbol tells how many 0 bits occurred between consecutive 1 bits, to handle this long 0 runs, the symbol consisting of all 1 bits means that the true distance is $(2^k - 1)$ plus the value of the following symbol (or symbols).

RLE algorithm is very easy to implement and does not require much CPU horsepower. RLE compression is only efficient with files that contain lots of repetitive data such as in text files, if they contain lots of spaces for indenting and line-art images that contain large white or black areas are more suitable. Computer generated color image such as architectural drawings, can also give fair compression ratios. RLE compression can be used in the Tagged Image File Format TIFF and PDF file format.

An example of RLE, consider a screen containing plain black text on a solid white background. There will be many long runs of white pixels in the blank space, and many short runs of pixels within the text. Let us take a hypothetical single scene line with B representing a black pixel and W representing white pixels:

WWWWWWWWWWWBWWWWWWWWWWWWWBBBWWWWWWWWWWWWWWWWWWWWWW
WWWWWWWBWWWWWWWWWWWWWWW

, the RLE data compression algorithm to this hypothetical scan line will be

11W1B12W3B24W1B14W

, where the RLE represents the original 66 characters in only 18 characters. RLE is well suited to palette based iconic images. The Joint Photographic Experts Group JPEG for compression of still images, uses the RLE effectively on the coefficients that remain after transforming and quantizing. RLE is used in Fax machines (combined with other techniques into Modified Huffman coding). It is relatively efficient because most faxed documents are mostly white space, with occasional interruptions of black. Data that have long sequential runs of bytes (such as lower quality sound sample) can be RLE compressed after applying a predictive filter such as delta encoding (section 10.12.7).

Problem 10.14

A Run Length Encoding RLE, encodes long binary bit strings containing mostly zeros, based on each k-bit symbol tells how many 0 bits occurred between consecutive 1 bits. Encode the following bit string:
0001000001000000100000000000001000001000100000001101000000101
, using 3-bit symbols and state how much saving percentage of bits ?

Solution

The bit string consists of runs of length 3, 5, 6, 14, 6, 3, 7, 0, 1, 5, and 1. It would be Run Length Encoded using 3-bit symbols as

011101110111111000110011111000000001101001

, and for a saving of bits equals 34% .

10.2.5. Lempel-Ziv-Wekh Encoding

In the computer data, there is no control over the data statistics and the data to be stored could be instructions, images, tables, text files and so on. Each having their own code value distributions. In this case, a coder relying on fixed source statistics will be completely inadequate. Lempel-Ziv-Wekh LZW codes was originally developed for text compression by building up a conversion table between frequent long source data strings and short transmitted data codes at both coder and decoder. When a sequence of symbols matches a sequence stored in the dictionary, an index is sent rather than the symbol sequence itself. If no match is found, the sequence of symbols is sent without being coded and the dictionary is updated. The image is encoded by processing its pixels in a left to right and top to bottom manner and each successive gray level value is concatenated with a variable [46].

At the start of the LZW coding process, a codebook (dictionary) containing the source symbols to be coded, is constructed where for 8-bit monochrome images. The first 256 words " 2^8 " of the dictionary are assigned to the gray value 0, 1, 2, 3, ... , 255. The encoder sequentially examines the image's pixels, where gray-level sequences that are not in the dictionary are placed in algorithmically determined (text unused) locations. As an example, consider the following 4×4, 8-bit image:

$$
\begin{array}{cccc}
39 & 39 & 126 & 126 \\
39 & 39 & 126 & 126 \\
39 & 39 & 126 & 126 \\
39 & 39 & 126 & 126 \\
\end{array}
$$

, and 512 word dictionary is assumed, Table 10.2. The image is encoded by processing its pixels in a left-to-right, and top-to-bottom manner. Each successive gray level value is concatenated with a variable-column of Table 10.2, called the currently recognized sequence.

The advantages of the Lempel-Ziv-Wekh coder over the other coding are: it requires no prior information about the input data stream, it can comprises the input stream in one single pass , simplicity and allowing fast execution , and it can consume less power and use smaller silicon size than the other coding such as the Huffman coding.

Table 10.2. Dictionary code word LZW encoding

Dictionary entry	Dictionary code word
0	0
1	1
2	2
...	...
39	39
...	...
126	126
...	...
555	555
39 39	556
39 126	777
126 126	558
39 39 126	559
126 126 39	260
126 39 39	261
39 126 126	262
39 39 126 126	363
... ...	...
... ...	511

10.12.6. Predictive Coding

Predictive data reduction (low bit rate) has many principles in common with digital modulation techniques and the two predictors in the encoder and decoder are identical, Fig.10.30. At the encoder, the previous sample (predictor output) is subtracted from the input data (actual next sample) to produce a residual (prediction error) which is a data reduction stored or transmitted. At the decoder, the previous sample (predictor output) is added to the input prediction error (residual), to recreate the original value of the sample, which also becomes the next predictor input in the decoder [8,10].

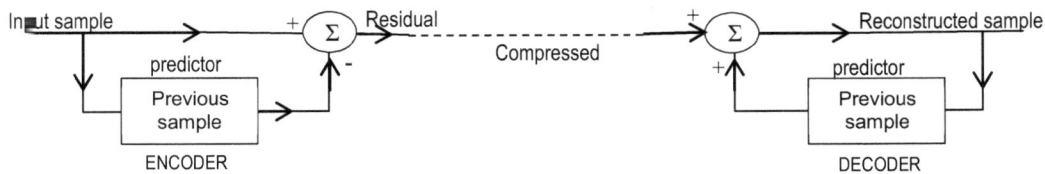

Fig.10.30. Block diagram of encoder and decoder using simple predictor.

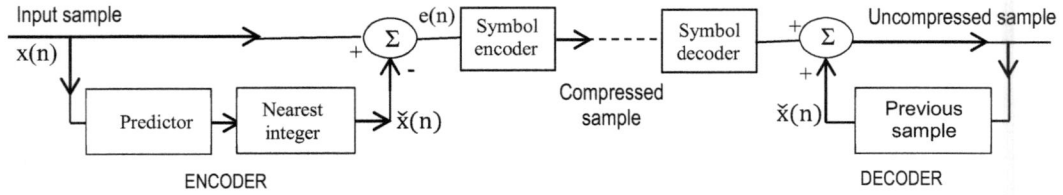

Fig.10.31. Lossless predictive coding model, coder and decoder.

Fig.10.31. shows a lossless predictive coding of an image data compression, where the compressed image from the encoder and decompressed image from the decoder. If $x(n)$ is the actual pixel value and $\check{x}(n)$ is the predicted value of that pixel. Based on eliminating the inter-pixel redundancy in an image, the new information is defined as the difference between the actual $x(n)$ and the predicted value $\check{x}(n)$ of that pixel where in the encoder (compressor), the compressed output (error) is given by

$$e(n) = x(n) - \check{x}(n)$$

, and in the decoder (expander), the decompressed output (reconstructed sample) is given by

$$x(n) = e(n) + \check{x}(n)$$

, also in the decoder, the decompressed output becomes the next predictor input and the most general form is given by

$$\check{x}(n) = \sum_{i=1}^{m} \alpha_i \, x(n-i)$$

, where the most simple form of $\check{x}(n)$ is given by $x(n-1)$.

The most common choice in optimization of parameters α_i, is the root mean square criterion which is also called the autocorrelation criterion. In this method, the expected value of the squared error $E[e^2(n)]$, is minimized, which yields the following equation

$$\sum_{i=1}^{m} \alpha_i \, R(i-j) = -\, R(j) \qquad , \quad 1 \le j \le m$$

, where $R(i)$ is the autocorrelation function of the signal $x(n)$ defined as

$$R(i) = E\{ \, x(n) \, x(n-i) \, \}$$

, and E is the expected value.

Another more general approach is to minimize the sum of the squares of the errors defined in the form

$$e(n) = x(n) - \check{x}(n) = x(n) - \sum_{i=1}^{m} \alpha_i \, x(n-i)$$

, or equivalently

$$e(n) = -\sum_{i=0}^{m} \alpha_i \, x(n-i)$$

, where the optimization problem searching over all α_i must now be constrained with $\alpha_0 = -1$.

In the digital communication, the input data of the encoder may be Differential Pulse Code Modulation DPCM or sigma delta modulation, where the lossy predictive coding. Also, the Linear predictive coding LPC is a tool used mostly in audio signal processing and speech processing for representing the spectral envelope of a digital signal of speech in a compressed form, using the information of a linear predictive model. LPC is one of the most powerful speech analysis techniques, and one of the most useful methods for encoding good quality speech at a low bit rate and provides extremely accurate estimates of speech parameters. Predictive coding is the most powerful and cost-effective review technology on the market, enabling attomeys to instantly find and automatically code documents [46].

10.12.7. Delta Encoding (Delta Compression)
Delta Encoding is a way of storing or transmitting data in the form of differences between sequential data rather than complete files, more generally, this is known as data differencing. Delta encoding is sometimes called delta compression, particularly where archival histories of changes are required such as in software projects. The differences are recorded in discrete files called "deltas" or "diffs". The Greek letter "Δ" is used to denote the change in a variable where the term delta encoding refers.

If the original data stream is   17 19 24 21 15 10 89 85 86 86 85 84 85 83 90 87 86 .. .. ..
The delta encoded will be   17 2 5 -3 -6 -5 79 -4 1 0 -1 -1 1 -2 7 -3 -1 .. .. ..

The first value in the encoded file is the same as the first value in the original file. Thereafter, each sample in the encoded file is the difference between the current and the last sample in the original file.

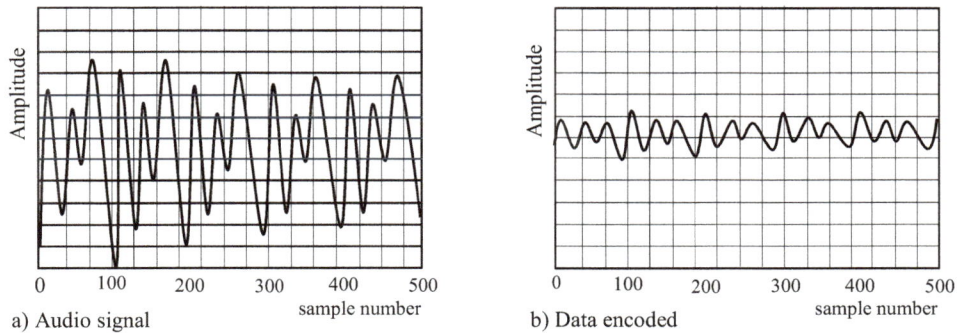

a) Audio signal   b) Data encoded

Fig.10.32. a) Audio signal digitized to 8 bits, b) The delta encoded version.

Delta encoding can be used for data compression when the values in the original data are smooth, that is, there is typically only a small change between adjacent values. This is not the case for ASCII text and executable code; however, it is common when the file represents a signal. Fig.6.32a. shows a segment of an audio signal, digitized to 8 bits, with each sample between -127 and 127. Fig.10.32b. shows the data encoded version of this signal. The key feature is that the delta encoded signal has a lower amplitude than the original signal. If the original signal is not changing, or is changing in a straight line, delta encoding will result in runs of samples having the same value. Correspondingly, delta encoding followed by Huffman, or followed by run-length encoding, is a common strategy for compressing signals. One drawback of delta encoding is that, to know the value of the signal at any point in time, one has to know all the previous values [8,10,46].

The idea used in delta encoding can be expanded into the more complicated technique of the Linear Predictive Coding LPC. To understand LPC, imagine that the first 99 samples from the input signal have been encoded, and we are about to work on sample number 100. We then ask ourselves: based on the first 99 samples, what is the most likely value for sample 100 ?. In delta encoding, the answer is that the most likely value for sample 100 is the same as the previous value, sample 99. This expected value is used as a reference to encode sample 100. That is, the difference between the sample and the expectation is placed in the encoded file. LPC expands on this by making a better guess at what the most probable value is. This is done by looking at the last several samples, rather than just the last sample. The algorithms used by LPC are similar to recursive filters, making use of the z-transform and other intensively mathematical techniques.

10.12.8. Drawbacks of Compression Techniques

Since the compression technique reduces the redundancy and may remove it from signals. The redundancy is essential for making data resistant to errors. It is true that compressed data are more sensitive to errors than the uncompressed data. Thus the transmission systems using compressed data must incorporate more powerful error correction strategies and avoid compression techniques which are notoriously sensitive. The compression technique using LZW table codes are very sensitive to bit errors in the transmission of a table value results in bit errors every time that table location is accessed. This is known as error propagation. The compression technique using variable length codes are also sensitive to bit errors such as Huffman code. The perceptive (lossy) coders introduce noise, where in cascaded system, the second codec could be confused by the noise due to the first coder. Another practical drawback of the compression systems, is that they are largely generic in structure and the same hardware can be operated at a variety of compression ratios. Clearly the higher the compression ratio the cheaper the system will be to operate so there will be economic pressure to use high compression ratios [46].

10.12.9. Audio Compression

Audio compression has become well entrenched in consumer and professional digital audio products such as the Compact Disc CD, Digital Audio Tap DAT, the Mini-Disc MD, the Digital Compact Cassette DCC, Digital Versatile Disc DVD, Digital Audio Broadcasting DAB, and Moving Picture Experts Group MPEG audio layer 3 (MP3) distribution on the internet. In addition, speech compression for telephony, and in particular, cellular telephone, required to preserve bandwidth and enable the battery life has spawned a large number of speech compression standards. Different algorithms are applied to speech signals and to signals of wider bandwidth consumer entertainment. Audio and speech compression schemes can be conveniently partitioned into applications reflecting some measure of acceptable quality, Table 10.3, [48,49].

Table 10.3. Typical parameters values for three classes of audio signals

	Frequency Range	Sampling Rate	PCM bits/ Sample	PCM Bit-rate
Telephone Speech	300-3400 Hz	8 kHz	8	64 kb/s
Wideband Speech	60-7000 Hz	16 kHz	14	224 kb/s
Wideband Audio	10-20000 Hz	48 kHz	16	768 kb/s

In audio compression, telephony speech processing, the Adaptive Differential Pulse Code Modulation ADPCM codec, encodes sample by sample, predicting the value of each sample from the reconstructed speech of previous samples using an adaptive feedback predictor. It accepts toll-quality 8-bit linear, A-law, or μ-law input sampled speech at 64 Kbits/sec, and it outputs a compressed speech at rates of 16, 24, 32, and 40 Kbits/sec. The encoder uses a decoder in its feedback path to analyze and modify algorithm parameters in order to minimize the reconstruction errors. While in audio compression, wideband speech coding, a compression of significant improvement in quality over that of telephone quality speech and is closer to broadcast quality speech and music signals [8].

On the other hand, speech coders using linear predictive filters (Codebook Excited Linear Predictive Coding CELP) can provide high quality encoded speech at rates above 16 Kbits/sec, but they degrade quickly at lower rates. The Linear Predictive Coding LPC, encoders can be modified to obtain high quality speech compression at rates on the order of 4.8 to 9.6 Kbits/sec. Linear Predictive Coding LPC techniques are used in the mobile systems designed to the American Interim Standard IS-95.

10.12.9.1. MPEG Audio Layers I, II, III

The International Standards Organization ISO and Moving Picture Experts Group MPEG audio coding standard describes audio compression for synchronized audio to accompany the compressed video known MPEG. It combines features of Masking pattern adapted Universal Sub-band Integrated Coding And Multiplexing MUSICAM, and Adaptive Spectral Perceptual Entropy Coding ASPEC. It consists of three layers (codes) of increasing complexity and improving subjective performance. It operates with input sampling rates of 32, 44.1, and 48 KHz, and it outputs bit rates per monophonic channel between 32 and 192 Kbits/sec, or per stereophonic channel between 64 and 384 Kbits/sec. The standard supports single channel mode, stereo mode, dual channel mode (for bilingual audio programs), and an optional joint stereo mode.

MPEG-1 audio offers three compatible layers. The objective of these MPEG layers, to obtain a good tradeoff between quality and bit-rate. Each succeeding layer is able to understand the lower layers and offering more complexity in the psychoacoustic model and better compression for a given level of audio quality and with increased compression effectiveness, accompanied by extra delay. Layer I quality can be quite good provided a comparatively high bit-rate is available. Digital Audio Tape DAT typically uses layer I at around 192 Kbits/sec. Layer II has more complexity, and was proposed for use in Digital Audio Broadcasting DAB. Layer III (MP3) is most complex, and was originally aimed at audio transmission over Integrated Services Digital Network ISDN lines. Most of the complexity increase, is at the encoder, not at the decoder, Fig.10.33.

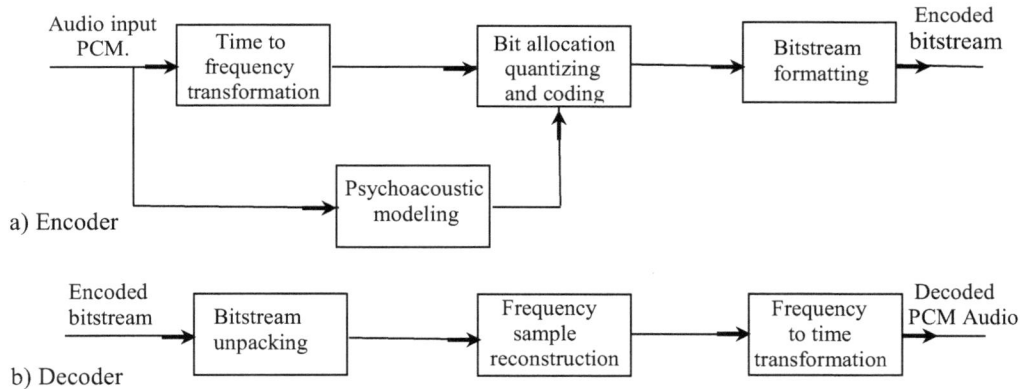

Fig.10.33. MPEG audio: a) Encoder, b) Decoder.

In MPEG layer III (MP3) standard, a higher frequency resolution is achieved that closely matches critical frequency-resolution bandwidths of the human auditory process. This enhanced partition is achieved by using Modified Discrete Cosine Transform MDCT [8].

10.12.10. Video Compression

The television converts the three dimensional moving image into a series of still pictures, taken at the same frame rate. The two dimensional image is scanned as a series of lines to produce a single voltage varying with time which can be recorded or transmitted. Europe, the Middle East and the Russian use the scanning standard of 625 lines/50 Hz, whereas the USA and Japan use 525 lines/60 Hz. Fig.10.34. shows some of the basic types of analog color video. Each of these types can exist in a variety of line standard. Since the practical color cameras generally have three separate sensors, one for each primary color, the Red-Green-Blue RGB system will exist at some stage inside the camera. RGB system consists of three parallel signals, each having the same spectrum, and is used where the highest accuracy is needed, often for production of still pictures, such as the Computer Aided Design CAD displays [46].

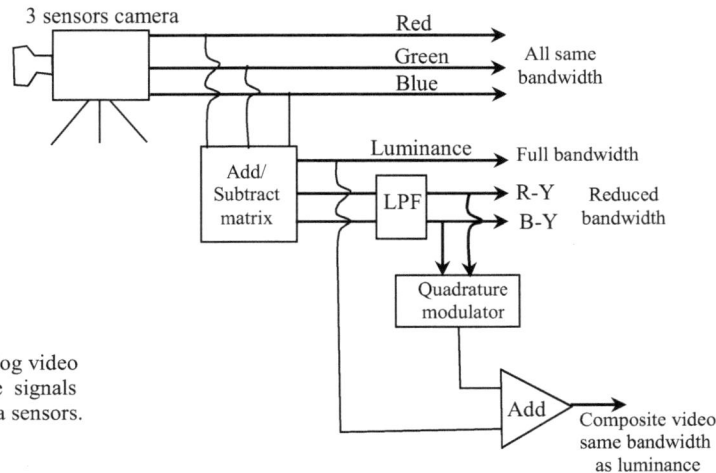

Fig.10.34.
The major types of analog video , Red, Green, and Blue signals emerge from the camera sensors.

Some saving of bandwidth can be obtained by using color difference technique. The human eye relies on brightness to convey detail, and much less resolution is needed in the color information. R, G, and B, are matrixed together to form a luminance (and monochrome compatible) signal Y which has full bandwidth. The matrix also produces two color difference signals, R-Y and B-Y, but these do not need the same bandwidth as Y, one half or one quarter will suffice, depending on the application. Color difference signals represent an early application of perceptive (lossy) coding: a saving in bandwidth is obtained by expressing the signals according to the way the eye operates.

For the color television broadcast in a single channel, the PAL, SECAM and NTSC systems interleave into the spectrum of a monochrome signal, a subcarrier which carries two color difference signals of restricted bandwidth. As the bandwidth required for composite video is no greater than that of luminance, it can be regarded as a form of compression performed in the analog domain. The subcarrier is intended to be invisible on the screen of a monochrome television set. A subcarrier based color system is generally referred to as composite video, the modulated subcarrier is called chroma.

In the image signal processing, technology has presented us with low-cost, high resolution color printers, scanners, cameras, and monitors, enabling us to capture and present images for commerce and entertainment. The storage and transport of these images rely quite heavily on source coding to reduce the demands on bandwidth and memory [50,51]. An example, selecting 8.5 by 11.0 inch sheet of paper containing an image resolved at 300 pixels per inch. Then the image contains $8.5\times300\times11.0\times300$, or 8.4×10^6 picture elements. If the picture is full color with three colors per element, each described by 8-bits words, the image contains 2×10^8 bits which is equivalent to 4.6×10^6 6-character ASCII words. For comparison with other image formats, the single frame of a high definition television image contains approximately 1.8×10^6 pixels, the standard television image contains approximately 0.33×10^6 pixels, and the high-end computer monitor contain from 1.2 to 3.1×10^6 picture elements [8]. Two primary compression schemes namely: Joint Photographic Experts Group JPEG and Moving Picture Experts Group MPEG.

10.12.10.1. Joint Photographic Experts Group JPEG

JPEG is the common name given for the digital compression of continuous-tone still images. The JPEG standard is introduced in the early of 1990`s to encode and store the still images. JPEG is primarily known as a transform based lossy compression scheme which permits errors in the signal construction. The levels of the errors are restricted to be below the perception threshold of a human observer. JPEG supports three modes of operation related to the Discrete Cosine Transform DCT: sequential DCT, progressive DCT, and Hierarchical DCT, as well as a lossless mode using differential prediction and entropy coding of the prediction error. DCT is a numerical transform related to the DFT for obtaining the spectral decomposition of even symmetric sequences. When the input sequence is even symmetric, there is no need for the sine components of the transform, hence the DCT can replace the FFT [8].

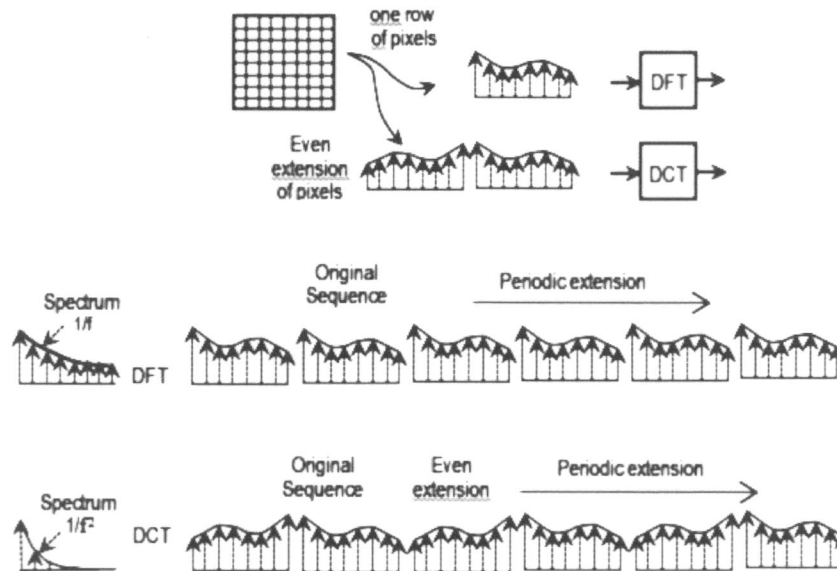

Fig.10.35. Spectral decay and periodic extension of time series by DFT and DCT.

An example, consider the two dimensional 8 by 8 DCT, Fig.10.35. This DCT is a separable transform that can be written as a double sum over the two dimensions. The separable DCT performs eight point DCT's in each direction, hence the basic building block is a single eight point DCT. The first question to address is, why use a DCT rather than DFT ?. The answer is related to the interaction between the sampling theorem and Fourier transforms. Sampling in one domain induces periodicity in the other domain. When samples in a time series, its spectrum becomes periodic. On the other hand, when sampling a spectrum of a time series, the time series is periodically extended. This resulting periodic extension is labeled by periodogram. For the DFT, the periodic extension of the original data, Fig.10.35 exhibits discontinuities at the boundaries that limit the rate of spectral decay in the spectrum to 1/f. But for the DCT, the even extension of the data by reflecting the data about one of its boundaries when this data is periodically extended, Fig.10.35, and the discontinuities no longer reside in the amplitude of the data but rather in its first derivative, so that the rate of spectral decay increases to $1/f^2$. The faster rate of the spectral decay means that there will be a reduced number of significant spectral terms. A second advantage to the DCT is that since the data is even symmetric, its transform is also real and symmetric and hence has no need for the odd symmetric basis terms, the sine functions, then all the discrete cosine transform multiplication are real, this lowers the number of the required multiplications, as compared to the discrete Fourier transform [10]. On the other hand, the drawback of the DCT is the blocking effect which made it impossible to follow the ridge lines after reconstruction, where in the personal Identification ID fingerprint compression standard, the DCT did not perform well at high compression ratios because it produced severe blocking effects which made it impossible to follow the ridge lines in the reconstruction of the fingerprints [22]. This drawback disappear in the wavelet transform due to its property of retaining the details present in the data, where JPEG 2000.

Since the amplitude of an image is highly correlated over short spatial intervals, the DCT of a 8 by 8 block of pixels is dominated by the dc terms and further contains relatively few significant terms. A typical set of amplitudes and their DCT is suggested in Fig.10.36. Note that the spectral terms fall at least as fast as $1/f^2$, Fig.10.36, and that most of the high frequency terms are essentially zeros. The spectrum is passed to a quantizer that uses standard quantization tables to assign bits to the spectral terms in proportion to their amplitude, as well as their psycho-visual contribution. Different quantization tables are used for the luminance and chrominance components [8].

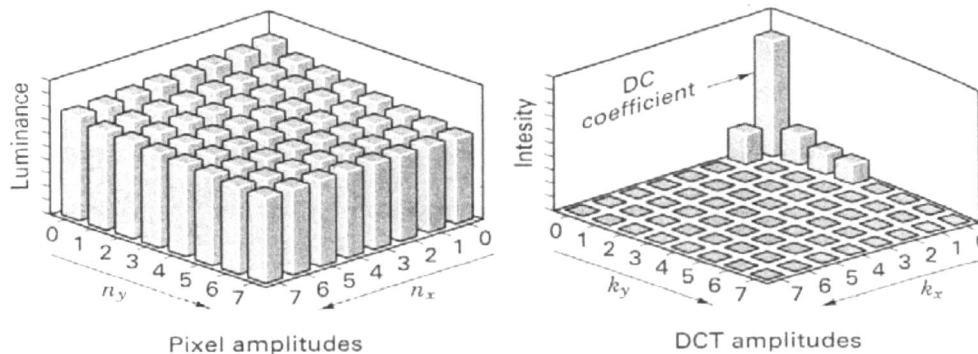

Pixel amplitudes   DCT amplitudes

Fig.10.36. Pixel and DCT amplitudes describing same 8 by 8 block of pixels.

To take the advantage of the large number of zero-valued entries in the quantized DCT, the spectral addresses of the DCT are scanned in a zigzag pattern, Fig.10.37, the zigzag pattern assures a long run of zeros. This improves the coding efficiency of the run length Huffman code describing the spectral samples.

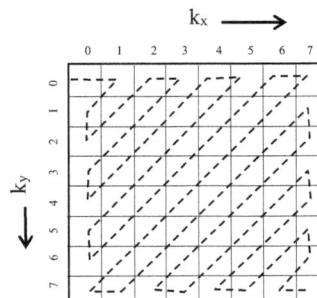

Fig.10.37. Zigzag scan pattern for DCT spectral terms.

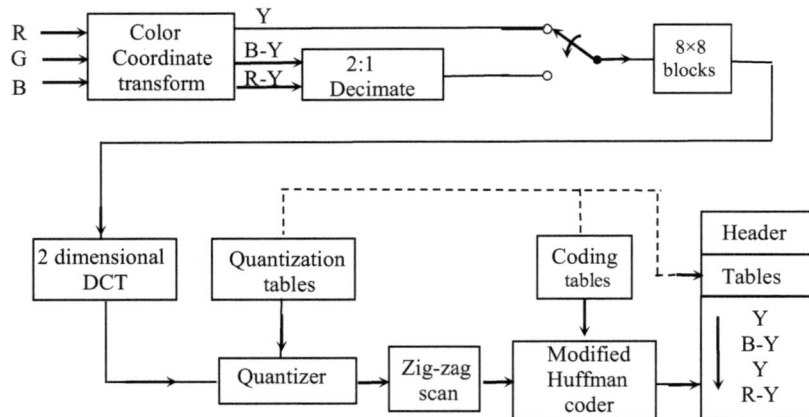

Fig.10.38. Block diagram of JPEG encoder.

Fig.10.38 presents the block diagram of a JPEG encoder. The signal delivered to the encoder is normally presented as raster scanned and sampled primary additive colors, Red-Green-Blue RGB. The color planes are transformed to luminance Y and chrominance components $0.564 \times (B-Y)$, and $0.713 \times (R-Y)$, using a variation of the color difference transformation developed for color TV. This matrixed mapping is given by

$$\begin{vmatrix} Y \\ B-Y \\ R-Y \end{vmatrix} = \begin{vmatrix} 0.299 & 0.582 & 0.114 \\ -0.169 & -0.331 & 0.500 \\ 0.500 & -0.419 & -0.081 \end{vmatrix} \begin{vmatrix} R \\ G \\ B \end{vmatrix}$$

, where the Y component is formed to reflect the human eye sensitivity to the primary colors. Both analog and digital color TV takes the advantage of the acuity difference to deliver the additional color components in a significantly reduced bandwidth. NTSC delivers all the three colors in a 0.5 MHz bandwidth, rather than the 4.2 MHz bandwidth required for the luminance component. JPEG similarly takes advantage of the color acuity difference and down samples the color difference components 2 to 1 in the scan x-direction, but not in the successive line y-direction. The luminance and down sampled color difference signals are sequentially presented as 8 by 8 blocks to the 2 dimensional DCT, Fig.10.38. The outputs of the DCT are quantized by the appropriate table and then zigzag scanned for presentation to the Huffman coder. JPEG uses the Huffman coder to encode the ac coefficients of the 2 dimensional DCT, but since the dc components are highly correlated over adjacent blocks, uses differential encoding for the dc component. The decoder, of course, reverses these operations to form an image [46].

During JPEG image reconstruction, the decoder can operate sequentially, starting in the upper left corner of the image and forming 8 by 8 pixel blocks as they arrive. This is the sequential mode of JPEG. But in the progressive coding mode, the image is first assembled in 8 by 8 blocks formed by only the dc component in each block. This is a very quick progress that presents a coarse but recognizable preview image, a process often seen on the internet when down loading the Graphic Interchange Format GIF files which has delivered the only dc components at the beginning of the data transfer. The image is then updated with each 8 by 8 block formed from the dc component and the first two dc components which are the next set of data delivered to the decoder. Finally, the image is updated at full resolution formed by the full set of coefficients associated with each 8 by 8 block.

While in the hierarchical coding, the image is encoded and decoded as overlapped frames. A low resolution, down sampled 4-to-1 in each direction, image is encoded using the DCT and quantized coefficient processing to form the first frame. The image formed by this frame is up-sampled and compared with a higher resolution, down sampled (2 to 1 in each direction) version of the original image and the difference, representing an error in the formation of the image, is again coded as an MPEG image. The two frames formed by the two layers of coding are used to form a composite image that is up-sampled and compared with the original image. The difference between the original image and the two lower level resolution reconstructions, is formed at the highest resolution and JPEG-encoded once again. This process is useful for delivering images with successively higher quality reconstruction in a way similar to the progressive coding [8].

On the other hand, JPEG-2000 is a JPEG initiative to define a new image coding system, addressing internet and mobile applications. This system offers low bandwidth, multiple resolution, error resilience, image security, and low complexity. It is based on wavelet compression algorithms, and relative to JPEG, it offers improved compression efficiency with multiple resolution capabilities [52]. JPEG-2000, in addition to being a great way to efficiently store highest quality images, also it has a great many benefits to offer web graphics, especially full color images with transparent backgrounds.

10.12.10.2. Joint Photographic Experts Group JPEG-2000

JPEG-2000 is an emerging standard for still image compression, where digital imagery becomes more common place and of higher quality. Image compression must not only reduce the necessary storage and bandwidth requirements, but also allow extraction for editing, processing, and targeting particular devices and applications. JPEG-2000 image compression system has a rate distortion advantage over the original JPEG. JPEG-2000 also allows extraction of different resolutions, pixel fidelities, regions of interest, components, and more, all from a single compression bitstream, this allows an application to manipulate or transmit only the essential information for any target device from the JPEG-2000 compressed source image. JPEG-2000 has a long list of features, a subset of which are:

   i. State-of-the-art low bit rate compression performance.
   ii. Progressive transmission by quality, resolution, component, or spatial locality.
   iii. Lossy and lossless compression.
   iv. Random (spatial) access to the bitstream.
   v. Pan and zoom (with decompression of only a subset of the compressed data).
   vi. Compressed domain processing such as rotation and cropping.
   vii. Region of interest coding by progression.
  viii. Limited memory implementations.

JPEG-2000 project was motivated by Ricoh`s submission of the Compression with Reversible Embedded Wavelets CREW algorithm [29], and the Set Partitioning In Hierarchical Trees SPIHT algorithm [30], to an earlier standardization effort for the lossless and near-lossless compression, known as JPEG-LS. JPEG-2000 provides better rate distortion performance, for any given rate, than the original JPEG standard. While the original JPEG provided different methods of generating progressive bitstreams, with JPEG-2000 the progression is simply a matter of the order the compressed bytes are stored in a file. Furthermore, the progression can be changed, additional quantization can be done, or a server can respond only with the data desired by a client, all without decoding the bitstream.

There are four basic dimensions of progression in the JPEG-2000 bitstream: resolution, quality, spatial location, and component where different types of progression are achieved by ordering of packets within the bitstream. JPEG-2000 does not explicitly define a method of subsampling color components as the original JPEG does, where the original JPEG provides subsampling on color components as a means to reduce computational complexity, and because it provides quantization the human visual system is unlikely to notice.

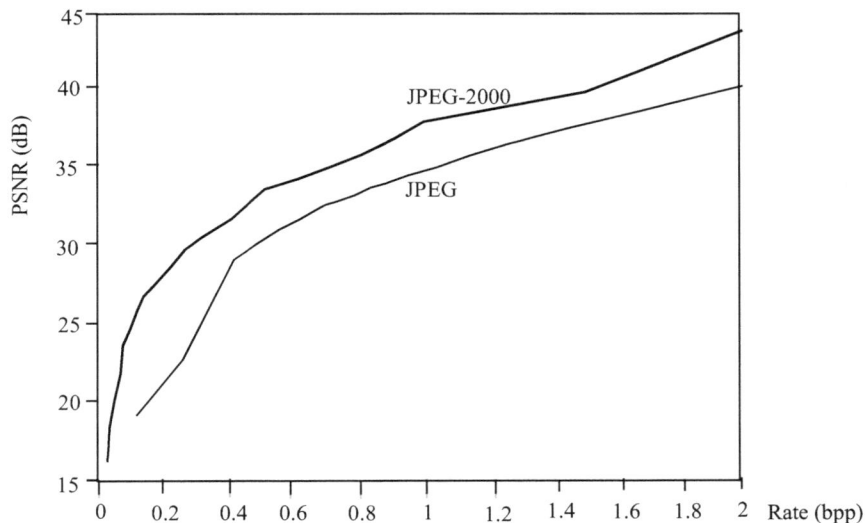

Fig.10.39. Rate-distortion performance for JPEG and JPEG-2000 on the SCID bike image.

Fig.10.39. provides rate-distortion performance for the original JPEG mode and JPEG-2000 mode for the bike image (grayscale, 2048 by 2560) from the Standard Color Image Data SCID test set. The original JPEG mode is Progressive Discrete Cosine Transform P-DCT with optimized Huffman tables while the JPEG-2000 is Progressive Discrete Wavelet Transform P-DWT. JPEG-2000 results are significantly better than the JPEG where JPEG-2000 provides only a few dB improvement. Although JPEG-2000 provides significantly lower distortion for the same bitrate, the computational complexity is significantly higher. The speed of the JPEG-2000 code should increase over time with implementation optimization, but the multi-pass bitplane context model and the arithmetic entropy coder will prevent any software implementation from reaching the speed JPEG obtains with the DCT and Huffman coder. JPEG-2000 also requires more memory than JPEG, but not as much as might be expected [52].

10.12.10.3. Moving Picture Experts Group MPEG

MPEG is a set of standards designed to support coding of moving pictures and associated audio, for digital storage media at up to 1.5 Mbits/sec. MPEG-1 is designed (year 1991) to permit full motion video recordings on Compact Disc CD players originally designed for stereo audio playback. MPEG-2 (year 1995) is generic coding of moving pictures and associated audio, addressed greater input-output format flexibility, data rates, and system considerations such as transport and synchronization, and topics neglected in MPEG-1. MPEG-2 supports variations of digital TV covering digitized video that matches existing analog formats with selectable quality through Digital Video Disc DVD, and High Definition Tele-Vision HDTV with different aspect ratio, line rate, pixel span, scan conversion options, and various re-sampling options for the color difference components. MPEG-4 is designed (year 1999) to permit full motion internet video. MPEG-7 (year 2002) is designed for object descriptions. MPEG-21 (year 2005) is designed for Digital Rights Management DRM.

10.12.10.4. Moving Picture Experts Group MPEG-2

Concerning the basic theory of operation of the simplest version of MPEG-2. MPEG compresses a sequence of moving images by taking advantage of the high correlation that exists between successive pictures of a moving picture. MPEG constructs the following three types of pictures:

i. The intra-pictures (I-frames) contains the all full image information (standard JPEG algorithm).
ii. The predicted-pictures (P-frames) contains the difference image information from past I or P frames (standard MPEG algorithm).
iii. The Bi-directionally predicted pictures (B-frames) contains the difference information from image on either side (from past and future I or P frames), (standard JPEG algorithm).

In MPEG, every M-th picture in a sequence can be fully compressed using a standard JPEG algorithm; these are the I-pictures. The process then compares successive I-pictures and identifies portions of the image that have moved. The image sections that didn`t move are carried forward in time to intermediate pictures by the decoder memory. The process then selects a subset of intermediate pictures, then predicts (via linear interpolation between I-pictures), and corrects the location of the image sections that are found to have moved. These predicted and corrected images are P-pictures. The pictures between the I-pictures and the P-pictures are the B-pictures that incorporate the stationary image sections uncovered by moving sections. The relative position of these pictures is shown in Fig.10.40. Note that P and B-pictures are permitted but not required, and their quantity is each variable. A sequence can be formed without any P and B-pictures, but a sequence containing only P or B-pictures cannot exist [8].

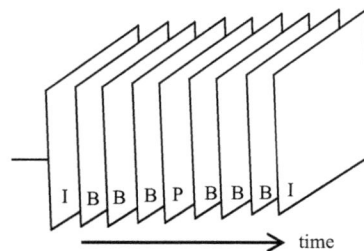

Fig.10.40. Sequence of pictures in MPEG compression.

The I-pictures are compressed as if they were JPEG images. This compression is applied to four contiguous 8 by 8 blocks called a macro block. The macro blocks can be down sampled for subsequent compression of the chrominance components. The macro blocks and their down sampling options are shown in Fig.10.41. Compression of the I-frame is performed without reference to any earlier or later pictures in the frame sequence. The distance in sequence count between I-pictures is adjustable, and it can be as small as I so that the I-pictures are adjacent, or as large as reconstruction memory would allow. Editing cuts in a sequence of pictures and local program insertion can only occur at I-pictures. Since one-half second is acceptable time accuracy for executing such editing, then the distance between I-pictures is usually limited to approximately 15 image pictures for the NTSC standard of 30 pictures per second, or 12 image pictures for the British standard Phase Alternating Lines PAL of 25 pictures per second [8].

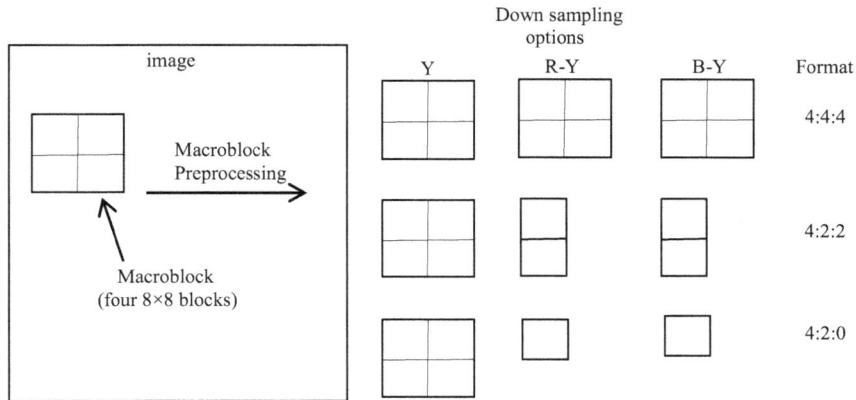

Fig.10.41. Preprocessing of macroblock for chrominance down sampling.

The first processing step performed by MPEG is the task of determining which macro blocks have moved between I-pictures. This is determined by carrying each macro block from one I-frame forward to the next and performing a two dimensional cross-correlation in the neighborhood of its original location. Motion vectors are determined for each macro block that identifies the direction and the amount of movement of each shifted macro block. Macro blocks that have not shifted are stationary in the image pictures between the I-pictures and can be carried forward in the intermediate image pictures.

The next processing step in MPEG is to form the P-frame between the I-pictures. Assuming that the shifted macro blocks have moved linearly in time between the two positions identified in the first processing step. Each macro block is placed at its predicted location on the P-frame and is cross-correlated in its neighborhood to determine the true location of the macro block in the P-frame. The difference between the predicted and true position of the macro block is an error in the prediction and this error is compressed using the DCT and is used to correct the P-frame. The same information is forwarded to the decoder so that it can correct its predictions. Fig.10.42. presents the macro block shift between I-pictures and an intermediate P-picture [8].

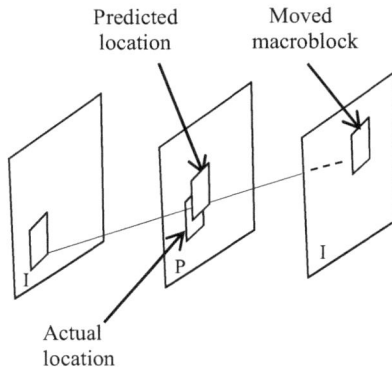

Fig.10.42. Macro block movement between I-pictures and P-picture.

B-pictures are image pictures located between I-pictures and P-pictures. In these pictures the motion vectors move the shifted macro blocks linearly through time to their bi-directional-interpolated positions at each successive B-frame in the sequence. I-pictures require the maximum amount of data to describe their DCT compressed contents. P-pictures require less data, having only to describe the error pixels in the motion predicted macro blocks in the frame. The rest of the pixels in the frame are carried forward in memory from the previous I-frame. B-pictures are the most efficient pictures in the set. They only have to linearly shift and correct the pixels newly covered and uncovered due to the motion of macro blocks on that image frame [8].

Reconstruction of the images at the decoder requires that the sequence of images be delivered in the order necessary for appropriate processing. For instance, since the computation of B-pictures require information from the I and P or between the P-and P-pictures on either side, the I and P pictures must be delivered first. The following is an example of the required ordering of images at the input and output of the encoder and decoder [8].

Picture Order at Encoder Input

1	2	3	4	5	6	7	8	9	10	11	12	13
I_n	B_1	B_2	P_1	B_3	B_4	P_2	B_5	B_6	I_{n+1}	B_1	B_2	P_1

Encoded Picture Order at Encoder Output and Decoder Input

1	2	3	4	5	6	7	8	9	10	11	12	13
I_n	P_1	B_1	B_2	P_1	B_3	B_4	I_{n-1}	B_5	B_6	P_1	B_1	B_2

Picture Order at Decoder Output

1	2	3	4	5	6	7	8	9	10	11	12	13
I_n	B_1	B_2	P_1	B_3	B_4	P_2	B_5	B_6	I_{n+1}	B_1	B_2	P_1

10.12.11. Principles Behind Compression

The uncompressed multimedia (graphic, audio, video) data requires considerable storage capacity and transmission bandwidth despite the rapid progress in mass storage density, processor speeds and digital communication system performance, demand for data storage capacity and data transmission bandwidth continues to outstrip the capabilities of available technologies. The recent growth of data intensive multimedia-based web applications have not only the need for more efficient ways to encode signals and images but have made compression of such signals central to storage and communication technology.

Fig.10.43. shows the MPEG channel/Multiplexing, where the video and audio signals are encoded while the data signal is formatted, then all are to be in the form of Elementary Stream ES, and they are packetized to be in the form of Packet Elementary Stream PES. While the tables are Programed Specific Info PSI. The three PES and the PSI are multiplexed to be in the form of Program Stream PS.

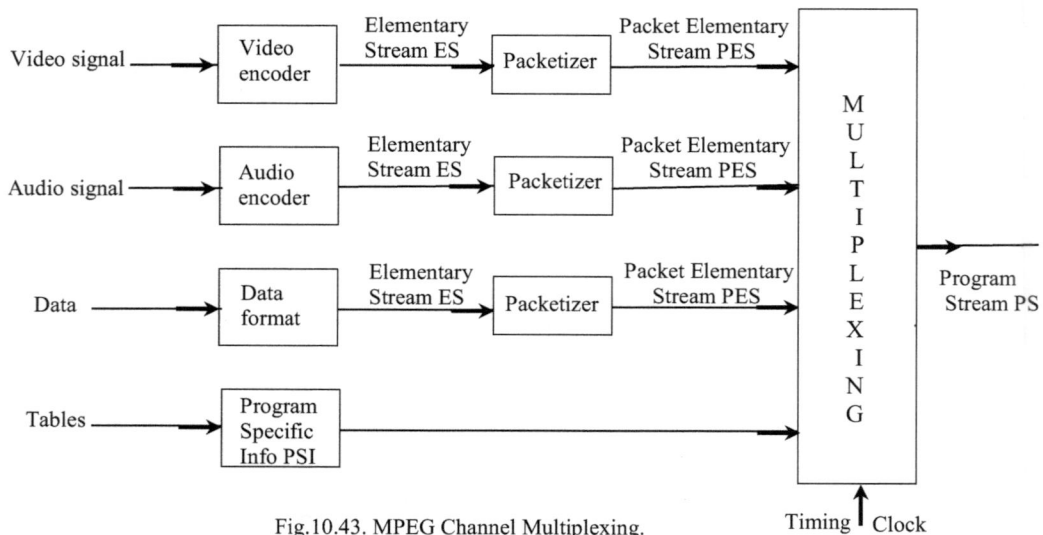

Fig.10.43. MPEG Channel Multiplexing.

Table 10.4. shows the multimedia data types and uncompressed storage space, transmission bandwidth and transmission time required. These information clearly illustrates the need for sufficient storage space, large transmission bandwidth and long transmission time for image, audio and video data. At the present state technology, the multimedia data is compressed before its storage and transmission, and decompress it at the receiver for playback. For example, for a compression ratio of 32:1 the space, bandwidth and the transmission time requirements can be reduced by a factor 32, with acceptable quality [53-56].

Table 10.4. Multimedia Data

Multimedia Data	Size/duration	Bits/pixel or Bits/sample	Uncompressed Size (bytes)	Transmission Bandwidth (bits)	Transmission time 28.8 K modem
Page of Text	11``×8.5``	Varying Resolution	4-8 K bytes	32-64 K bits/page	1.1-2.2 secs
Telephone quality speech	10 seconds	8 bits/s	80 K bytes	64 K bits/sec	22.2 secs
Grayscale Image	512×512	8 bpp	262 K bytes	2.1 M bits/image	1 min 13 secs
Color Image	512×512	24 bpp	786 K bytes	6.29 M bits/image	3 min 39 secs
Medical Image	2048×2048	12 bpp	5.16 M bytes	41.3 M bits/image	23 min 54 secs

A common characteristic of most images is that the neighboring pixels are correlated and therefore contain redundant information. The foremost task then is to find less correlated representation of the image. Two fundamental components of compression are redundancy and irrelevancy reduction. Redundancy reduction aims at removing duplication from the signal source (image/video). Irrelevancy reduction omits parts of the signal that will not be noticed by the signal receiver namely Human Visual System HVS. In general, three types of redundancy can be identified: 1) Spatial redundancy or correlation between neighboring pixel, 2) Spectral redundancy or correlation between different color planes or spectral bands, and 3) Temporal redundancy or correlation between adjacent frames in a sequence of images (in video applications).

Image compression research aims at reducing the number of bits needed to represent an image by removing the spatial and spectral redundancies as much as possible. In many different fields, digitized images are replacing conventional analog images as photograph or x-rays. The volume of data required to describe such images greatly slow transmission and makes storage prohibitively costly. The information contained in images must, therefore, be compressed by extracting only visible elements, which are then encoded. The quantity of data involved is thus reduced substantially. The fundamental goal of image compression is to reduce the bit rate for transmission or storage while maintaining an acceptable fidelity or image quality.

One of the most successful applications of wavelet methods is the transform-based image compression (coding). The overlapping nature of the wavelet transform alleviates blocking artifacts, while the multimedia character of the wavelet decomposition leads to superior energy compaction and perceptual quality of the decompressed image. Furthermore, the multi-resolution transform domain means that wavelet compression methods degrade much more gracefully than the block-DCT methods as the compression ratio increases. Since a wavelet basis consists of functions with both short support (for high frequencies) and long support (for low frequencies), large smooth areas of an image may be represented with very few bits, and detail added where it is needed [53,57].

Wavelet-based coding [53] provides substantial improvements in picture quality at higher compression ratios. Over the past few years, a variety of powerful and sophisticated wavelet-based schemes for image compression, have been developed and implemented. Because of the many advantages, wavelet-based compression algorithms are the suitable candidates for the new JPEG-2000 standard [54]. Such a coder operates by transforming the data to remove redundancy, then quantizing the transform coefficients (a lossy step), and finally entropy coding the quantizer output. The loss of information is introduced by the quantization stage which intentionally rejects less relevant parts of the image information. Because of their superior energy compaction properties and correspondence with the human visual system, wavelet compression methods have produced superior objective and subjective results [55].

With wavelets, a compression rate of up to 1:300 is achievable. Wavelet compression allows the integration of various compression techniques into one algorithm. With lossless compression, the original image is recovered exactly after decompression. Unfortunately, with images of natural scenes, it is rarely possible to obtain error-free compression at a rate beyond 2:1. Much higher compression ratios can be obtained if some error, which is usually difficult to perceive, is allowed between the decompressed image and the original image. This is lossy compression. In many cases, it is not necessary or even desirable that there be error-free reproduction of the original image. In such a case, the small amount of error introduced by lossy compression may be acceptable. Lossy compression is also acceptable in fast transmission of still images over the internet [56].

10.12.12. Moving Picture Experts Group MPEG-4 International Standard (MP4)

Moving Picture Experts Group MPEG-4 is a file format that is commonly used to store media types defined by the ISO/IEC Moving Picture Experts Group, though it can store other media types as well. The files of this format usually have extension MP4. The start point design of the MP4 file format was Apple's QuickTime file format, having been improved in many ways. Today, MPEG-4 differs markedly from its predecessor. MPEG-4 allows each streaming over the internet, multiplexing of multiple video and audio streams in one file, variable frame-and bit-rates, subtitles and still images. MP4 is being quite often used as the alternative to MP3 on an apple iPod and iTunes, MP4 is still not as widely used in computer and hardware players as the MP3 in spite of the higher quality of the Advanced Audio Coding AAC codec.

The following kinds of data are recommended (for compatibility reasons) to be embedded in MPEG-4: a) Video: MPEG-4, MPEG-2, and MPEG-1, b) Audio: MPEG-4 AAC, MP3, MP2, MPEG-1 part 3, MPEG-2 part 3, Codebook Excited Linear Predictive CELP coding (speech), Transform-domain weighted interleave Vector Quantization TwinVQ (very low bitrates), and Structured Audio Orchestra Language SAOL describes a set of tools that will be the next standard for computer music, audio for gaming, streaming internet music/sound, and other multimedia applications, c) Pictures: JPEG and Portable Network Graphics PNG, d) Subtitles: MPEG-4 Timed Text, and/or xmt/bt text format (means that subtitles have to be translated into xmt/bt), and e) Systems: Allowing animation, interactivity and DVD-like menus. Also some file extensions used on the files that contain data in the PM4 format are: a) MP4: Official extension, for audio, video and advanced content files, b) M4A: For audio-only files, can safely be renamed to MP4, c) M4P: FairPlay protected files, d) MP4V and M4V: Video-only (sometimes also used for raw MPEG-4 video streams not in the MP4 container format), e) 3GP and 3G2: Using by 3G mobile phones, may also store content not specified directly in the MP4 specification (based on recommendation H.263).

10.12.13. Transform-domain weighted interleave Vector Quantization TwinVQ in MPEG-4

Transform-domain weighted interleave Vector Quantization TwinVQ is an audio compression technique developed by Nippon Telegraph and Telephone Corporation NTT Human Interface Laboratories, year 1994, (now Cyber Space Laboratories) [58-61]. The compression technique has been used in both standardized and proprietary designs. In the context of MPEG-4 audio (MPEG-4 part 3). TwinVQ is an audio codec optimized for audio coding at ultra-low bitrates around 8 Kbit/s, published as subpart of ISO/IEC (year 1999), MPEG-4 audio version 1 [62-66].

TwinVQ type is based on a general audio transform coding scheme integrated with the Advanced Audio Coding ACC frame work in the spectral flattening module and the weighted interleave vector quantization module. TwinVQ reportedly has high coding gain for low bit rate and potential robustness against channel errors and packet loss, since it does not use any variable length coding and adaptive bit allocation. Some commercialized products such as Metasound (Voxware products) [67-68], SoundVQ (Yamaha products) [69-71], and SolidAudio (Hagiwara products) are also based on TwinVQ technology but the configurations are different from MPEG-4 TwinVQ [63].

The proprietary audio compression format TwinVQ developed by Nippon Telegraph and Telephone Corporation NTT Human Interface Laboratories [72-73] and marketed bt Yamahaunder SoundVQ [70]. Also NTT offered a TwinVQ demonstration software for non-commercial purposes, NTT TwinVQ encoder and TwinVQ player codec AP1 and header file format [74-75]. TwinVQ uses Twin vector quantization. The proprietary TwinVQ codec supports constant bit rate encoding at 80, 96, 112, 128, 160 and 192 Kbit/s. It was claimed that TwinVQ files are about 30 to 35% smaller than MP3 files of adequate quality. For example, a 96 Kbit/s TwinVQ file allegedly has roughly the same quality as a 128 Kbit/s MP3 file. The higher quality is achieved at the cost of higher processor usage. Yamaha marketed TwinVQ as an alternative to MP3 but the format never became very popular. This could be attributed to the proprietary nature of the format-third party software was scarce and there was no hardware support. Also the encoding was extremely slow and there was not much music available in TwinVQ format. As other MP3 alternatives emerged TwinVQ quikly became obsolete. The proprietary version of TwinVQ can be also used for speech compression. Compression technology specifically designed to handle voice compression was published by NTT. NTT TwinVQ implementation supported sampling frequencies from 8 KHz or 11.025 KHz and bit rate from 8 Kbit/s [71], [76-79].

10.12.14. Comparison of MPEG-4 (H.264) and JPEG-2000 Video Compression

In the past decade, digital video recording and distribution has been made much more viable by the introduction of new video compression and decompression techniques. Analog video applications have traditionally typically needed large investments in infrastructure to be able to provide any useful recording and distribution functionality (even at very low channel counts), even with the older generation of standard definition PAL/SECAM video. With the introduction of high definition video, much higher data rates need to be recorded and moved around in the digital domain meaning new technologies have had to be developed in order to accommodate these higher specification video streams. In addition to hardware level innovations such as Gigabit transceivers which allow much more data to be transmitted across single channels (rather than the blunt instrument of just adding more parallel infrastructure), hardware (or software) based coder/decoders (codecs) can help ease the issues of system development, not only by reducing the bandwidth required for each channel of video but also by reducing the storage requirements. This of course has benefits in terms of development time as well as overall cost for any given system [80].

The two more popular technologies are: MPEG-4 part 10 (also known as Advanced Video Coding AVC and ITU-T based on recommendation H.264), and JPEG-2000. Both these codecs have distinct practical advantages and disadvantages in relation to each other. In their codec technology, the core concept behind any video codec is that in any given picture there is redundancy in the image (regarded as invisible to the human eye) can be removed reducing the amount of data required to represent the picture, at least up to a certain point. After this point a compressed then decompressed (reconstructed) image will contain certain artifacts, the nature of these artifacts differs depending upon the algorithm used to compress and decompress the image, as well as the image contents. Depending on the application, one codec's artifacts may be preferable over those of another. Also, limitations on the bandwidth of any network being used are a very important consideration. If an existing network infrastructure is in place, it must be remembered that it will not take many channels of high definition video to completely overwhelm even a Gigabit Ethernet network, and directly related to this are storage requirements. Some applications can require that several days worth of video is stored, in which case multiple Terabytes of hard disk may be required, which has a related system cost. This can be not only in terms of the cost of the storage media itself, but the ability to remove the media, perhaps for debrief (for instance from a flight mission camera to a ground station playback PC). Solid State Derive SSD storage is becoming considered more frequently for this sort of application and the cost/ density/reliability aspects of that all have an influence as well. It is important to remember that all codecs have their upsides and downsides. In considering any single algorithm for use the system architect must be aware of all factors and also that it is unlikely that any one codec will be completely ideal for its intended application.

The MPEG-4 part 10 was introduced in year 2003, is slightly newer codec. MPEG stands for the Motion Picture Experts Group, and as the name suggests, the MPEG-4 codec is designed to encode and decode motion pictures, as opposed to single frames (such as a still picture). MPEG-4 is known as a lossy codec as the reconstructed images will always be inferior to the original pictures. The key principle behind MPEG-4 encoding is the idea of the codec working on what is known as a Group Of Pictures GOP. These consist of three types of mage. The first type, the I-frame is a stand-alone compressed frame, and there is one I-frame at the beginning of each GOP. Between this are two further types of image which predict the change (motion) between each I-frame. The first level of predicted frame is the P-frame, which contains the difference between the current and preceding frame. Secondly, the B-frame contains the difference between the current frame and both the preceding and following frames. So, where before compression each frame is transmitted in full, after compression the bandwidth of the images is reduced as only a subset of information from each I-frame is required. To reduce the bandwidth for any given moving image, the length of the GOP is increased. As the GOP is increased the bandwidth will fall, but so will the quality of the images as deeper prediction is required. Furthermore as the number of I-frames in a video stream is reduced, the stream becomes less easily editable. Fig.10.44. shows a graphic representation of this [80]. This has as effect that is more pronounced depending upon the source material.

For images where frame to frame changes are very small (such as a camera pointed at a sunset, or where there are large areas of the same color, or some one speaking to a camera where only his/her mouth is moving) then inter-frame changes are very small, leading to a good compression ratio and more accurate reconstruction at the decoder. For action movies, where there are many sharp cuts from scene to scene and fast moving objects, the inter-frame prediction starts to break down as the changes between successive frames are greater. This is one of the reasons why MPEG-4 part 10 is popular in video conferencing applications as there is little frame to frame motion meaning very low streaming bit rated can be achieved [80].

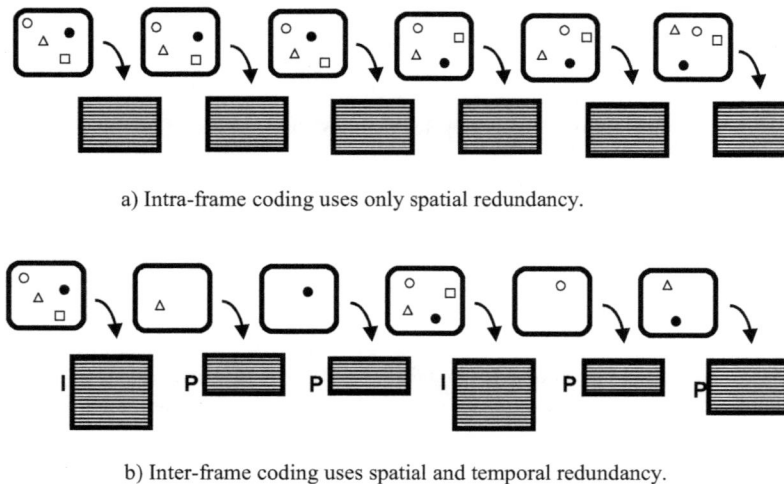

a) Intra-frame coding uses only spatial redundancy.

b) Inter-frame coding uses spatial and temporal redundancy.

Fig.10.44. Intra and Inter-frame coding example.

The JPEG-2000 is a successor to the earlier Joint Photographic Experts Group JPEG standard introduced in the early 1990`s to encode and store still images, and in its motion JPEG-2000 incarnation, moving images. Unlike MPEG-4`s use of both intra- and inter-frame encoding, JPEG employs only intra-frame encoding. This means that each frame is compressed individually, and no attention is paid by the encoder to either previous or following frames, and so it is not predictive in any way. In contrast to MPEG-4, JPEG-2000 can be operated in both a lossless as well as a lossy mode. The Discrete Wavelet Transform DWT at the heart of the algorithm can be based on either a reversible filter (lossless) or a non-reversible filter which is lossy but provides a higher compression rate for any given material.

Because each frame is individually compressed, JPEG-2000 offers a number of advantages over MPEG-4. For example, the latency to compress a frame is shorter (as the codec does not have to rely on generating forward and reverse differences between frames). In terms of video editing capabilities, as no frames are dropped during encoding there is a direct correlation between each frame of encoded and decoded video, although with lossy compression the image quality of the reconstructed frame will be reduced. The disadvantages of this of course is that JPEG-2000 has to encode and transmit each frame individually, unlike the MPEG-4 algorithm, which sends only the differences between a number of frames, ultimately meaning that the bandwidth requirements to transmit JPEG-2000 can be higher, with an increase in storage needs. JPEG-2000 is more resilient to errors in transmission than MPEG-4, a small loss of data in a JPEG-2000 stream will be far less noticeable than the corresponding loss in an MPEG-4 stream.

Also, JPEG-2000 is computationally much more complex to decode than MPEG-4. This means that though JPEG-2000 can be decoded in software (for example on a generic PC CPU running a software application), this can place quite a burden on the host CPU, leaving it less capable of running other tasks, or in extreme cases leading to frames being dropped. This means that most of the time it is desirable to accelerate JPEG-2000 using a hardware-based decoder.

Table 10.5. Intra and inter-frame coding example

File Size (MB)	MPEG-4			JPEG-2000		
	Comp.Factor 15	Comp.Factor 30	Comp.Factor 45	30 Mbps	20 Mbps	10 Mbps
Stock 1 Video: High Entropy	451	67	17	1100	730	378
Stock 2 Video: Medium Entropy	387	56	19	1000	714	375
Synthesized Graphics and Text	81	35	17	1100	702	349

Table 10.5. shows the intra and inter-frame coding example of the more popular technologies MPEG-4 part 10 (Advanced Video Coding AVC, H.264) and the JPEG-2000. For the MPEG-4 codec test, a compression setting for the codec 15, 30 and 45. At a compression setting of 15, with MPEG-4 the reconstructed images are almost lossless to the naked eye-no macro-blocking can be seen and motion is smooth. At a setting of 30, artifacts start to introduce themselves as a mosaic effect (restricted to certain parts of the image) and the motion between frames is just a little more juddery in compression to the lowest setting, but the effects are subtle and the video is still highly watchable. Predictably, at the highest compression setting 45, the replayed video is very heavily pixelated into larger macro-blocks, frame to frame motion is very poor and while it is still easy enough to make out what the images are it cannot claim to be high quality [80]. Where there have been encoding errors on playback this is largely noticeable in that parts of the picture are not reconstructed at all. Interestingly, the file sizes for all three video sources seem to converge as the compression quality moves towards maximum compression.

In JPEG-2000, Table.10.5, the main quality control is the number of Megabits per second (Mbps), so Table 10.5. shows the JPEG-2000 tests 30, 20 and 10 Mbps were used on the same material as the MPEG-4 test. In terms of quality JPEG-2000 fares extremely well at 30 Mbps, with the reconstructed video being nearly identical to the source, with only the smallest amount of fidelity lost which is almost invisible to the naked eye. At 20 Mbps quality was only reduced slightly but there is uniform noise over the entire image-however, the picture is still fairly high quality with most of clarity maintained, especially around the edges of objects. Down at 10 Mbps the noise that appears at 20 Mbps becomes much more pronounced, becoming a uniform watery like effect across each frame. Unlike the MPEG-4 at its lowest setting this picture is still very watchable, and it is still possible to make out distinct objects and their edges [80].

It is quite clear that in terms of reduced requirements for both network streaming and storage, MPEG-4 is the winner. However, JPEG-2000 is the winner in terms of picture quality. JPEG-2000 is far more resilient in terms of transmission or encoding error than MPEG-4 and because of its frame-by frame nature can be easily manipulated in its decompressed form - MPEG-4 is digital VHS meaning little can be done with it in terms of editing. An interesting point here is also in terms of legal use – in some countries MPEG-4 video is inadmissible as evidence due to its inter-frame encoding whereas JPEG-2000 has been successfully used as evidence in a number of legal cases worldwide.

Overall, although the comparison of the two codecs cannot be done directly due to their different compression methods (meaning that compression quality settings between the two never directly correlate) it can be seen that JPEG-2000 compression appears to degrade in terms of the reconstructed pictures much more gracefully between the lowest and highest compression settings, but the price that is paid in terms of storage and network bandwidth (for network transmission or streaming) purposes is relatively high. Conversely, though MPEG-4 is lower quality, it has many advantages in the network/storage domain, even though the reconstructed video may be of poorer quality and the end user has less ability to edit it.

In some applications both codecs have their place. As an example, in a recent application a video recording and distribution system required that remote Closed Circuit Tel-Vision CCTV sensors streamed live video to remote operators over a Wireless Area Network WAN. The WAN was highly bandwidth limited (it was already in place, and time shared between many users), however, the highest quality of video recording needed to be stored for later analysis. The solution to this was to provide both codecs in parallel, an MPEG-4 encoder streaming live video to the operators over the network while a local Redundant Array of Inexpensive Disks RAID array stored the high quality JPEG-2000 video [80].

Problems

1. Evaluate the Discrete Fourier Transform DFT of the following data sequences, consider the sampling period T_s equals unity:

    i. $x(n) = 0 , 0 , 1 , 0$ ii. $x(n) = 0 , 1 , 0 , 0$

    iii. $x(n) = 1 , 0 , 0 , 1$ iv. $x(n) = 1 , 0 , 0 , 1 , 0 , 0$

    vi. $x(n) = 1 , 1 , 0 , 0 , 0 , 0$

2. Evaluate the Inverse Discrete Fourier Transform IDFT of the following data sequences, consider the sampling period T_s equals unity:

    i. $X(k) = 1 , 1 , 0 , 0$ ii. $X(k) = 1 , j , -j , 0$

    iii. $X(k) = 0 , 1 , 0 , 1 , 0 , 1$ vi. $X(k) = 1 , 0 , 0 , 0 , 1 , 0 , 0 , 0$

3. Based on the following Discrete Fourier Transform DFT pairs of problems 10.1, and 10.4

$$1, 1, 0, 0 \;\rightleftharpoons\; 2 , (1 - j) , 0 , (1 + j)$$

, and

$$1, 1, 1, 1 \;\rightleftharpoons\; 4 , 0 , 0 , 0$$

Prove that their addition and subtraction can be given by the following Discrete Fourier Transform DFT pairs

$$2 , 2 , 1 , 1 \;\rightleftharpoons\; 6 , (1 - j) , 0 , (1 + j)$$

, and

$$0 , 0 , -1 , -1 \;\rightleftharpoons\; -2 , (1 - j) , 0 , (1 + j) \qquad\qquad \text{respectively.}$$

4. Based on the following Discrete Fourier Transform DFT pair

$$0 , 1 , 0 , 1 \;\rightleftharpoons\; 2 , 0 , -2 , 0$$

, and using shifting property. Find:

    i. The discrete Fourier transform of the data array sequence " $0 , 1 , 0 , 1$ ", shifted by 17 samples.

    ii. The discrete Fourier transform of the data array sequence " $0 , 1 , 0 , 1$ ", shifted by 18 samples.

5. Evaluate the following Discrete Fourier Transform DFT circular convolution signal $x_{12}(n)$ given by , consider the sampling period T_s equals unity:

    i. $x_{12}(n) = 1 , 3 , 2 , 4 \otimes 2 , 3 , 1$ ii. $x_{12}(n) = 1 , 2 , 3 , 4 \otimes 3 , 2 , 1$

    iii. $x_{12}(n) = 1 , 3 , 2 , 4 \otimes 2 , 3 , 1 , 4$

6. Evaluate the Discrete Cosine Transform of the one dimensional data array sequence "0, 1, 0, 1" , consider the sampling period T_s equals unity.

7. Evaluate the Inverse Discrete Cosine Transform of the one dimensional data array sequence "2 , 0 , -2 , 0" , consider the sampling period T_s equals unity.

8. Evaluate the Discrete Cosine Transform DCT of the following two dimensional data array sequence, consider the sampling period T_s equals unity.

    i. $x(n_1, n_2) = \begin{vmatrix} 1 & 0 \\ 0 & 1 \end{vmatrix}$ ii. $x(n_1, n_2) = \begin{vmatrix} 0 & 1 \\ 0 & 1 \end{vmatrix}$ iii. $x(n_1, n_2) = \begin{vmatrix} 1 & 0 \\ 1 & 0 \end{vmatrix}$

9. Evaluate the Inverse Discrete Cosine Transform IDCT of the following two dimensional data array sequence, consider the sampling period T_s equals unity.

    i. $X(k_1, k_2) = \begin{vmatrix} 2 & 0 \\ 0 & 2 \end{vmatrix}$ ii. $X(k_1, k_2) = \begin{vmatrix} 2 & 0 \\ -2 & 0 \end{vmatrix}$ iii. $X(k_1 , k_2) = \begin{vmatrix} 2 & 0 \\ 2 & 0 \end{vmatrix}$

10. Sketch the Huffman coding tree building of a quantized data contain four messages m_1, m_2, m_3, and m_4, have the probabilities of occurrence of 0.02, 0.25, and 0.73 respectively, and then
    i. Deduce the code word of each message.
    ii. Find the average length of this Huffman coding.
    iii. Find the entropy of the source data.
    iv. Find the code efficiency.
    v. Find the redundancy.
    vi. Find the compression ratio if each message is represented by 2 fixed bits before compression.

11. Sketch the Huffman coding tree building of a quantized data contain seven messages m_1, m_2, m_3, m_4, m_5, m_6, and m_7, have the probabilities of occurrence of 0.005, 0.005, 0.04, 0.1, 0.2, 0.3, and 0.35 respectively, and then
    i. Deduce the code word of each message.
    ii. Find the average length of this Huffman coding.
    iii. Find the entropy of the source data.
    iv. Find the code efficiency.
    v. Find the redundancy.
    vi. Find the compression ratio if each message is represented by 3 fixed bits before compression.

12. Sketch the Huffman coding tree building of a quantized data contain four messages m_1, m_2, m_3, and m_4, have the probabilities of occurrence of 0.05, 0.2, 0.35, and 0.4 respectively, and then
    i. Deduce the code word of each message.
    ii. Find the average length of this Huffman coding.
    iii. Find the entropy of the source data.
    iv. Find the code efficiency.
    v. Find the redundancy.
    vi. Find the compression ratio if each message is represented by 2 fixed bits before compression.

13. Sketch the Huffman coding tree building of a quantized data contain eight messages m_1, m_2, m_3, m_4, m_5, m_6, m_7, and m_8, have the probabilities of occurrence of 0.02, 0.04, 0.09, 0.1, 0.12, 0.13, 0.14, and 0.36 respectively, and then
    i. Deduce the code word of each message.
    ii. Find the average length of this Huffman coding .
    iii. Find the entropy of the source data.
    iv. Find the code efficiency.
    v. Find the redundancy.
    vi. Find the compression ratio if each message is represented by 3 fixed bits before compression.

14. Sketch the Huffman coding tree building of a quantized data contain six messages m_1, m_2, m_3, and m_4, have the probabilities of occurrence of 0.1, 0.1, 0.1, 0.1, 0.2, and 0.4 respectively, and then
    i. Deduce the code word of each message.
    ii. Find the average length of this Huffman coding.
    iii. Find the entropy of the source data.
    iv. Find the code efficiency.
    v. Find the redundancy.
    vi. Find the compression ratio if each message is represented by 3 fixed bits before compression.

15. Why compression techniques (coding) is necessary in video and audio digital communication systems ?

16. Compare between lossy and lossless compression (coding) in the following parameters :
    i. Compression ratio.   ii. Entropy and redundancy.
    iii. Applications.   iv. State examples for each.

17. What are the drawbacks of the compression technique (coding) in audio and video digital communication systems ?

18. Compare between the intra-coded and inter-coded video compression categories techniques. State examples.

19. Explain, How does the Fast Fourier Transform FFT refer to several sufficient algorithms for the Discrete Fourier transform DFT.

20. What are the drawbacks of the Discrete Fourier Transform DFT and state How does the Discrete Cosine Transform DCT overcome these drawbacks ?

21. Sketch a block diagram and state the equations to show, How can the Fast Fourier Transform FFT algorithm be used to handle the computation of the Inverse DFT and the DFT.

22. What are the drawbacks of the Discrete Cosine Transform DCT and state How does the Discrete Wavelet Transform DWT overcome these drawbacks ?

23. State the time domain admissibility condition in the wavelet transform. Why this condition must be achieved. What is the effect of getting more vanishing higher moments of the wavelet analysis function ?

24. In the Discrete Wavelet Transform DWT, What are the main conditions to get the inversely wavelet transformed original signal.

25. Explain using sketches and equations the Mallat tree three levels of decomposition and reconstruction in the Discrete Wavelet Transform DWT, illustrate How will this analysis be efficient and accurate to give multi resolution analysis.

26. In the Discrete Wavelet Transform DWT, the time scale representation of the digital signal is obtained using digital filtering techniques. Explain using sketches and state the equations of three decomposition and reconstruction levels of the discrete signal filtering techniques.

27. Explain, What is meant by dyadic technique in Discrete Wavelet Transform DWT, illustrate your answer using equations, and What are the advantages of this technique ?

28. What are the drawbacks of the Discrete Wavelet Transform DWT and state How does the Contourlet Transform CT overcome these drawbacks.

29. Explain the Pyramidal Directional Filter Bank PDFB of the Contourlet Ttransform CT, state How the Laplacian Pyramid LP does the decomposition of the images into sub-bands and How does the Directional Filter Bank DFB analyze each detail image ?

30. Joint Photographic Experts Group JPEG supports three modes of operation related to the Discrete Cosine Transform DCT: Sequential DCT, Progressive DCT, and Hierarchical DCT. Compare between these modes algorithms.

31. In Joint Photographic Experts Group JPEG, What are the reasons of using Discrete Cosine Transform DCT rather than Discrete Fourier Transform DFT ?, illustrate your answer using sketches and equations.

32. In Moving Picture Experts Group MPEG audio layer 3 (MP3). There are three classes of audio signals: telephone speech, wideband speech, and wideband audio. Compare between these three classes in the following parameters:
    i. Frequency range.   ii. Sampling rate.
    iii. PCM bits/sample.   iv. PCM bit rate.

33. Joint Photographic Experts Group JPEG-2000, is a JPEG initiative to define a new image coding system, addresses internet, and mobile applications. What are the advantages of JPEG-2000, and compare between JPEG and JPEG-2000, in the following parameters:
    i. Types of compression (coding) algorithm.   ii. Compression efficiency.
    iii. Resolution capabilities.   iv. Rate-distortion performance.

34. State the three types of pictures to construct MPEG-2. Sketch the sequence of these pictures, and what are the reasons of using this sequence ?

35. Compare between the spatial redundancy, spectral redundancy and the temporal redundancy, in the multi-media data compression technique.

36. Compare between MPEG-4 and JPEG-2000, in the following parameters:
    i. Picture quality.   ii. Encoding error.
    iii. Storage and network bandwidth.

37. State the main advantages of the Transform-domain weighted interleave Vector Quantization TwinVQ file format in MPEG-4.

Appendix A

Mathematical Tables and Properties

Table A.1. Summary of Fourier Transform FT Properties

Property	Mathematical Description
	Let $\quad x(t) \rightleftharpoons X(f)$
Linearity	$a_1 x_1(t) + a_2 x_2(t) \rightleftharpoons a_1 X_1(f) + a_2 X_2(f)$, where a_1 and a_2 are constants
Duality	$X(-t) \rightleftharpoons x(f)$
Scaling	$x(at) \rightleftharpoons \dfrac{1}{\lvert a \rvert} X(\dfrac{f}{a})$, where the scalar "a" is positive real constant ($a \geq 0$).
Shifting	$x(t \mp t_o) \rightleftharpoons X(f)\, e^{\mp j2\pi f\, t_o}$, where t_o is the time shifting. , and $\quad x(t)\, e^{\pm j2\pi f_c t} \rightleftharpoons X(f \mp f_c)$, where f_c is the frequency shifting.
Differentiation	$\dfrac{dx(t)}{dt} \rightleftharpoons j2\pi f\, X(f) \qquad$ (in the time domain) $, \quad -j2\pi t\, x(t) \rightleftharpoons \dfrac{dX(f)}{df} \qquad$ (in the frequency domain)
Integration	$\displaystyle\int_{-\infty}^{t} x(\tau)\, d\tau \rightleftharpoons \dfrac{X(f)}{j2\pi f} \qquad$ (Zero Boundary Condition) $, \quad \dfrac{x(t)}{-j2\pi t} \rightleftharpoons \displaystyle\int_{-\infty}^{f} X(\lambda)\, d\lambda \qquad$ (Zero Boundary Condition)
Convolution	$x_1(t)\, x_2(t) \rightleftharpoons \displaystyle\int_{-\infty}^{\infty} X_1(\lambda)\, X_2(f-\lambda)\, d\lambda \quad$ (in the f domain) $, \displaystyle\int_{-\infty}^{\infty} x_1(\tau)\, x_2(t-\tau)\, d\tau \rightleftharpoons X_1(f)\, X_2(f) \qquad$ (in the t domain)
Area under x(t)	$\displaystyle\int_{-\infty}^{\infty} x(t)\, dt = X(0) \qquad$ (frequency domain origin value)
, Area under X(f)	$\displaystyle\int_{-\infty}^{\infty} X(f)\, df = x(0) \qquad$ (time domain origin value)
Conjugate	$x^*(t) \rightleftharpoons X^*(-f)$

Table A.2. Summary of some Discrete Fourier Transform DFT Properties

Property	Mathematical Description
Property	Let $\quad x(nT_s) \rightleftharpoons X(kf_s)$, where $\; n = 0, 1, 2, \dots, (N-1), k = 0, 1, 2, \dots, (N-1)$, and N denote the number of frequency samples contained in an interval f_s.
Linearity	$a_1\, x_1(nT_s) + a_2\, x_2(nT_s) \rightleftharpoons a_1\, X_1(kf_s) + a_2\, X_2(kf_s)$, where a_1 and a_2 are constants.
Shifting	$x[(n \mp i)T_s] \rightleftharpoons X(kf_s)\; e^{\mp j2\pi\, ki/N}$, or $\quad x[n \mp (i \bmod N)T_s] \rightleftharpoons X(kf_s)\; e^{\mp j2\pi\, ki/N}$, where iT_s is the time shifting of the data array sequence $x(nT_s)$, and $\quad x(nT_s)\, e^{\pm j2\pi\, ni/N} \rightleftharpoons X[(k \mp i)f_s)]$, or $\quad x(nT_s)\, e^{\pm j2\pi\, ni/N} \rightleftharpoons X[k \mp (i \bmod N)f_s]$, where if_s is the frequency shifting of the data array sequence $X(kf_s)$.
Convolution in t domain	$T_s \displaystyle\sum_{i=0}^{2N-1} x_1(iT_s)\; x_2[(n-i)T_s] \rightleftharpoons X_1(kf_s)\; X_2(kf_s)$ $, n = 0, 1, 2, \dots, (2N-1)$

A.1. Some Fourier Transform Pairs

$$A \operatorname{rect}\left(\frac{t}{T}\right) \rightleftharpoons AT \operatorname{sinc}(fT)$$

$$A \operatorname{tri}\left(\frac{t}{T}\right) \rightleftharpoons AT \operatorname{sinc}^2(fT)$$

$$e^{-\pi t^2} \rightleftharpoons e^{-\pi f^2}$$

$$e^{-t^2} \rightleftharpoons \sqrt{\pi}\; e^{-\pi^2 f^2}$$

$$A \operatorname{rect}\left(\frac{t}{T}\right) \cos(2\pi f_c t) \rightleftharpoons \tfrac{1}{2} AT \operatorname{sinc}[T(f - f_c)] + \tfrac{1}{2} AT \operatorname{sinc}[T(f + f_c)]$$

$$e^{\mp t} u(\pm t) \rightleftharpoons \frac{1}{1 \pm j2\pi f}$$

$$e^{-|t|} \rightleftharpoons \frac{2}{1 + (2\pi f)^2}$$

$$e^{-|at|} \rightleftharpoons \frac{2a}{a^2 + (2\pi f)^2}$$

$$\frac{1}{1 \mp j2\pi t} \rightleftharpoons e^{\mp f} U(\pm f)$$

$$\frac{2}{1 + (2\pi t)^2} \rightleftharpoons e^{-|f|}$$

$$AW \operatorname{sinc}(Wt) \rightleftharpoons A \operatorname{rect}\left(\frac{f}{w}\right)$$

$$AW \operatorname{sinc}^2(Wt) \rightleftharpoons A \operatorname{tri}\left(\frac{f}{w}\right)$$

$$e^{-2t} u(t) \rightleftharpoons \frac{1}{2} \frac{1}{(1 + j\pi f)}$$

$$e^{2\pi t} u(-t) \rightleftharpoons \frac{1}{2\pi} \frac{1}{(1 - jf)}$$

$$\frac{1}{1 - jt} \rightleftharpoons 2\pi\, e^{-2\pi f} u(f)$$

$$e^{-(t-5)} u(t - 5) \rightleftharpoons \frac{1}{1 + j2\pi f}\, e^{-j2\pi f 5}$$

$$, \qquad e^{-(t-7)}\, u(t-5) \;\rightleftharpoons\; \frac{e^2}{1+j2\pi f}\, e^{-j2\pi f 5}$$

$$, \qquad 5\,\text{rect}\left[\frac{t+3}{4}\right] \;\rightleftharpoons\; 20\,\text{sinc}(4f)\, e^{j2\pi f 3}$$

$$, \qquad 6\,\text{tri}\left(\frac{t}{5}\right) e^{j2\pi 10^6 t} \;\rightleftharpoons\; 30\,\text{sinc}^2[5(f-10^6)]$$

$$, \qquad e^{\mp t}\, e^{\pm j2\pi 10^6 t}\, u(\pm t) \;\rightleftharpoons\; \frac{1}{1\pm j2\pi(f\mp 10^6)}$$

$$, \qquad e^{-t}\sin(2\pi f_c t)\, u(t) \;\rightleftharpoons\; \frac{2\pi f_c}{(1+j2\pi f)^2+(2\pi f_c)^2}$$

$$, \qquad t\,\text{rect}\left(\frac{t}{2}\right) \;\rightleftharpoons\; j\,\frac{2\pi f\cos(2\pi f)-\sin(2\pi f)}{2\pi^2 f^2}$$

$$, \qquad t\,e^{-t}\, u(t) \;\rightleftharpoons\; \frac{1}{(1+j2\pi f)^2}$$

$$, \qquad \frac{1}{(N-1)!}\,\frac{1}{(RC)^{N-1}}\, t^{N-1}\, e^{-t}\, u(t) \;\rightleftharpoons\; \frac{1}{(1+j2\pi f\, RC)^N} \qquad , \quad N>1$$

$$, \qquad t\,e^{-|t|} \;\rightleftharpoons\; -j\,\frac{8\pi f}{[1+(2\pi f)^2]^2}$$

$$, \qquad t^2\, e^{-\pi t^2} \;\rightleftharpoons\; \frac{1}{2\pi}\,[2\pi f^2-1]\, e^{-\pi f^2}$$

$$, \qquad \int_{-\infty}^{t} \tau\, e^{-|\tau|}\, d\tau \;\rightleftharpoons\; \frac{1}{j2\pi f}\,\frac{-j8\pi f}{[1+(2\pi f)^2]^2}$$

$$, \qquad \delta(t) \;\rightleftharpoons\; 1$$

$$, \qquad 1 \;\rightleftharpoons\; \delta(f)$$

$$, \qquad \cos(2\pi f_c t) \;\rightleftharpoons\; \frac{1}{2}\,[\,\delta(f-f_c)+\delta(f+f_c)\,]$$

$$, \qquad \sin(2\pi f_c t) \;\rightleftharpoons\; \frac{1}{2j}\,[\,\delta(f-f_c)-\delta(f+f_c)\,]$$

$$, \qquad \text{sgn}(t) \;\rightleftharpoons\; \frac{1}{j\pi f}$$

$$, \qquad -\frac{1}{j\pi t} \;\rightleftharpoons\; \text{sgn}(f)$$

$$, \qquad -\frac{1}{j\pi t}\, e^{-j2\pi t5} \;\rightleftharpoons\; \mathrm{sgn}(f+5)$$

$$, \qquad j\pi\, \mathrm{sgn}(t) \;\rightleftharpoons\; \frac{1}{f}$$

$$, \qquad \frac{1}{t} \;\rightleftharpoons\; -j\pi\, \mathrm{sgn}(f)$$

$$t \;\rightleftharpoons\; \frac{1}{(-j2\pi)}\frac{d\,\delta(f)}{df}$$

$$, \qquad \frac{1}{(j2\pi)}\frac{d\,\delta(t)}{dt} \;\rightleftharpoons\; f$$

$$, \qquad \mathrm{sgn}(t+2) \;\rightleftharpoons\; \frac{1}{j\pi f}\, e^{j2\pi f2}$$

$$, \qquad j\pi\, \mathrm{sgn}(t)\, e^{-j2\pi t3} \;\rightleftharpoons\; \frac{1}{3+f}$$

$$, \qquad 6\pi\, e^{-6\pi t}\, u(t) \;\rightleftharpoons\; \frac{1}{3+jf}$$

$$, \qquad u(t) \;\rightleftharpoons\; \frac{1}{j2\pi f} + \frac{\delta(f)}{2}$$

$$, \qquad -\frac{1}{j2\pi t} + \frac{\delta(t)}{2} \;\rightleftharpoons\; U(f)$$

$$, \qquad u(t+2) \;\rightleftharpoons\; \frac{1}{j2\pi f}\, e^{j2\pi f2} + \frac{\delta(f)}{2}$$

$$, \qquad \int_{-\infty}^{t} x(\tau)\, d\tau \;\rightleftharpoons\; \frac{X(f)}{j2\pi f} + \frac{X(0)\delta(f)}{2}$$

$$, \qquad \int_{-\infty}^{t} \delta(\tau)\, d\tau \;\rightleftharpoons\; \frac{1}{j2\pi f} + \frac{\delta(f)}{2}$$

$$, \qquad \frac{1}{1+jt} \;\rightleftharpoons\; 2\pi\, e^{2\pi f}\, u(-f)$$

$$, \qquad \frac{1}{1+t} \;\rightleftharpoons\; -j\pi\, \mathrm{sgn}(f)\, e^{j2\pi f}$$

$$, \qquad \frac{t}{1+jt} \;\rightleftharpoons\; j\left[\, 2\pi\, e^{2\pi f}\, u(-f) - \delta(f)\,\right]$$

$$[\,\delta(t) - e^{-t}\, u(t)\,] \;\rightleftharpoons\; \frac{j2\pi f}{1+j2\pi f}$$

$$, \quad \int_{-\infty}^{t} e^{-\pi\tau^2} d\tau \quad \rightleftharpoons \quad \frac{e^{-\pi f^2}}{j2\pi f} + \frac{\delta(f)}{2}$$

$$, \quad erf(t) \quad \rightleftharpoons \quad \left[\frac{e^{-2(\pi f)^2}}{j2\pi f} + \frac{\sqrt{\pi}}{2} \delta(f) \right]$$

$$, \quad erfc(t) \quad \rightleftharpoons \quad \left[-\frac{e^{-2(\pi f)^2}}{j2\pi f} + (1 - \frac{\sqrt{\pi}}{2}) \delta(f) \right]$$

$$, \quad \sum_{m=-\infty}^{\infty} x(t - mT_o) \quad \rightleftharpoons \quad \frac{1}{T_o} \sum_{m=-\infty}^{\infty} X(\frac{n}{T_o}) \, \delta(f - \frac{n}{T_o})$$

$$, \quad \frac{1}{T_o} \sum_{n=-\infty}^{\infty} X(\frac{n}{T_o}) \, e^{j2\pi \frac{n}{T_o} t} \quad \rightleftharpoons \quad \frac{1}{T_o} \sum_{n=-\infty}^{\infty} X(\frac{n}{T_o}) \, \delta(f - \frac{n}{T_o})$$

$$, \quad \sum_{m=-\infty}^{\infty} x(t - mT_o) \quad \rightleftharpoons \quad \frac{1}{T_o} \sum_{n=-\infty}^{\infty} X(\frac{n}{T_o}) \, \delta(f - \frac{n}{T_o})$$

$$, \quad \sum_{m=-\infty}^{\infty} \delta(t - mT_o) \quad \rightleftharpoons \quad \frac{1}{T_o} \sum_{n=-\infty}^{\infty} \delta(f - \frac{n}{T_o})$$

$$, \quad \frac{1}{T_o} \sum_{m=-\infty}^{\infty} R(\tau - mT_o) \quad \rightleftharpoons \quad \frac{1}{T_o^2} \sum_{n=-\infty}^{\infty} \left| X(\frac{n}{T_o}) \right|^2 \delta(f - \frac{n}{T_o})$$

, the autocorrelation function of the energy signal $R(\tau)$ and its energy spectral density $\Psi(f)$ form a Fourier transform pair

$$R(\tau) \quad \rightleftharpoons \quad \Psi(f)$$

, the autocorrelation function of the power signal $R_p(\tau)$ and its power spectral density $\Omega(f)$ form a Fourier transform pair

$$R_P(\tau) \quad \rightleftharpoons \quad \Omega(f)$$

, the cross-correlation function of two energy signals $R_{12}(\tau)$ and their cross energy spectral density $\Psi_{12}(f)$ form Fourier transform pair

.

$$R_{12}(\tau) \quad \rightleftharpoons \quad \Psi_{12}(f)$$

, and the cross-correlation function of two power signals $R_{12p}(\tau)$ and their cross power spectral density $\Omega_{12}(f)$ form Fourier transform pair

$$R_{12p}(\tau) \quad \rightleftharpoons \quad \Omega_{12}(f)$$

A.2. Some Convolution Functions

$$u(t) \otimes u(t) = t\, u(t)$$

$$u(t) \otimes u(t) \otimes u(t) = u(t) \otimes t\, u(t) = \tfrac{1}{2}\, t^2\, u(t)$$

$$t\, u(t) \otimes t\, u(t) = \frac{1}{6}\, \tau^3\, u(t)$$

$$\text{rect}(t) \otimes \text{rect}(t) = \text{tri}(t)$$

$$\text{rect}(t) \otimes t^2 = t^2 + \frac{1}{12}$$

$$e^{-t}\, u(t) \otimes u(t) = [\, 1 - e^{-t}\,]\, u(t)$$

$$e^{-t}\, u(t) \otimes t\, u(t) = t + e^{-t} - 1$$

$$e^{-t}\, u(t) \otimes e^{-t}\, u(t) = t\, e^{-t}\, u(t)$$

$$e^{-3t}\, u(t) \otimes e^{-t} = \tfrac{1}{2}\, e^{-t}$$

$$e^{-t}\, u(t) \otimes e^{-2t}\, u(t) = 2\, e^{-3t/2}\, \sinh(t/2)\, u(t)$$

$$t^{n+1}\, u(t) \otimes u(t) = \frac{1}{n+1}\, t^{n+1}\, u(t) \qquad , n = 1, 2, 3, \dots$$

$$e^{-\pi t^2} \otimes e^{-\pi t^2} = \frac{1}{\sqrt{2}}\, e^{-\pi t^2/2}$$

$$\text{rect}(f)\, e^{-j2\pi f} \otimes e^{-j2\pi f} = e^{-j2\pi f}$$

$$\frac{1}{1 - j2\pi t} \otimes \frac{1}{1 - j\pi t} = \frac{\tfrac{2}{3}}{1 - j\,\tfrac{2}{3}\,\pi t}$$

$$2\, \text{sinc}(2t - 8) \otimes 14\, \text{sinc}(7t + 14) = 4\, \text{sinc}(2t - 4)$$

$$16\, \text{sinc}(8f) \otimes 18\, \text{sinc}^2(4f) = 16\, \text{sinc}(8f) + 16\, \text{sinc}^2(4f)$$

$$e^{-t}\, u(t) \otimes \cos(t) = 0.707\, \cos(t - 45^\circ)$$

$$\sin(2\pi 10^6 t) \otimes \frac{1}{\pi t} = -\cos(2\pi 10^6 t)$$

$$\frac{1}{2f}\, U(f) \otimes U(f) = \tfrac{1}{2}\, \log(f)$$

$$e^{-2t}\, u(t) \otimes u(t) = \frac{1}{4}\, \text{sgn}(t) - \frac{1}{2}\, e^{-2t}\, u(t) + \frac{1}{4}$$

$$\delta(2t - 1) \otimes u(t + 2) = \frac{1}{4}\, \text{sgn}\left(t + \frac{3}{2}\right) + \frac{1}{4} = \frac{1}{2}\, u\left(t + \frac{3}{2}\right)$$

$$\cos(2t + \theta) \otimes \frac{1}{\pi t} = \sin(2t + \theta)$$

$$e^{-2t}\, u(t) \otimes \text{sgn}(t) = \tfrac{1}{2}\, \text{sgn}(t) - e^{-2t}\, u(t)$$

$$, \quad e^{-t} \, \text{rect}\left[\frac{t - \frac{T}{2}}{T}\right] \otimes e^{t} \, \text{rect}\left[\frac{t - \frac{T}{2}}{T}\right] = \begin{cases} \sinh(t) & \text{for } 0 < t < T \\ \sinh(2T - t) & \text{for } T < t < 2T \end{cases}$$

$$, \quad \text{rect}(t) \otimes u(t) = \begin{cases} 0 & \text{for } \quad t \leq -\frac{1}{2} \\ t + \frac{1}{2} & \text{for } -\frac{1}{2} \leq t \leq \frac{1}{2} \\ 1 & \text{for } \quad t \geq \frac{1}{2} \end{cases}$$

$$, \quad \text{rect}(t) \otimes t\, u(t) = \begin{cases} 0 & \text{for } \quad t \leq -\frac{1}{2} \\ \frac{1}{2} t^2 + \frac{1}{2} t + \frac{1}{8} & \text{for } -\frac{1}{2} \leq t \leq \frac{1}{2} \\ t & \text{for } \quad t \geq \frac{1}{2} \end{cases}$$

$$, \quad e^{-t} u(t) \otimes \text{rect}(t) = \begin{cases} 0 & \text{for } \quad t \leq -\frac{1}{2} \\ 1 - 0.61\, e^{-t} & \text{for } -\frac{1}{2} \leq t \leq \frac{1}{2} \\ 1.04\, e^{-t} & \text{for } \quad t \geq \frac{1}{2} \end{cases}$$

$$, \quad \text{tri}(t) \otimes u(t) = \begin{cases} \frac{1}{2} t^2 + t + \frac{1}{2} & \text{for } -1 \leq t \leq 0 \\ -\frac{1}{2} t^2 + t + \frac{1}{2} & \text{for } \quad 0 \leq t \leq 1 \\ 1 & \text{for } \quad t \geq 1 \end{cases}$$

$$, \quad \text{sgn}(t) \otimes \text{rect}(t) = \begin{cases} -1 & \text{for } \quad t \leq -\frac{1}{2} \\ 2t & \text{for } -\frac{1}{2} \leq t \leq \frac{1}{2} \\ 1 & \text{for } \quad t \geq \frac{1}{2} \end{cases}$$

$$, \quad \text{tri}(t) \otimes \text{rect}(t) = \begin{cases} \frac{1}{2} t^2 + 1.5\, t + \left(\frac{9}{8}\right) & \text{for } -3/2 \leq t \leq -\frac{1}{2} \\ \left(\frac{3}{4}\right) - t^2 & \text{for } -\frac{1}{2} \leq t \leq \frac{1}{2} \\ \frac{1}{2} t^2 - 1.5\, t + \left(\frac{9}{8}\right) & \text{for } \quad \frac{1}{2} \leq t \leq 3/2 \end{cases}$$

, and

$$100 \otimes 111 = \begin{cases} A^2\, t & \text{for } 0 \leq t \leq T \\ 2\, A^2\, T - A^2\, t & \text{for } T \leq t \leq 4T \\ A^2\, t - 6\, A^2\, T & \text{for } 4T \leq t \leq 6T \end{cases}$$

A.5. Some Area Integrals

$$\int_{-\infty}^{\infty} \frac{1}{1 \mp j2\pi f} \, df = \frac{1}{2}$$

$$\int_{-\infty}^{\infty} e^{\mp t} u(\pm t) \, dt = 1$$

$$\int_{-\infty}^{\infty} (1 - at) e^{-bt} u(t) \, dt = \frac{1}{b^2} [b - a]$$

$$\int_{-\infty}^{\infty} e^{-\pi x^2} \, dx = 1$$

$$\int_{-\infty}^{\infty} e^{-x^2} \, dx = \sqrt{\pi}$$

$$\int_{-\infty}^{\infty} e^{-(\pi x)^2} \, dx = \sqrt{\frac{2}{\pi}}$$

$$\int_{-\infty}^{\infty} \text{sinc}(x) \, dx = 1$$

$$\int_{-\infty}^{\infty} \text{sinc}^2(x) \, dx = 1$$

$$\int_{-\infty}^{\infty} \text{sinc}^2(t) e^{-j\pi t} \, dt = \frac{1}{2}$$

$$\int_{-\infty}^{\infty} \frac{\sin(\lambda)}{\lambda} \, d\lambda = \pi$$

$$\int_{-\infty}^{\infty} \frac{\sin^2(\lambda)}{\lambda^2} \, d\lambda = \pi$$

$$\int_{-\infty}^{\infty} t^2 e^{-\pi t^2} \, dt = \frac{1}{2\pi}$$

$$\int_{-\infty}^{\infty} t^2 e^{-t^2} \, dt = \frac{\sqrt{\pi}}{2}$$

$$\int_{-\infty}^{\infty} e^{-(at)^2 + bt} \, dt = e^{(b/2a)^2} \sqrt{\frac{\pi}{a}} \qquad , \quad a > 0$$

$$\int_{-\infty}^{\infty} e^{\mp j2\pi ft} \, dt = \delta(f)$$

, and
$$\int_{-\infty}^{\infty} e^{-|at|} \, dt = \frac{2}{a}$$

Table. A.3. Hilbert Transform Pairs

Time function	Hilbert transform
$\cos(2\pi f_c t)$	$\sin(2\pi f_c t)$
$\sin(2\pi f_c t)$	$-\cos(2\pi f_c t)$
$x(t) \otimes y(t)$	, or $\begin{array}{c} x(t) \otimes \hat{y}(t) \\ \hat{x}(t) \otimes y(t) \end{array}$
$m(t)\,e^{j2\pi f_c t}$, m(t) is a low-pass band limited signal of W Hz , f_c is much higher than W Hz.	$-j\,m(t)\,e^{j2\pi f_c t}$
$m(t)\,c(t)$, m(t) is the low-pass signal , c(t) is the carrier wave	$m(t)\,\hat{c}(t)$
$\dfrac{d}{dt}\,x(t)$	$\dfrac{d}{dt}\,\hat{x}(t)$
$\delta(t)$	$\dfrac{1}{\pi t}$
$\dfrac{1}{t}$	$-\pi\,\delta(t)$
$\text{rect}(\dfrac{t}{T})$	$-\dfrac{1}{\pi}\,\ln\left[\dfrac{t-\frac{1}{2}T}{t+\frac{1}{2}T}\right]$
$\text{sinc}(t)$	$\dfrac{1-\cos(\pi t)}{\pi t}$
$\dfrac{1}{1+t^2}$	$\dfrac{t}{1+t^2}$
constant	zero
$\dfrac{1-\cos(\pi t)}{\pi t}$	$-\text{sinc}(t)$
$\dfrac{t}{1+t^2}$	$-\dfrac{1}{1+t^2}$

Appendix B

Trigonometric Polynomials

B.1. Harmonic and Polynomial Representation

Theorem B.1. DeMoivre`s Theorem

$$Z = r\, e^{j\theta}$$

, then
$$Z^n = [\, r\, e^{j\theta}]^n = r^n\, e^{jn\theta}$$
$$= r^n\, [\cos(n\theta) + j\sin(n\theta)] \qquad\qquad \text{Euler's identity}$$

Theorem B.2.
The cosine of an argument with an integer factor n can be expressed as a polynomial in the cosine without the integer factor, where

$$\cos(n\theta) = \sum_{k=0}^{\lfloor\frac{n}{2}\rfloor} \sum_{m=0}^{k} (-1)^{k+m} \binom{n}{2k}\binom{k}{m} [\cos(\theta)]^{n-2(k-m)} \qquad (B.1)$$

, and
$$\sin(n\theta) = \sum_{k=0}^{\lfloor\frac{n}{2}\rfloor} \sum_{m=0}^{k} (-1)^{k+m} \binom{n}{2k}\binom{k}{m} [\sin(\theta)]^{n-2(k-m)} \qquad (B.2)$$

, where the standard Binomial coefficients are: $\binom{n}{2k} = \dfrac{n!}{(n-2k)!\,2k!}$, and $\binom{k}{m} = \dfrac{k!}{(k-m)!\,m!}$

Proof:
$$\cos(n\theta) = \text{Re}[\cos(n\theta) + j\sin(n\theta)] = \text{Re}[\,e^{jn\theta}\,] = \text{Re}[(e^{j\theta})^n] = \text{Re}[(\cos(\theta) + j\sin(\theta))^n]$$

$$= \text{Re}\left[\sum_{k}^{n} \binom{n}{k} [\cos(\theta)]^{n-k} [j\sin(\theta)]^k\right] = \text{Re}\left[\sum_{k}^{n} j^k \binom{n}{k} [\cos(\theta)]^{n-k} \sin^k(\theta)\right]$$

$$= \text{Re}\left[\sum_{k\in 0,4,\dots,n} \binom{n}{k} [\cos(\theta)]^{n-k} \sin^k(\theta) + j \sum_{k\in 1,5,\dots,n} \binom{n}{k} [\cos(\theta)]^{n-k} \sin^k(\theta)\right.$$
$$\left. - \sum_{k\in 2,6,\dots,n} \binom{n}{k} [\cos(\theta)]^{n-k} \sin^k(\theta) + (-j) \sum_{k\in 3,7,\dots,n} \binom{n}{k} [\cos(\theta)]^{n-k} \sin^k(\theta)\right]$$

$$= \sum_{k\in 0,4,\dots,n} \binom{n}{k} [\cos(\theta)]^{n-k} \sin^k(\theta) - \sum_{k\in 2,6,\dots,n} \binom{n}{k} [\cos(\theta)]^{n-k} \sin^k(\theta)$$

$$= \sum_{k\in 0,2,4,6,\dots,n} \binom{n}{k} (-1)^{k/2} [\cos(\theta)]^{n-k} \sin^k(\theta)$$

$$= \sum_{k=0}^{\lfloor\frac{n}{2}\rfloor} \binom{n}{2k} (-1)^k [\cos(\theta)]^{n-2k} [\sin(\theta)]^{2k}$$

$$= \sum_{k=0}^{\lfloor\frac{n}{2}\rfloor} \binom{n}{2k} (-1)^k [\cos(\theta)]^{n-2k} [1 - \cos^2(\theta)]^k$$

$$= \left[\sum_{k=0}^{\lfloor\frac{n}{2}\rfloor} \binom{n}{2k} (-1)^k [\cos(\theta)]^{n-2k}\right]\left[\sum_{m=0}^{k} \binom{k}{m} (-1)^m [\cos(\theta)]^{2m}\right]$$

$$= \sum_{k=0}^{\lfloor\frac{n}{2}\rfloor} \sum_{m=0}^{k} (-1)^{k+m} \binom{n}{2k}\binom{k}{m} [\cos(\theta)]^{n-2(k-m)} \qquad (B.1)$$

, also the sine of an argument with an integer factor n can be expressed as a polynomial in the sine without the integer factor, where

$$\sin(n\theta) = \cos(n\theta - \frac{\pi}{2}) = \cos[n(\theta - \frac{\pi}{2n})]$$

$$= \sum_{k=0}^{\lfloor \frac{n}{2} \rfloor} \sum_{m=0}^{k} (-1)^{k+m} \binom{n}{2k} \binom{k}{m} \left[\cos[(\theta - \frac{\pi}{2})]\right]^{n-2(k-m)}$$

$$= \sum_{k=0}^{\lfloor \frac{n}{2} \rfloor} \sum_{m=0}^{k} (-1)^{k+m} \binom{n}{2k} \binom{k}{m} [\sin(\theta)]^{n-2(k-m)} \tag{B.2}$$

Example B.1.
The monomials $\cos(5\theta)$ and $\sin(5\theta)$ as polynomials in $\cos(\theta)$ and $\sin(\theta)$ respectively, are

$$\cos(5\theta) = 16\cos^5(\theta) - 20\cos^3(\theta) + 5\cos(\theta) \tag{B.3}$$
$$\sin(5\theta) = 16\sin^5(\theta) - 20\sin^3(\theta) + 5\sin(\theta) \tag{B.4}$$

Proof:

1. proof using DeMoivre's Theorem

$$\cos(5\theta) + j\sin(5\theta) = e^{j5\theta} = [e^{j\theta}]^5 = [\cos(\theta) + j\sin(\theta)]^5$$

$$= \sum_{k=0}^{5} \binom{5}{k} [\cos(\theta)]^{5-k} [j\sin(\theta)]^k$$

$$= \binom{5}{0}[\cos(\theta)]^{5-0}[j\sin(\theta)]^0 + \binom{5}{1}[\cos(\theta)]^{5-1}[j\sin(\theta)]^1$$
$$+ \binom{5}{2}[\cos(\theta)]^{5-2}[j\sin(\theta)]^2 + \binom{5}{3}[\cos(\theta)]^{5-3}[j\sin(\theta)]^3$$
$$+ \binom{5}{4}[\cos(\theta)]^{5-4}[j\sin(\theta)]^4 + \binom{5}{5}[\cos(\theta)]^{5-5}[j\sin(\theta)]^5$$

$$= 1\cos^5(\theta) + j5\cos^4(\theta)\sin(\theta) - 10\cos^3(\theta)\sin^2(\theta)$$
$$-j10\cos^2(\theta)\sin^3(\theta) + 5\cos(\theta)\sin^4(\theta) + j1\sin^5(\theta)$$

$$= [\cos^5(\theta) - 10\cos^3(\theta)\sin^2(\theta) + 5\cos(\theta)\sin^4(\theta)]$$
$$+j[5\cos^4(\theta)\sin(\theta) - 10\cos^2(\theta)\sin^3(\theta) + 1\sin^5(\theta)]$$

$$= [\cos^5(\theta) - 10\cos^3(\theta)[1 - \cos^2(\theta)] + 5\cos(\theta)[1 - \cos^2(\theta)]^2]$$
$$+j[5[1 - \sin^2(\theta)]^2\sin(\theta) - 10[1 - \sin^2(\theta)]\sin^3(\theta) + \sin^5(\theta)]$$

$$= [\cos^5(\theta) - 10[\cos^3(\theta) - \cos^5(\theta)] + 5[\cos(\theta) - 2\cos^3(\theta) + \cos^5(\theta)]$$
$$+j[5\sin(\theta) - 2\sin^3(\theta) + \sin^5(\theta) - 10[\sin^3(\theta) - \sin^5(\theta)] + \sin^5(\theta)]$$

$$= [16\cos^5(\theta) - 20\cos^3(\theta) + 5\cos(\theta)] + j[16\sin^5(\theta) - 20\sin^3(\theta) + 5\sin(\theta)]$$

, where the real part is Eq.(B.3) and the imajinary part is Eq. (B.4)

2. proof using cosine harmonic to cosine polynomial formula, Eq.(B.1) yields

$$\cos(5\theta) = \sum_{k=0}^{\lfloor \frac{5}{2} \rfloor} \sum_{m=0}^{k} (-1)^{k+m} \binom{5}{2k} \binom{k}{m} \cos^{5-2(k-m)}(\theta)$$

$$= \sum_{k=0}^{2} \sum_{m=0}^{k} (-1)^{k+m} \binom{n}{2k} \binom{k}{m} \cos^{5-2(k-m)}(\theta)$$

$$= (-1)^0 \binom{5}{0}\binom{0}{0}\cos^5(\theta) + (-1)^1 \binom{5}{2}\binom{1}{0}\cos^3(\theta) + (-1)^2 \binom{5}{2}\binom{1}{1}\cos^5(\theta)$$

$$+ (-1)^2 \binom{5}{4}\binom{2}{0}\cos^1(\theta) + (-1)^3 \binom{5}{4}\binom{2}{1}\cos^3(\theta) + (-1)^4 \binom{5}{4}\binom{2}{2}\cos^5(\theta)$$

$$= (1)(1)\cos^5(\theta) - (10)(1)\cos^3(\theta) + (10)(1)\cos^5(\theta) + (5)(1)\cos(\theta)$$
$$- (5)(2)\cos^3(\theta) + (5)(1)\cos^5(\theta)$$

$$= (1 + 10 + 5)\cos^5(\theta) + (-10 - 10)\cos^3(\theta) + 5\cos(\theta)$$

$$= 16\cos^5(\theta) - 20\cos^3(\theta) + 5\cos(\theta)$$

Example B.2. Harmonic to Polynomial conversion

n   cos(nθ)   polynomial in cos(θ)

0   cos(0θ) = 1
1   cos(1θ) = $\cos^1(\theta)$
2   cos(2θ) = $2\cos^2(\theta) - 1$
3   cos(3θ) = $4\cos^3(\theta) - 3\cos(\theta)$
4   cos(4θ) = $8\cos^4(\theta) - 8\cos^2(\theta) + 1$
5   cos(5θ) = $16\cos^5(\theta) - 20\cos^3(\theta) + 5\cos(\theta)$
6   cos(6θ) = $32\cos^6(\theta) - 48\cos^4(\theta) + 18\cos^2(\theta) - 1$
7   cos(7θ) = $64\cos^7(\theta) - 112\cos^5(\theta) + 56\cos^3(\theta) - 7\cos(\theta)$

Proof:
For $\cos(2\theta)$, Eq. (B.1) yields

$$\cos(2\theta) = \sum_{k=0}^{\lfloor\frac{2}{2}\rfloor}\sum_{m=0}^{k}(-1)^{k+m}\binom{2}{2k}\binom{k}{m}[\cos(\theta)]^{2-2(k-m)}$$

$$= \sum_{k=0}^{1}\sum_{m=0}^{k}(-1)^{k+m}\binom{2}{2k}\binom{k}{m}[\cos(\theta)]^{2-2(k-m)}$$

$$= (-1)^0\binom{2}{0}\binom{0}{0}\cos^2(\theta) + (-1)^1\binom{2}{2}\binom{1}{0}\cos^0(\theta) + (-1)^2\binom{2}{2}\binom{1}{1}\cos^2(\theta)$$

$$= (1)(1)\cos^2(\theta) - (1)(1) + (1)(1)\cos^2(\theta)$$

$$= 2\cos^2(\theta) - 1$$

, for $\cos(3\theta)$, Eq. (B.1) yields

$$\cos(3\theta) = \sum_{k=0}^{\lfloor\frac{3}{2}\rfloor}\sum_{m=0}^{k}(-1)^{k+m}\binom{3}{2k}\binom{k}{m}[\cos(\theta)]^{3-2(k-m)}$$

$$= (-1)^0\binom{3}{0}\binom{0}{0}\cos^3(\theta) + (-1)^1\binom{3}{2}\binom{1}{0}\cos^1(\theta) + (-1)^2\binom{3}{2}\binom{1}{1}\cos^3(\theta)$$

$$= \binom{3}{0}\binom{0}{0}\cos^3(\theta) - \binom{3}{2}\binom{1}{0}\cos^1(\theta) + \binom{3}{2}\binom{1}{1}\cos^3(\theta)$$

$$= (1)(1)\cos^3(\theta) - (3)(1)\cos^1(\theta) + (3)(1)\cos^3(\theta)$$

$$= (1 + 3)\cos^3(\theta) - 3\cos(\theta)$$

$$= 4\cos^3(\theta) - 3\cos(\theta)$$

, for $\cos(4\theta)$, Eq. (B.1) yields

$$\cos(4\theta) = \sum_{k=0}^{\lfloor\frac{4}{2}\rfloor}\sum_{m=0}^{k}(-1)^{k+m}\binom{4}{2k}\binom{k}{m}[\cos(\theta)]^{4-2(k-m)}$$

$$= \sum_{k=0}^{2}\sum_{m=0}^{k}(-1)^{k+m}\binom{4}{2k}\binom{k}{m}[\cos(\theta)]^{4-2(k-m)}$$

$$= (-1)^{0+0}\binom{4}{2\times 0}\binom{0}{0}\cos^{4-(0-0)}(\theta) + (-1)^{1+0}\binom{4}{2\times 1}\binom{1}{0}\cos^{4-2(1-0)}(\theta)$$

$$+ (-1)^{1+1}\binom{4}{2\times 1}\binom{1}{1}\cos^{4-2(1-1)}(\theta) + (-1)^{2+0}\binom{4}{2\times 2}\binom{2}{0}\cos^{4-2(2-0)}(\theta)$$

$$+ (-1)^{2+1}\binom{4}{2\times 2}\binom{2}{1}\cos^{4-2(2-1)}(\theta) + (-1)^{2+2}\binom{4}{2\times 2}\binom{2}{2}\cos^{4-2(2-2)}(\theta)$$

$$= (1)(1)\cos^4(\theta) - (6)(1)\cos^2(\theta) + (6)(1)\cos^4(\theta) + (1)(1)\cos^0(\theta)$$

$$-(1)(2)\cos^2(\theta) + (1)(1)\cos^4(\theta)$$

$$= (1 + 6 + 1)\cos^4(\theta) + (-6 - 2)\cos^2(\theta) + (1)\cos^0(\theta)$$

$$= 8\cos^4(\theta) - 8\cos^2(\theta) + 1$$

, for $\cos(5\theta)$, Eq. (B. 1) yields

$$\cos(5\theta) = \sum_{k=0}^{\left\lfloor\frac{5}{2}\right\rfloor} \sum_{m=0}^{k} (-1)^{k+m} \binom{5}{2k} \binom{k}{m} [\cos(\theta)]^{5-2(k-m)}$$

$$= \sum_{k=0}^{2} \sum_{m=0}^{k} (-1)^{k+m} \binom{5}{2k} \binom{k}{m} [\cos(\theta)]^{5-2(k-m)}$$

$$= (-1)^0 \binom{5}{0}\binom{0}{0} \cos^5(\theta) + (-1)^1 \binom{5}{2}\binom{1}{0} \cos^3(\theta) + (-1)^2 \binom{5}{2}\binom{1}{1} \cos^5(\theta)$$

$$+ (-1)^2 \binom{5}{4}\binom{2}{0} \cos^1(\theta) + (-1)^3 \binom{5}{4}\binom{2}{1} \cos^3(\theta) + (-1)^4 \binom{5}{4}\binom{2}{2} \cos^5(\theta)$$

$$= (1)(1) \cos^5(\theta) - (10)(1) \cos^3(\theta) + (10)(1) \cos^5(\theta) + (5)(1) \cos^1(\theta)$$

$$- (5)(2) \cos^3(\theta) + (5)(1) \cos^5(\theta)$$

$$= (1 + 10 + 5) \cos^5(\theta) + (-10 - 10) \cos^3(\theta) + (5) \cos(\theta)$$

$$= 16 \cos^5(\theta) - 20 \cos^3(\theta) + 5 \cos(\theta)$$

, for $\cos(6\theta)$, Eq. (B. 1) yields

$$\cos(6\theta) = \sum_{k=0}^{\left\lfloor\frac{6}{2}\right\rfloor} \sum_{m=0}^{k} (-1)^{k+m} \binom{6}{2k} \binom{k}{m} [\cos(\theta)]^{6-2(k-m)}$$

$$= \sum_{k=0}^{3} \sum_{m=0}^{k} (-1)^{k+m} \binom{6}{2k} \binom{k}{m} [\cos(\theta)]^{6-2(k-m)}$$

$$= (-1)^0 \binom{6}{0}\binom{0}{0} \cos^6(\theta) + (-1)^1 \binom{6}{2}\binom{1}{0} \cos^4(\theta) + (-1)^2 \binom{6}{2}\binom{1}{1} \cos^6(\theta)$$

$$+ (-1)^2 \binom{6}{4}\binom{2}{0} \cos^2(\theta) + (-1)^3 \binom{6}{4}\binom{2}{1} \cos^4(\theta) + (-1)^4 \binom{6}{4}\binom{2}{2} \cos^6(\theta)$$

$$+ (-1)^3 \binom{6}{6}\binom{3}{0} \cos^0(\theta) + (-1)^4 \binom{6}{6}\binom{3}{1} \cos^2(\theta) + (-1)^5 \binom{6}{6}\binom{3}{2} \cos^4(\theta)$$

$$+ (-1)^6 \binom{6}{6}\binom{3}{3} \cos^6(\theta)$$

$$= (1)(1) \cos^6(\theta) - (15)(1) \cos^4(\theta) + (15)(1) \cos^6(\theta) + (15)(1) \cos^2(\theta)$$

$$- (15)(2) \cos^4(\theta) + (15)(1) \cos^6(\theta) - (1)(1) \cos^0(\theta) + (1)(3) \cos^2(\theta)$$

$$- (1)(3) \cos^4(\theta) + (1)(1) \cos^6(\theta)$$

$$= (1 + 15 + 15 + 1)\cos^6(\theta) - (15 + 30 + 3))\cos^4(\theta) + (15 + 3)\cos^2(\theta) - (1) \cos^1(\theta)$$

$$= 32 \cos^6(\theta) - 48 \cos^4(\theta) + 18 \cos^2(\theta) - 1$$

, and for $\cos(7\theta)$, Eq. (B. 1) yields

$$\cos(7\theta) = \sum_{k=0}^{\left\lfloor\frac{7}{2}\right\rfloor} \sum_{m=0}^{k} (-1)^{k+m} \binom{7}{2k} \binom{k}{m} [\cos(\theta)]^{7-2(k-m)}$$

$$= \sum_{k=0}^{3} \sum_{m=0}^{k} (-1)^{k+m} \binom{7}{2k} \binom{k}{m} [\cos(\theta)]^{7-2(k-m)}$$

$$= (-1)^0 \binom{7}{0}\binom{0}{0} \cos^7(\theta) + (-1)^1 \binom{7}{2}\binom{1}{0} \cos^5(\theta) + (-1)^2 \binom{7}{2}\binom{1}{1} \cos^7(\theta)$$

$$+ (-1)^2 \binom{7}{4}\binom{2}{0} \cos^3(\theta) + (-1)^3 \binom{7}{4}\binom{2}{1} \cos^5(\theta) + (-1)^4 \binom{7}{4}\binom{2}{2} \cos^7(\theta)$$

$$+ (-1)^3 \binom{7}{6}\binom{3}{0} \cos^1(\theta) + (-1)^4 \binom{7}{6}\binom{3}{1} \cos^3(\theta) + (-1)^5 \binom{7}{6}\binom{3}{2} \cos^5(\theta)$$

$$+ (-1)^6 \binom{7}{6}\binom{3}{3} \cos^7(\theta)$$

$$= (1)(1)\cos^7(\theta) - (21)(1) \cos^5(\theta) + (21)(1)\cos^7(\theta) + (35)(1)\cos^3(\theta)$$

$$- (35)(2)\cos^5(\theta) + (35)(1) \cos^7(\theta) - (7)(1) \cos^1(\theta) + (7)(3)\cos^3(\theta)$$

$$- (7)(3)\cos^5(\theta) + (7)(1) \cos^7(\theta)$$

$$= (1 + 21 + 35 + 7)\cos^7(\theta) - (21 + 70 + 21))\cos^5(\theta) + (35 + 21)\cos^3(\theta)$$

$$- (7) \cos^1(\theta)$$

$$= 64 \cos^7(\theta) - 112 \cos^5(\theta) + 56 \cos^3(\theta) - 7 \cos(\theta)$$

Theorem B.3.

A polynomial in $\cos^n(\theta)$ can be expressed as a linear combination of $\cos(n\theta)$, where

$$\cos^n(\theta) = \frac{1}{2^n} \sum_{k=0}^{n} \binom{n}{k} \cos[(n-2k)\theta] \tag{B.5}$$

$$= \begin{cases} \dfrac{1}{2^n} \binom{n}{n/2} + \dfrac{1}{2^{n-1}} \displaystyle\sum_{k=0}^{\frac{n}{2}-1} \binom{n}{k} \cos[(n-2k)\theta] & \text{for n even} \quad\text{(B.6)} \\[4ex] \dfrac{1}{2^{n-1}} \displaystyle\sum_{k=0}^{\lfloor\frac{n}{2}\rfloor} \binom{n}{k} \cos[(n-2k)\theta] & \text{for n odd} \quad\text{(B.7)} \end{cases}$$

Proof:

$$\cos^n(\theta) = \left[\frac{e^{j\theta} + e^{-j\theta}}{2}\right]^n$$

$$= \text{Re}\left[\left[\frac{e^{j\theta} + e^{-j\theta}}{2}\right]^n\right]$$

$$= \text{Re}\left[\frac{1}{2^n} \sum_{k=0}^{n} \binom{n}{k} e^{j(n-k)\theta} e^{-jk\theta}\right]$$

$$= \text{Re}\left[\frac{1}{2^n} \sum_{k=0}^{n} \binom{n}{k} e^{j(n-2k)\theta}\right]$$

$$= \text{Re}\left[\frac{1}{2^n} \sum_{k=0}^{n} \binom{n}{k} \{\cos[(n-2k)\theta] + j\sin[(n-2k)\theta]\}\right]$$

$$= \text{Re}\left[\frac{1}{2^n} \sum_{k=0}^{n} \binom{n}{k} \cos[(n-2k)\theta] + j\frac{1}{2^n} \sum_{k=0}^{n} \binom{n}{k} \sin[(n-2k)\theta]\right]$$

$$= \frac{1}{2^n} \sum_{k=0}^{n} \binom{n}{k} \cos[(n-2k)\theta] \tag{B.5}$$

$$= \begin{cases} \dfrac{1}{2^n} \binom{n}{n/2} + \dfrac{1}{2^{n-1}} \displaystyle\sum_{k=0}^{\frac{n}{2}-1} \binom{n}{k} \cos[(n-2k)\theta] & \text{for n even} \quad\text{(B.6)} \\[4ex] \dfrac{1}{2^{n-1}} \displaystyle\sum_{k=0}^{\lfloor\frac{n}{2}\rfloor} \binom{n}{k} \cos[(n-2k)\theta] & \text{for n odd} \quad\text{(B.7)} \end{cases}$$

Example B.3. Polynomial to Harmonic conversion

n $\cos^n(\theta)$ harmonic expansion

0        $\cos^0(\theta) = 1$

1        $\cos^1(\theta) = \cos(\theta)$

2        $\cos^2(\theta) = \dfrac{\cos(2\theta) + 1}{2}$

3        $\cos^3(\theta) = \dfrac{\cos(3\theta) + 3\cos(\theta)}{2^2}$

4        $\cos^4(\theta) = \dfrac{\cos(4\theta) + 4\cos(2\theta) + 3}{2^3}$

5        $\cos^5(\theta) = \dfrac{\cos(5\theta) + 5\cos(3\theta) + 10\cos(\theta)}{2^4}$

6        $\cos^6(\theta) = \dfrac{\cos(6\theta) + 6\cos(4\theta) + 15\cos(2\theta) + 10}{2^5}$

7        $\cos^7(\theta) = \dfrac{\cos(7\theta) + 7\cos(5\theta) + 21\cos(3\theta) + 35\cos(\theta)}{2^6}$

Proof:
For $\cos^0(\theta)$, Eq.(B. 5) yields

$$\cos^0(\theta) = \frac{1}{2^0} \sum_{k=0}^{0} \binom{0}{k} \cos[(0 - 2k)\theta]$$

$$= \binom{0}{0} \cos[(0 - 2 \times 0)\theta]$$

$$= 1$$

, for $\cos^1(\theta)$, Eq.(B. 5) yields

$$\cos^1(\theta) = \frac{1}{2^1} \sum_{k=0}^{1} \binom{1}{k} \cos[(1 - 2k)\theta]$$

$$= \frac{1}{2^1} \binom{1}{0} \cos[(1 - 2 \times 0)\theta] + \frac{1}{2^1} \binom{1}{1} \cos[(1 - 2 \times 1)\theta]$$

$$= \frac{1}{2} \cos(\theta) + \frac{1}{2} \cos(-\theta)$$

$$= \cos(\theta)$$

, for $\cos^2(\theta)$, Eq.(B. 5) yields

$$\cos^2(\theta) = \frac{1}{2^2} \sum_{k=0}^{2} \binom{2}{k} \cos[(2 - 2k)\theta]$$

$$= \frac{1}{2^2} \left[\binom{2}{0} \cos[(2 - 2 \times 0)\theta] + \binom{2}{1} \cos[(2 - 2 \times 1)\theta] + \binom{2}{2} \cos[(2 - 2 \times 2)\theta] \right]$$

$$= \frac{1}{2^2} [\cos(2\theta) + 2\cos(0\theta) + \cos(-2\theta)]$$

$$= \frac{1}{2^2} [\cos(2\theta) + 2 + \cos(-2\theta)]$$

$$= \frac{1}{2} [\cos(2\theta) + 1]$$

, for $\cos^3(\theta)$, Eq.(B. 5) yields

$$\cos^3(\theta) = \frac{1}{2^3} \sum_{k=0}^{3} \binom{3}{k} \cos[(3-2k)\theta]$$

$$= \frac{1}{2^3} \left[\binom{3}{0} \cos[(3-2\times 0)\theta] + \binom{3}{1} \cos[(3-2\times 1)\theta] + \binom{3}{2} \cos[(3-2\times 2)\theta \right.$$
$$\left. + \binom{3}{3} \cos[(3-2\times 3)\theta] \right]$$

$$= \frac{1}{2^3} \left[1 \cos(3\theta) + 3 \cos(1\theta) + 3 \cos(-1\theta) + 1 \cos(-3\theta) \right]$$

$$= \frac{1}{2^3} \left[\cos(3\theta) + 3 \cos(\theta) + 3 \cos(\theta) + \cos(3\theta) \right]$$

$$= \frac{1}{2^2} \left[\cos(3\theta) + 3 \cos(\theta) \right]$$

, for $\cos^4(\theta)$, Eq.(B. 5) yields

$$\cos^4(\theta) = \frac{1}{2^4} \sum_{k=0}^{4} \binom{4}{k} \cos[(4-2k)\theta]$$

$$= \frac{1}{2^4} \left[\binom{4}{0} \cos[(4-2\times 0)\theta] + \binom{4}{1} \cos[(4-2\times 1)\theta] + \binom{4}{2} \cos[(4-2\times 2)\theta] \right.$$
$$\left. + \binom{4}{3} \cos[(4-2\times 3)\theta] + \binom{4}{4} \cos[(4-2\times 4)\theta]] \right]$$

$$= \frac{1}{2^4} \left[1 \cos(4\theta) + 4 \cos(2\theta) + 6 \cos(0\theta) + 4 \cos(-2\theta) + 1 \cos(-4\theta] \right]$$

$$= \frac{1}{2^3} \left[\cos(4\theta) + 4 \cos(2\theta) + 3 \right]$$

, for $\cos^5(\theta)$, Eq.(B. 7) yields

$$\cos^5(\theta) = \frac{1}{2^{5-1}} \sum_{k=0}^{\lfloor \frac{5}{2} \rfloor} \binom{5}{k} \cos[(5-2k)\theta]$$

$$= \frac{1}{16} \sum_{k=0}^{2} \binom{5}{k} \cos[(5-2k)\theta]$$

$$= \frac{1}{16} \left[\binom{5}{0} \cos(5\theta) + \binom{5}{1} \cos(3\theta) + \binom{5}{2} \cos(\theta) \right]$$

$$= \frac{1}{16} \left[\cos(5\theta) + 5 \cos(3\theta) + 10 \cos(\theta) \right]$$

, for $\cos^6(\theta)$, Eq.(B. 6) yields

$$\cos^6(\theta) = \frac{1}{2^6} \binom{6}{6/2} + \frac{1}{2^{6-1}} \sum_{k=0}^{\frac{6}{2}-1} \binom{6}{k} \cos[(6-2k)\theta]$$

$$= \frac{1}{2^6} \binom{6}{3} + \frac{1}{2^5} \sum_{k=0}^{2} \binom{6}{k} \cos[(6-2k)\theta]$$

$$= \frac{1}{64} 20 + \frac{1}{32} \left[\binom{6}{0} \cos(6\theta) + \binom{6}{1} \cos(4\theta) + \binom{6}{2} \cos(2\theta) \right]$$

$$= \frac{1}{32} \left[\cos(6\theta) + 6 \cos(4\theta) + 15 \cos(2\theta) + 10 \right]$$

, and for $\cos^7(\theta)$, Eq.(B. 7) yields

$$\cos^7(\theta) \;=\; \frac{1}{2^{7-1}} \sum_{k=0}^{\left\lfloor \frac{7}{2} \right\rfloor} \binom{7}{k} \cos[(7-2k)\theta]$$

$$=\; \frac{1}{64} \sum_{k=0}^{2} \binom{7}{k} \cos[(7-2k)\theta]$$

$$=\; \frac{1}{64} \sum_{k=0}^{2} \binom{7}{k} \cos[(7-2k)\theta]$$

$$=\; \frac{1}{64} \left[\binom{7}{0} \cos(7\theta) + \binom{7}{1} \cos(5\theta) + \binom{7}{2} \cos(3\theta) + \binom{7}{3} \cos(\theta) \right]$$

$$=\; \frac{1}{64} \left[\cos(7\theta) + 7\cos(5\theta) + 21\cos(3\theta) + 35\cos(\theta) \right]$$

, then theorem B.3. show that a polynomial in $\cos^n(\theta)$ can be expressed as a linear combination of $\cos(n\theta)$.

Appendix C

Trigonometric Identities

C.1. Definitions

Consider x is a real variable , then

$$\cos(x) = \frac{e^{jx} + e^{-jx}}{2} \quad , \quad \sin(x) = \frac{e^{jx} - e^{-jx}}{2j} \quad , \quad \tan(x) = \frac{\sin(x)}{\cos(x)} = \frac{e^{jx} - e^{-jx}}{j(e^{jx} - e^{-jx})}$$

$$, \quad \cos(x)\sec(x) = 1 \quad , \quad \sin(x)\csc(x) = 1 \quad , \quad \tan(x)\cot(x) = 1$$

C.2. Circular (Triangle) Functions

$$e^{jx} = \cos(x) + j\sin(x)$$
$$, \qquad e^{-jx} = \cos(x) - j\sin(x) \qquad \text{(Euler's theorem)}$$
$$, \qquad \sin^2(x) + \cos^2(x) = 1$$
$$, \qquad 1 + \tan^2(x) = \sec^2(x)$$
$$, \qquad 1 + \cot^2(x) = \csc^2(x)$$

, if x and y are real variables, $x > y$, *then*

$$2\cos(x)\cos(y) = \cos(x - y) + \cos(x + y)$$
$$, \quad 2\sin(x)\sin(y) = \cos(x - y) - \cos(x + y)$$
$$, \quad 2\sin(x)\cos(y) = \sin(x + y) + \sin(x - y)$$
$$, \quad 2\cos(x)\sin(y) = \sin(x + y) - \sin(x - y)$$

$$, \quad \sin(x + y) = \sin(x)\cos(y) + \cos(x)\sin(y)$$
$$, \quad \sin(x - y) = \sin(x)\cos(y) - \cos(x)\sin(y)$$
$$, \quad \cos(x + y) = \cos(x)\cos(y) - \sin(x)\sin(y)$$
$$, \quad \cos(x - y) = \cos(x)\cos(y) + \sin(x)\sin(y)$$

$$, \quad \sin(x) + \sin(y) = 2\sin[(x+y)/2]\cos[(x-y)/2]$$
$$, \quad \sin(x) - \sin(y) = 2\cos[(x+y)/2]\sin[(x-y)/2]$$
$$, \quad \cos(x) + \cos(y) = 2\cos[(x+y)/2]\cos[(x-y)/2]$$
$$, \quad \cos(x) - \cos(y) = -2\sin[(x+y)/2]\sin[(x-y)/2]$$

$$\sin(2x) = 2\sin(x)\cos(x)$$
$$, \quad \cos(2x) = \cos^2(x) - \sin^2(x)$$
$$, \quad \tan(2x) = \frac{2\tan(x)}{1 - \tan^2(x)}$$

$$, \quad \cos^2(x) = \tfrac{1}{2}[1 + \cos(2x)]$$
$$, \quad \sin^2(x) = \tfrac{1}{2}[1 - \cos(2x)]$$
$$, \quad \tan^2(x) = \frac{1 - \cos(2x)}{1 + \cos(2x)}$$

$$\sin(3x) = 3\sin(x) - 4\sin^3(x)$$
$$, \quad \cos(3x) = 4\cos^3(x) - 3\cos(x)$$
$$, \quad \tan(3x) = \frac{3\tan(x) - \tan^3(x)}{1 - 3\tan^2(x)}$$

, $$\sin^3(x) = \frac{3}{4}\sin(x) - \frac{1}{4}\sin(3x)$$

, $$\cos^3(x) = \frac{3}{4}\cos(x) + \frac{1}{4}\cos(3x)$$

, $$\sin^4(x) = \frac{3}{8} - \frac{1}{2}\cos(2x) + \frac{1}{8}\cos(4x)$$

, $$\cos^4(x) = \frac{3}{8} + \frac{1}{2}\cos(2x) + \frac{1}{8}\cos(4x)$$

, $$\cos(x + \frac{\pi}{2}) = -\sin(x)$$

, $$\cos(x - \frac{\pi}{2}) = \sin(x)$$

, $$\sin(x + \frac{\pi}{2}) = \cos(x)$$

, $$\sin(x - \frac{\pi}{2}) = -\cos(x)$$

, and $A\cos(x) - B\sin(x) = R\cos(x + \theta)$

, where $R = \sqrt{A^2 + B^2}$, $\theta = \tan^{-1}[B/A]$

, $A = R\cos(\theta)$, and $B = R\sin(\theta)$

C.3. Inverse Circular Functions

If $y = \sin^{-1}(x)$, then $x = \sin(y)$

, if $y = \cos^{-1}(x)$, then $x = \cos(y)$

, and if $y = \tan^{-1}(x)$, then $x = \tan(y)$

Appendix D

Mathematical Techniques and Identities

D.1. Differential Calculus

D.1.1. Definition

$$\frac{d}{dt}[x(t)] = \lim_{\Delta t \to 0} \frac{x(t + \Delta t)) - x(t)}{\Delta t}$$

D.1.2. Differentiation Rules

$$\frac{d}{dt}[x(t)\,y(t)] = x(t)\frac{dy(t)}{dt} + y(t)\frac{dx(t)}{dt} \qquad \text{products}$$

$$, \qquad \frac{d}{dt}\left[\frac{x(t)}{y(t)}\right] = \frac{y(t)\frac{dx(t)}{dt} - x(t)\frac{dy(t)}{dt}}{y^2(t)} \qquad \text{quotient}$$

$$, \text{and} \qquad \frac{d}{dt}[x(y(t))] = \frac{dx(t)}{dy(t)}\frac{dy(t)}{dt} \qquad \text{(chain rule)}$$

D.1.3. Derivative Table

$$\frac{d}{dt}[x^n(t)] = n\,x^{n-1}(t)$$

$$, \quad \frac{d}{dt}[\sin(t)] = \cos(t) \qquad , \quad \frac{d}{dt}[\sin(at)] = a\,\cos(at)$$

$$, \quad \frac{d}{dt}[\cos(t)] = -\sin(t) \qquad , \quad \frac{d}{dt}[\cos(at)] = -a\,\sin(at)$$

$$, \quad \frac{d}{dt}[\tan(t)] = \sec^2(t) \qquad , \quad \frac{d}{dt}[\tan(at)] = a\,\sec^2(at)$$

$$, \quad \frac{d}{dt}[\operatorname{cosec}(t)] = -\operatorname{cosec}(t)\cot(t), \quad \frac{d}{dt}[\operatorname{cosec}(at)] = a\,\operatorname{cosec}(at)\cot(at)$$

$$, \quad \frac{d}{dt}[\sec(t)] = \sec(t)\tan(t) \qquad , \quad \frac{d}{dt}[\sec(at)] = a\,\sec(at)\tan(at)$$

$$, \quad \frac{d}{dt}[\cot(t)] = -\operatorname{cosec}^2(t) \qquad , \quad \frac{d}{dt}[\cot(at)] = -a\,\operatorname{cosec}^2(at)$$

$$, \quad \frac{d}{dt}[\sin^{-1}(t)] = \frac{1}{\sqrt{1-t^2}} \qquad , \quad \frac{d}{dt}[\sin^{-1}(at)] = \frac{a}{\sqrt{1-(at)^2}}$$

$$, \quad \frac{d}{dt}[\cos^{-1}(t)] = -\frac{1}{\sqrt{1-t^2}} \qquad , \quad \frac{d}{dt}[\cos^{-1}(at)] = -\frac{a}{\sqrt{1-(at)^2}}$$

$$, \quad \frac{d}{dt}[\tan^{-1}(t)] = \frac{1}{1+t^2} \qquad , \quad \frac{d}{dt}[\tan^{-1}(at)] = \frac{a}{1+(at)^2}$$

$$, \quad \frac{d}{dt}[\operatorname{cosec}^{-1}(t)] = -\frac{1}{t\sqrt{t^2-1}} \quad , \quad \frac{d}{dt}[\operatorname{cosec}^{-1}(at)] = -\frac{a}{at\sqrt{(at)^2-1}}$$

$$, \quad \frac{d}{dt}[\sec^{-1}(t)] = \frac{1}{t\sqrt{t^2-1}} \qquad , \quad \frac{d}{dt}[\sec^{-1}(at)] = \frac{a}{at\sqrt{(at)^2-1}}$$

$$, \quad \frac{d}{dt}[\cot^{-1}(t)] = -\frac{1}{1+t^2} \qquad , \quad \frac{d}{dt}[\cot^{-1}(at)] = -\frac{a}{1+(at)^2}$$

$$, \qquad \frac{d}{dt}[e^t] = e^t \qquad , \qquad \frac{d}{dt}[e^{at}] = a\, e^{at}$$

$$, \qquad \frac{d}{dt}[a^t] = a^t\, \ln(a) \qquad , \qquad \frac{d}{dt}[a^{bt}] = b\, a^{bt}\, \ln(a)$$

$$, \qquad \frac{d}{dt}[\ln(t)] = \frac{1}{t} \qquad , \qquad \frac{d}{dt}[\log_a(t)] = \frac{1}{t}\log_a e$$

, and

$$\frac{d}{dt}\left[\int_{a(t)}^{b(t)} x(\tau,t)\,d\tau\right] = x(b(t),t)\,\frac{db(t)}{dt} - x(a(t),t)\,\frac{da(t)}{dt} + \int_{a(t)}^{b(t)} \frac{\partial}{\partial t}[x(\tau,t)]\,d\tau \quad \text{(Leibniz's rule)}$$

D.2. Indeterminate Forms

If $\lim\limits_{t \to a} x(t)$ is of the form: $\dfrac{0}{0}$, $\quad \dfrac{\infty}{\infty}$, $\quad 0 \times \infty$, $\quad \infty - \infty, 0^0$, $\quad \infty^0$, $\quad$ and 1^∞.

, then $\qquad \lim\limits_{t \to a} x(t) = \lim\limits_{t \to a} \dfrac{N(t)}{D(t)} = \lim\limits_{t \to a} \dfrac{dN(t)/dt}{dD(t)/dt} \qquad$ (L'Hospital's rule)

, where $N(t)$ and $D(t)$ are the numerator and the denominator of the function $x(t)$, and $N(a) = 0$, and $D(a) = 0$.

D.3. Integral Calculus

D.3.1. Definition

$$\int x(t)\,dt = \lim_{\Delta t \to 0}\left[\sum_n [x(n\,\Delta t)]\,\Delta t\right]$$

D.3.2. Integration Techniques

D.3.2.1. Change in variable. Let $v = u(t)$

$$\int_a^b x(t)\,dt = \int_{u(a)}^{u(b)} \left[\left.\frac{x(t)}{dv/dt}\right|_{\text{at } t=u^{-1}(v)}\right] dt$$

D.3.2.2. Integration by parts

$$\int_a^b x(t)\,dy(t) = [\,x(t)\,y(t)\,]_a^b - \int_a^b y(t)\,dx(t)$$

D.3.2.3. Integral Tables

D.3.2.3.1. Indefinite Integrals

$$\int t^n\,dt = \frac{t^{n+1}}{n+1} \qquad\qquad , \quad \text{and } n \geq 0$$

$$, \quad \int \frac{1}{t}\,dt = \ln|t|$$

$$, \quad \int \frac{1}{t^n}\,dt = \frac{-1}{(n-1)\,t^{n-1}} \qquad , \quad \text{and } n > 1$$

$$, \quad \int (a + bt)^n \, dt = \frac{(a + bt)^{n+1}}{b(n + 1)} \qquad , \text{ and } n \geq 0$$

$$, \quad \int \frac{1}{a + bt} \, dt = \frac{1}{b} \ln |a + bt|$$

$$, \quad \int \frac{1}{(a + bt)^n} \, dt = \frac{-1}{(n - 1) \, b \, (a + bt)^{n-1}} \qquad , \text{ and } n > 1$$

$$, \quad \int \frac{1}{at^2 + bt + c} \, dt = \begin{cases} \dfrac{2}{\sqrt{4ac - b^2}} \tan^{-1} \left(\dfrac{2at + b}{\sqrt{4ac - b^2}} \right) & \text{for } 4ac > b^2 \\[2ex] \dfrac{1}{\sqrt{b^2 - 4ac}} \ln \left[\dfrac{2at + b - \sqrt{b^2 - 4ac}}{2at + b + \sqrt{b^2 - 4ac}} \right] & \text{for } b^2 > 4ac \\[2ex] \dfrac{-2}{2at + b^2} & \text{for } 4ac = b^2 \end{cases}$$

$$, \quad \int \frac{t}{at^2 + bt + c} \, dt = \frac{1}{2a} \ln |at^2 + bt + c| - \frac{b}{2a} \int \frac{1}{at^2 + bt + c} \, dt$$

$$, \quad \int \frac{1}{a^2 + t^2} \, dt = \frac{1}{a} \tan^{-1} \left(\frac{t}{a} \right)$$

$$, \quad \int \frac{1}{a^2 + b^2 t^2} \, dt = \frac{1}{ab} \tan^{-1} \left(\frac{bt}{a} \right)$$

$$, \quad \int \frac{t}{a^2 + t^2} \, dt = \frac{1}{2} \ln(a^2 + t^2)$$

$$, \quad \int \frac{t^2}{a^2 + t^2} \, dt = t - a \tan^{-1} \left(\frac{t}{a} \right)$$

$$, \quad \int \frac{1}{(a^2 + t^2)^2} \, dt = \frac{t}{2a^2(a^2 + t^2)} + \frac{1}{2a^3} \tan^{-1} \left(\frac{t}{a} \right)$$

$$, \quad \int \frac{t}{(a^2 + t^2)^2} \, dt = \frac{-1}{2(a^2 + t^2)}$$

$$, \quad \int \frac{t^2}{(a^2 + t^2)^2} \, dt = \frac{-t}{2(a^2 + t^2)} + \frac{1}{2a} \tan^{-1} \left(\frac{t}{a} \right)$$

$$, \quad \int \frac{1}{(a^2 + t^2)^3} \, dt = \frac{t}{4a^2(a^2 + t^2)^2} + \frac{3t}{8a^4(a^2 + t^2)} + \frac{3}{8a^5} \tan^{-1} \left(\frac{t}{a} \right)$$

$$, \quad \int \frac{t^2}{(a^2 + t^2)^3} \, dt = \frac{-t}{4(a^2 + t^2)^2} + \frac{t}{8a^2(a^2 + t^2)} + \frac{1}{8a^3} \tan^{-1} \left(\frac{t}{a} \right)$$

$$, \quad \int \frac{t^4}{(a^2 + t^2)^3} \, dt = \frac{a^2 t}{4(a^2 + t^2)^2} - \frac{5t}{8(a^2 + t^2)} + \frac{3}{8a} \tan^{-1} \left(\frac{t}{a} \right)$$

$$, \quad \int \frac{1}{(a^2 + t^2)^4} \, dt = \frac{t}{6a^2(a^2 + t^2)^3} + \frac{5t}{24a^4(a^2 + t^2)^2} + \frac{5t}{16a^6(a^2 + t^2)} + \frac{5}{16a^7} \tan^{-1} \left(\frac{t}{a} \right)$$

$$\int \frac{t^2}{(a^2+t^2)^4}\, dt = \frac{-t}{6(a^2+t^2)^3} + \frac{t}{24a^2(a^2+t^2)^2} + \frac{t}{16a^4(a^2+t^2)} + \frac{1}{16a^5}\tan^{-1}\left(\frac{t}{a}\right)$$

$$\int \frac{t^4}{(a^2+t^2)^4}\, dt = \frac{a^2 t}{6(a^2+t^2)^3} - \frac{7t}{24(a^2+t^2)^2} + \frac{t}{16a^2(a^2+t^2)} + \frac{1}{16a^3}\tan^{-1}\left(\frac{t}{a}\right)$$

$$\int \frac{1}{a^4+t^4}\, dt = \frac{1}{4a^3\sqrt{2}}\ln\left(\frac{t^2+at\sqrt{2}+a^2}{t^2-at\sqrt{2}+a^2}\right) + \frac{1}{2a^3\sqrt{2}}\tan^{-1}\left(\frac{at\sqrt{2}}{a^2-t^2}\right)$$

, and

$$\int \frac{t^2}{a^4+t^4}\, dt = -\frac{1}{4a\sqrt{2}}\ln\left(\frac{t^2+at\sqrt{2}+a^2}{t^2-at\sqrt{2}+a^2}\right) + \frac{1}{2a\sqrt{2}}\tan^{-1}\left(\frac{at\sqrt{2}}{a^2-t^2}\right)$$

D.3.2.3.2. Trigonometric functions

$$\int \cos(t)\, dt = \sin(t)$$

$$\int t\cos(t)\, dt = \cos(t) + t\sin(t)$$

$$\int t^2 \cos(t)\, dt = 2t\cos(t) + (t^2 - 2)\sin(t)$$

$$\int \sin(t)\, dt = -\cos(t)$$

$$\int t\sin(t)\, dt = \sin(t) - t\cos(t)$$

, and

$$\int t^2 \sin(t)\, dt = 2t\sin(t) - (t^2 - 2)\cos(t)$$

D.3.2.3.3. Exponential functions

$$\int e^t\, dt = e^t$$

$$\int e^{at}\, dt = \frac{e^{at}}{a} \qquad\qquad , \quad a \text{ is real or complex}$$

$$\int t\, e^{at}\, dt = e^{at}\left[\frac{t}{a} - \frac{1}{a^2}\right] \qquad , \quad a \text{ is real or complex}$$

$$\int t^2\, e^{at}\, dt = e^{at}\left[\frac{t^2}{a} - \frac{2t}{a^2} + \frac{2}{a^3}\right] \qquad , \quad a \text{ is real or complex}$$

$$\int t^3\, e^{at}\, dt = e^{at}\left[\frac{t^3}{a} - \frac{3t^2}{a^2} + \frac{6t}{a^3} - \frac{6}{a^4}\right] \qquad , \quad a \text{ is real or complex}$$

$$\int e^{at}\sin(t)\, dt = \frac{e^{at}}{a^2+1}\left[a\sin(t) - \cos(t)\right]$$

, and

$$\int e^{at}\cos(t)\, dt = \frac{e^{at}}{a^2+1}\left[a\cos(t) - \sin(t)\right]$$

D.3.2.3.4. Definite Integrals

$$\int_0^\infty \frac{t^{m-1}}{1+t^n}\, dt = \frac{\pi/n}{\sin(\frac{m\pi}{n})} \qquad , \qquad n > m > 0$$

$$, \quad \int_0^\infty t^{\alpha-1}\, e^{-t}\, dt = \Gamma(\alpha) \qquad , \qquad \alpha > 0$$

, where $\quad \Gamma(\alpha+1) = \alpha\,\Gamma(\alpha)$

$, \qquad\qquad\qquad \Gamma(1) = 1$

$, \qquad\qquad\qquad \Gamma(\tfrac{1}{2}) = \sqrt{\pi}$

, and $\qquad\qquad \Gamma(n) = (n-1)!$ for n is positive integer.

$$, \quad \int_0^\infty (1-at)\, e^{-bt}\, dt = \frac{1}{b^2}\,[\,b-a\,]$$

$$, \quad \int_0^\infty e^{-(at)^2}\, dt = \sqrt{\pi}\,/2a \qquad , \quad a > 0$$

$$, \quad \int_0^\infty t^{2n}\, e^{-at^2}\, dt = \frac{1 \times 3 \times 5 \,\dots\dots\, (2n-1)}{2^{a+1}\, a^n}\, \sqrt{\pi/a}$$

$$, \quad \int_0^\infty e^{-at} \cos(bt)\, dt = \frac{a}{a^2+b^2} \qquad , \quad a > 0$$

$$, \quad \int_0^\infty e^{-at} \sin(bt)\, dt = \frac{b}{a^2+b^2} \qquad , \quad a > 0$$

$$, \quad \int_0^\infty e^{-(at)^2} \cos(bt)\, dt = \frac{\sqrt{\pi}}{2a}\, e^{-(b/2a)^2} \qquad , \quad a > 0$$

$$, \quad \int_0^\infty t^{\alpha-1}\, \cos(bt)\, dt = \frac{\Gamma(\alpha)}{b^\alpha} \cos\left(\frac{\pi\alpha}{2}\right) \qquad , \quad 0 < \alpha < 1 \;, \; b > 0$$

$$, \quad \int_0^\infty t^{\alpha-1}\, \sin(bt)\, dt = \frac{\Gamma(\alpha)}{b^\alpha} \sin\left(\frac{\pi\alpha}{2}\right) \qquad , \quad 0 < \alpha < 1 \;, \; b > 0$$

$$, \quad \int_0^\infty t\, e^{-at^2}\, I_k(bt)\, dt = \frac{1}{2a}\, e^{b^2/4a}$$

, where $\quad I_k(bt) = \dfrac{1}{\pi}\displaystyle\int_0^\pi e^{bt\cos(\theta)}\,\cos(k\theta)\,d\theta$

$$, \quad \int_0^\infty \frac{\sin(t)}{t}\, dt = \frac{\pi}{2}$$

$$, \quad \int_0^\infty \frac{\sin^2(t)}{t^2}\, dt = \frac{\pi}{2}$$

$$, \quad \int_0^\infty \frac{\cos(at)}{b^2+t^2}\, dt = \frac{\pi}{2b}\, e^{-ab} \qquad , \quad a > 0, \; b > 0$$

$$, \quad \int_0^\infty \frac{t\sin(at)}{b^2+t^2}\, dt = \frac{\pi}{2}\, e^{-ab} \qquad , \quad a > 0, \; b > 0$$

D.4. Series Expansions

D.4.1. Finite Series

$$\sum_{n=1}^{N} n = \frac{N(N+1)}{2}$$

$$, \sum_{n=1}^{N} n^2 = \frac{N(N+1)(2N+1)}{6}$$

$$, \sum_{n=1}^{N} n^3 = \frac{N^2 (N+1)^2}{4}$$

$$, \sum_{n=0}^{N} t^n = \frac{t^{N+1} - 1}{t - 1}$$

$$, \sum_{n=0}^{N} e^{j(\theta + n\phi)} = \frac{\sin[(N+1)\phi/2]}{\sin(\phi/2)} e^{j[\theta + \left(\frac{n\phi}{2}\right)]}$$

$$, \sum_{k=0}^{N} \binom{N}{k} t^{N-k} \tau^k = (t+\tau)^N$$

, where the standard Binomial coifficient $\binom{N}{k} = \dfrac{N!}{(N-k)! \; k!}$

$$, \sum_{n=0}^{N-1} \cos(n\theta) = \frac{\sin\left(\frac{N\theta}{2}\right) \cos\left[\frac{(N-1)\theta}{2}\right]}{\sin\left(\frac{\theta}{2}\right)}$$

, and

$$, \sum_{n=0}^{N-1} \sin(n\theta) = \frac{\sin\left(\frac{N\theta}{2}\right) \sin\left[\frac{(N-1)\theta}{2}\right]}{\sin\left(\frac{\theta}{2}\right)}$$

D.4.2. Infinite Series

$$x(t) = \sum_{n=0}^{\infty} \left[\frac{x^{(n)}(a)}{n!}\right] (t-a)^n \qquad \text{(Taylor`s series)}$$

$$, \qquad x(t) = \sum_{n=-\infty}^{\infty} C_n e^{j2\pi \frac{n}{T_o} t} \qquad , \qquad -T_o/2 \le t \le T_o/2 \qquad \text{(Fourier series)}$$

$$, \text{where} \qquad C_n = \frac{1}{T_o} \int_{-T_o/2}^{T_o/2} x_p(t) \, e^{-j2\pi \frac{n}{T_o} t} \, dt$$

$$, \qquad e^t = \sum_{n=0}^{\infty} \frac{t^n}{n!}$$

$$, \qquad \sin(t) = \sum_{n=0}^{\infty} \frac{(-1)^n t^{2n+1}}{(2n+1)!}$$

, and

$$\cos(t) = \sum_{n=0}^{\infty} \frac{(-1)^n t^{2n}}{(2n)!}$$

Appendix E

Hyperbolic Functions

E.1. Definitions

Consider x is a real variables, then

$$\cosh(x) = \frac{e^x + e^{-x}}{2} \quad , \quad \sinh(x) = \frac{e^x - e^{-x}}{2} \quad , \quad \tanh(x) = \frac{\sinh(x)}{\cosh(x)} = \frac{e^x - e^{-x}}{e^x + e^{-x}}$$

$$, \quad \cosh(x)\,\mathrm{sech}(x) = 1 \quad , \quad \sinh(x)\,\mathrm{cosech}(x) = 1 \quad , \quad \tanh(x)\,\coth(x) = 1$$

E.2. Hyperbolic Identities

$$e^x = \cosh(x) + \sinh(x)$$
,
$$e^{-x} = \cosh(x) - \sinh(x)$$
,

,
$$\cosh^2(x) - \sinh^2(x) = 1$$
,
$$1 - \tanh^2(x) = \mathrm{sech}^2(x)$$
,
$$\coth^2(x) - 1 = \mathrm{cosech}^2(x)$$

, if x and y are real variables, $x > y$, *then*

,
$$\sinh(x + y) = \sinh(x)\cosh(y) + \cosh(x)\sinh(y)$$
,
$$\sinh(x - y) = \sinh(x)\cosh(y) - \cosh(x)\sinh(y)$$
,
$$\cosh(x + y) = \cosh(x)\cosh(y) + \sinh h(x)\sinh(y)$$
,
$$\cosh(x - y) = \cosh(x)\cosh(y) - \sinh(x)\sinh(y)$$

,
$$2\cosh(x)\cosh(y) = \cosh(x + y) + \cosh(x - y)$$
,
$$2\sinh(x)\sinh(y) = \cosh(x + y) - \cosh(x - y)$$
,
$$2\sinh(x)\cosh(y) = \sinh(x + y) + \sinh(x - y)$$
,
$$2\cosh(x)\sinh(y) = \sinh(x + y) - \sinh(x - y)$$

,
$$\cosh(x) + \cosh(y) = 2\cosh[(x + y)/2]\ \cosh[(x - y)/2]$$
,
$$\cosh(x) - \cosh(y) = 2\sinh[(x + y)/2]\ \sinh[(x - y)/2]$$
,
$$\sinh(x) + \sinh(y) = 2\sinh[(x + y)/2]\ \cosh[(x - y)/2]$$
,
$$\sinh(x) - \sinh(y) = 2\cosh[(x + y)/2]\ \sinh[(x - y)/2]$$

,
$$\tanh(x + y) = \frac{\tanh(x) + \tanh(y)}{1 + \tanh(x)\tanh(y)}$$

,
$$\tanh(x - y) = \frac{\tanh(x) - \tanh(y)}{1 - \tanh(x)\tanh(y)}$$

,
$$\cosh(2x) = \cosh^2(x) + \sinh^2(x)$$
,
$$\sinh(2x) = 2\sinh(x)\cosh(x)$$
,
$$\tanh(2x) = \frac{2\tanh(x)}{1 + \tanh^2(x)}$$

,
$$\cosh^2(x) = \tfrac{1}{2}\,[\cosh(2x) + 1]$$
,
$$\sinh^2(x) = \tfrac{1}{2}\,[\cosh(2x) - 1]$$

$$\sinh(3x) = 4\sinh^3(x) + 3\sinh(x)$$
$$\cosh(3x) = 4\cosh^3(x) - 3\cosh(x)$$
,

$$\sinh^3(x) = -\frac{3}{4}\sinh(x) + \frac{1}{4}\sinh(3x)$$

$$\cosh^3(x) = \frac{3}{4}\cosh(x) + \frac{1}{4}\cosh(3x)$$

E.3. Inverse Hyperbolic Functions

If $\quad y = \sinh^{-1}(x) \quad$, then $\quad x = \sinh(y)$

, if $\quad y = \cosh^{-1}(x) \quad$, then $\quad x = \cos(y)$

, and if $\quad y = \tanh^{-1}(x) \quad$, then $\quad x = \tanh(y)$

E.4. Differentiation of Hyperbolic and Inverse Hyperbolic Functions

$$\frac{d}{dt}[\sinh(t)] = \cosh(t)$$

$$\frac{d}{dt}[\cosh(t)] = \sinh(t)$$

$$\frac{d}{dt}[\tanh(t)] = \operatorname{sech}^2(t)$$

$$\frac{d}{dt}[\coth(t)] = -\operatorname{cosech}^2(t)$$

$$\frac{d}{dt}[\operatorname{sech}(t)] = -\operatorname{sech}(t)\tanh(t)$$

$$\frac{d}{dt}[\operatorname{cosech}(t)] = -\operatorname{cosech}(t)\coth(t)$$

$$\frac{d}{dx}[\sinh^{-1}(x)] = \frac{1}{\sqrt{x^2+1}} \qquad x \geq 1$$

$$\frac{d}{dx}[\cosh^{-1}(x)] = \frac{1}{\sqrt{x^2-1}} \qquad x \geq 1$$

$$\frac{d}{dx}[\tanh^{-1}(x)] = \frac{1}{1-x^2} \qquad |x| < 1$$

E.5. Relation between Hyperbolic Functions and Circular Functions

$$\cosh(ix) = \cos(x) \qquad , \qquad \sinh(ix) = i\sin(x)$$

$$\cos(ix) = \cosh(x) \qquad , \qquad \sin(ix) = i\sinh(x)$$

Appendix F

Logarithms

The word "Logarithm" is made up by Scottish mathematician John Napier (1550-1617) from two Greek words, the first word is "Logos" which mean "ratio", and the second word is "arithmos" which mean "number", and the two words together means "ratio number".

F.1. The logarithms obey the following rules

    i. $\log_a(a) = \log_b(b) = \log_c(c) = 1$

    ii. $\log_b(a) \times \log_c(b) = \log_c(a)$

    iii. $\log_c(a \times b) = \log_c(a) + \log_c(b)$

    iv. $\log_c\left(\dfrac{a}{b}\right) = \log_c(a) - \log_c(b)$

    v. $\log_c\left(\dfrac{1}{a}\right) = -\log_c(a)$

    vi. $\log_c(a^b) = b\log_c(a)$

F.2. The common logarithm, when the base is 10, where

$$\log_{10}(0) = 0 \qquad , \qquad \log_{10}(10) = 1$$
$$\log_{10}(2) = 0.30103 \qquad , \qquad \log_{10}(e) = 0.4343$$

F.3. The natural (Napierian) logarithm, when the base is e, where $\log_e(..)$, and

$$e = \lim_{n \to \infty}[1 + \frac{1}{n}]^n$$

, and due to Maclaurin`s theory, e is given by

$$e = 1 + [1 + \frac{1}{2!} + \frac{1}{3!} + \frac{1}{4!} + \ldots]$$

, or equivalently

$$e = 2.71828$$

. , where

$$\log_e(0) = 0 \qquad , \qquad \log_e(e) = 1$$
$$\log_e(2) = 0.693147 \qquad , \qquad \log_e(10) = 2.303$$

F.4. Logarithm to base 2, where

$$\log_2(0) = 0 \qquad , \qquad \log_2(2) = 1$$
$$\log_2(10) = 0.303 \qquad , \qquad \log_2(e) = 1.442695$$

F.5. Numerical examples

    i. $\log_{10}(a) = \log_e(a) \times \log_{10}(e) = 0.4343\log_e(a)$

    ii. $\log_e(a) = \log_{10}(a) \times \log_e(10) = 2.303\log_{10}(a)$

    iii. $\log_2(a) = \log_{10}(a) \times \log_2(10) = 0.303\log_{10}(a)$

    iv. $\log_2(a) = \log_e(a) \times \log_2(e) = 1.442695\log_e(a)$

Glossary

Symbols

$\alpha(f)$	gain or loss
α_i	optimized parameter, and parameter (root mean square criterion)
$\beta(f)$	phase angle of transfer function
$\beta(f_c)$	phase angle between input and output carrier vector of a system
γ	redundancy
τ	time axis, (varying time delay of correlation process)
τ_p	phase delay (carrier delay)
τ_g	group delay (envelope delay)
$\delta(t)$	Dirac delta function (unit impulse)
Δ	Greek letter used to denote the change in a variable
Δf	incremental spacing between two successive discrete frequencies equal $1/T_o$
$\Psi(f)$	Energy Spectral Density ESD
$\Psi_i(f)$	input energy spectral density
$\Psi_o(f)$	Output Energy Spectral Density OESD
$\Psi_{12}(f)$	Cross Energy Spectral Density CESD
$\varphi(t)$	phase angle
η	code efficiency
φ_c	phase of carrier waveform
$\Omega(f)$	Power Spectral Density PSD
$\Omega_{12}(f)$	Cross Power Spectral Density CPSD
$\otimes$	convolution process notation
^oC	degree Celsius
$\Psi(f,\tau)$	spectrogram of the Short Time Fourier Transform STFT
$\Psi(s,\tau)$	scalogram of the Wavelet transform
$\rightleftharpoons$	convenient symbol of Fourier transform pair
| |	magnitude of the complex quantity contained within
Re[ ]	real part of the complex quantity
Im[ ]	imaginary part of the complex quantity
ln[ ]	natural logarithm of the quantity contained within
$\log_c(a)$	logarithm of a to the base c
x^*	complex conjugate of the complex quantity x
F[ ]	Fourier transform operation
F^{-1}[ ]	Inverse Fourier transform operation
Ω	ohm

Recommended unit prefixes

Prefixes	Multiples and submultiples
Tera	10^{12}
Giga	10^{9}
Mega	10^{6}
Kilo	10^{3}
milli	10^{-3}
micro	10^{-6}
nano	10^{-9}
pico	10^{-12}

Functions

1. Rectangular functionj

$$A \, \mathrm{rect}(\frac{t}{T}) = \begin{cases} A & -T/2 < t < T/2 \\ 0 & |t| > T/2 \end{cases}$$

2. Triangle function

$$A \, \mathrm{tri}(\frac{t}{T}) = \begin{cases} \left[A + \dfrac{A}{T} t\right] & -T/2 < t < 0 \\ \left[A - \dfrac{A}{T} t\right] & 0 < t < T/2 \\ 0 & |t| > T/2 \end{cases}$$

3. Unit step function

$$u(t) = \begin{cases} 1 & t > 0 \\ \tfrac{1}{2} & t = 0 \\ 0 & t < 0 \end{cases}$$

, and in terms of Dirac delta function the unit step function is given by

$$u(t) = \int_{-\infty}^{t} \delta(\tau) \, d\tau$$

4. Signum function

$$\mathrm{sgn}(t) = \begin{cases} 1 & t > 0 \\ 0 & t = 0 \\ -1 & t < 0 \end{cases}$$

5. Dirac delta function $\delta(t)$, where

$$\int_{-\infty}^{\infty} \delta(t) dt \;=\; 1 \qquad \text{at} \quad t = 0$$
$$\qquad\qquad\qquad = \; 0 \qquad \text{at} \quad t \neq 0$$

, and in terms of unit step function the Dirac delta function is given by

$$\delta(t) = \frac{d\,u(t)}{dt}$$

6. Sinc function

$$\mathrm{sinc}(t) = \frac{\sin(\pi t)}{\pi t}$$

7. Sine integral function

$$Si(u) = \int_{0}^{u} \frac{\sin(x)}{x} dx$$

8. Error function

$$\mathrm{erf}(t) = \frac{1}{\sqrt{2\pi}} \int_{-\infty}^{t} e^{-x^2/2} \, dx$$

9. Complementary error function

$$\mathrm{erfc}(t) = 1 - \mathrm{erf}(t) = \frac{1}{\sqrt{2\pi}} \int_{t}^{\infty} e^{-x^2/2} \, dx$$

10. Bessel function of the first kind of order n

$$J_n(x) = \frac{1}{2\pi} \int_{-\pi}^{\pi} e^{j[x\sin(\theta)-n\theta]} \, d\theta$$

11. Modified Bessel function of the first kind of zero order $I_0(x) = \dfrac{1}{2\pi} \displaystyle\int_{-\pi}^{\pi} e^{x\cos(\theta)} \, d\theta$

Series expansions

1. Exponential series

$$e^x = 1 + x + \frac{x^2}{2!} + \frac{x^3}{3!} + \frac{x^4}{4!} + \frac{x^5}{5!} + \cdots$$

,

$$e^{-x} = 1 - x + \frac{x^2}{2!} - \frac{x^3}{3!} + \frac{x^4}{4!} - \frac{x^5}{5!} + \cdots$$

,

$$e^{jx} = 1 + jx + \frac{(jx)^2}{2!} + \frac{(jx)^3}{3!} + \frac{(jx)^4}{4!} + \frac{(jx)^5}{5!} + \cdots$$

, and

$$e^{-jx} = 1 - jx + \frac{(jx)^2}{2!} - \frac{(jx)^3}{3!} + \frac{(jx)^4}{4!} - \frac{(jx)^5}{5!} + \cdots$$

2. Logarithmic series

$$\log_{10}(1 + x) = x - \frac{x^2}{2!} + \frac{x^3}{3!} - \frac{x^4}{4!} + \frac{x^5}{5!} - \cdots$$

3. Circular (Traingle) series

$$\cos(x) = 1 - \frac{x^2}{2!} + \frac{x^4}{4!} - \frac{x^6}{6!} + \cdots$$

,

$$\sin(x) = x - \frac{x^3}{3!} + \frac{x^5}{5!} - \frac{x^7}{7!} + \cdots$$

,

$$\tan(x) = x + \frac{x^3}{3} + 2\frac{x^5}{15} + 17\frac{x^7}{315} + \cdots \qquad \text{for} \quad |x| < \pi/2$$

,

$$\csc(x) = \frac{1}{x} + \frac{1}{6}x + \frac{7}{360}x^3 + \frac{31}{15120}x^5 + \cdots \qquad \text{for} \quad 0 < |x| < \pi$$

,

$$\sec(x) = 1 + \frac{1}{2}x^2 + \frac{5}{24}x^4 + \frac{61}{720}x^6 + \cdots \qquad \text{for} \quad 0 < |x| < \pi$$

, and

$$\cot(x) = \frac{1}{x} - \frac{1}{3}x - \frac{1}{45}x^3 - \frac{2}{945}x^5 - \cdots \qquad \text{for} \quad 0 < |x| < \pi$$

4. Inverse circular series

$$\sin^{-1}(x) = x + \frac{x^3}{6} + 3\frac{x^5}{40} + \frac{15}{336}x^7 + \cdots$$

, and

$$\tan^{-1}(x) = x - \frac{x^3}{3} + \frac{x^5}{5} - \cdots \qquad , \quad |a| < 1$$

5. Hyperbolic series

$$\cosh(x) = 1 + \frac{x^2}{2!} + \frac{x^4}{4!} + \frac{x^6}{6!} + \cdots$$

, and

$$\sinh(x) = x + \frac{x^3}{3!} + \frac{x^5}{5!} + \frac{x^7}{7!} + \cdots$$

6. Sinc series

$$\text{sinc}(x) = 1 - \frac{1}{3!}(\pi x)^2 + \frac{1}{5!}(\pi x)^4 - \frac{1}{7!}(\pi x)^6 + \cdots$$

7. Taylor series

$$f(x) = f(a) + \frac{f'(a)}{1!}(x-a) + \frac{f''(a)}{2!}(x-a)^2 + \cdots + \frac{f^n(a)}{n!}(x-a)^n + \cdots$$

, where
$$f^n(a) = \left.\frac{d^n(x)}{dx^n}\right|_{\text{at } x=a}$$

8. MacLaurin series

$$f(x) = f(0) + \frac{f'(0)}{1!}x + \frac{f''(0)}{2!}x^2 + \cdots + \frac{f^n(0)}{n!}x^n + \cdots$$

, where
$$f^n(0) = \left.\frac{d^n(x)}{dx^n}\right|_{\text{at } x=0}$$

9. Binomial series

$$(1+x)^n = 1 + nx + \frac{n(n-1)x^2}{2!} + \cdots \qquad , \qquad |nx| < 1$$

10 Binomial coefficient

$$\binom{k}{m} = \frac{k!}{(k-m)!\,m!}$$

Abbreviations

"a"   positive real constant (a $\geq$ 0)

a_0 average value of an arbitrary periodic signal $x_p(t)$

a_n real coefficient of Fourier series expansion

ac   alternating current

a(t)   natural envelope

$a_i(n)$ approximations wavelet coefficients

arg[ ]   phase angle

A   constant amplitude

A_c amplitude of carrier wave.

ADPCM   Adaptive Differential Pulse Code Modulation

AM   Amplitude Modulation

AND   AND gate

ASCII   American Standard Code for Information Interchange

ASK   Amplitude Shift Keying

ASPEC   Adaptive Spectral Perceptual Entropy Coding

b_n real coefficient of Fourier series expansion

B   channel frequency bandwidth

B_1 skirt steepness bandwidth

B_2 pass-bandwidth

B-Y   color difference signal

c(t)   carrier wave

C   maximum channel capacity bits per second,

C_n complex coefficient of Fourier series expansion

CCITT   Consultative Committee of International Telegraph and Telephone

CCTV   Closed Circuit Tel-Vision

CD   Compact Disc

CELP   Codebook Excited Linear Predictive coding

CESD   Cross Energy Spectral Density

CMF   Conjugate Mirror Filter

CPSD   Cross Power Spectral Density

CPU   Central Processing Unit

CR   Compression Ratio

CREW   Compression with Reversible Embedded Wavelets

CT   Contourlet Transform

CW   Continuous Wave

CWT   Continuous Wavelet Transform

$d_i(n)$ details wavelet coefficients

dB   deci-Bell

dc   direct current

diffs   differences are recorded in discrete files called "deltas"

DAB   Digital Audio Broadcasting

DAT   Digital Audio Tap

DCC   Digital Compact Cassette

DCT   Discrete Cosine Transform

DFB   Directional Filter Bank

DFT   Discrete Fourier Transform

DM   Delta Modulation

DRM   Digital Rights Management

DSB-SC   Double Side Band-Suppressed Carrier

DSB-TC   Double Side Band-Transmitted Carrier

DSP   Digital Signal Processing

DVD   Digital Versatile Disc

DVM   Directional Vanishing Moment

DWT   Discrete Wavelet Transform

$e(n)$ Discrete compressed output error
$erf(t)$ error function
$erfc(t)$ error complementary function
E   Energy content
E_i input energy content of system
E_o output energy content of system
$E[..]$ Expected value
ECG   Electro-Cardio-Gram signals
EM   Electro-Magnetic
ES   Elementary Stream
ESD   Energy Spectral Density

f frequency domain axis, (varying frequency of convolution process in the frequency domain)
f_c carrier frequency (frequency shifting)
f_s sampling frequency
$F[\ ]$ Fourier transform operation
$F^{-1}[\ ]$ inverse Fourier transform operation
Fax   Facsimile
FDCT   Fast Discrete Cosine Transform
FFT   Fast Fourier Transform
FM   Frequency Modulation
FSK   Frequency Shift Keying
FWT   Fast Wavelet Transform

G_o low-pass analysis filter of wavelet transform
G_1 low-pass synthesis filter of wavelet transform
Gbits/s   Giga bits per second
GHz   Giga Hz
GIF   Interchange Format Files
GSM   Global System for Mobile communication

$h(t)$ impulse response
$h_{eq}(t)$ equivalent impulse response
$h_N(t)$ equivalent impulse response of N sections
$h_w(t)$ weighted impulse response
$h(t)_{ILPF}$ impulse response of the ideal lowpass filter
$h(t)_{IHPF}$ impulse response of the ideal highpass filter
$h(t)_{IBPF}$ impulse response of the ideal bandpass filter
$h(t)_{IBSF}$ impulse response of the ideal bandstop filter
H_0 high-pass analysis filter of wavelet transform
H_1 high-pass synthesis filter of wavelet transform
Hz   Hertz
$H(f)$ transfer function
$H_{eq}(f)$ equivalent transfer function
$H_N(f)$ equivalent transfer function of N sections
$H_w(f)$ weighted transfer function
$H(m)$ entropy of source data
$H(F)_{ILPF}$ transfer function of the ideal lowpass filter
$H(F)_{IHPF}$ transfer function of the ideal highpass filter
$H(F)_{IBPF}$ transfer function of the ideal bandpass filter
$H(F)_{IBSF}$ transfer function of the ideal bandstop filter
$|H(f)|$ amplitude of transfer function
HVS   Human Visual System
H.261   Video codecs recommendation (video codec for audio visual services at p×64 kbits/sec, and p ≤ ±15) for compression of video telephony and teleconferencing
H.263   Video codecs recommendation (Hybrid of interframe compression that makes use of temporal redundancy and intraframe transform coding to use spatial redundancy) as a compression advantage of low bitrate communications, video conferencing and video telephony applications
H.264   Video codecs recommendation (Advanced Video Coding AVC and ITU-T) for compression of video telephony and teleconferencing

i	index (integer number)
i(t)	current waveform
i Mod N	notation means that, to divide i by N and retain the remainder only
IBPS	Ideal Band-Pass Systems
IEC	International Electro-technical Commission
ILPS	Ideal Low-Pass Systems
ISDN	Integrated Services Digital Networks
IS-95	American Interim Standard
ISO	International Standard Organization
I(t)	Inphase component of complex envelope

j	$\sqrt{-1}$
J	Laplacian pyramid level of contourlet transform
JPEG	Joint Photographic Experts Group

k	accounts the change in amplitude, (number of subbands $k = 2^\ell$ with wedge-shaped frequency partitioning) of contourlet transform
k_B	Boltzmann's constant (1.38×10^{-23} Joule/Kelvin),
Kbit/s	Kilo bits per second
KHz	Kilo Hertz

ℓ	level binary tree structured of decomposition directional filter bank of contourlet transform
ℓ_j	number of directional filter bank decomposition levels
L	average length of Huffman code
L_i	code word length of message m_i
L_f	The fixed code length of the messages before compression
LED	Light Emitting Diode
LP	Laplacian Pyramid
LPC	Linear Predictive Coding
LZW	Lempel Ziv Wekh

m	integer number
m(t)	baseband (low frequency) signal
m`(t)	first derivative of m(t)
m``(t)	second derivative of m(t)
$m_s(t)$	discrete time sampling signal
Mbps	Megabits per second
MD	Mini-Disc
MDCT	Modified DCT
M(f)	Fourier transform (spectrum) of m(t)
MHz	Miga Hz
Modem	Modulator/demodulator
MPEG	Moving Picture Experts Group (Moving Picture Experts Group)
MP3	Moving Picture Experts Group MPEG layer III audio standard
MP4	Moving Picture Experts Group MPEG part 10 (Advanced Video Coding AVC and ITU-T based on recommendation H.264)
MUSICAM	Masking pattern adapted Universal Subband Integrated Coding And Multiplexing

n	integer number
N	number of data sequence
N_1	bandwidth of discrete spectrum of $x_1(n)$
N_2	bandwidth of discrete spectrum of $x_2(n)$
N_c	bandwidth sequence of circular convolution $x_{12}(n)$
N_{th}	thermal noise power
n/T_o	discrete frequencies of an arbitrary power signal $x_p(t)$ of period T_0
NTSC	National Television Standards Committee
NTT	Nippon Telegraph and Telephone corporation

OFDM	Orthogonal Frequency Division Multiplexing

P_{av}	average power of periodic signal $x_p(t)$
p_i	probability of message m_i
$p_{in}(t)$	instantaneous dissipated power
PAM	Pulse Amplitude Modulation
PAL	Phase Alternating Line
PCM	Pulse Code Modulation
PDF	Portable Document Format, is an open standard for electronic document exchange
PDFB	Pyramidal Directional Filter Bank of contourlet transform
PES	Packet Elementary Stream
PM	Phase Modulation
PPM	Pulse Position Modulation
PSK	Phase Shift Keying
PSD	Power Spectral Density
PSI	Program Specific Info
PWM	Pulse Width Modulation
QAM	Quadrature Amplitude Modulation
$Q(t)$	Quadrature component of complex envelope
r_1	pass-band ripples level (Fresnel ripples)
r_2	sidelobe ripples level (Gibbs ripples).
$rect(t)$	rectangular function
R	Resistance
R_s	source date rate
R_r	Reduced data rate
$R(\tau)$	autocorrelation function of energy signal
$R_i(\tau)$	input autocorrelation function
$R_o(\tau)$	output autocorrelation function
$R_p(\tau)$	autocorrelation function of power signal
$R_{12}(\tau)$	cross-correlation function of energy signal
$R_{12p}(\tau)$	cross-correlation function of power signals
RADAR	Radio Detection And Ranging
RC	Resistance-Capacitance circuit
RF	Radio Frequency
RAID	Redundant Array of Inexpensive Disks
RL	Resistance-Inductance circuit
RLC	Resistance-Inductance-Capacitance circuit
RLE	Run Length Encoding
R-Y	color difference signal
SAOL	Structured Audio Orchestra Language
$s(t)$	transmitted modulated signal
$\hat{s}(t)$	Hilbert transform of $s(t)$
$sgn(t)$	signum waveform
$sinc(t)$	sinc waveform
$sinc^2(t)$	sinc squared waveform
$sgn(f)$	signum spectrum
S	signal power
SC	Suppressed Carrier
SECAM	Sequential Color with Memory
SF	Shape Factor
SSD	Solid State Derive SSD
$S(f)$	Fourier transform (spectrum) of $s(t)$
$Si(u)$	sine integral function
S/N_{th}	channel signal to noise power ratio
SNR	Signal to noise power ratio
SCID	Standard Color Image Data SCID
SSB	Single Side Band
STFT	Short Time Fourier Transform
SPIHT	Set Partitioning In Hierarchical Trees
SAW	Surface Acoustic Wave

t	time domain axis, (varying time of convolution process in the time domain)
$tri(t)$	triangle waveform
t_o	accounts the delay in transmission system (time shifting)
T	pulse width of rectangular function
T_o	finite period of an arbitrary power signal
T_d	constant delay
T_b	bit duration
T_s	sampling period.
T_{emp}	absolute temperature (Kelvin degrees = 273 + °C),
Telex	Telegraph exchanger
TC	Transmitted Carrier
TIFF	Tagged Image File Format
$u(t)$	unit step function
UMTS	Universal Mobile Telecommunication System
UWB	Ultra Wide Band
$v(t)$	voltage waveform
VoIP	Voice over Internet Protocol
W	frequency bandwidth (maximum frequency) of m(t)
WAN	Wireless Area Network
$W_{s,\tau}(t)$	wavelet analysis function
$W^*_{s,\tau}(t)$	complex conjugate of $W_{s,\tau}(t)$
WT	Wavelet Transform
$x(t)$	arbitrary complex energy signal (unperiodic signal)
$x`(t)$	first derivative of x(t)
$x``(t)$	second derivative of x(t)
$\tilde{x}(t)$	complex-envelope of x(t)
$x_+(t)$	pre-envelope of x(t)
$\hat{x}(t)$	Hilbert transform of x(t)
$\breve{x}(n)$	predicted value of the discrete signal x(n)
$x^*(t)$	complex conjugate of x(t)
$x_p(t)$	arbitrary complex power signal (periodic signal)
$x_+(t)$	pre-envelope of x(t)
$x_s(t)$	sampled signal
$x(nT_s)$	discrete waveform
$x_p^*(t)$	complex conjugate of an arbitrary power signal $x_p(t)$
$x_{12}(t)$	convolution function of $x_1(t)$ and $x_2(t)$
$x_{12}(n)$	circular convolution function of $x_1(n)$ and $x_2(n)$
$[x_p(t)]_{real}$	real value of an arbitrary complex power signal
$x(n_1,n_2)$	two dimensional discrete N×N matrix waveform
$X(f)$	input spectrum of a system
$X_s(f)$	spectrum of discrete signal
$X_+(f)$	pre-envelope spectrum
$X(kf_s)$	discrete spectrum
$X(n/T_o)$	discrete spectrum of an arbitrary power signal $x_p(t)$
$X^*(n/T_o)$	complex conjugate of $x_p(t)$
$X^*(f)$	complex conjugate of X(f)
$X_1(kf_s)$	discrete Fourier transform of $x_1(nT_s)$
$X_2(kf_s)$	discrete Fourier transform of $x_2(nT_s)$
$X(k_1,k_2)$	DCT of two dimensional N×N matrix $x(n_1,n_2)$.
$y(t)$	output waveform of a system
Y	luminance (monochrome compatible) signal
$Y(f)$	output spectrum of a system
$z(t)$	output waveform of a system
$Z(f)$	output spectrum (Fourier transform) of a system

REFERENCES

[1] Simon Haykin, "Communication Systems", John Wiley & Sons, Inc., 1983, Second Edition, ISBN 81-224-0164-3

[2] George Kennedy, "Electronic Communication Systems", McGraw-Hill, Inc., 1977, Second Edition, ISBN 0-07-034052-8

[3] Taub, Herbert, and D.L.Schilling, "Principles of Communication Systems", McGraw-Hill, Inc., New York, 1971.

[4] B. P. Lathi, "Communication Systems", Fourth Edition, John Wiley & Sons, Inc., 1978, ISBN 0-85226-505-0

[5] Peyton Z. Peebles, Jr, "Communication Systems Principles", Addison-Wesely, Inc., Canada, 1971, ISBN 0-201-05758-1

[6] Leon W. Couchi II, "Digital and Analog Communication Systems", Prentice–Hall International, Inc, 1997, Fifth Idition, USA, ISBN 0-13-599028-9

[7] Alan V. Oppenheim, Alan S.Willsky, S.Hamid Nawab, "Signals & Systems", Prentice–Hall International, Inc, 1997, USA, ISBN 0-13-651175-9

[8] Bernard Sklair, "Digital Communications", Prentice–Hall, International, Inc, Second Edition, 2000, USA, ISBN 0-13-084788-7

[9] James W. Nilsson, "Electric Circuits", Addison-Wesely, Inc., 1983, ISBN 0-201-06238-0

[10] B. P. Lathi, "Modern Digital and Analog Communication Systems", Third Edition, Exford University Press, Inc. 1998.

[11] Kh. El-Shennawy, "New Design Criterion for Improving the Performance of SAW Bandpass Digital Signal Processing in Communication Systems", IEEE Transactions on Instrumentation and Measurement, IM-Vol.50, Dec. 2001, pp.1796-1800.

[12] Kh. El-Shennawy, M. Mousa,"Transfer Function of an Ideal Theoretical Distorsionless Surface Acoustic Wave Filters", IEEE Africon`96 Conference, Stellenbosch, South Africa, Sept.24-27, 1996, pp.528-531.

[13] Kh. El-Shennawy, A.Shafik, M. Zaghloul,"The Transfer Function of the Distorsionless SAW Duplexer for IS`95, GSM, and UMTS Mobile Systems, 25th National Radio Science Conference, NRSC 2008, Tanta University, Egypt, C24-1, March, 2007.

[14] Kh. El-Shennawy, G. Ph. Fiani, M. B. Tayel,"Closed Form Solutions for Voltage Pulse Response of Open, Shorted, and Loaded Distributed RC Thin Film Structure", IEEE Transactions on Circuits and Systems, vol.CAS - 38, Dec.1991, pp.1567-1571.

[15] Kh. El-Shennawy,"Closed Form Solutions for Step Input Voltage of Uniformly Distributed Inductively Loaded RC Thin Film Structure", The International Association of Science and Technology for Development, IASTED-Symposium, Modeling, Identification and Control, Innsbruk, Austria, Feb.1992, pp.530-533

[16] Kh. El-Shennawy,"Effect of Pulse Width Duration on the Closed Form Solutions for Voltage Pulse Response of Open and Shorted Distributed RC Thin Film Structure", The International Association of Science and Technology for Development, IASTED-Symposium, Modeling, Identification and Control, Innsbruk, Austria, Feb. 1993, pp.74-76.

[17] Kh. .El-Shennawy, G. Ph. Fiani, M. B. Tayel,"Transient Waveform Response of Distributed Loaded RC Thin Film Structure for Voltage Step Function Using Laplace Transformation", the 8th Annual IEEE International ASIC Conference and Exhibit, Circuits and Systems Society, Austin, Texas, Sept.1995, pp.153 -156.

[18] Kh. El-Shennawy, "Simulation of Open and Shorted RC Interconnects Waveforms for Voltage Step Function with Low and High Transient Response", Microelectronics Journal, Elsevier Science, Volume 29, No.8, August 1998, pp.559-564.

[19] Simon Haykin, "Communication Systems", John Wiley & Sons, Inc., 1994, Third Edition , ISBN 0-471-57176-8

[20] Ahmed, N. and K. R. Rao, "Orthogonal Transforms for Digital Signal Processing", Springer-Verlag New York, Inc., Secaucus, HJ, 1975.

[21] Hsiung, W., S. Chen and S.Frlick, "A Fast Computational Algorithm for the Discrete Cosine Transform, IEEE Transactions on Communications, 25(9), 1004-1009, 1977.

[22] H. W. Park and Y. L. Lee,"A Postprocessing Method for Reducing Quantization Effects in Low Bit-Rate Moving Picture Coding", IEEE Transaction on Circuits and Systems, Video Technology , Vol. 9, pp.161-171, 1999.

[23] S. Mallat. "A Wavelet Tour of Signal Processing", 2nd Edition, Academic Press, 1999.

[24] J. S. Bendat and A. G. Piersol, "Randam Data: Analysis and Measurement Procedures", 3rd Edition, JohnWiley & Sons, Inc. 2000.

[25] D. Gabor, "Theory of Communication", Journal of IEE, vol.93, No.3, pp.429-457, 1946.

[26] Gross.A. and Morlet J., " Decomposition of Hardy Functions into Square Integrable Wavelets of Constant Shape", SIAM J. Applied Mathematics, Vol. 15, pp.723-736, 1984.

[27] Mallat S., "A.Theory for Multiresolution Signal Decomposition: The Wavelet Decomposition", IEEE Transaction PAMI Vol. 11, pp.674-693, 1989.

[28] Mallat S., "Multifrequency Channel Decomposition of Images and Wavelet Models", IEEE Transaction Pattern Anal. Mach. Intell. 11, pp.674-693, 1989.

[29] A. Zandi, J. D. Allen, E. L. Schwartz, and M. Boliek, " CREW: Compression with reversible embedded wavelets", Proceding of IEEE Data Compression Conference, Snowbird, Utah , pp. 212-221, March 1995.

[30] LI Yi and GONG Jianhua,"Global Terrain Data Organization and Compression Methods", The International Archives of the Photogrammetry, Remote Sensing and Spatial Information Sciences, Vol. XXXVII, Part B5, Beijing 2008.

[31] R. H. Bamberger and M. J. T. Smith, "A Filter Bank for the Directional Decomposition of Images: Theory and Design", IEEE Transaction on Signal Processing, Vol. 40, No.4, pp.882-893, Apr.1992, Thessaloniki, Greece.

[32] M. N. Do and M. Vetterli, "Pyramidal Directional Filter Banks and Curvelets", in 2001 Proc. of IEEE Int. Conference on Image Processing, Vol.3, pp.158-161, Thessaloniki, Greece.

[33] D. D. Y. Po and M. N. Do, "Directional Multiscale Modeling of Images Using the Contourlet Transform", IEEE Transaction on Image Processing, to appear, June 2006.

[34] P. J. Burt and E. H. Adelson, "The Laplacian Pyramid as a Compact Image Code", IEEE Transaction. on Communication, vol.31, no.4, pp.532-540, 1983.

[35] M. N. Do, "Directional Multiresolution Image Representation", Ph.D. Thesis, EPFI, Lausanne, Switzerland, Dec.2001.

[36] A. L. Cunha, J. Zhou, and Minh. N. Do, " The Nonsubsampled Contourlet Transform: Theory, Design, and Applications", IEEE Transaction on Image Processing, May 19, 2005, pp. 1-30.

[37] E. J. Candes and D. L. Donoho, "Curvelets – a surprisingly effective nonadaptive representation for objects with edges", in Curve and Surface Fitting, A. Cohen, C. Rabut, and L. L. Schumaker, Eds. Saint-Malo:Vanderbilt University Press, 1999.

[38] M. Vetterli, "Wavelets: approximation and compression", IEEE Signal Proc. Mag., pp.59-73, Sept.2001.

[39] E. J. Candes and D. L. Donoho, "Ridgelets: a key to higher-dimensional intermittency?", Phil.Trans. R. Soc. Lond. A., pp.2495-2509, 1999.

[40] M. N. Do and M. Vetterli, "The Contourlet Transform: An Efficient Directional Multiresolution Image Representation", IEEE Transaction on Image Processing, Vol.14, No.12, pp.2091-2106, Dec.2004.

[41] C. M. Brislawn, J.N.Bardley and R.J.Onyschczak and T.Hopper, "The FBI Compression Standard for Digitized Fingerpoint Images", in 1996 Proc. SPIE. Vol. 2847. Pp.344-355.

[42] A. Gersho and R. M. Gray, "Vector Quantization and Signal Compression", Boston, MA: Klower, 1992.

[43] S. Esakkirajan, T. Veerakumar, V. Senthil Murugan and R. Sudhakar, "Fingerprint Compression Using Contourlet Transform and Multistage Vector Quantization", International Journal of Biomedical Sciences, Volume 1, Number 2, 2006.

[44] M. Antonini, M. Barlaud, P. Mathieu, and I. Daubechies, "Image Coding Using Wavelet Transform", IEEE Transaction on Image Processing, pp.205-220. Apr.1992.

[45] B. H. Juang and A. H. Gray, "Multiple Stage Vector Quantization for Speech Coding", in 1992 Proc.IEEE int.Conference Acoust, Speech, Signal Processing, Paris, France, pp.597-600.

[46] John Watkinson, "Compression in Video and Audio", British Library of Cataloguing in Publishing Data, 1995, ISBN 0 240 51394 0.

[47] D. A. Huffman,"A Method for the Construction of Minimum Redundancy Codes", Proc.IRE, 40, 1098-1101, 1952.

[48] Cox, R., "Three New Speech Coders from the ITU Cover a Range of Applications", IEEE Comm. Mag., Vol. 35, No. 9, Sept. 1997, pp.40-47.

[49] Noll, P., "Wideband Speech and Audio Coding", IEEE Comm. Mag., Vol.31, No 11, Nov. 1993, pp.34-44.

[50] Solari, S., Digital Video and Audio Compressions, McGraw-Hill, New York, 1997.

[51] Rzeszewski, T., Digital Video: Concepts and Applications Across Industries, IEEE Press, 1995.

[52] Ebrahimi, T., Santa Cruz, D., Christopoulos, C., Askelof, J., Larsson, M., "JPEG-2000 Still Image Coding Versus other Standards", SPIE International Symposium, 30 July – 4 August 2000, Special Session on JPEG-2000, San Diego, CA., USA.

[53] Volkmer, H., "On the Regularity of Wavelets", IEEE Transactions on Information Theory, 38, 872-876, 1992.

[54] http://www.cipr.rpi.edu/research/SPIHT.

[55] Bower, B.V., "Low-Bit-Rate Image Compression Evaluations", Proc. Of SPIE, Orlando, FL, April 4-9, 1994.

[56] Strang, G. and Nguyen, T., "Wavelets and Filter Banks, Wellesley-Cambridge Press, Wellesley, MA, 1996, http://www-math.mit.edu/~gs/books/wfb.html.

[57] K.P.Soman and K.I.Ramachandran, "Insight into Wavelets from Theory to Practice, Prentice Hall, India, New Delhi, 2002, ch.9.

[58] Nippon Telegraph and Telephone Corp. (1995), "R & D Activities of NTT`s Research and Development Headquarters in 1994-An Integral Multimedia Capability-Compression Encoding of Music with TwinVQ (archived website)", Archived from the original on 1997-10-09, Retrieved 2010-08-06.

[59] Nippon Telegraph and Telephone Corp. (1996), " Welcome to the home of TwinVQ! (archived website) (Japanese)", Archived from the original on 2000-08-30, Retrieved 2010-08-06.

[60] " AES E-Library-Transform-Domain Weighted Interleave Vector Quantization (TwinVQ) ", Audio Engineering Society, 1996, Retrieved 2010-08-06.

[61] " Our Research of Audio",Nippon Telegraph and Telephone Corporation NTT Human Interface Laboratories, 1997, Archived from the original on 1999-01-28, Retrieved 2010-08-06.

[62] ISO (1999), " ISO/IEC 14496-3, 1999-Information Technology-Coding of Audio-Visual Objects-part 3: Audio", ISO, Retrieved 2009-10-09.

[63] D. Thom, H. Purnhagen and the MPEG Audio Subgroup (1998-10), " MPEG Audio FAQ Version 9-MPEG-4-An Introduction to MPEG-4 Audio", chiariglione.org, Retrieved 2009-10-06.

[64] ISO/IEC JTC 1/SC 29/WG 11 (1999-07) (PDF), ISO/IEC 14496-3/Amd,1-Final Committee Draft-MPEG-4 Audio Version 2, Retrieved 2009-10-07.

[65] Heiko Purnhagen (2001-06-01), "The MPEG-4 Audio Standard: Overview and Applications", Heiko Purnhagen, Retrieved 2009-10-07.

[66] ISO/IEC JTC 1/SC 29/WG 11 N2203 (1998-03), "MPEG-4 Audio (Final Committee Draft 14496-3)", Heiko Purnhagen , Retrieved 2009-10-07.

[67] Business Wire (1996-12-11), "Voxware Expands Technology Offerings & Signs Licensing Agreement with NTT", the free library, Retrieved 2009-10-06.

[68] Business Wire (1997-05-13), "IBM Licenses Voxware`s MetaSound Technology for Use in Multimedia Products", the free library, Retrieved 2009-10-06.

[69] Yamaha Corporation (2000), "Yamaha SoundVQ", Archive.org, Archived from the original on 2003-02-27, Retrieved 2009-10-06.

[70] Yamaha Corporation (1997), "Yamaha SoundVQ", Archived from the original on 1998-12-07, Retrieved 2010-08-06.

[71] Nippon Telegraph and Telephone NTT-East Multimedia Business Department (2000-03-31), "About TwinVQ", Archive.org, Archived from the original on 2000-08-17, Retrieved 2009-10-06.

[72] Music Compression Technology "TwinVQ" (archived website) (Japanese)", 1996, Archived from the original on 1997-06-27, Retrieved 2010-08-06.

[73] " About TwinVQ (archived website) (Japanese)", 1997, Archived from the original on 1997-07- 25, Retrieved 2010-08-06.

[74] Nippon Telegraph and Telephone NTT-East Multimedia Business Department (2008), "TwinVQ Software", Archive.org, Archived from the original on 2008-04-19, Retrieved 2009-10-07.

[75] Nippon Telegraph and Telephone NTT-East Multimedia Business Department (2002), "TwinVQ-Libraries and Sample Programs", Archive.org, Archived from the original on 002-12-27, Retrieved 2009-10-07.

[76] Nippon Telegraph and Telephone NTT-East Multimedia Business Department (2000), "TwinVQ FAQ", Archive.org, Archived from the original on 2000-08-19, Retrieved 2009-10-06.

[77] Nippon Telegraph and Telephone NTT (1998-03-24), "TwinVQ (archived website)", Archive.org, Archived from the original on 1998-04-30, Retrieved 2009-10-06.

[78] MultimediaWiki (2009), "VQF", MultimediaWiki, Retrieved 2009-10-07.

[79] " TwinVQ F.A.Q. (archived website) (Japanese)", 1997, Archived from the original on 1997-07- 25, Retrieved 2010-08-06.

[80] Technology white paper, "A Comparison of MPEG-4 (H.264) and JPEG-2000 Video Compression and Decompression Algorithms", Curtiss Wright Control, Embedded Computing, Video Distribution Systems, Homepage at: www.cwcembedded,com/video_distribution_system.htm

INDEX

A

Absolute temperature, 6
Acoustic, 13
Advanced Audio Coding AAC, 244
ASPEC, 234
ADPCM codec, 234
Admissibility condition, 214
Advanced Video Coding AVC, 244
Aliasing distortion, 195
Aliasing free, 217
A-law, μ-law, 234
Amplitude spectrum, 23
Amplitude distortion, 154
Amplitude Modulation AM, 4
American Interim Standard IS-95, 234
Analog pulse modulation, 5
Analog phase shift keying, 5
Analog signals, 13
Anti-causal signal, 13
AND gate, 13
Applications of WT, 219
Applications of compression Technique, 227
Area property, 66
Average length, 228
Average value, 21, 104
Average power, 10
Area property, 66
Associative law of algebra, 54
ASCII, 233, 236
Audio compression, 234
Audio video system, 3, 4
Autocorrelation function, 116
Autocorrelation function of energy signals, 117
Autocorrelation function of power signals, 124

B

Band limited, 5
Band-pass signal, 14
Bandwidth, 5, 6
    ... of low-pass, 147
    ... of band-pass, 147
Band-pass envelope, 177
Band-pass system, 180
Bidirectional pictures, 240
Bilingual audio programs, 234
Boltzmann`s constant, 6
Bite natural order, 205
Bite reversed order, 205
Biorthogonal wavelet filter, 217
Block-DCT, 209, 243
Blocking effects, 209, 237
Butterfly, 205
Byte, 204
Byte force computation, 204

C

Carrier wave, 4
Carrier delay, 190
Cascaded systems, 143
Causal systems, 146
Causal signals, 13
Causality condition, 146
Closed Circuit Tel-Vision CCTV, 247

Central Processing Unit CPU, 219
Channel, 3, 5
Channel capacity, 5
Chroma, 236
Chrominance components, 238, 241
Circular convolution, 202
Classification of signals, 6
Coder, 13
Code word length, 228
Codec, 227
Coding gain, 227
CELP coding, 234
Codec, 227
Code efficiency, 228
Code gain, 227
Code word, 229
Coiflet wavelet function, 218
Color difference components, 238
Communication system, 3
Communication channel, 3
Commutative law of algebra, 54
Compact Disc CD, 234
Compander, 227
Complex multiplication, 206
Complex addition, 206
Compression, 219
Compression ratio, 227
Compression principle, 227
Compression efficiency, 239
Compressor, 232
Complex-envelope, 178
Complementary error probability function, 100
Composite video, 235
Computational efficiency, 204
Computer Aided Design CAD, 235
Compression techniques, 227
Conjugate property, 68
Conjugate Mirror Filter CMF, 217
Contourlet transform, 193, 222
Continuous modulation, 4
Continuous spectrum, 34
Continuous wave, 4
Contamination, 4
Convolution property, 53
Correlation function, 113, 116
Cross-correlation function, 128
Cross-correlation function of energy signals, 129
Cross-correlation function of Power signals, 134
Cross spectral density, 129
Cross spectral density of energy signals, 129
Cross spectral density of power signals, 135
CREW, 220
Critical sampling, 223

D

Daubechies wavelet function, 218
Demodulation, 4
Deterministic signal, 7
Delta function, 7, 8, 81
Delay, 13
Delta compression, 233
Delta encoding, 233
Delta modulation, 5
Delay distortion, 154